THIN FILMS—
INTERDIFFUSION AND REACTIONS

THE ELECTROCHEMICAL SOCIETY SERIES

THE CORROSION HANDBOOK
Edited by Herbert H. Uhlig

MODERN ELECTROPLATING, THIRD EDITION
Edited by Frederick A. Lowenheim

THE ELECTRON MICROPROBE
Edited by T. D. McKinley, K. F. J. Heinrich, and D. B. Wittry

CHEMICAL PHYSICS OF IONIC SOLUTIONS
Edited by B. E. Conway and R. G. Barradas

HIGH-TEMPERATURE MATERIALS AND TECHNOLOGY
Edited by Ivor E. Campbell and Edwin M. Sherwood

ALKALINE STORAGE BATTERIES
S. Uno Falk and Alvin J. Salkind

THE PRIMARY BATTERY (*in Two Volumes*)
Volume I Edited by George W. Heise and N. Corey Cahoon
Volume II Edited by N. Corey Cahoon and George W. Heise

ZINC-SILVER OXIDE BATTERIES
Edited by Arthur Fleischer and J. J. Lander

LEAD-ACID BATTERIES
Hans Bode
Translated by R. J. Brodd and Karl V. Kordesch

THIN FILMS—INTERDIFFUSION AND REACTIONS
Edited by J. M. Poate, K. N. Tu, and J. W. Mayer

THE CORROSION MONOGRAPH SERIES

Edited by R. T. Foley, N. Hackerman, C. V. King, F. L. LaQue, and
H. H. Uhlig

THE STRESS CORROSION OF METALS
H. L. Logan

CORROSION OF LIGHT METALS
H. P. Godard, M. R. Bothwell, R. L. Kane, and W. B. Jepson

THE CORROSION OF COPPER, TIN, AND THEIR ALLOYS
H. Leidheiser

CORROSION IN NUCLEAR APPLICATIONS
Warren C. Berry

HANDBOOK ON CORROSION TESTING AND EVALUATION
Edited by William H. Ailor

MARINE CORROSION
F. L. LaQue

THIN FILMS—
INTERDIFFUSION AND REACTIONS

Edited by

J. M. POATE
BELL LABORATORIES
MURRAY HILL, NEW JERSEY

K. N. TU
IBM THOMAS J. WATSON RESEARCH CENTER
YORKTOWN HEIGHTS, NEW YORK

J. W. MAYER
CALIFORNIA INSTITUTE OF TECHNOLOGY
PASADENA, CALIFORNIA

Sponsored by

THE ELECTROCHEMICAL SOCIETY, INC.
Princeton, New Jersey

A WILEY-INTERSCIENCE PUBLICATION

JOHN WILEY & SONS

NEW YORK • CHICHESTER • BRISBANE • TORONTO

Copyright © 1978 by John Wiley & Sons, Inc.

Library of Congress Cataloging in Publication Data:

Main entry under title:

Thin films.

"A Wiley-Interscience publication."
Includes bibliographical references.
1. Thin films. 2. Semiconductors. I. Poate, J. M., 1940–
II. Tu, King-ning, 1937– III. Mayer, James W.,
1930–

TK7871.15.F5T44   621.3817'1   77-25348
ISBN 0-471-02238-1

Printed in the United States of America.

10 9 8 7 6 5 4 3 2 1

Contributors

J. E. E. BAGLIN, *IBM Research Center, Yorktown Heights, New York*

D. R. CAMPBELL, *IBM Research Center, Yorktown Heights, New York*

F. M. D'HEURLE, *IBM Research Center, Yorktown Heights, New York*

D. GUPTA, *IBM Research Center, Yorktown Heights, New York*

P. S. HO, *IBM Research Center, Yorktown Heights, New York*

J. K. HOWARD, *IBM System Products Division, East Fishkill, New York*

S. S. LAU, *California Institute of Technology, Pasadena, California*

J. W. MAYER, *California Institute of Technology, Pasadena, California*

J. O. McCALDIN, *California Institute of Technology, Pasadena, California*

T. C. McGILL, *California Institute of Technology, Pasadena, California*

S. M. MYERS, *Sandia Laboratories, Albuquerque, New Mexico*

E. H. NICOLLIAN, *Bell Laboratories, Murray Hill, New Jersey*

J. C. PHILLIPS, *Bell Laboratories, Murray Hill, New Jersey*

J. M. POATE, *Bell Laboratories, Murray Hill, New Jersey*

R. ROSENBERG, *IBM System Products Division, East Fishkill, New York*

A. K. SINHA, *Bell Laboratories, Murray Hill, New Jersey*

M. J. SULLIVAN, *IBM System Products Division, East Fishkill, New York*

K. N. TU, *IBM Research Center, Yorktown Heights, New York*

W. F. vAN DER WEG, *Philips Research Laboratories, Amsterdam, The Netherlands*

Preface

This monograph is devoted almost exclusively to phenomena associated with interdiffusion or reactions between thin films. We outline the general aims and contents in Chapter 1, "An Overview." The subject has developed rapidly over the past five years and shows every sign of continuing to expand. When is the right time to try and review such a complex and evolving subject? We feel that the understanding in many of the areas has reached a sufficient level of maturity to warrant detailed exposition. We decided to concentrate on areas where interdiffusion or reactions play a dominant role. Therefore, much of the volume is concerned with structures used in contacts with semiconductors, where these phenomena are recognized as important factors.

Loss of coherence can be a deficiency in a contributed volume. We discussed the book at length with our colleagues, both before and during writing, to provide the coherence required to present the many facets of thin film phenomena. The thin film topics cover such a wide area that we included sufficient background information so that the book will be useful to research workers, teachers, and students in many disciplines.

We are indebted to our fellow contributors who agreed to partake in this venture and then cheerfully acquiesced to our editorial comments; to Dr. Newton Schwartz and the Electrochemical Society for encouraging us to compile the monograph; to our colleagues and management at Bell Laboratories, IBM, and Caltech who gave us considerable support and stimulation; and to Joyce Otis, Edith Nitchie, and Carol Norris for secretarial work.

J. M. POATE
K. N. TU
J. W. MAYER

Murray Hill, New Jersey
Yorktown Heights, New York
Pasadena, California

January, 1978

Contents

THIN FILMS—
INTERDIFFUSION AND REACTIONS

1

AN OVERVIEW

J. M. Poate

Bell Laboratories, Murray Hill, New Jersey

K. N. Tu

IBM Thomas J. Watson Research Center, Yorktown Heights, New York

J. W. Mayer

California Institute of Technology, Pasadena, California

Thin films play a dominant role in modern technology. Their most successful and important use arises in the all-pervasive technology of integrated circuits. They also appear as vital elements in such diverse areas as solar energy conversion devices and superconducting elements. An essential criterion of the thin film structures in these applications is that they maintain structural integrity. In the world of bulk or large-scale structures, interdiffusion or reaction on the 100 Å scale can generally be ignored. This is not the case for thin film structures, where pronounced reaction or interdiffusion can occur over these distances even at room temperature. It is only recently that experimental techniques have been available to determine mass transport on this scale. The general theme of this book will be to present the techniques and review the various diffusion or reaction phenomena.

The problems that arise in trying to untangle thin film reactions can assume

1

forbidding proportions. Film thicknesses range from 10 Å to several microns. Compositions and structures have to be evaluated before and after reaction. Once the physical and chemical parameters have been determined, models of the reactions and structures can be constructed. Despite the inherent complexities of the systems, it will become apparent from the following chapters that in many cases a single process or parameter will dominate. Structures or physical behavior of remarkable simplicity can result.

A nice example of the underlying simplicity, to be amply demonstrated later, is the reaction of metal films on Si to produce thin film metal silicides. Although equilibrium phase diagrams typically predict many phases, usually only one phase nucleates and grows in thin film reactions. Moreover, these solid phase reactions, which occur at temperatures far below the melting point of silicon or metal, produce well-defined planar silicides.

It is quite obvious that much of the incentive for thin film reaction studies originates from the demands of technology. These demands can only increase as tighter constraints are placed on film materials and dimensions. Indeed, in the field of integrated circuits we are approaching the time when films will be tailored to dimensions close to an atomic monolayer. Technological incentives aside, it will be apparent from the following chapters that the subject is also generating considerable scientific excitement. Much of the interest centers around the relationship of the behavior of matter in the bulk to that of interfacial or thin film behavior. Understanding the materials science in this submicron world may lead to significant developments in other areas.

The contents of this book are influenced by our bent toward electronic materials and our use of ion beams for materials analysis or modification. To keep the length of the book within reasonable bounds, we have excluded fields such as superconducting films, amorphous films, and superlattices produced by molecular beam epitaxy. Although these areas overlap quite well with many of the general concepts in thin film technology, they are not based upon the interdiffusion and reactions which constitute the theme of this volume. We now outline the intent of the various chapters.

TECHNOLOGICAL IMPERATIVES (CHAPTER 2)

This chapter sets the framework and flavor for the rest of the book. It delineates, within the context of Si device technology, most of the areas that will be presented. We can give some idea of the thrust of the chapter and the progress that has been made in solid state electronics with the following illustrations. Figure 1.1 shows the original point contact Ge transistor of Bardeen and Brattain, while Figure 1.2 shows one of the developments in integrated circuits. These photographs help demonstrate the complete convergence to planar thin film technologies over the past 25 years or so.

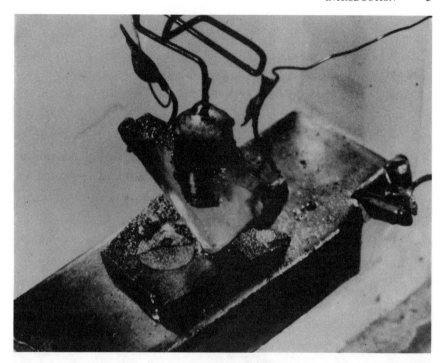

Figure 1.1. The original point contact transistor of Bardeen and Brattain (J. Bardeen and W. H. Brattain, Phys. Rev., **75**, 1208, 1949). The device consisted of two pointed gold contacts, less than 50 μm apart, on one side of a Ge wafer. Photo courtesy of Bell Laboratories.

Rosenberg, Sullivan, and Howard concentrate on the properties and reactions of metal films on Si integrated circuits. They trace, for example, some of the developments of Al metallization. The behavior of the Al thin film metallurgy, in such areas as solid phase reactions with Si and electromigration, had to be understood before successful devices could be constructed.

The importance of solid phase reactions to the electronics industry was graphically demonstrated by the appearance of the "purple plague" when thin films of Al and Au were contacted together. The most distressing symptom of this disease was the loss of mechanical strength at the metallic bond. Aluminum and Au form the intermetallic $AuAl_2$ at relatively low temperatures with a characteristic purple color, and it is well known that intermetallic phases are often brittle. This problem was the precursor of a series of interesting and perplexing solid phase reactions that have dominated the approaches used in applying thin film technology to the field of integrated circuits. Chapter 2 gives a feeling for some of the problems and their solutions.

Figure 1.2.   Charge coupled device shift register, 16 bit n-channel, with poly-Si and Al gates of 5 μm linewidth. Photo courtesy of Hewlett-Packard Laboratories.

UNDERSTANDING INTERFACES (CHAPTERS 3 AND 4)

The key to understanding thin film reactions and most of the phenomena discussed in this book lies in the physical and chemical nature of the interface. Theoretical treatments are scarce in this area. This is understandable when it is realized that only recently has the vacuum solid interface, a system of much less complexity, been subjected to detailed investigation or calculation.

Two areas can be discerned: the metallurgy or chemistry of the interface, and the electronic or electrochemical properties of the interface. Chapters 3 and 4 tackle various aspects. In Chapter 3, Phillips stresses the need for chemical and physical models of interfacial regions and speculates on a broad range of interfacial phenomenon. He nicely details the excitement and potential in this area. In Chapter 4, McCaldin and McGill address the specific and important problem of the semiconductor–conductor interface. They are primarily concerned with the roles played by the anion in a compound semiconductor at the interface.

EXPERIMENTAL TECHNIQUES (CHAPTERS 5 AND 6)

Thin films are unique microstructural entities. They can be formed in single crystal, polycrystalline, or amorphous layers. Their microstructure will evidently dominate many of the reactions and phenomena of interest. Metal films, for example, are usually deposited as polycrystalline layers; consequently, many of the phenomena such as electromigration and interdiffusion may be dominated by diffusion and transport along the grain boundaries. Chapters 5 and 6 discuss methods of characterizing thin film structures.

It is our belief that much of the progress in the science of thin films is due not only to the sophisticated deposition techniques but also to the recent developments and accessibility of many analytical techniques. These techniques come from such diverse fields as nuclear physics and surface science. The marriage of these disciplines has led to many exciting developments.

Tu and Lau in Chapter 5 summarize the immense fields of film deposition and characterization. They place primary emphasis on the types of thin film structure that can be fabricated and on the analytical techniques of characterizing the surface morphology and internal microstructure.

The content of this book deals primarily with reaction or interdiffusion across interfaces. Such a process is illustrated in Fig. 1.3, which shows an SEM picture of the formation of a thin Pd silicide layer located 1 μm beneath the surface of amorphous Si. Mayer and Poate in Chapter 6 discuss the various depth-profiling techniques which have the requisite depth resolution to allow measurement of reaction kinetics that occur on the submicron scale.

GRAIN BOUNDARY PHENOMENON (CHAPTERS 7 AND 8)

One of the remarkable features of thin film reactions is that they can occur at temperatures well below those characteristic of bulk processes. It is now generally accepted that diffusion and transport along grain boundaries can dominate many thin film reaction processes. The width of the grain boundaries is approximately 5 to 10 Å, and the grain size itself is typically of the order of a few hundred to a few thousand Ångstroms. Both the experimental evaluation and theoretical description of grain boundary diffusion in thin films require considerable sophistication because of the complicated physical structures.

In Chapter 7 Gupta, Campbell, and Ho present the current understanding of grain boundary diffusion. They concentrate upon tracer or self-diffusion studies and relate experimental depth profiles to theoretical models. An indication of the substantial progress that has been made in this area are the links that the authors establish between thin film and bulk grain boundary diffusion phenomena. The import of that chapter is to show that we have now

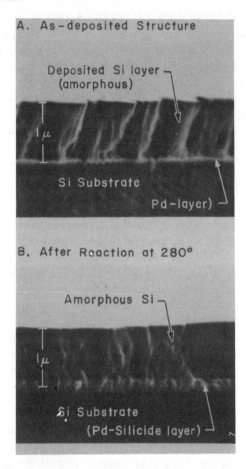

Figure 1.3.   Scanning electron microscope photographs of an edge view of a cleaved sample of 1 μm Si deposited on 0.02 μm Pd on a Si substrate: (A) as-deposited sample; (B) sample after heat treatment at 280°C to form the phase Pd₂Si. From J. W. Mayer, J. M. Poate, and K. N. Tu, *Science*, **190**, 228 (1975).

reached the stage where quantitative assessments can be made of the role of grain boundary diffusion phenomena.

An important area where grain boundary transport effects are crucial is that of electromigration. The critical role of electromigration in device technology is highlighted in the field of large-scale integrated circuits where current densities in conductors can reach 10^6 A cm^{-2}. The destructive effect of such current densities is shown in Figure 1.4a and 1.4b. Voids and hillocks are formed in the metal film owing to the migration of metal atoms under high current density conditions. D'Heurle and Ho in Chapter 8 review the

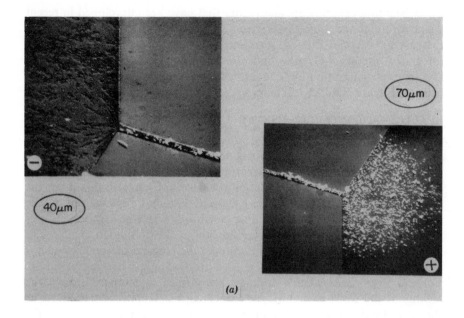

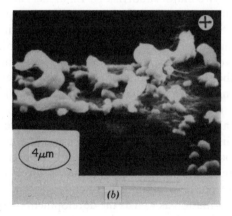

Figure 1.4. (a) Voids (negative terminal) and hillocks (positive terminal) in a single-crystal Al + 3 wt % Mg thin film after electromigration testing for 900 h at 234°C with a current density of 4×10^6 A cm^{-2}. (b) Hillocks (in greater detail) in Al + 3 wt % Mg after 16,000 h at 176°C with a current density of 4×10^6 A cm^{-2}. From A. Gangulee and F. M. d'Heurle, *Thin Solid Films*, **25**, 317 (1975).

many beautiful experiments and theory that have been developed in recent years. Although the subject of electromigration is over 100 years old, only in the past 10 years has it been established as a significant source of failure of integrated circuits. D'Heurle and Ho discuss the failure models used to predict electromigration lifetimes as well as evaluating the various methods used to increase conductor lifetimes.

REACTION AND INTERDIFFUSION IN LAYERED STRUCTURES (CHAPTERS 9 TO 12)

Perhaps the most striking and intriguing area of thin film reactions is the large-scale transport of material across interfaces at low temperatures. It was the advent of the profiling techniques discussed in Chapter 6 that made the observation of this mass transport feasible. Figure 1.5 shows one of the manifestations of this behavior. Silicon can diffuse in Al films at temperatures well below the eutectic, and severe erosion of Si at contacts may occur. Although such low-temperature reactions have plagued the semiconductor industry, many of these reactions are used to advantage in device fabrication. For example, stable silicide contacts are formed at temperatures in the vicinity of $400°C$. These compound layers can be exceptionally uniform, as shown in Figure 3. The exceptional uniformity of these layers permits the possibility of thin film engineering whereby knowledge of the reaction kinetics permits fabrication of precise structures.

The field of metal–metal reactions covered in Chapter 9 by Baglin and Poate lends itself to two major subdivisions. First, there are systems that form series of solid solutions; here grain boundary diffusion is the primary transport mechanism. Second, there are systems that form intermetallic compounds; here formation of layered phases is often the dominant feature. The results of these investigations impact on many areas of thin film technology; among these are the complex conductor metallization schemes on integrated circuits, formation of superconducting thin films as well as contacts, and diffusion barriers in solar energy convertors.

Whereas in metal–metal systems polycrystalline films are almost invariably encountered, the discussion in Chapter 10 by Tu and Mayer on silicide formation deals with polycrystalline metal films on single crystal silicon. The systematics of silicide formation have reached a level of maturity where general patterns of phase formations have emerged. The existence of these patterns of phase formation as well as the development of marker techniques have given rise to a physical understanding of the reaction process. Moreover, the simplicity of the pattern of silicide formation is such that silicide-forming systems can be used with confidence in technology.

Compared with the field of silicide formation, the study of metal–compound semiconductor reactions, presented in Chapter 11 by Sinha and

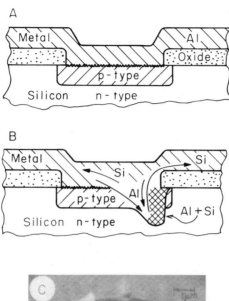

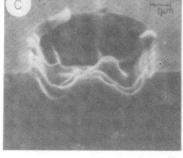

Figure 1.5. Schematic diagrams and scanning electron microscope photographs of the erosion of Si that can occur when heat treating Al films in contact with Si: (A) as-deposited structure with Al in contact with *p*-type layer of Si *p–n* junction; (B) after heat treatment at temperatures around 400°C solid phase reactions occur between Al and Si; (C) view of the Si surface after the Al layer is removed. From R. W. Bower, *Appl. Phys. Lett.*, **23**, 99 (1973).

Poate, is still in its infancy. Large area Ohmic contacts to GaAs microwave devices were achieved by a brute force approach that involved considerable consumption of the underlying semiconductor. Such erosion cannot be tolerated in the GaAs large-scale integrated circuits presently under development. For this application it will be necessary to have controlled and uniform reactions between the contact metal and semiconductor. Chapter 11 describes some of the approaches and advances that have been made in this area. It appears that the gross features of the reaction or interdiffusion may be explained in terms of the electronegativities of the conductor metal. This chapter should be read in connection with Chapter 3 on the semiconductor–conductor interface.

An obvious development of the concepts of mass transport and uniform layer growth presented in Chapters 9 to 11 is the growth of planar epitaxial layers. Lau and van der Weg in Chapter 12 detail the experimental approach to solid phase epitaxy. Although the field is new, it is included in this book because the concepts, experimental techniques, and physical processes are intimately tied to the work of the previous chapters. The formation of large-area epitaxial layers has now been achieved through an understanding of interface properties. The limiting case of epitaxial growth, which is discussed in the chapter, is the recrystallization of amorphous layers formed by implanting Si atoms into single crystal Si or Ge.

ELECTRICAL CHARACTERISTICS OF INTERFACES (CHAPTER 13)

The previous chapters have been mainly concerned with the physical or metallurgical structure of interfaces. From a technological standpoint, of course, the electrical characteristics of the semiconductor contact are of overwhelming importance. Nicollian and Sinha in Chapter 13 present the effect of interfacial reactions on the electrical characteristics of metal–semiconductor contacts. Two types of electrical contacts exist; the Ohmic contact, in which the metal–semiconductor contact offers minimum resistance to current flow for either polarity of applied voltage, and the Schottky barrier contact, which offers low resistance to current flow in one direction and high resistance in the other. In this latter case, the most important property of the contact is the potential barrier between metal and semiconductor. Chapter 13 describes both the factors that determine this potential barrier and how the current–voltage characteristics are in turn determined by the potential barrier. This description also covers the influence of a thin intervening oxide layer between the metal electrode and the semiconductor. The presence of this thin oxide layer can have a strong influence on the contact characteristics.

IMPLANTATION METALLURGY (CHAPTER 14)

One of the major themes of this book has been the microstructural properties of layered structures in the micron range. A significant new development appears to be in the use of ion beams to fabricate layered structures of unusual properties. The implantation process, which is not limited by the usual constraints of equilibrium thermodynamics, can lead to the formation of metastable phases. Ion-implanted metal layers are discussed in Chapter 14 by Myers.

There is little doubt that interesting and significant information on non-equilibrium metallurgy and diffusion properties will emerge. For example, it has now been shown that there is a close connection between the classical "splat-cooled" alloys and equivalent ion-implanted systems. One may also have confidence that the modification of surface layers of metals by implantation will have the same impact on such diverse areas as corrosion and wear resistance as the use of implantations for doping in Si device technology.

We believe that there are strong connections between thin film reactions and implantation metallurgy. There are the obvious ones of the submicron dimensions and the fact that the same analytical tools have to be used to characterize the layers. Further, one uses equilibrium phase diagrams in trying to systematize the various processes. In both systems the growth of phases is controlled by nucleation and diffusion processes. Myers indeed shows that implantation can be employed to obtain diffusion and phase boundary parameters.

SUMMARY

The stage is set for both basic and applied studies of interdiffusion and reactions in thin film studies. Both the analysis techniques and the practical applications are at hand. The chapters in this book establish the background and current understanding of these studies. It is an exciting area. From the 100-year-old field of electromigration to the newly developed field of implantation metallurgy, new concepts and new insights have emerged. The intent of the monograph is to present some of these developments and to point the directions for future work.

2

EFFECT OF THIN FILM
INTERACTIONS ON
SILICON DEVICE TECHNOLOGY

R. Rosenberg, M. J. Sullivan, and J. K. Howard

IBM System Products Division, East Fishkill, New York

2.1 INTRODUCTION

The basic objective of the text in which this chapter appears is to provide insight into the fundamentals of the interactions between polycrystalline thin films in temperature ranges where defects such as dislocations, grain boundaries, or interfaces dominate. The kinetics of such interactions, the physical modeling, and illustrations of the modern analytic methods available are given in detail in other chapters; here they will be addressed in depth only when some detail may be required for clarity. The following discussion is concerned chiefly with some examples for which such interactions have a major bearing on the fabrication or stability of device technologies.

Applications in which multilayered thin films are critical parts of the physical structure are as widespread as the various types of films, ranging from optical wave guides and mirrors, solar absorbers, and superconducting devices to the vast electronics industry, in which large-scale integration (LSI) has come of age. Examples of the various important interactions can be found in the silicon device industry, for which much of the work has been developed, and where degradation modes are best understood, characterized, and improved upon. The subjects covered are metal–silicon contact reactions; Schottky barrier formation and instability; metal–metal interdiffusion, related problems, and diffusion barrier studies; electromigration in polycrystalline metal lines, width effects, and high current-carrying structures; polysilicon gate MSOS structural and electrical studies; and metal film corrosion.

2.2   INSTABILITY OF SCHOTTKY BARRIER CONTACTS

An important consequence of low-temperature diffusion and the resulting interlayer interactions is the instability produced in SBDs (Schottky barrier diodes), either with single-layer Al–Si or with more complex Al–silicide–Si metallurgies. Schottky diodes are solid–state devices formed by placing a metal in intimate contact with a lightly doped semiconductor. These devices find many applications in the semiconductor industry, for example, as "clamps" to prevent transistors from going into saturation and as discrete devices in logic and memory arrays. The disadvantage of using SBDs is that the electrical characteristics are extremely sensitive to the conditions of the interface between metal and semiconductor. Contamination, oxide layers, and/or metallurgical reactions at this interface can cause major variations in diode behavior.

Typically, an SBD is fabricated by chemical vapor deposition of a thin epitaxial layer of lightly doped Si onto a heavily doped Si substrate. The Si wafer is then oxidized, and contact windows are defined photolithographically. The oxide is chemically etched to expose the contacts, and the contacts are chemically cleaned immediately before they are loaded into the vacuum system for metal deposition. For aluminum metallization on Si, the diodes are sintered after deposition to allow the Al to reduce any native SiO_2 in the contact. This procedure should eliminate the effects of contaminants and oxides. However, during the sinter and subsequent processing such as glassing, metallurgical reactions occur by grain–boundary diffusion of Si in Al and alter diode behavior. This review will deal mainly with instability in Al- and Si-doped Al diodes, with extension to reactions of Al with metal silicide contacts.

2.2.1 Pure Al SBDs

2.2.1.1 Penetration. As was noted above, diodes fabricated with Al metallization in contact with Si must be sintered (typical sintering temperatures are 400 to 500°C) to allow the Al to reduce the native SiO_2 in the contacts. This results in what is called "alloy penetration" (1–3). The Al–Si equilibrium phase diagram shows an appreciable solid-state solubility of Si in Al at the processing temperatures used. During sintering, Si from the single-crystal substrate diffuses into the Al through grain-boundary paths to satisfy the solubility. Concomitantly, Al diffuses to the substrate surface; this is the "penetration." Since the Al metallization usually extends some distance from the contact hole, a large volume of Al is available to act as a semi-infinite sink for the Si. As Figure 2.1 shows, the Si is usually dissolved from a number of specific sites within the contact. These sites are thought to correspond to local defects in the native SiO_2, which locally expose the Si substrate to the Al metallization during sintering. Also, for asymmetrical Al coverage of the contact, the penetration is greatest at the edge of the contact nearest the semi-infinite sink; the Al away from the sink is quickly saturated with Si. The extent of the penetration is greater at higher temperatures because diffusion rates are higher and there is greater solubility of Si in Al. All of these features are illustrated in Figure 2.2 which shows an Al–Si contact after 1 hour at 400°C. The Al metallization has been etched away for examination, and the Si contact is surrounded by SiO_2.

The mechanism described above results in a diode contact composed of a number of Al spikes extending into the Si. These metal spikes locally increase the electric fields in the Si near the sharp tips of the spikes. It has been shown (4) that the site of the penetration pit is the site of the diode breakdown within the contact. Figure 2.3 shows the $I–V$ characteristics of an Al–Si diode after heat treatment (450°C, 1 h) and an SEM micrograph of the contact after Al removal. Also shown is a picture of the anode and cathode contacts, obtained by use of a TV microscope sensitive to infrared radiation. The bright spots (D and E) in Figure 2.3b are caused by infrared photoemission at the location of breakdown, where the avalanche current is the greatest. A comparison of Figures 2.3b and 2.3c shows that these points coincide exactly with the alloy penetration pits.

2.2.1.2 Epitaxial Layers. Another complicating phenomenon in the Al–Si system occurs because of thermal cycles needed in the process. Since the solubility of Si in Al decreases with decreasing temperature, the Al becomes supersaturated with Si during cooling from the sintering temperature; the magnitude of supersaturation is greatest adjacent to the contact

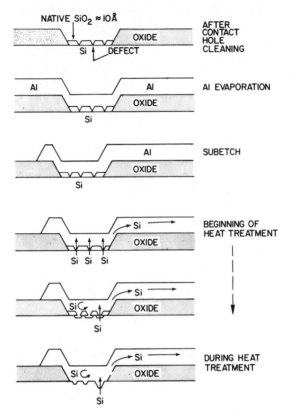

Figure 2.1.   Formation of alloy penetration pits. Courtesy of T. M. Reith.

from which the Si was originally depleted. During cooling, the Si precipitates from solution in the Al, forming particles at the Al grain boundaries, and at the same time forms an epitaxial layer on the Si in the contact (5–10). The Si particles and the epitaxial layer become saturated with Al, which acts as a p-type dopant; the resulting contact consists of a layer of p-type Si on n-type Si. The effect of the p-layer can be determined by solving Poisson's equation for the situation shown in Figure 2.4, after Card (11).

The magnitude of the potential at the maximum, ϕ_m, and the location of the maximum potential, x_m, are given by

$$\phi_m = \frac{-qN_D}{2\varepsilon}\left(1+\frac{N_D}{N_A}\right)w^2 \tag{1}$$

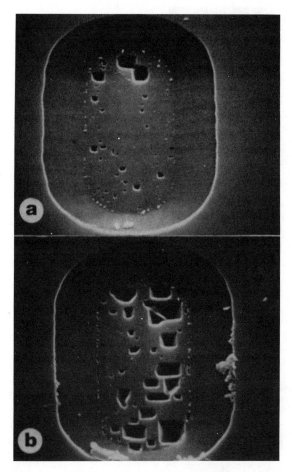

Figure 2.2. Scanning electron micrographs, 4500×, of alloy penetration pits formed after sintering for 1 h at 400°C: (a) narrow Al land extending from top of contact; (b) wide Al land extending to right of contact. Courtesy of T. M. Reith.

$$x_m = -w \frac{N_D}{N_A} \qquad (2)$$

$$w = \left[\left(1 + \frac{N_A}{N_D} \right) a^2 + w_0^2 \right]^{1/2} - a \qquad (3)$$

where w is the depletion width, w_0 is the zero bias depletion width for $a = 0$ (no p-layer), and all other symbols have their usual significance.

This result indicates that the barrier height will increase with increasing

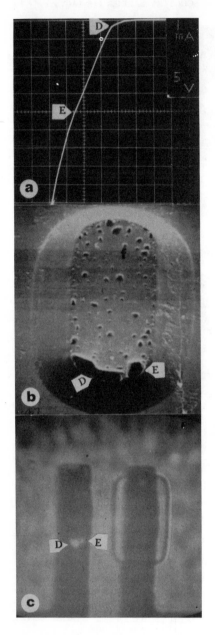

Figure 2.3. Effect of alloy penetration on SBD electrical behavior: (a) reverse-bias *I–V* curve; (b) scanning electron micrograph, 4400×, of contact after Al removal; (c) infrared TV photograph showing location of diode breakdown, D, E. From reference 4.

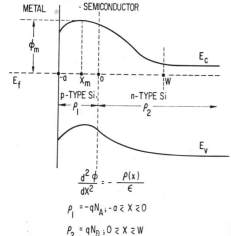

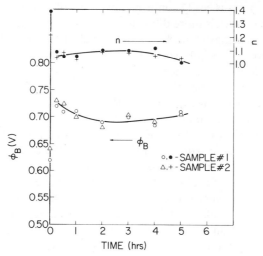

Figure 2.4. Energy band diagram of an Al-to-Si contact with an epitaxially crystallized p-layer. From reference 11.

p-layer thickness a and/or increasing p-layer doping N_A. Both of these parameters are increased by increased sintering temperature. References 12 to 16 report that the barrier height does indeed increase as a result of higher sintering temperatures.

The effect of sintering time at constant temperature (400°C) on the barrier height and ideality of Al–Si diodes has been investigated by Sullivan (17); see Figure 2.5. The initial drop in barrier height is caused by the formation of

Figure 2.5. Barrier height and ideality of Al-to-Si SBDs as a function of sintering time at 400°C. From reference 17.

penetration pits. This effect is compensated for by the formation of a p-layer at longer sintering times, which causes the barrier height to increase.

2.2.2 Silicon-Doped Al SBDs

In the presence of shallow diffused junctions, the alloy penetration discussed in the preceding section may short-circuit the junction and destroy the device. To prevent this, the Al metallization is presaturated with Si by evaporation of a thin Si layer on top of the Al before sintering. The Si can also be placed as a layer in the middle of the Al film, or be codeposited with it. The thicknesses of the Al and Si layers are adjusted to yield about 1 wt % Si. This level of Si will prevent alloy penetration up to approximately 550°C. However, the Al is supersaturated with Si at all processing temperatures below 550°C, and therefore the epitaxial p-layer formation discussed in the preceding section will take place during the entire heating cycle, as well as during cooling.

  2.2.2.1 Epitaxial Mesas.   As was noted earlier, after contact holes are opened to the Si substrate, a thin layer of native SiO_2 is still present on the Si. During heat treatment after Al and Si deposition, the Al reduces the SiO_2. While this is taking place, however, the Si overlay diffuses into the Al; then, as was the case for the alloy penetration, a reaction occurs with the Si exposed through the local defects in the SiO_2. The Al is supersaturated with Si, and epitaxial Si growth takes place rather than alloy penetration. After all of the native SiO_2 has been reduced, the epitaxial growth occurs over the entire contact, and the final structure, shown in Figure 2.6, consists of Al-doped Si epitaxial "mesas" on a continuous Al-doped Si epitaxial layer. The presence of the epitaxial mesas is easily confirmed by direct observation; the presence of the continuous layer is more difficult to demonstrate because it is so thin, only a few hundred angstroms. Reith and Schick (12) prepared heat-treated, Si-doped Al diodes specially so that the Al metallization over half of the contact was removed *after* heat treatment. Figure 2.7a shows an SEM emission-mode micrograph of the contact after half of the Al has been removed; the epitaxial mesas are clearly visible. In the electron beam-induced current (EBIC) mode, the SEM display shows a bright spot at the location of a junction, where the built-in potential prevents the recombination of electron–hole pairs. Figure 2.7b shows the contact in the EBIC mode; the area between the mesas is bright, indicating the presence of a p-layer on the n-type Si substrate. Figure 2.7c is a superposition of Figures 2.7a and 2.7b. After a very short Si etch, the thin p-layer has been removed (Fig. 2.7d). The epitaxial mesas have been disturbed very little. From etch rate and electron beam penetration considerations, these authors estimate that the p-layer is about 200 to 400 Å thick. As was the case with pure Al SBDs, the effect of the p-layer is to increase the barrier height (see Fig. 2.7e), and the calculation of the potential given in the preceding section

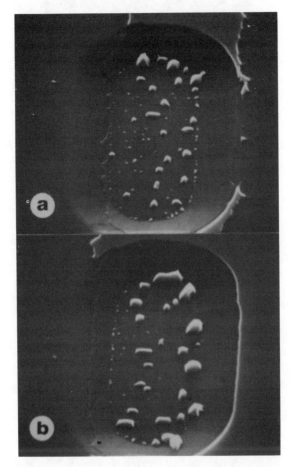

Figure 2.6.   Scanning electron micrographs, 4500 ×, of epitaxial mesas formed after sintering for 1 h at 400°C: (*a*) narrow Al–Si land extending from top of contact; (*b*) wide Al–Si land extending to right of contact. Courtesy of T. M. Reith.

applies exactly. We will forego discussion of the detailed electrical characteristics of the junction, which are outside the scope of this chapter and which are to be published elsewhere shortly (17).

2.2.2.2   *Aging.*   The presence of the *p*-layer can lead to an aging effect in Si-doped Al SBDs. Reith (18) describes an experiment in which diodes were sintered for 1 h at 500°C and then quickly cooled to room temperature. This treatment resulted in *p*-layer formation, and the quick cooling held the doping

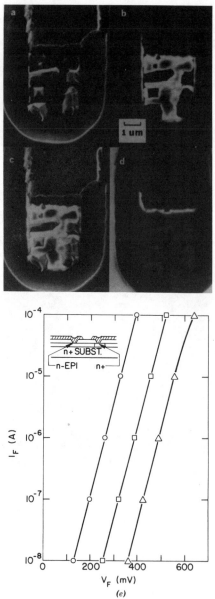

Figure 2.7. SEM analysis of SBD structures: (*a*) emission-mode SEM micrograph; (*b*) EBIC-mode SEM micrograph; (*c*) double exposure of emission-mode and EBIC-mode micrographs; (*d*) double exposure after etching for 3 s in $HNO_3:H_2O:HF$; (*e*) forward *I–V* characteristics of Al and Al–Si SBDs. (○) Al, 400°C for 1 h; $\phi = 0.68$ V, $n = 1.06$. (□) Al–Si, 400°C for 1 h; $\phi = 0.79$ V, $n = 1.06$. (△) Al–Si, 500°C for 1 h; $\phi = 0.89$ V, $n = 1.07$. From reference 12.

of the p-layer at the solubility of Al in Si at $500°C$ ($\sim 10^{18}$ to 10^{19} cm^{-3}). The diodes were then sintered at temperatures between $225°$ and $300°C$. The barrier heights of the diodes were monitored as a function of time at temperature. As Figure 2.8 shows, the barrier height varies linearly with $t^{1/2}$ before

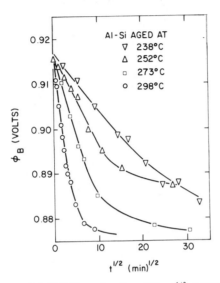

Figure 2.8.   Barrier height of Al–Si SBD as a function of (time)$^{1/2}$ at several aging temperatures. From reference 18.

the effect saturates. The variation is attributed to the precipitation of the Al dopant from solid solution in the epitaxial Si layer, where it is electrically active, to sites where it is electrically inactive, such as dislocations and stacking faults (Al is not active electrically unless it is in substitutional solid solution). This implies that the change in barrier height is due to a change in N_A. For illustration, the equation for the maximum potential due to the presence of the p-layer, Eq. 1, is rewritten with Eq. 3 substituted for the depletion width:

$$\phi_m = \frac{q}{2\varepsilon} \left\{ \left[\frac{(N_A + N_D)^2}{N_A} a^2 + \frac{N_A + N_D}{N_A} N_D w_0^2 \right] \right.$$
$$\left. -2a \left[\frac{(N_A + N_D)^3}{N_A^2} N_D a^2 + \left(\frac{N_A + N_D}{N_A} \right)^2 N_D^2 w_0^2 \right]^{1/2} + \frac{N_A + N_D}{N_A} N_D a^2 \right\} \quad (4)$$

For $N_A \gtrsim 10 N_D$ and $a > 300$ Å, which are reasonable for this situation, only the first term on the right side of Eq. 4 is significant, and therefore the equation reduces to

$$\phi_m \simeq k N_A \qquad (5)$$

This is illustrated in Figure 2.9, which shows that the barrier height is directly proportional to N_A in the region of interest: $N_A > 10^{18}$ cm^{-3}. In Figure 2.10, the squares of the slopes obtained from Figure 2.8 are plotted versus re-

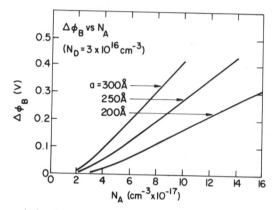

Figure 2.9. Change in barrier height as a function of *p*-layer doping and thickness as determined from Eq. 4. Courtesy of T. M. Reith.

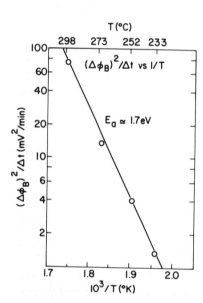

Figure 2.10. Square of slopes from Figure 2.8 vs. $1/T$ from which the activation energy for the process is found. From reference 18.

ciprocal temperature, to yield an activation energy for the process of 1.7 eV. This value is about half of the value for Al diffusion through bulk single-crystal Si, which is reasonable for the highly disordered epitaxial layers, in which Al can diffuse along defects such as dislocations.

In this situation, an SBD acts as a sensor to determine diffusion data at low temperatures and extremely low concentrations of diffusing species.

2.2.3 Metal Silicide Formation

In circuit applications, the SBD parameter that is usually of the most interest is the forward voltage drop at some specified current level. With LSI, where the chip affords little room to vary the SBD area, the desired voltage drop is achieved by adjusting the barrier height of the diode, most often by using different metals in contact with the single-crystal Si substrate.

During process temperature excursions, most of the metals commonly used with Si to form SBDs will react with the substrate to form metal silicides (19–24), with the notable exception of Al. Silicide formation is advantageous for contamination reduction, since the critical metal–semiconductor rectifying interface is within the Si substrate, where it has not been exposed to the environment.

In working toward an understanding of silicide formation, the parameters most often studied have been minimum formation temperature, kinetics, diffusing species and diffusion mechanism, phases formed, and phase stability. A great deal of this work has been done by use of nuclear backscattering and x-ray diffraction methods; Mayer and Tu (19) provide an excellent review of these experiments.

2.2.3.1 Al–PtSi Reaction.

Platinum and palladium react to form silicides at very low temperatures, about 200°C. The activation energy for the process is 1.1 to 1.5 eV. This suggests a grain-boundary diffusion mechanism. Mayer and Tu (19), however, believe that bulk diffusion is taking place because the activation energy, at least for Pd_2Si, remains constant up to 700°C, indicating no change in mechanism. Also Pd_2Si has a very open structure that should make diffusion easy.

Platinum silicide (PtSi) is an important silicide because it has the highest reported barrier height to n-type Si (24–33) and therefore readily makes good Ohmic contact to relatively lightly doped p-type Si (the barrier height of any material to an n-type semiconductor plus that to a p-type semiconductor equals the semiconductor bandgap). SBDs of this type are usually fabricated by first opening contact holes through the SiO_2 to the Si substrate and then evaporating or sputtering Pt (32). The contacts are then sintered at 550 to 600°C, and PtSi forms in the contact holes only. The unreacted Pt is then chemically removed. The interconnection metallization, here Al, is then deposited over the PtSi. During subsequent thermal processing steps, however, the Al reacts with the PtSi, and the characteristics of the SBDs change (33). The barrier height is about 0.85 V at the outset; during sintering at temperatures from 350 to 450°C, it drops to 0.60 V and then increases to an Al-like value of 0.72 V. The same changes occur at all testing temperatures; only the time scale changes. Hosack (33) suggests that these observations can be explained by noting that the initial barrier height is simply that of PtSi to n-type Si. During sintering, the Al reacts with the PtSi to form $PtAl_2$ and also Si, which dissolves in the Al. The barrier height remains at the initial value until the $PtAl_2$ reaction front touches the Si substrate, and then drops to 0·6 V; this is assumed to be the $PtAl_2$–Si barrier height. Further sintering, suggests Hosack, might result in continued Al diffusion, with an Al layer forming at the Si surface to account for the increase of the barrier height to 0.72 V. Finally, Hosack determined an activation energy for the reaction of Al with PtSi by forming Al islands on blanket PtSi. The reaction length parallel to the substrate around the Al islands was optically monitored at temperatures from 426 to 550°C. The kinetics showed a $t^{1/2}$ dependence, indicating a diffusion-limited reaction with an activation energy of 0.77 eV.

This reaction has also been investigated by Sullivan et al. (34). Figure 2.11 shows the variation of barrier height and the ideality of a group of SBDs sintered for different times at 340°C. The general features of the barrier height variation are the same as those described above; however, microsections, SIMS, RBS, and AES have all failed to reveal the formation of an Al layer at the Si interface after extended sintering. In all cases, the $PtAl_2$ seems to

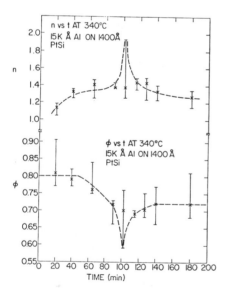

Figure 2.11. Variation of barrier height and ideality of PtSi–Si SBDs as a function of sintering time at 340°C. Variation is due to the reaction of PtSi with Al. From reference 34.

remain at the Si interface, although possibly an Al layer, too thin to be measured, has formed between the PtAl$_2$ and the Si.

This evidence, and the manner in which the ideality reaches a maximum at the same time that the barrier height is at the minimum, suggest the following alternative mechanism for the observed variations in the barrier height. As before, the barrier height is that of PtSi to Si at the outset and remains the same until the PtAl$_2$ reaction front touches the Si surface. With further reaction, however, it is now assumed by Sullivan et al. (34) that PtAl$_2$ and Al have about the same barrier height, 0.72 V. As the PtAl$_2$ reaction front touches the Si surface, the diode is composed of a large-area SBD of high barrier height ($\phi_{PtSi} \simeq 0.85$ V) in parallel with a small-area SBD of lower barrier height ($\phi_{PtAl_2} \simeq 0.72$ V). Both diodes also have a series resistance component, chiefly because of the resistance of the epitaxial layer, as Figure 2.12a shows. The figure also gives the expression for the forward voltage drop, including the series resistance term. Figure 2.12b shows ln I_F versus V_F for different diode areas—curve 1 for diodes of 100% PtSi and curve 2 for diodes of 100% PtAl$_2$. Curve 3 is the intermediate case, mentioned above, in which the PtAl$_2$ reaction front has just touched the Si surface, resulting in parallel diodes.

At low currents the voltage drop due to the series resistance of the small-area, low-barrier-height diode is negligible; therefore almost all of the current

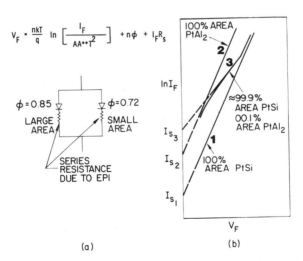

Figure 2.12. Parallel diode model to explain the electrical behavior shown in Figure 2.11: (a) parallel diodes with series resistance due to epitaxy; (b) forward I–V characteristics showing barrier height as determined from saturation current I_s. (1) $\phi \approx 0.85$ V; $n \approx 1$. (2) $\phi \approx 0.70$ V; $n \approx 1.7$. From reference 34.

flows through this part of the diode, and curves 2 and 3 are almost coincident. At high current, the voltage drop due to series resistance of the low-barrier-height diode is large because the area is small. The current takes the path of least resistance, which is the diode with high barrier height and large area. In this current region, curve 3 coincides with curve 1. At intermediate currents, there is a smooth transition from one behavior to the other, and the net diode $I-V$ characteristics are as given by curve 3. The overall mechanism during the reaction is a transition from curve 1 to curve 2 via an intermediate situation in which the parallel diodes create an artificially low barrier height (determined by $I-V$ measurements) because of the very large apparent ideality.

Sullivan et al. have also examined the kinetics of the Si-doped Al–PtSi interaction and found them identical, within the limits of their measurements, to those of pure Al and PtSi. A consequence of this work, which has great importance for shallow contacts, is that the reaction of Al-Si with PtSi is the only successful method for achieving an Al-like Schottky barrier without some formation of epitaxial mesas or Al-doped silicon layers.

2.3   INTERDIFFUSION BETWEEN METAL LAYERS

Interdiffusion between metal layers in integrated circuit technology is necessary for strong adhesive bonding, phase-in of needed physical and electrical properties, and transition from one medium to another. Often, fabrication at temperatures and other conditions chosen on the basis of bulk transport data has led to poor yield and excessive mixing. The major causes of degradation are the comparatively small sizes of grains in thin films and the short distances that need to be traversed by diffusants, so that what appear as anomalous surface-dominated diffusion profiles in bulk samples represent the typical film behavior. The fundamentals of these effects are discussed elsewhere in the present book and are not to be pursued further here. In the following discussion, we shall concentrate on examples illustrating possible yield implications and describe the effectiveness of intermetallic diffusion barriers in cases for which interdiffusion must be kept at a minimum.

2.3.1   Gold–Transition Metal Layers

In applications for which gold is used as a conducting metallurgical or non-corrosive layer, an underlayer must be provided for adhesion (35). Usually, thin films of metals such as titanium, chromium, or tantalum make satisfactory transition layers. Many applications, however, require subsequent processing at 400 to 500°C, which has serious effects on the gold–metal layers. Experience with the Au–Cr system can serve to illustrate the nature of the problem. Munitz and Komem (36), for example, in the process of examining gold films, used a 300 Å Cr adhesive layer on an oxide substrate. Annealing of the film resulted in gross changes in the top surface of the gold and local delamination from the oxide.

This led to a study of Au–Cr layers annealed in air and vacuum between 280 and 450°C by resistivity change, SEM, and TEM methods. The resistivity data for air annealing are shown in Figure 2.13 for an 8000 Å film. At tem-

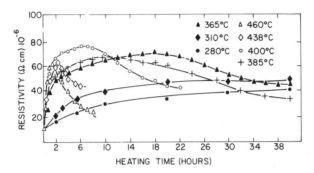

Figure 2.13.   Resistivity of Cr–Au layers as a function of time for Au films 8000 Å thick. From reference 36.

peratures above 350°C, resistivity rises rapidly to values about 10 times the original and then undergoes a slow decay. This is similar to results reported earlier by Rairden et al. (37), in which maximum resistivity was observed. The same behavior was found for 500 Å Au films, except that resistivity increased faster. TEM (in the thin Au samples) and SEM results were obtained with time at 385°C. The sequence of events appeared to be as follows: (1) grain boundary grooving was observed in networks whose mean diameter was much larger than the Au grain size; (2) Cr_2O_3 particles were observed in Au grain boundary intersections, and a thin polycrystalline Cr_2O_3 layer was present over the entire surface after annealing for about the same time as was required to reach the maximum resistivity; (3) the grain boundary grooves evolved into open channels arranged in closed loops, and single crystals of Cr_2O_3 started to grow in the center of each loop at the expense of the earlier particles; (4) the Cr_2O_3 crystals grew larger until the Cr layer was completely gone, at which time the resistivity approached the original values.

Several observations can be made from the data. The presence of grain boundary grooving and Cr_2O_3 particles at boundary intersections, and the coincidence of the resistivity maximum (lattice saturation) with the appearance of surface Cr_2O_3, indicate that significant lattice diffusion and grain boundary diffusion are occurring simultaneously. Indicative of the grain boundary effects are the slower rates of increase of resistivity found in thicker

Au films, where the grains are significantly larger; lattice-dominated kinetics would not show dependence on grain size. A Fisher-type (38) boundary diffusion process appears to be active in which the lattice becomes saturated with Cr diffusing from the boundary areas as well as from the Au–Cr inter-face; thus, Cr enters the lattice at higher rates than in the depletion process that occurs after saturation, in which, basically, Cr at the top surface is removed by oxidation through the thin native oxide. Holloway (39) used Auger methods for profiling to study the top oxide, which produced high-resistance electrical contacts.

It can be postulated that the oxidation process itself plays an important role in the kinetics by acting as a sink for the Cr ions diffusing to the surface. This could, in effect, increase the flux by setting the concentration of free Cr at zero at the external surface. Resistivity curves for air- and vacuum-annealed samples are shown in Figure 2.14, where two characteristics of the vacuum sample can be noted: (1) the resistivity does not decrease by Cr depletion once the maximum level has been achieved and (2) the rate of resistivity increase is lower. The value of the maximum resistivity, about 88 $\mu\Omega$-cm, is approxi-mately equal to that calculated from the solubility level of Cr in Au at the annealing temperature and the published values for resistivity of Cr in Au.

The open channels observed are difficult to interpret. Mechanisms were proposed suggesting stress relief and free energy lowering as driving forces for the formation of the large Cr_2O_3 particles by long-range surface diffusion of Cr and accumulation of vacancies in loops around the particles to form the grooves and channels. As a separate consequence of the reaction, the channels could result in direct shorting to the substrate of an overlayer deposited onto the Au film.

The results of this work illustrate a wide variety of modes in which metal layers can be degraded and thus should provide fruitful paths for analysis and additional studies.

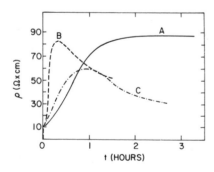

Figure 2.14. Resistivity of Au films heat treated at 450°C: (A) 500 Å layer in vacuum; (B) 500 Å ($t_{Cr}/t_{Au} \sim 0.6$); (C) 8000 Å ($t_{Cr}/t_{Au} \sim 0.4$) layer in air. From reference 36.

2.3.2   Aluminum–Polysilicon Interactions

With the increasing importance of polysilicon gate FET structures in the electronics industry, it has become of interest to consider another instance in which low-temperature interdiffusion can adversely affect device integrity. In certain configurations, gate structures can consist of polysilicon gate contacts with conducting aluminum overlays directly over the gate oxide. Metallurgical changes in this channel region can lead to changes in the electrical characteristics. Work by Nakamura et al. (40), for example, has illustrated complete mixing of Al and polysilicon layers after a nominally low stress of about $\frac{1}{2}$ h at 400°C, which in an FET environment would effectively create a metal gate situation. For thin oxides, this could result in the degradation of reliability, in accordance with the widely held view that polysilicon provides more breakdown-resistant gate structures (41). The following set of experiments (42) was undertaken to illustrate the phenomenon further and to provide some idea of the electrical breakdown characteristics to be expected.

Experiments were performed on capacitor structures of the MSOS type (Al–polySi–SiO$_2$–Si) with layers of 250 to 500 Å SiO$_2$, 0.1 to 0.5 μm polysilicon, and 1 μm Al. Stress testing was done at 4×10^6 V cm^{-1} and 250°C; pretest anneals were done at 400°C, usually for 1 h. The effects of doping and high-temperature argon annealing of polysilicon layers were also investigated.

The time dependence of the reaction was monitored by capacitance change. The capacitance of the initial unreacted layered structure was characteristic of the SiO$_2$–polysilicon layers, about 40 pF for the 5000 Å polysilicon layer. As mixing occurred and aluminum began to contact the SiO$_2$, the capacitance rose to about 120 pF, typical of MOS values. Figure 2.15 shows the change

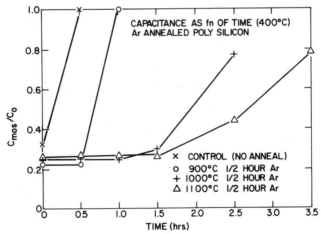

Figure 2.15. Time dependence of the Al–polysilicon reaction, monitored by capacitance change, as a function of prior Ar annealing. Courtesy of P. L. Garbarino and J. R. Gardiner.

with time at 400°C as a function of the prior argon annealing temperature of the polysilicon. The unannealed sample shows massive restructuring after only 0.5 h, as was previously reported (40), and the anneal at 1100°C led to a significant decrease in reactivity. The major effect of annealing would appear to be the increase in grain size of the polysilicon. This would be consistent with the previous findings (40) that crystallization by precipitation of silicon dissolved in the aluminum during the initial stages of the interaction depends on the relative free energy of the crystallites with respect to the energy of the underlying polysilicon layer. The smaller the grain size of the layer, the greater the probability that crystallites will be stable in the aluminum. For example, in Al–single-crystal Si contacts, no crystallites are found in the Al layer. The structure of the reacted layers, shown in Figure 2.16, is similar to that reported by Nakamura et al. (40), that is, large crystallites of silicon over the porous polysilicon underlayer.

Typical dielectric breakdown results are shown in Figure 2.17, in which samples were stress tested at room temperature before and after being annealed for 30 min at 400°C. The unannealed sample (Fig. 2.17a) has a sharp breakdown characteristic, about 10^7 V cm^{-1} at about $+53.5$ V. After

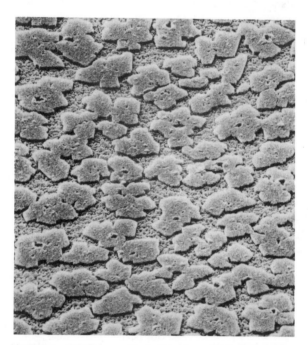

Figure 2.16. Aluminum–polysilicon reaction after 1 h at 400°C exposes large crystallites of Si over the porous polysilicon underlayer. Courtesy of P. L. Garbarino and J. R. Gardiner.

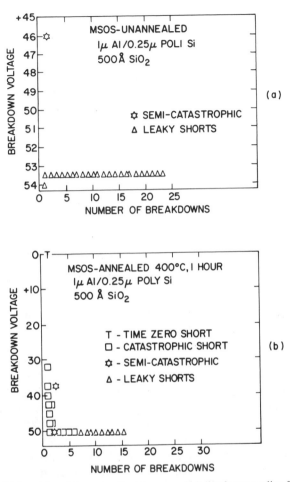

Figure 2.17. Dielectric breakdown results: (*a*) unannealed; (*b*) after annealing for 1 h at 400°C. Courtesy of P. L. Garbarino and J. R. Gardiner.

annealing (Fig. 2.17*b*), many lower voltage breakdowns occur, as is typical for standard aluminum-gate MOS structures. Once the first breakdowns have occurred, however, the time-dependent behavior at 250°C, 4×10^6 V cm^{-1}, is relatively independent of anneal, oxide thickness, or structure. This is shown in Figure 2.18, in which, except for the initial number of early fails, the curves are parallel for MOS and MSOS samples. Only in the thinnest oxide layer, 250 Å, did annealed MSOS and MOS differ in the number of early fails; the latter is especially susceptible. From these results, the conclusion is that

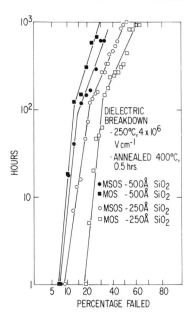

Figure 2.18. Time-dependent breakdown for MSOS and MOS structures. Courtesy of P. L. Garbarino and J. R. Gardiner.

annealing at 400°C causes a higher incidence of lower-voltage breakdowns and early fails but that long-term reliability failure rates are about the same in annealed, unannealed, and MOS controls.

One other point of interest pertains to phosphorus-doped polysilicon layers. The capacitance change during annealing at 400°C was inhibited to the point that the aluminum–polysilicon reaction took about 1.5 to 2 h to reach completion without argon anneal. Breakdown results, shown in Figure 2.19 for a 250 Å oxide layer, indicate that a 0.5 h anneal at 400°C produced no change in early fails and that only after 1.5 h did the curves look like those in Figure 2.18 for 250 Å oxide layers. The mechanism is not clear at this time and is being explored further; certainly the absence of any increase in grain size suggests that a diffusion barrier effect may be present.

2.3.3 Diffusion Barriers

From the above interactions and from the preceding sections on ohmic and Schottky contact degradation, it has become increasingly clear that barriers to diffusion must be provided between active layers. For single elemental barriers such as those described in references 42 to 45, restriction of interdiffusion is related basically to the diffusion coefficients through the barrier, which, as was shown in the Au–Cr study above for Cr barriers, is not very effective when temperatures are in the range of 400°C. The most effective barrier should be intermetallic compounds formed either between the barrier metal

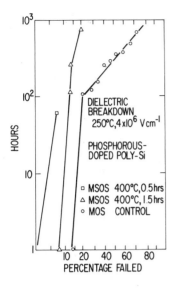

10^3

10^2

HOURS

10

1

DIELECTRIC
BREAKDOWN
250°C, 4×10^6 V cm^{-1}

PHOSPHOROUS-
DOPED POLY-Si

□ MSOS 400°C, 0.5 hrs
△ MSOS 400°C, 1.5 hrs
○ MOS CONTROL

10 20 40 60 80
PERCENTAGE FAILED

Figure 2.19. Time-dependent breakdown for phosphorus-doped polysilicon structures.

and one of the layers or between the layers themselves. The diffusion characteristics could then be significantly different from those in the base metal layers. To illustrate the formation of compounds as barriers to Al, two examples will be used—Cr and Hf—which represent parabolic and nonparabolic compound growth behavior. A more complete investigation was made by Howard et al. (46).

The Cr data will be extracted from the work of Howard et al. (46), in which the sample configuration shown in Figure 2.20 was used as the diffusion couple and the interdiffusion and compound formation kinetics were measured by resistivity change, nuclear backscattering, and x-ray diffraction methods. Backscattering was used to show that $CrAl_7$ formed symmetrically at both interfaces A and B. As the reaction proceeded, the top Al layer was entirely consumed and the $CrAl_7$ was converted to Cr_2Al_{11}. The kinetics of the reaction were monitored by the sheet resistance method, which made possible sensitive determination of residual Al thickness under the conditions of limited solubility. If the compound has thickness X and forms a fraction X/X_0 of the total thickness, then

$$\frac{X}{X_0} = \frac{\Delta G_s(t)}{\Delta G_s(t=\infty)} \simeq \frac{R_s(t) - R_s(t=0)}{R_s(t) R_s(t=0)} \tag{6}$$

where $\Delta G_s(t)$ is the change in conductance decrease with time, $\Delta G_s(t=\infty)$ is the total change in conductance, and R_s is the sheet resistance. Plotting $\Delta G_s(t)$ versus time gives the rate at which the compound layer forms. As

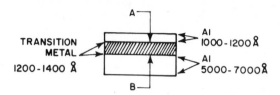

Figure 2.20.   Sample thickness configuration used in Al–transition metal kinetic studies. From reference 46.

Figure 2.21 shows, the results over a range of temperature are consistent with parabolic diffusion, $\Delta G_s(t) \propto t^{1/2}$, indicating a planar reaction process controlled by diffusion through the forming compound. An Arrhenius plot of the data gave an activation energy of about 1.91 eV, which most probably represents transport of Al through $CrAl_7$ (46). In summary, the formation of $CrAl_7$ and the resultant growth are quite representative of a physical barrier whose effectiveness increases with time by thickness increase. Implications with respect to electromigration-resistant configurations will be discussed in a later section.

The nature of the Al–Hf interaction is distinctly different and represents perhaps the ideal class of barrier, whose effectiveness is not thickness dependent but rather is immediate (i.e., diffusivity $\rightarrow 0$). This system has been

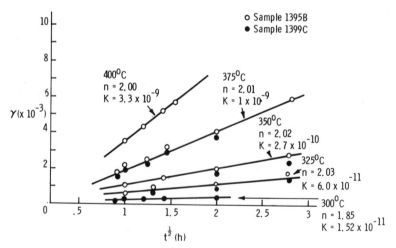

Figure 2.21.   Plot of γ (proportional to compound thickness) vs. $t^{1/2}$ for $CrAl_7$ (○) and Cr_2Al_{11} (●). From reference 46.

examined in detail by Lever et al. (47) and Smith et al. (48). The samples used were similar to that in Figure 2.20, that is, thick Al underlayer (6000 Å), thin Hf (1000 Å), and thin Al (1000 Å). Depth profiles were obtained by use of backscattering and Auger electron spectroscopy (AES) combined with Xe^+ sputtering. The composition profiles before and after annealing for 2 h at $400°C$ are shown by the AES scans of Figures 2.22a and 2.22b, respectively, where A and B are the original Al–Hf interfaces. Compound formation at A is inferred from the flattened regions in the Al and Hf profiles. The absence of this flattening and the formation of long tails in the Hf profile at B indicate that what is happening is not the formation of a compound by a planar reaction process but rather a diffusion of Hf into the thick Al layer. The structure of the layers was determined by TEM after ion thinning to different levels. The top interface (A) was converted to $HfAl_3$, but the bottom structure consisted of needle-like precipitates of $HfAl_3$ at Al grain boundaries (Fig. 2.23). Clearly, the symmetrical behavior of the Al–Cr system was not repeated with Al–Hf.

The most surprising finding was that the residual Al layer on top, after the initial reaction, remains stable through the rest of the interdiffusion process. The aluminum needed to continue formation of the compound is supplied from the bottom layer, below B, and the growth interface, $HfAl_3$–Hf, moves toward B. The reaction stops when the Hf layer is depleted. A schematic representation of the cross section is shown in Figure 2.24. From these results it is apparent that neither Al nor Hf can diffuse readily through the $HfAl_3$ phase.

The kinetics of the Al–Hf reaction are not parabolic. Howard et al. (46), after comparing the kinetics for many systems by the resistance method, divided compounds into two types: those with high melting point, such as $HfAl_3$ and $ZrAl_3$, and those with lower melting point, such as $CrAl_7$ and $PdAl_3$. For the latter compounds, diffusion and parabolic behavior are evidenced. For the former, however, the nonparabolic kinetics lead to speculation (46) that compound growth occurs by Harrison's type C diffusion (49), in which only grain boundary penetration occurs, with minimal lattice contribution. This is consistent with the short-time conductance data for $HfAl_3$, the backscattering data that show an irregular growth interface, and the work of Baird (50) illustrating the relationship between the irregular interface and nonparabolic kinetics with preferential grain boundary compound formation.

Poor deposition conditions can lead to excessive oxidation of the transition metals (Cr, Ta, Ti, etc). In some cases this can inhibit or prevent compound formation with Al.

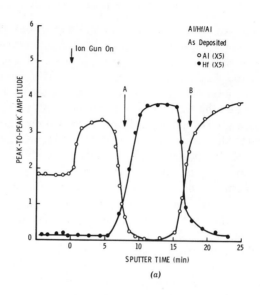

(a)

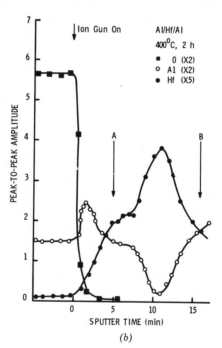

(b)

Figure 2.22. Depth profile obtained by Auger spectroscopy, expressed as Auger peak-to-peak amplitude vs. sputtering time for (*a*) unannealed samples and (*b*) samples annealed at 400°C for 2 h. From reference 47.

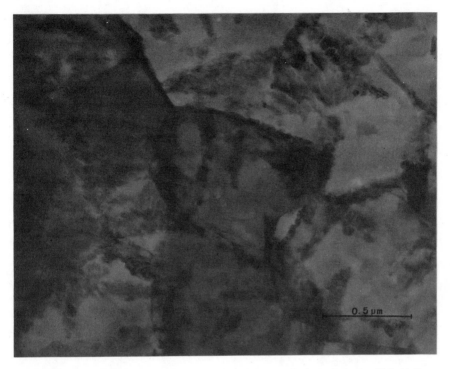

Figure 2.23. Needle-shaped precipitates of $HfAl_3$ at Al grain boundaries in thick Al film, below B in Figure 2.20. From reference 47, courtesy of P. J. Smith.

2.4   ELECTROMIGRATION

2.4.1   Conductor Metallurgy

The trend in integrated circuit technology is toward smaller device structures and for greater packing density on the chip and the circuit level. A smaller device geometry requires a reduction in the linewidth for metal interconnections. Several investigators (51–52) have shown that the median failure time for electromigration is dependent on the linewidth of the conductor. For Al or Al–Cu metallization structures, the median failure time was reduced by a factor of 10 or more when the linewidth was reduced from 10–15 μm to 1–2 μm.

A new method for improving the electromigration capacity of Al or Al–Cu metallization for integrated circuit applications was recently reported (53). A brief synopsis of the data is given here. For linewidths of 1 to 2 μm, the median lifetime can be increased to as much as 100 times that of Al–Cu. The distribution in the failure times is comparable to or better than that of Al–Cu. The vast improvement is due to the formation of intermetallic compounds by

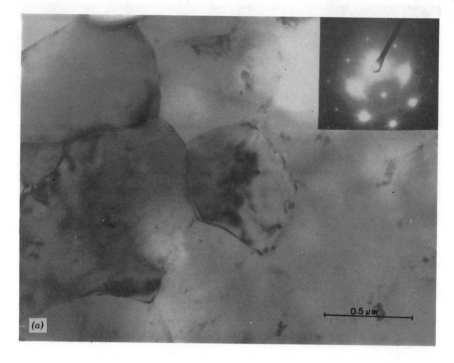

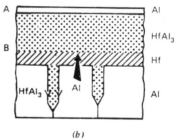

Figure 2.24.   TEM analysis of Al–Hf–Al structure: (*a*) transmission electron micrograph show-ing unreacted Al at A in Figure 2.24(*b*), courtesy of P. J. Smith; (*b*) schematic of Al–Hf–Al reaction.

reacting Al or Al–Cu with various transition metals. Life test data for narrow lines of Al, Al–Cu, and Al–Cu–Si fabricated and tested in the same laboratory are included for comparison. Uniform films were processed with the test stripes to determine the extent of compound formation during annealing. The properties of Al–transition metal compounds have been discussed elsewhere (46–48).

The substrates for metal film deposition were oxidized silicon wafers. These were covered with a film of electron beam-sensitive photoresist and then exposed to yield a lift-off pattern of the test structures. The films were prepared by codeposition of Al and Cu from two sources to yield about 5000 Å of Al–4% Cu (wt %). During the same pumpdown, a layer of the desired transition metal 500 to 600 Å thick was deposited; the layered structure was completed by adding another 5000 Å Al–Cu. The test stripe, shown in Figure 2.25, consists of a 254-μm-long stripe of variable width; the structure was

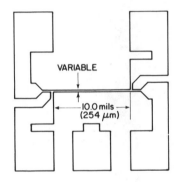

Figure 2.25. Test stripe configuration for electromigration evaluation of Al, Al–Cu, and Al–Cu/intermetallic structures. From reference 53.

designed for measurement with a four-point probe. After film deposition and lift-off, the stripes were passivated with a 2 μm layer of SiO_2. The wafers were diced and mounted on TO-5 headers; the outside contacts were made by wire bonding. Figure 2.26 shows optical photographs and scanning electron micrographs of the stripes 1 to 2 μm wide. The effective linewidth (μm) was calculated from the ratio $R_s L/R$, where R_s is the sheet resistance ($\Omega/\square$) of a uniform metal film heat treated under the same conditions as the test stripes, L is the line length (μm), and R is the measured stripe resistance (Ω).

The electromigration tests were conducted with a current density of 1×10^6 A cm^{-2} (about 20×10^{-3} A) and a temperature of 250°C. The tests were continued until 50% or more of the stripes had failed. A failure was registered when an actual open region occurred or the stripe resistance exceeded a predetermined level. The cumulative failure data were found to fit a lognormal probability plot; the median failure time and standard deviation were computed from the graph.

In Figure 2.27 the median failure times for Al–Cu intermetallic structures (HfAl$_3$, TaAl$_3$, etc.) are plotted with the data of Scoggan et al. (52) for comparison. The Al, Al–4% Cu and Al–6% Cu data are in good agreement. The failure times for lines 1 to 2 μm wide (Al–Cu intermetallic), however, are

Figure 2.26.   Analysis of a conductor stripe 1 to 2 μm wide: (a) optical photograph (2500 $\times$);
(b) scanning electron micrograph, before glassing. From reference 53.

considerably greater than even those for the 10μm-wide Al–Cu structures. The standard deviation of the log-normal distribution also shows a width dependence. In Figure 2.28 the values for Al, Al–Cu, and Al–Cu intermetallic structures are plotted with the data of Scoggan et al. TiAl$_3$, TaAl$_3$, and CrAl$_7$ yield values comparable to the values for Al–Cu. HfAl$_3$ exhibits a much lower σ value, possibly because of its unique microstructure, which consists of inter-granular precipitates (47, 48).

A more complete discussion of the effect of microstructure on electromigra-

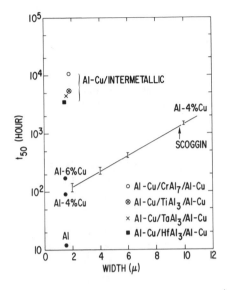

Figure 2.27. Median failure times for Al–Cu/intermetallic structures (HfAl$_3$, TiAl$_3$, TaAl$_3$, CrAl$_7$) are compared to those of Al and Al–Cu. From reference 53.

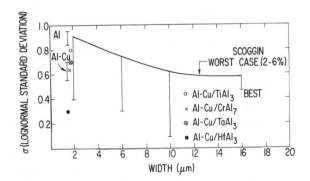

Figure 2.28. Standard deviation in failure times for Al, Al–Cu, and Al–Cu/intermetallic structures.

tion will be published in the future (53). Here the results can be summarized as follows:

1. Intermetallic layers of CrAl$_7$ or TiAl$_3$ provide an excellent diffusion barrier for Al, and block voids from penetrating the stripe. However, the inter-metallic layer forms a structurally coherent electrical contact between the top and bottom layers of Al or Al–Cu.

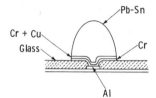

Figure 2.29.   Terminal metallurgy structure for integrated circuit contact. From reference 55.

2. HfAl$_3$ and, to a lesser extent, TaAl$_3$ form at Al grain boundaries and reduce the electromigration rate; also they act as a barrier to void growth.

2.4.2   Contact Metallurgy

Two important physical processes, liquid metal embrittlement and grain boundary melting, are related to the existence of liquid layers at grain boundaries in polycrystalline materials. The formation and movement of these liquid zones in thin films results from an interaction between the solder pad and the conductor stripe (54, 55). The terminal metallurgy in integrated circuit contacts is depicted in Figure 2.29. The Cr layer has two roles: it provides adhesive connection between the 95 : 5% Pb–Sn solder pad and the Al stripe, and during reflow it serves as the main barrier to diffusion between the molten solder and the stripe. Under certain processing conditions, however, the Cr layer is not continuous, and solder–stripe interactions can occur. In the case of Al–Sn reaction, the direction of liquid migration is toward the positive terminal. X-Ray stress topography was found to be a very sensitive and nondestructive method of tracking the zone migration.

A (220) x-ray topograph (Mo$K\alpha$ radiation) of a sample before testing (Fig. 2.30) shows the outline of a film 7580 Å thick superimposed on an image of dislocations in the silicon substrate. The film, because of its negligible absorption, appears transparent to the incident x-ray beam, but the edges are visible because of stress risers at the stripe boundaries. The Pb–Sn dot, because it totally absorbs the x-ray beam, is displayed as a null contrast region (arrow 1). After 66 h of testing at 150°C (0.37×10^6 A cm^{-2}), the alloy zone was detected (arrow 2) on the anode side of the dot (Fig. 2.30b). After 16 h of additional testing at 120°C (0.34×10^6 A cm^{-2}), the alloy zone continued to move toward the anode. A traveling microscope examination of Figure 2.30d (Mo$K\alpha$ radiation) revealed a 2080 μm displacement from the location in Figure 2.30c, which corresponds to a velocity of 130 μm h^{-1}. When the testing was continued at 150°C (0.37×10^6 A cm^{-2}), the stripe failed because of local film melting.

In parts b and c of Figure 2.30, the contrast between the Al stripe and the

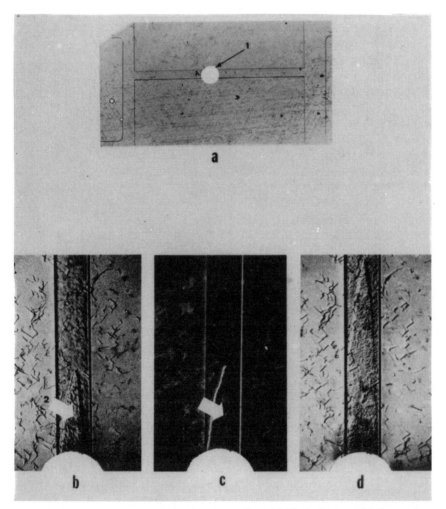

Figure 2.30. X-Ray stress topographs: (a) stripe outline with Pb–Sn dot (arrow) before testing; (b) after 66 h of testing; (c) contrast reversal at boundaries between alloy zone and stripe; (d) alloy zone movement toward anode after 16 h of additional testing. From reference 55.

alloy inclusion is reversed. This effect, a consequence of anomalous transmission, can be used to deduce the sign of the stress in the alloy region (55). The concentration of Sn in the zone was determined by point count analysis with an electron microprobe (Fig. 2.31). A sample tested at $150°C$ (0.39×10^6 A cm^{-2}) was found to show zone spreading in the positive terminal (Fig. 2.31, insert). The Pb–Sn dot was chemically removed and examined with an

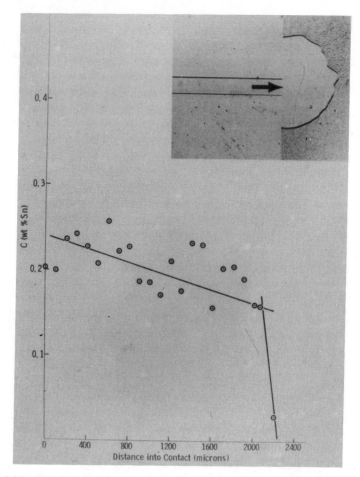

Figure 2.31.   Distribution of Sn across alloy zone (insert), as determined by point count analysis with an electron microscope. From reference 55.

electron microprobe; the SnLα scan was started at the interface of stripe and contact and proceeded in a straight line across the zone. The Sn concentration was found to be about 0.2 to 0.3 wt % at the interface and remained nearly constant across the zone; at the edge of the zone, about 2100 μm in, it dropped abruptly. Since the solid solubility of Sn in Al is estimated to be 10^{-3} wt % at the eutectic temperature, the measured concentration reflects the increased solubility of Sn in the liquid phase.

A direct image showing the relationship between grain structure and alloy was obtained by transmission electron microscopy. Specimens were prepared from the stripe of a sample similar to the one in Figure 2.30. Since the samples were viewed in a 200 kV electron microscope, the 6000 Å thick film could be examined without thinning. To protect the film from etch damage during the chemical removal of the silicon substrate, the samples were prepared with a 700 Å thick layer of silicon nitride (etchant barrier) separating the Al from the SiO_2 layer. In the electron micrograph from the stripe region (Fig. 2.32), the increased electron transmission at the grain boundaries shows that the boundaries are preferentially dissolved as the liquid zone is swept along the stripe by electromigration.

The experimental results demonstrate that conductor stripes can melt

Figure 2.32.   Transmission electron micrograph of stripe area, showing local grain boundary dissolution (arrow) due to zone migration. From reference 55.

during accelerated testing of integrated circuit devices. Stripe failure was tracked to the formation of Al–Sn eutectic at the grain boundaries (55); catastrophic melting was noted in some cases and attributed to excessive anode Joule heating due to thinned grain boundaries. The liquid migration rate, as controlled by electromigration, was measured and related to the product term DZ_1^*, where D is the diffusion coefficient in the liquid and Z_1^* is the effective charge.

2.5 IN-PROCESS CORROSION OF ALUMINUM ALLOY FILMS

One of the technological impacts of modifying aluminum alloy conductors to gain electromigration resistance is the possible introduction of a new set of reliability and degradation problems. A problem that caused concern but failed to materialize had to do with the use of copper dopants in the region of silicon contacts, where poisoning by copper is well known. Fortunately, the segregation coefficient of copper between polycrystalline Al and single-crystal silicon was high enough to prevent silicon contamination. However, another problem related to corrosion of Al–Cu lines (56) caused high yield loss and necessitated significant changes in film deposition and cleaning procedures. The appearance of the degradation in a circuit pattern is shown in Figure 2.33, in which specific lines appear to have been attacked. Figure 2.34 shows SEM photographs of an attacked area before and after ion etching of the top surface, in which intergranular fissures are shown to be prevalent and the damage appears to be more toward the bottom of the stripe. This can be seen better in Figure 2.35, where the damage becomes denser as ion milling proceeds from top to bottom of a stripe area. The TEM photograph in Figure 2.36 shows the open regions to be confined to grain boundaries surrounding large grains of Al–Cu and $CuAl_2$ (θ phase) particles. By selective area diffraction, some of the surface particles were identified as mainly $AlCl_3$, $AlOCl$, and $6CuO–Cu_2O$.

The intergranular nature of the damage and the presence of aluminum chlorides as corrosion products leads to the conclusion that electrochemical cell reactions similar to those found in bulk Al–4% Cu alloys are occurring in the films. In bulk, the attack is accelerated in the boundary regions by the presence of $CuAl_2$ particles surrounded by zones of copper depletion in the adjacent grains. The electrode potentials in the structure are

	Potential, V
Al + 4% Cu solid solution	−0.69
$CuAl_2$ (θ phase)	−0.73
Al + 2% Cu solid solution	−0.75
Al	−0.85

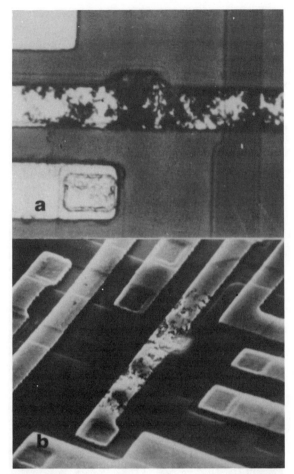

Figure 2.33. Localized attack of interconnection metallurgy: (*a*) optical micrograph; (*b*) scanning electron micrograph of same area. From reference 56.

From these values, regions of low copper appear anodic with respect to the base alloy and therefore are preferentially attacked in the presence of an electrolytic solution. With chlorine present, the reaction forming Al^{3+} ions is greatly accelerated (57). Briefly, the source of chlorine was found to be related to cleaning procedures in which either trichlorethylene or Freon* was used in conjunction with water and alcohol. It is well established that

*Registered trademark, E. I. du Pont de Nemours.

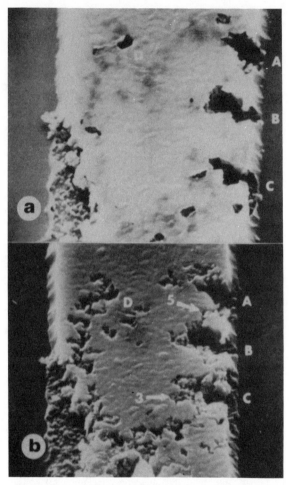

Figure 2.34. Scanning electron micrograph of attack area before (*a*) and after (*b*) ion etching of top surface. From reference 56.

water contamination of chlorinated solvents leads to hydrolysis reactions, which free chloride ions.

In a completely homogeneous film, electrochemical reactions would not be expected. In the present case, however, the deposition method was a major contributor to nonhomogeneous copper distribution. Figure 2.37 shows copper profiles through the thickness of films, obtained by an ion microprobe

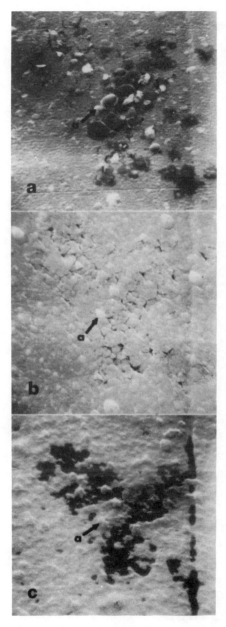

Figure 2.35. Scanning electron micrographs showing intergranular fissuring: (*a*) as deposited; (*b*) after ion etching about one third of the Al–Cu metal: (*c*) after thinning all but about 500 Å. From reference 56.

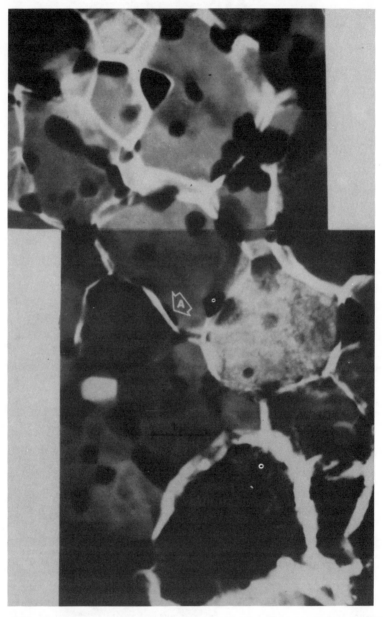

Figure 2.36. Transmission electron micrograph shows open regions at grain boundaries surrounding CuAl₂ particles and large grains of Al–Cu. From reference 56.

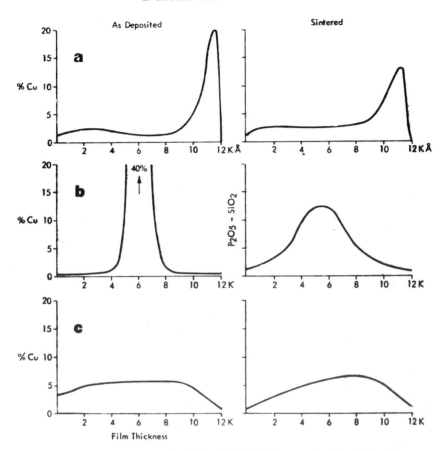

Figure 2.37. Ion microprobe profile of Cu in Al–Cu thin films as a function of deposition process: (a) Al–Cu single-source alloy; (b) layered Al–Cu–Al; (c) codeposition of Al–Cu from two sources. From reference 56.

method. The alloy evaporation method (single source) generally used to generate Al–Cu films is characterized by very high copper concentrations at the substrate interface ($\sim 20\%$), even after postdeposition sintering at $400°$C. The bulk of the film is relatively low in copper and contains a low density of $CuAl_2$ particles. These films are especially susceptible in the region near the high copper content, where wide variations in the composition of the various phases are to be found. Figures 2.34 and 2.35 illustrate this by showing the attack to be denser near the substrate interface. Other methods of deposition were attempted, notably an Al–Cu–Al sandwich deposition and a low-temperature, two-source codeposition; these gave the profiles shown in

Figure 2.37. Both proved to be less susceptible—the sandwich structure because of anodic protection of the alloy by Al on both sides and the codeposition because of the uniformity of copper concentration.

In summary, the grain boundary nature of diffusion and precipitation in thin films, the manner in which the films are deposited, and the chemical environment to which the films are exposed are all important considerations when it becomes necessary to introduce multielement metallization into device structures.

REFERENCES

1. P. A. Totta and R. P. Sopher, *IBM J. Res. Dev.*, **13**, 226 (1967).
2. I. E. Price and L. A. Berthoud, *Solid-State Electron.*, **16**, 1303 (1973).
3. C. H. Lane, *Metall. Trans.*, **1**, 713 (1970).
4. J. F. White, IBM, East Fishkill, New York, private communication.
5. J. O. McCaldin and H. Sankur, *Appl. Phys. Lett.*, **20**, 171 (1972).
6. J. Sankur, J. O. McCaldin, and J. Devaney, *Appl. Phys. Lett.*, **22**, 64 (1973).
7. G. Ottaviani, D. Sigurd, V. Marello, J. W. Mayer, and J. O. McCaldin, *J. Appl. Phys.*, **45**, 1730 (1974).
8. D. Sigurd, G. Ottaviani, H. J. Arnal, and J. W. Mayer, *J. Appl. Phys.*, **45**, 1740 (1974).
9. J. S. Best and J. O. McCaldin, *J. Appl. Phys.*, **46**, 4071 (1975).
10. R. L. Boatright and J. O. McCaldin, *J. Appl. Phys.*, **47**, 2260 (1976).
11. H. C. Card, in *Metal-Semiconductor Contacts*, Inst. of Physics Conf. Ser. No. 22, Manchester, England (1974).
12. T. M. Reith and J. D. Schick, *Appl. Phys. Lett.*, **25**, 524 (1974).
13. K. Chino, *Solid-State Electron.*, **16**, 119 (1973).
14. H. C. Card, *Solid-State Commun.*, **16**, 87 (1975).
15. J. Basterfield, J. M. Shannon, and A. Gill, *Solid-State Electron.*, **18**, 290 (1975).
16. H. C. Card and K. E. Singer, *Thin Solid Films*, **28**, 265 1975.
17. M. J. Sullivan, IBM, East Fishkill, New York, to be published.
18. T. M. Reith, *Appl. Phys. Lett.*, **28**, 3 (1976).
19. J. W. Mayer and K. N. Tu, *J. Vac. Sci. Technol.*, **11**, 86 (1974).
20. K. N. Tu, *Appl. Phys. Lett.*, **27**, 221 (1975).
21. R. W. Bower and J. W. Mayer, *Appl. Phys. Lett.*, **20**, 359 (1972).
22. C. J. Kircher, *Solid-State Electron.*, **14**, 507 (1970).
23. R. W. Bower, D. Sigurd, and R. E. Scott, *Solid-State Electron.*, **16**, 1461 (1973).
24. D. Kahng and M. P. Lepselter, *Bell Syst. Tech. J.*, **44**, 1525 (1965).
25. M. P. Lepselter, *Bell Syst. Tech. J.*, **45**, 233 (1966).
26. F. B. Koch and P. B. Byrnes, *J. Electrochem. Soc.*, **117**, 262c (1970).
27. H. N. S. Lee, F. B. Koch, and W. R. Costello, *J. Electrochem. Soc.*, **118**, 72c (1971).
28. H. Muta and D. Shinoda, *J. Appl. Phys.*, **43**, 2413 (1972).
29. A. K. Sinha, R. B. Marcus, T. T. Sheng, and S. E. Haszko, *J. Appl. Phys.*, **43** (1972).
30. J. M. Poate and T. C. Tisone, *Appl. Phys. Lett.*, **24**, 8 (1974).
31. A. Hiraki, M. A. Nicolet, and J. W. Mayer, *Appl. Phys. Lett.*, **18**, 178 (1971).
32. R. M. Anderson and T. M. Reith, *J. Electrochem. Soc.*, **122**, 1337 (1975).
33. H. H. Hosack, *J. Appl. Phys.*, **44**, 8 (1973).
34. M. J. Sullivan, T. M. Reith, C. Goldsmith, W. K. Chu, and M. Mitchell, to be published.
35. D. M. Mattox, *Thin Solid Films*, **18**, 173 (1973).
36. A. Munitz and Y. Komem, *Thin Solid Films*, **37**, 171 (1976).

37. J. R. Rairden, C. A. Neugebauer, and R. A. Sigsbee, *Metall. Trans.*, **2**, 719 (1971).
38. J. Fisher, *J. Appl. Phys.*, **22**, 74 (1951).
39. P. Holloway, D. E. Amos, and G. C. Nelson, *J. Appl. Phys.*, **47**, 3769 (1976).
40. K. Nakamura, M. A. Nicolet, J. W. Mayer, R. J. Blattner, and C. A. Evans, Jr., *J. Appl. Phys.*, **46**, 4678 (1975).
41. C. Osburn, private communication.
42. Test results obtained in conjunction with P. Garbarino and J. Gardiner, IBM, East Fishkill, New York.
43. T. C. Tisone and J. Drobek, *J. Vac. Sci. Technol.*, **9**, 271 (1972).
44. T. C. Tisone and S. S. Lau, *J. Appl. Phys.*, **45**, 1667 (1974).
45. R. W. Bower, *Appl. Phys. Lett.*, **23**, 99 (1973).
46. J. K. Howard, R. F. Lever, and P. J. Smith, *J. Vac. Sci. Technol.*, **13**, 68 (1976).
47. R. F. Lever, J. K. Howard, W. K. Chu, and P. J. Smith, *J. Vac. Sci. Technol.*, **14**, 158 (1977).
48. P. J. Smith, J. K. Howard, W. K. Chu, and R. F. Lever, *Proc. 34th Ann. Elec. Micros. Soc. Amer.*, G. W. Bailey, Ed., Miami Beach, Florida (1976) p. 634.
49. L. G. Harrison, *Trans. Faraday Soc.*, **57**, 1191 (1961).
50. J. D. Baird, *J. Nucl. Energy*, **11**, 88 (1960).
51. B. N. Agarawala, M. Attardo, and A. Ingraham, *J. Appl. Phys.*, **41**, 3945 (1970).
52. G. A. Scoggan, B. N. Agarawala, P. P. Peressini, and A. Brouillard, *IEEE Proc. 13th Annual Conf. on Reliability Physics*, Las Vegas, Nevada (1975), p. 51.
53. J. K. Howard, J. F. White, and P. Ho, submitted to *J. Appl. Phys.*
54. J. K. Howard, *J. Appl. Phys.*, **44**, 1997 (1973).
55. J. K. Howard, *J. Appl. Phys.*, **46**, 1910 (1975).
56. P. A. Totta, *J. Vac. Sci. Technol.*, 13, 26 (1976).
57. M. G. Fontana and N. D. Greene, *Corrosion Engineering*, McGraw-Hill, New York (1967), pp. 49–66.

3

CHEMISTRY AND PHYSICS
OF SOLID–SOLID INTERFACES

J. C. Phillips

Bell Laboratories, Murray Hill, New Jersey

3.1 INTRODUCTION

For many years research on thin films and interfaces has been dominated by pragmatic considerations developed to overcome the many materials problems which arise at solid–solid interfaces. Today, however, there is a growing literature of well-defined experiments on the structure and properties of interfaces that suggests the need for chemical and physical models of interfacial regions. How do the structures of these regions relate to known bulk and surface structures? Can we define an interfacial phase diagram, and what are the limits of its definition? How would such interfacial phase diagrams be related to bulk and surface phase diagrams? Can structural and thermodynamic models be used to interpret the directions and degrees of interfacial reactions? We do not have the answers to these and many other questions which are important to many scientists working on interfaces, but there are some general ideas that seem likely to be useful in this area. We present them here, together with discussion of some specific examples where progress has already taken place, in the hope that the reader will find them useful in his own work.

The classical literature on liquid–liquid and liquid–solid interfaces has been summarized by Davies and Rideal (1). Unless a chemical reaction takes place, the interaction at the interface is likely to be weak compared to the other interactions present, and the systems should be close to thermal equilibrium. The situation is much more complex at solid–solid interfaces, which may be far from equilibrium and which may also be influenced substantially by constraints (internal strain fields). Indeed, the most general and important characteristic of a solid–solid interface is whether it is epitaxial or non-epitaxial. Both cases are of great practical importance, but the epitaxial case is more simple. In the nonepitaxial case at least one of the solids is usually polycrystalline, and interfacial behavior is strongly affected by grain boundaries.

3.2   NEARLY IDEAL INTERFACES

One of W. Shockley's rules for research on complex problems is to "try simplest cases first." It seems reasonable to assume that the simplest interface will be the one between two elements that form the most nearly ideal solid solution. In Figure 3.1 we show the phase diagram for Cu–Ni alloys (cubic close-packed structures miscible in all proportions). When this diagram is analyzed by regular solution theory (2), the liquidus curve is fitted, and there is some discrepancy between the calculated and experimental solidus curves, which is possibly caused by intrinsic limitations of thermal analysis of cooling curves. In any case, Figure 3.1 presents one of the most favorable cases (another is Ge–Si) among the elemental systems. The lattice constants of Cu and Ni differ by $\Delta a/a = 3\%$, and the difference ΔX in Pauling electronegativity is 0.1, while the heats of melting $\delta(\Delta H)$ (see Fig. 3.1) differ by about 0.05 eV at.$^{-1} \sim 1$ kcal g-at.$^{-1}$. In the Ge–Si case, $\Delta a/a = 0.04$, $\Delta X = 0.0$ (Pauling) or 0.06 (Phillips-Van Vechten), and $\delta(\Delta H) \sim 3.5$ kcal g-at.$^{-1}$. In Figure 3.1 at $x = \frac{1}{2}$, $T(\text{liquidus}) - T(\text{solidus}) = \Delta T = 20°$ K, whereas for Ge–Si at $x = \frac{1}{2}$, $\Delta T = 100°$ K, which shows the expected qualitative trend. We discuss later some recent metallurgical models which attempt to assess quantitatively the relative importance of strain, ionic, and vibrational entropy energies on bulk phase diagrams.

Most elemental systems are not miscible in all proportions, and in most cases one metallic element is not soluble in another over a range of more than a few percent. The most elaborate studies of this situation have been based

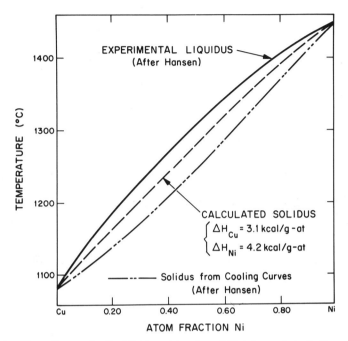

Figure 3.1.   Phase diagram for Cu–Ni alloys, taken from J. Steininger, *J. Appl. Phys.*, **41**, 2713 (1970).

on an idea originally put forward by Darken and Gurry (3), which is illustrated in Figure 3.2. The atomic radius and Pauling electronegativity of each element B dissolved in A are mapped in a Cartesian plot, and a circle with $\Delta r/r = 0.15$ and $\Delta X = 0.4$ defines with 75% success solates B that are more than 5% soluble in A. Incidentally, this plot provides a good check on $X(A)$, and at one time it was used to modify the value of $X(Ag)$ originally suggested by Pauling from molecular heats of formation.

3.3   SINGLE-CRYSTAL FILMS

The development of single-crystal films has been summarized in an excellent article by Matthews (4), from which (for the reader's convenience) much of the following discussion is taken. While film growth usually begins with islands, monolayer growth can occur when the contact angle θ of overgrowth on substrate is zero. (Gibbs defined θ in terms of the surface free energies σ_s of substrate, σ_0 of overgrowth, and σ_i of the interface by $\sigma_s = \sigma_i + \sigma_0 \cos \theta$, so that $\theta = 0$ when $\sigma_i + \sigma_0 \leqslant \sigma_s$.) This is true monolayer growth [e.g., Pb on Ag, because $\sigma(Pb) \ll \sigma(Ag)$], but monolayer-like growth can also occur for a

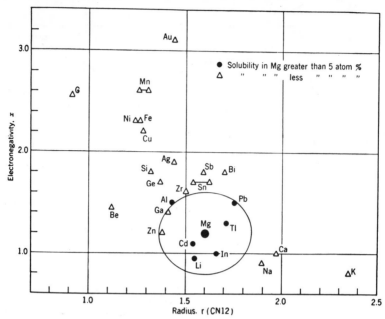

Figure 3.2. Darken-Gurry plot for solubilities of elements in Mg. Taken from reference 3.

number of kinetic reasons unrelated to thermodynamic equilibrium conditions; for example, during deposition some diffusion takes place, enough to grow a monolayer but not too much compared to the deposition rate, so that the system never reaches equilibrium. Some examples of this kind are Pt on Au, and Co on Cu or Ni. We have here very simple examples of the importance of kinetic rather than thermodynamic considerations in determining interfacial morphology.

3.4 MISFIT DISLOCATIONS

Motion of steps across a surface can produce a single-crystal film in much the same way that single crystals are grown from single-crystal substrates of the same material. In the latter case growth at steps can be contrasted with growth nucleated by impurities (pyramidal growth); at least in the case of silicon vapor deposition this can be an intrinsic process (5), although of course most growth centers will be dislocation lines or impurities unless special care is taken. However, in the film case, as the thickness of the film increases, the strain energy at the interface also increases. In the classic paper on accommodation of misfit, Frank and Van der Merwe (6) found that the nature of the strain misfit for ideal (abrupt) interfaces between isostructural films and substrates depends on $\delta a/a$ (lattice misfit), film thickness t, and dislocation geometry. Although the principles behind misfit dislocation relief of mechanical strain are straightforward, often the agreement between theory and experiment is only semiquantitative.

The results are as follows: (a) For $\delta a/a \gtrsim 0.2$, even monolayer films will contain vacancies or interstitials relative to the substrate lattice, that is, these are misfit dislocations even in the monolayer, which has the structure and lattice constant of the bulk material (e.g., Pb on Ag). (b) Smaller misfits (between 0.02 and 0.04) will be accommodated by strain up to film thicknesses t of about 3 to 15 Å (several monolayers), after which misfit dislocations form (e.g., Ni on Cu and Pt on Au). In practice a larger fraction of the misfit energy appears as strain, that is, there are fewer misfit dislocations than predicted by theory. This is possibly a nonequilibrium effect which could be expected from step-growth kinetics.

There are many ways in which film growth can be nucleated, and in favorable cases some aspects of the kinetics of nucleation can be inferred from the distribution and orientation of nuclei. Matthews (4) reviews simple nucleation mechanisms, nucleation at surface steps, reduction of misfit energies for small nuclei, reduction of growth rates for misoriented nuclei, and the coalescence of nuclei and accompanying defect formation (strains, stacking faults, dislocations). All of these effects show the importance of kinetic mechanisms even in the growth of nearly ideal films.

An interesting example of competing growth mechanisms is provided by recent studies (7) of combined island and layer growth of epitaxial Cu on Ni. Here we have very thin (one or two monolayer) films on which islands are superposed; effectively the two Cu "phases" (islands and planar layers) coexist because of kinetic conditions. The situation is entirely similar to pyramidal growth on Si from the vapor phase (5) as shown in Figure 3.3.

Compound epitaxial films have been studied extensively for semiconductor device applications. One of the most studied examples are AlAs–GaAs alloys on GaAs. This system is very favorable; through further alloying, for example, with GaP, the strain misfit can be tuned to a very low value (of the order of 10^{-4}) either at the growth temperature or at lower temperatures. The perfection of the electronic structure of the interface can often be inferred in general terms from electrical measurements or from the sharpness of optical transitions between electron and hole states trapped in wells between two interfaces. Because of their nearly ideal character as well as their practical importance, these interfaces have been prepared to a level of perfection not established for previous "favorable" systems (such as Cu–Ni). Thus in the case of nearly ideal metal–metal interfaces, it has been suggested that misfit can be graded by motion of dislocations away from the interface (8) or by cross doping; these problems have been greatly reduced or eliminated in compound semiconductor films produced by molecular beam epitaxy (9).

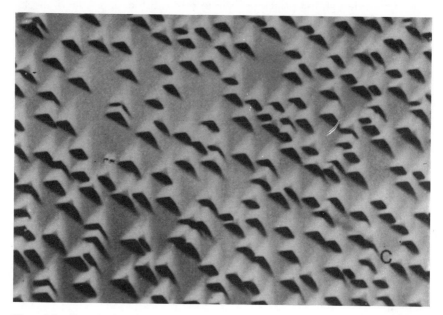

Figure 3.3. Growth of pyramidal islands on Si by vapor phase epitaxy. Taken from reference 5.

3.5 THERMODYNAMICS OF COMPLEX INTERFACIAL PHASES

We have seen in the preceding section that, at least for sufficiently thin films, for sufficiently small misfit, and for sufficiently simple phase diagrams, the growth of heteroepitaxial films (Cu on Ni) can closely resemble the growth of homoepitaxial films (Si on Si) under favorable conditions. We now turn our attention to situations where the phase diagram is more complex. As an example of a situation which has received much attention, consider the Pt–Si system whose bulk phase diagram is shown in Figure 3.4. Many phases of different stoichiometry are formed, and at lower temperatures these may undergo further structural transformations.

Many transition metal–silicon interfaces have been formed by depositing transition metal films of thickness ~ 2–3×10^{-6} cm on Si substrates and heating the samples until a solid–solid reaction takes place to form a transition metal silicide[10] (see Chapter 10). This reaction is exothermic, and so it is perhaps not surprising that it can take place rapidly at temperatures several hundred degrees lower than the eutectic temperature of the bulk phase diagram. The reaction product depends on the annealing temperature and period and certainly involves kinetic factors. Moreover, when the thin film phase can choose between several structures, none of which is isostructural to

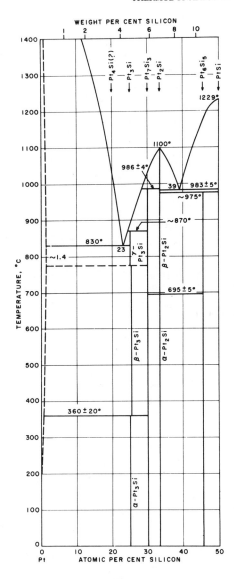

Figure 3.4. Partial phase diagram for the Pt–Si system. Taken from F. A. Shunk, *Constitution of Binary Alloys*, Second Supplement, McGraw-Hill, New York (1969).

the substrate, the interfacial misfit energy must be considered in discussing interfacial reaction kinetics.

The foregoing statement of the problem may serve to indicate some of the complexities generated at solid–solid interfaces by multiphase systems. At this writing these complexities are by no means resolved, but data are accumulating, and there is some hope that systematic phenomenologic models can be developed. For example, we show in Figure 3.5 the Schottky barrier heights ϕ_{Bn} of a variety of transition metal silicides on n-type silicon; these are seen to correlate extremely well (10) with the heats of formation of the intermetallic compound normalized per transition metal atom per formula unit. One would expect some correlation between these barrier heights (which include a dielectric double-layer term) and the interfacial dipole moment; the latter, in turn, is related to the charge transferred between the transition metal and the silicon. This charge transferal is related to the Pauling electronegativity difference ΔX, which is defined in terms of heats of formation. For the very small differences involved here, the heat of formation constitutes a more accurate measure of the charge distribution at the interface than any derived quantity (e.g., ΔX, which is based on averages over many other structurally distinct systems such as molecules).

An interesting feature of Figure 3.5 is that it enables one to determine which particular alloy is present at a given interface. (Thus, one identifies the Pt_2Si phase, which was prepared by Pt deposition on a substrate heated to several hundred $^\circ$C, rather than by annealing a room temperature substrate to 600°C, which gives PtSi.) Another interesting feature is the effect of quasi-epitaxial order on the interface; this order was anticipated in previous structural studies (11)* and is confirmed in Figure 3.5 for Pt–Si:Si by its anomalous barrier height. Finally, it is interesting to note that it was possible to arrest the usual reaction kinetics ($Pt+Si \to Pt_2Si \to PtSi$) at the Pt_2Si stage only for Si(111) surfaces (this is the stable or equilibrium surface but not for Si(100) or (110) substrates (which are inherently unstable). Thus, the reaction kinetics at the interface can be influenced substantially by surface anisotropy, even in the case of disordered interfaces.

Most transition metal–silicide:silicon interfaces are not epitaxial. At the same time many of the metals in question form metallic glasses when splat-cooled. This strongly suggests that there is an interfacial region, within 5 to 10 atomic layers of the actual interface, where the transition metal–silicide forms a glassy membrane. The study of such metallic glasses in bulk has shown that they are in general most stable at the composition of the lowest temperature eutectic.

---

*Sinha et al. (11) and other authors have found that PtSi can be quasi-epitaxial on Si(111), but the orientation of the PtSi film varies with its thickness and thermal history. The former is explicable in terms of temperature-dependent lattice constants of orthorhombic PtSi, which successively approximate the Si(111) geometry as the temperature is varied (equilibrium misfit energies). The latter is suggestive of differing growth rates, presumably via step motion, of different PtSi planes (kinetic effect). Note that by quasi-epitaxial we mean enough short-range order at the interface to generate electron diffraction spots in a thin film. In the metal part of a Schottky barrier, there could still be misfit dislocations without affecting most of the barrier properties. Note that a quasi-epitaxial film, like a true epitaxial film, has much less strain energy concentrated at the interface. Therefore, like a true epitaxial film, it is less subject to microscopic damage such as formation of cracks.

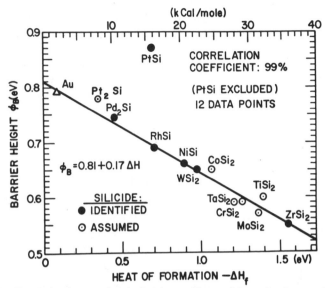

Figure 3.5. Correlation between barrier heights and heats of formation for transition metal silicides. From J. M. Andrews and J. C. Phillips, *Phys. Rev. Lett.*, **35**, 56 (1975).

If we now anneal the sample to accelerate a solid–solid reaction, a crystallization front of silicon atoms will diffuse from the glassy membrane into the transition metal. The composition of the first crystalline phase nucleated should then be that of the congruently melting phase closest in composition to the glassy membrane, as pointed out by Walser and Bené (12). Their model, which emphasizes the importance of reaction kinetics, appears to predict the first nucleation phases for many systems.

Evidence for the formation of glassy membranes at metal–semiconductor interfaces has also been obtained for the Al–Si system, where no compound is formed. Vitrification of the Al membrane region is observed by photoemission (13) as a sharpening of a band structure peak about 2 eV below E_F. The thickness of the membrane region is about four atomic layers.

3.6 KINETICS OF INTERFACIAL PHASE FORMATION

Some of the most interesting interfacial phase reactions may be dominated by kinetic considerations, as we have just seen. In this case we should attempt to analyze these reactions by techniques that have been developed to study bulk transitions dominated by kinetic effects. There is a wide class of such transitions in metals, called martensitic transitions. In insulators, similar effects are seen in materials called ferroelectric or ferroelastic. In all cases the strain energy produces long-range forces. These forces are cumulative in nature and are of the type characterized by theorists as nonlocal. This gives rise to "anomalous" behavior compared to "normal" phase transitions, where the forces are local and energies do not accumulate. Almost all currently fashionable theories of phase transitions assume dominance by local forces.

While we cannot hope for much help from the mathematicians on this problem, we can expect to benefit by systematic material analogies. Thus, island formation in thin films is analogous to microdomain formation in martensitic transformations. Some beautiful electron microscopy studies have been devoted to this phenomenon (14).

3.7 INTERDIFFUSION NORMAL TO INTERFACE

Here again a macroscopic approach seems appropriate. We have seen that at a metal–semiconductor interface, there is probably a glassy metallic interfacial phase some 10 to 20 Å thick. This glassy phase is similar to a liquid metal, and therefore the tendency of semiconductor atoms or impurities to diffuse into the interfacial region should be inferrable from the slopes of bulk liquidus–solidus alloy phase diagrams. This approach has recently been used successfully to predict surface segregation in alloys (15).

3.8 BULK, SURFACE, AND INTERFACIAL PHASE DIAGRAMS

Although the structures of surface and interfacial phases are determined in many cases by kinetic factors, we should assume that these structures are still close to thermal equilibrium. In some respects these structures should be described as metastable phases. The metastability is maintained by strain fields associated with bulk phases which act as three-dimensional substrates for the thin film surface and interfacial phases.

There is an extensive literature on metastable phases, especially of metallic alloys, which are generally produced by rapid quenching from the melt (16). Some general qualitative ideas about metastable phases, especially amorphous or glassy phases (17), have been developed. These suggest that amorphous metals should be analyzed in terms of random packing of spheres, while amorphous semiconductors are discussed in terms of random covalent networks.

While these qualitative ideas are probably correct in a general sense, they are weak in the sense that the Darken-Gurry model is weak, that is, they do not really identify in a quantitative way the structural elements as accurately as possible. Thus, I would like to suggest here that recent developments in the understanding of bulk phase diagrams and bulk structural patterns offer a path that, at least in principle, will produce more quantitative theoretical models. In particular, these models offer realistic possibilities for calculation of structural strain energies and hence of the effects of constraints.

The new models replace the isotropic central force classical models (hard spheres plus electronegativity corrections) with orbitally dependent radii r_s^α, r_p^α, and r_d^α for each element α. Different hybridization states (such as σ-hybridization, i.e., $s+p$ or $s+p+d$) are found to correspond to different combinations of the orbitally dependent radii (such as $r_s + r_p$ or $r_s + r_p + r_d$, etc.). These models have successfully differentiated the structures of a number of bulk phases (18, 19). The relative importance of ionic and covalent effects is treated correctly with these coordinates, and strain energies in distorted cubic structures (chalcopyrite, wurzite, CrB) correlate well with appropriate structural combinations of orbital radii (20). The coordinates separate layer and cubic structures (graphite, diamond) and are therefore well suited for discussing metastable thin film phases and possibly metastable bulk phases as well. This is an interesting direction for future theoretical work.

3.9 CONCLUSIONS

Most scientists tend to feel uncomfortable with inhomogeneous systems, where there are composition gradients, such as surface or interfacial segregation, islands, or microdomains. It may be, however, that with modern techniques of ion implantation and molecular beam epitaxy, kinetic inhomogeneities can be controlled and even designed. In that case we may be on the threshold of a new era in thin film technology. Many experimentalists already feel that this is the case. In this brief chapter, I have tried to persuade theorists to share their excitement.

REFERENCES

1. J. T. Davies and E. K. Rideal, *Interfacial Phenomena*, Academic Press, New York (1963); J. W. Cahn and J. E. Hilliard, *J. Chem. Phys.*, **28**, 258 (1958).
2. A. Lagier, *Rev. Phys. Appl.*, **8**, 259 (1973).
3. L. S. Darken and R. W. Gurry, *Physical Chemistry of Metals*, McGraw-Hill, New York (1953); J. T. Waber et al., *Trans. Metall. Soc. AIME*, **227**, 717 (1963).
4. J. W. Matthews, *Proc. Fourth Int. Vacuum Congr.*, (1968), p. 479.
5. M. Shimbo, J. Nishizawa, and T. Terasaki, *J. Cryst. Growth*, **23**, 267 (1974).
6. F. C. Frank and J. H. Van der Merwe, *Proc. Roy. Soc.*, **A198**, 216 (1949); M. D. Chinn and S. C. Fain, Jr., *Phys. Rev. Lett.*, **39**, 146 (1977).
7. J. W. Matthews, D. C. Jackson, and A. Chambers, *Thin Solid Films*, **26**, 129 (1975).
8. J. W. Matthews, *Scripta Metall.*, **8**, 505 (1974).
9. R. Dingle, A. C. Gossard, and W. Wiegmann, *Phys. Rev. Lett.*, **34**, 1327 (1975).
10. J. M. Andrews and M. P. Lepselter, *Solid State Electron.*, **13**, 1011 (1970).
11. A. K. Sinha et al., *J. Appl. Phys.*, **43**, 3637 (1972).
12. R. M. Walser and R. W. Bene, *Appl. Phys. Lett.*, **28**, 624 (1976).
13. G. Margaritondo, J. E. Rowe, and S. B. Christman, *Phys. Rev.*, **14B**, (October 15, 1976).
14. I. Cornelis, R. Oshima, H. C. Tong, and C. M. Wayman, *Scripta Metall.*, **8**, 133 (1974).
15. J. J. Burton and E. S. Machlin, *Phys. Rev. Lett.*, **37**, 1433 (1976); A. R. Miedema, *Philips Tech. Rev.*, **36**, 217 (1976).
16. A. K. Sinha, B. C. Giessen, and D. E. Polk, in *Treatise on Solid State Chemistry*, Vol. 3, N. B. Hannay, Ed., Plenum Press, New York (1976), pp. 1–88.
17. D. R. Uhlmann, *ibid.*, pp. 293–334.
18. J. St. John and A. N. Bloch, *Phys. Rev. Lett.*, **33**, 1095 (1974).
19. E. S. Machlin, P. Chow, and J. C. Phillips, *Phys. Rev. Lett.*, in press (1977).
20. J. R. Chelikowsky and J. C. Phillips, to be published.

4

THE SEMICONDUCTOR–CONDUCTOR INTERFACE*

J. O. McCaldin and T. C. McGill**

California Institute of Technology, Pasadena, California

4.1 INTRODUCTION

Semiconductors are made to interface with a variety of substances in modern technology. The more common of the interfacing substances are other semiconductors, metals, insulators, and vacuum ("no substance"). The perfection of the interface obtained varies greatly, depending on the terminating substance. The best interfaces are often with other semiconductors, for example, $Al_xGa_{1-x}As$–GaAs, where almost perfect lattice match and good continuity of electronic character can be obtained. Termination of a substrate semiconductor is likely to be less perfect against vacuum, as evidenced by the structural reconstruction that often occurs at the surface of the substrate and the large electronic barrier there. Terminations against conductors, the subject of this chapter, are perhaps the least perfect. With few exceptions, there

*Supported in part by the Office of Naval Research (L. Cooper and D. Ferry).
**Alfred P. Sloan Foundation Fellow.

is no possibility of lattice match, and furthermore the electronic characteristics to either side of the interface differ enormously. The principle exceptions are zero-gap semiconductors: they are semimetals at room temperature and they can afford almost perfect lattice match, for example, HgSe on CdSe.

Before beginning the main discussion, a few words about the role played by the anion of a semiconductor compound at the various interfaces. In the much-studied vacuum interface, Swank (1) found that the energy of the valence band maximum of a compound semiconductor substrate was usually determined by the anion of the compound. Subsequent model calculations by Van Vechten (2) and Nethercot (3) generally confirmed this result. In the case of lattice-matched semiconductor–semiconductor interfaces, the one very well-studied interface, $Al_xGa_{1-x}As$–GaAs, shows valence bands approximately in alignment (4) to either side of the interface between these two compounds with a common anion. Thus, the anion plays the same role as found by Swank for vacuum interfaces. Many semiconductor compounds at an interface with Au behave in a similar way, as is discussed in Sections 4.3 and 4.4. The possibility also exists of such anion-dominated behavior extending to semiconductor–electrolyte interfaces, as is discussed in Section V.

4.2 POTENTIAL DIFFERENCE ACROSS THE INTERFACE

A brief description is given here of the way a Schottky barrier arises near the surface of a semiconductor. The description is intended only for electrochemists who might otherwise not read the ensuing sections, couched as they are in the language of semiconductor electronics.

Just as the electrochemist focuses attention on the mobile species in electrolytes, namely, ions, so we focus attention on the mobile species in the semiconductor, namely, electronic carriers. In the semiconductor case the relevant electrochemical potential has a special name, the "Fermi level," designated E_F. In Figure 4.1 we show E_F in two isolated, electrically neutral substances. In this example the semiconductor happens to be p-type, and the metal happens to be relatively electropositive. When the two substances are brought into contact, as in Figure 4.2, the electrochemical potentials of the mobile species come to a common value. This entails a transfer of electrons from the metal to the semiconductor in this particular example, giving rise to a space charge layer. The disposition of the space charge layer depends on the Debye lengths, or screening lengths, in the two substances. For most semiconductors, the Debye length L_D is 10^{-5} cm or larger, whereas for metals and most electrolytes L_D is 10^{-7} cm or smaller. Consequently, the space charge layer lies almost entirely on the more weakly screened semiconductor side of the interface. Thus arises the barrier ϕ_p within the semiconductor.

The description just given applies best to semiconductors such as ZnS or CdS, the so-called "ionic" semiconductors. Semiconductors may be classified as either "ionic" or "covalent" according to simple criteria given by Kurtin,

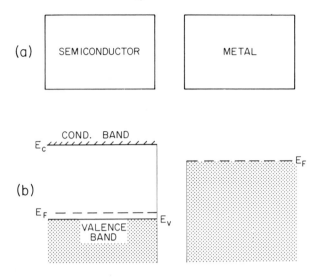

Figure 4.1. (*a*) A bulk semiconductor and a bulk metal, not in contact and not electrically connected. (*b*) Electron energies in these materials. In this example, the semiconductor is *p*-type and has a greater work function W_F than does the metal (i.e., the Fermi level E_F in the semiconductor lies below E_F in the metal).

McGill, and Mead (5). For "covalent" semiconductors, for example, GaAs and Si, the description given above must be modified to account for interface states. These electronic states are localized at the interface and absorb most of the charge transferred to equalize E_F, and thus do the screening. The density of interface states can often be greatly reduced, however, by suitable interface layers, usually oxides. Present-day Si MOS transistors depend upon this reduction of interface state density for their effectiveness. Other covalent semiconductors can be treated to reduce surface state densities substantially, though not yet as effectively as in the Si case (see, for example, reference 6 for the InP case and reference 7 for the GaAs case). We indicate such a beneficial interface layer that greatly reduces surface state densities in Figure 4.2 and shall invoke, from time to time, such a layer in subsequent discussion.

4.3   BARRIER ϕ_p DEPENDS ON SEMICONDUCTOR ANION

The rather extensive measurements available of barrier heights ϕ for various combinations of semiconductor and metal give rise from time to time to generalizations.* Besides the Kurtin, McGill, and Mead (5) rule mentioned

---

*For simplicity, the example of a *p*-type semiconductor with a barrier ϕ_p for holes is used in Figure 4.2. Barriers ϕ_n can be defined on *n*-type semiconductors in exactly analogous fashion, and furthermore, $\phi_p + \phi_n = E_g$, the energy gap of the semiconductor. We use the symbol ϕ without subscript to represent barriers of both types.

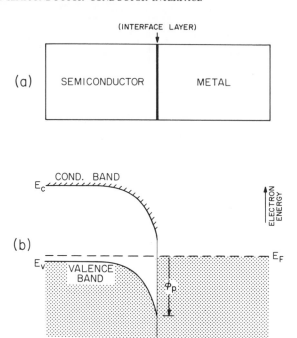

Figure 4.2. (a) The semiconductor and the metal now in contact (and thereby electrically connected). Occasionally an interface layer is introduced (e.g. a very thin oxide in solar cells) to decrease the number of surface states at the interface. (b) Electron energies after contact. The two materials are now at electronic equilibrium and therefore have the same E_F. The potential difference necessary to align E_F occurs almost entirely in the semiconductor, giving rise to the Schottky barrier ϕ_p.

earlier, there is the "two thirds" rule which applies to the most common semiconductors (8). More recently, it has been proposed (9, 10) that the barrier ϕ_p for the common III–V and II–VI semiconductors contacted by Au is fixed by the anion in the semiconductor compound. For example, InAs and GaAs have the same barrier height ϕ_p for holes, according to this rule. In arriving at this rule, the rather extensive data available were examined and are summarized in Figure 4.3. For discussion of the range of applicability, treatment of the data, etc., in this figure, consult references 9 and 10.

For clarity, we restate the main point in Figure 4.3 as follows. The anion in the semiconductor compound fixes the energy relationships at the interface with a Au contact. Thus, in Figure 4.2 the position of the valence band maximum E_v (and consequently of the conduction band minimum E_c) *at the interface* is fixed with respect to E_F by the particular anion present in the semiconductor. If the anion is As, for example, then $E_F - E_v \approx 0.45$ eV.

Besides indicating a relationship between semiconductors with the same

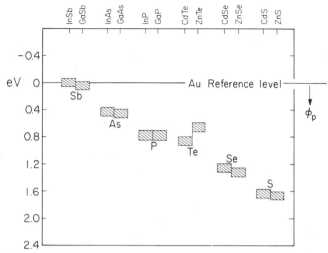

Figure 4.3. Schottky barrier height ϕ_p for Au contacts to the common III-V and II-VI semiconductors. The barrier ϕ_p is approximately the same for two compounds having the same anion.

anion, Figure 4.3 allows us to order the common binary semiconductors by anion. Although Figure 4.3 does not indicate an abscissa, it turns out that the electronegativity of the anion is suitable for abscissa [see (10)]. This behavior is reminiscent of the ordering of ions in electrochemical series.

4.4 THEORETICAL CONSIDERATIONS

The experimental results summarized in the previous section suggest a rather unexpected correlation. The energy required to remove an electron from the top of the valence band in the semiconductor and place it at the Fermi level in Au depends only on the anion in the III–V and II–VI semiconducting compounds. These compounds are all very similar. Structurally, they all are either zinc blende or wurzite, that is tetrahedrally coordinated, which suggests that covalent bonding plays an important role. The compounds with different cations but the same anion have rather similar ionicities (11, 12). For example, GaAs and InAs have ionicities of 0.310 and 0.357, respectively, on the Phillips scale (11, 12) and 0.26 and 0.26, respectively, on the Pauling scale.* Other sets of compounds with a common anion exhibit similar variations in ionicity.

The similarity of the ionicities in the common anion sets of compounds implies a similarity in charge transfer. Hence, we might expect that the position in energy of anion-derived states would not vary a great deal in going from one compound to the next.

*The values of the Pauling ionicity were taken from reference 11, Table A. The column labeled Pauling (1939) was used. See also reference 13.

Returning to the correlations found in barrier energies, we expect that anion-derived states can enter the determination of the value of the barrier energy in two ways: the maximum in the valence band is made up of anion-derived states, or there are anion-derived interface states that pin the metal Fermi level (14).

In the first case, we can argue that, since the state from which the electron is removed is an anion-derived state, its energy position should depend only on the anion. There is considerable evidence that the states at the top of the valence band are very anion-like. The valence band maxima of these materials are derived primarily from the p-like atomic states (15–17) of the anion. Chadi and Cohen (16) find in their tight binding calculation for GaAs and ZnSe that the valence band maximum is located within 0.2 eV of their value for the energy of the anion p-level. Using the same parameters as they used in their tight binding calculation, we computed the ratio of the square of the wave function on the anion to that on the cation. At the valence band maximum, we find that this ratio is 24 for GaAs and 60 or ZnSe. Recent values of the tight binding parameters (18) give ratios of 2.3 and 15 for GaAs and ZnSe, respectively. These variations indicate the type of errors that tight binding calculations are subject to, but both sets of values show that the valence band maximum is very anion-like.

However, there is other evidence for the strongly anionic character of the top of the valence band. A recent experimental study of the filled surface states on GaAs has shown that they are primarily anion derived and located near the valence band maximum (19). Another experimental measure which allows us to gauge the relative anion–cation contribution at the top of the valence band is the spin-orbit splitting of the band states at the top of the valence band (20). The spin-orbit splitting is a relativistic effect. The main contribution to the splitting comes from the core region of the atoms in the solid where the electrons are moving rapidly. In Figure 4.4, we have plotted the values of the spin-orbit splitting in the solid versus the atomic values for the anion and cation. This figure shows that splitting in the solid correlates with the anion splitting and shows little, if any, correlation with the atomic, cation splitting. Again this correlation suggests that the states at the top of the valence band are very anion-like.

In the case of anion-derived interface states, we might expect that these states would pin the Fermi level of the metal at a certain position. For if the Fermi level moved up or down, these states would fill or empty in such a way as to build up a dipole layer that would fix the position of the valence band edge with respect to the metal. If these interface states were positioned relative to the valence band maximum at an energy that depended largely on the anion and not on the cation, then we would observe the correlations reported here. At present, little is known about the systematic variation of interface

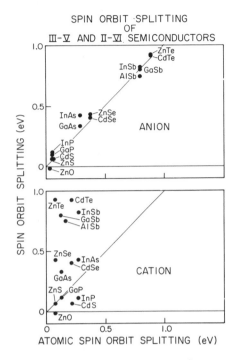

Figure 4.4. Spin orbit splitting energy at the top of the valence band vs. atomic spin orbit splitting between $p_{1/2}$ and $p_{3/2}$ states. In upper graph, the atomic anion splitting is used; in lower graph, the atomic cation splitting is used. The straight lines are lines of slope 1. Data obtained from reference 20.

states at metal–semiconductor interfaces. However, a theoretical research program initiated by Cohen and co-workers (14) promises to provide us with some of this information.

4.5 EXTENSION TO SEMICONDUCTOR–ELECTROLYTE INTERFACES

In Section 4.1, evidence from Swank (1) that the anion of the semiconductor compound fixes energy positions at an interface with vacuum was cited. In Section 4.3, recent evidence for a similar role of the anion at an interface with Au was cited. In this section, we suggest that the anion may also play such a role at an interface with an electrolyte.

One limitation to such a role for the semiconductor anion should be mentioned at the outset, namely, that such a viewpoint does not contemplate determining factors coming from the other side of the interface. For example, if Au is replaced by another metal in the Schottky barrier case, the barrier height may be affected. Similarly, in the electrolyte case, the influence of conditions in the electrolyte must be taken into account (21).

Nevertheless it is interesting to examine presently available data for semiconductor–electrolyte interfaces. These data come from Vanden Berghe and

Gomes (22) and from a compilation by Gleria and Memming (23). The data have been plotted in Figure 4.5 as a superposition on the previously discussed semiconductor–Au plot of Figure 4.3. The reference energies for the two cases, semiconductor–Au and semiconductor–electrolyte, have been set equal. This turns out to be reasonable because of the happenstance that an Au electrode immersed in aqueous electrolyte will be very nearly at its potential of zero charge, which is analogous to the flat-band condition in semiconductors, when at the potential of a saturated calomel reference electrode (24).

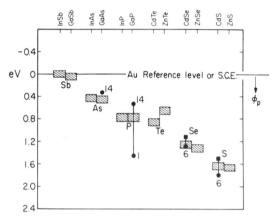

Figure 4.5.   Effect of an aqueous electrolyte contact, as contrasted with a Au contact, on the energy of the valence band maximum E_v at the surface of a semiconductor. The Au contact plot is identical to Figure 4.3. The superposed electrolyte contact data come from the compilation of Gleria and Memming (23), indicated by (●) data points with an associated pH number, and from the work of Vanden Berghe and Gomes (22), indicated by (■) data points.

The plot in Figure 4.5 suggests that in moderately alkaline solutions the barrier formed on a p-type semiconductor is about the same as when Au contacts the semiconductor. Thus, if indeed this turns out to be the case, the valence band energy E_v at the surface of the semiconductor would be the same with either contact. However, more data are needed before drawing such a conclusion. In particular, a crucial experiment is to compare two semiconductors with a common anion, for example, InP and GaP, in the same electrolyte to see whether they give rise to the same barrier height ϕ_p.

4.6   WHAT RANGE OF BARRIERS ϕ IS OBTAINABLE?

Having discussed semiconductor–conductor interfaces in some generality, we now turn to an application of current interest. How large and how small can the barriers on a given semiconductor be made by suitable choice of electronic conductor for contact?

This question is of considerable practical importance. If sufficiently large values of ϕ_p can be obtained for the semiconductor in Figure 4.2, the semiconductor becomes "inverted" at its surface, that is, $E_c \approx E_F$ at the surface to make the surface n-type. If, for example, the semiconductor in Figure 4.2 were a suitable light emitter, inverting its surface would permit minority carrier injection from the contact and hence light emission from this Schottky diode, that is, without resort to a conventional p–n junction. Similarly, Schottky barrier solar cells and field effect devices can benefit from the wider range of ϕ values resulting from a wider range of contact materials. What then are the contact materials available today?

Schottky barriers are almost always made today by evaporating an elemental metal onto a semiconductor substrate. The metals represent a majority of the elements in the periodic chart schematized in Figure 4.6,

Figure 4.6.   Periodic chart of the elements (with a few omissions for simplicity) divided according to electrical conductivity. The elements identified individually at the far right are very poor conductors but are highly electronegative.

hence there is a rather wide choice available. The elemental metals are ranked in Figure 4.7 according to their electronegativity. Electronegativity is a measure of how low the E_F will be located for a given metal in Figure 4.1. Thus, the electropositive metals such as Al and Mg will have their E_F posi-

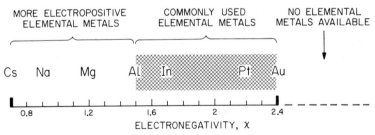

Figure 4.7.   Elemental metals arranged according to their Pauling electronegativity χ. The elemental metals commonly used to contact semiconductors vary from Al with $\chi = 1.5$ to Au with $\chi = 2.4$. Elemental metals with χ lower than 1.5 are available, but none with χ higher than 2.4.

tioned relatively high in Figure 4.1, whereas Au will have its E_F positioned low. It happens, as is suggested in Figure 4.7, that there is a multiplicity of electropositive metals, so that one can usually obtain large barriers ϕ_p on p-semiconductors. In practice, one may have to take steps, such as the insertion of an oxide interface layer as discussed in Section 4.2, to prevent chemical reaction between these very electropositive metals and the semiconductor substrate and to reduce surface state densities so that the desired band bending can be achieved. In principle, however, one should be able, for example, to fully invert p-type solar cell materials of current interest for use in Schottky barrier structures.

What then of the highly electronegative metals available to invert n-semiconductors? The most electronegative elemental metal is Au, but it is incapable of inverting many important n-semiconductors. Indeed, with a little thought one can see how to use Figure 4.3 to predict how far short Au comes from being able to fully invert n-type versions of the semiconductors listed at the top of the figure.

If, however, one is willing to work with binary-compound conductors in place of single-element conductors, then new possibilities for highly electronegative metallic contacts open up. For example, the elements individually labeled to the far right in Figure 4.6 are highly electronegative though very poor electrical conductors. By combining two such elements, however, reasonably good conductors can sometimes be produced, for example, $(SN)_x$, an equimolar polymer of sulfur and nitrogen. This material has been proved to be substantially more electronegative than Au when tested on the more ionic semiconductors (25).

Another possibility for the formation of highly electronegative contacts is to use the zero-gap semiconductors HgX, where X is one of the chalcogens S, Se, and Te. These materials are quite good conductors at room temperature, and are sometimes called semimetals. The Fermi level E_F, in such materials is necessarily positioned near the zero gap, that is, where the conduction and valence bands coincide. If such materials were to have their valence band maxima E_v positioned as in Figure 4.3, then indeed E_F would lie very low. Experimental results on HgSe recently available (26) do not show E_F lying so very low; evidently, when conduction and valence bands coincide, cation and anion derivation of the states discussed in Section 4.4 are no longer so distinct. Nevertheless, E_F in these materials does lie substantially below E_F in Au (26). Since the HgX materials are considerably more stable than $(SN)_x$, the prospects for these materials appear quite bright.

4.7 SUMMARY

The difference in electrostatic potential that develops across the interface between a semiconductor and a conductor, that is, the semiconductor band bending $\approx$ the Schottky barrier height ϕ, is described briefly for the non-specialist. Also discussed is the recent observation that ϕ is determined in many compound semiconductor–gold interfaces by the anion in the compound semiconductor. Theoretical reasoning for this special role of the anion as constrasted with that of the cation is presented, including correlations with spin-orbit splittings. Possible extension of the noted anion effects to semiconductor–electrolyte interfaces is suggested based on recent electrochemical studies. Finally, we discuss the range of barrier heights ϕ obtainable on a given semiconductor by utilizing various electronic conductors for contact. Considering first the elemental metals as conductors, there is an adequate repertoire of electropositive metals available, but not of electronegative metals. If binary compound conductors are considered, however, this limitation is eased. For example, HgSe, a zero-gap semiconductor which at room temperature is a rather good conductor, is expected, on the basis of the anion systematics just noted, to be quite electronegative. Indeed, HgSe and the polymer $(SN)_x$ are already among the new class of binary compounds that give more electronegative contact than the most electronegative elemental metal, Au.

ACKNOWLEDGMENT

The authors wish to thank Dr. Carl Wagner for helpful advice.

REFERENCES

1. R. K. Swank, *Phys. Rev.*, **153**, 844 (1967).
2. J. A. Van Vechten, *Phys. Rev.*, **187**, 1007 (1969).
3. Arthur H. Nethercot, Jr., *Phys. Rev. Lett.*, **33**, 1088 (1974).
4. R. Dingle, W. Wiegmann, and C. H. Henry, *Phys. Rev. Lett.*, **33**, 827 (1974).
5. S. Kurtin, T. C. McGill, and C. A. Mead, *Phys. Rev. Lett.*, **22**, 1433 (1969).
6. D. L. Lile, and D. A. Collins, *Appl. Phys. Lett.*, **28**, 554 (1976).
7. D. L. Lile, A. R. Clawson, and D. A. Collins, *Appl. Phys. Lett.*, **29**, 207 (1976).
8. C. A. Mead and W. G. Spitzer, *Phys. Rev.*, **134**, A713 (1964).
9. J. O. McCaldin, T. C. McGill, and C. A. Mead, *Phys. Rev. Lett.*, **36**, 56 (1976).
10. J. O. McCaldin, T. C. McGill, and C. A. Mead, *J. Vac. Sci. Technol.*, **13**, 802 (1976).
11. J. C. Phillips, *Rev. Mod. Phys.*, **42**, 317 (1970), and references contained therein.
12. J. C. Phillips, *Bonds and Bands in Semiconductors*, Academic Press, New York (1973), and references contained therein.
13. L. Pauling, *The Nature of the Chemical Bond*, 3rd ed., Cornell University Press, Ithaca, N.Y. (1960).
14. S. G. Louie, J. R. Chelikowsky, and M. L. Cohen, *J. Vac. Sci. Technol.*, **13**, 790 (1976).
15. G. G. Hall, *Phil. Mag.*, **43**, 338 (1952).
16. D. J. Chadi and M. L. Cohen, *Phys. Stat. Solidi*, **b68**, 405 (1975).
17. J. L. Birman, *Phys. Rev.*, **115**, 1493 (1959).
18. D. J. Chadi, private communication.
19. P. E. Gregory, W. E. Spicer, S. Ciraci, and W. A. Harrison, *Appl. Phys. Lett.*, **25**, 511 (1974).
20. M. Cardona, *Modulation Spectroscopy*, Academic Press, New York (1969), pp. 67–73, and references contained therein.

21. W. P. Gomes, private communication.

22. R. A. L. Vanden Berghe and W. P. Gomes, *Ber. Bunsen-Ges.*, **76**, 481 (1972).

23. M. Gleria and R. Memming, *J. Electroanal. Chem.*, **65**, 163 (1975).

24. See John O'M. Bockris, and Amulya K. N. Reddy, *Modern Electrochemistry*, Vol. 2, pp. 706–708. To convert the potentials in Table 7.6 to the SCE, subtract 0.242 V.

25. R. A. Scranton, J. B. Mooney, J. O. McCaldin, T. C. McGill, and C. A. Mead, *Appl. Phys. Lett.*, **29**, 47 (1976).

26. J. S. Best, J. O. McCaldin, T. C. McGill, C. A. Mead, and J. B. Mooney, *Appl. Phys. Lett.*, **29**, 433 (1976).

5

THIN FILM DEPOSITION
AND CHARACTERIZATION

K. N. Tu

IBM Thomas J. Watson Research Center, Yorktown Heights, New York

S. S. Lau

California Institute of Technology, Pasadena, California

5.1 INTRODUCTION

In the last decade, the application of thin film technology has greatly expanded. The active components of many kinds of devices used in computation and communication, for example, are now made of thin films. With the expansion there is a need to understand and improve the techniques of thin film deposition and characterization. In turn, the advances in thin film techniques have led to the design and fabrication of more sophisticated devices made of semiconducting, magnetic, optical, and low-temperature superconducting materials. Hence, the subject of thin film deposition and characterization has been frequently reviewed (1–6) to meet the fast pace of the technological development. Due to recent advances in vacuum and deposition techniques, the problems of reproducibility and producing thin film samples of comparable quality in different laboratories are no longer serious. Thin film characterization techniques such as ion backscattering, Auger electron spectroscopy, and glancing-angle x-ray diffraction are relatively recent, and new understanding of thin films has been gained from these techniques. Indeed, it is now possible to use thin film samples to observe interactions between various film components that are unique in thin film samples and to produce results that are sometimes not expected or obtained from bulk samples. For example, the present highest critical transition temperature of superconducting is obtained in Nb_3Ge thin films (7, 8), and new quantum states have been observed in man-made superlattices in thin films (9). Thus, it seems that a survey from the viewpoint of unique applications of thin films in material research is desired.

The objectives of this chapter are to review the capability of utilizing thin films for the investigations of structure- and composition-related physical properties of materials and to discuss the essential features of fabrication and characterization techniques needed for these investigations.

This chapter is intended to provide only a background knowledge of thin film deposition and characterization for those researchers who are interested in thin film physics to enable them to interact with specialists in these fields.

Those who need to know the details of the deposition and characterization techniques are referred to the articles listed in the reference section.

To maintain coherence and continuity in the discussion of thin film inter-diffusion and reactions throughout this book, we emphasize primarily structural characterization in this chapter, since the kinetics in thin films is extremely sensitive to the structure.

In the following sections, the unique microstructural and compositional variations in thin films is discussed first, followed by the discussion of deposition parameters in controlling these variations, and finally by methods and techniques to characterize thin film structures and compositions.

5.2 VARIATION OF THE MICROSTRUCTURE OF THIN FILMS

Many unique microstructures have been fabricated in thin film form. For example, Si thin layers have been made in the form of single-crystal, poly-crystal, and amorphous structures. By controlling the fabrication process, one can obtain a variety of microstructures in thin films. Many structure-related physical properties can therefore be investigated more suitably with thin film samples. In this section the various microstructures in thin films and their applications to structure-property related studies will be discussed.

5.2.1 Epitaxial Single-Crystal Films

Epitaxial thin film generally refers to a single-crystal overgrowth on a single-crystal substrate with which a fixed crystallographic relationship exists. When the overgrowth has the same composition as the substrate, it is called homo-epitaxy. When the overgrowth has a different composition than the substrate, it is called heteroepitaxy. Epitaxial single-crystal films generally contain defects such as dislocations, faults, twins, and substructures. Only in very rare cases are epitaxial layers free of defects.

Elemental semiconductor epitaxial layers can be obtained by vapor phase, liquid phase, and solid phase epitaxy (see Chapter 12). Compound semiconductor epitaxial layers are obtained mainly by liquid vapor and molecular beam epitaxy.

Silicon epitaxial thin layers are commonly used in Si device technology to form electrically active layers on the substrates. The junctions formed by epitaxy usually are better defined than those obtained by diffusion. On the other hand, epitaxial films formed by vapor phase epitaxy (VPE) usually contain relatively high defect density. They are ideal specimens for trans-mission electron microscopy investigations of defects such as dislocations and faults in these layers. For example, Booker (10, 11) used VPE Si thin layers for the characterization of defects such as dislocations, faults, and microtwins in both (100) and (111) oriented Si.

In the case of regrowth of amorphous Si layers on single-crystal substrates produced by ion implantation (12), high-density dislocations, well defined and relatively straight, can be obtained as illustrated in Figure 5.1 (see also Chapter 12). In principle, these samples can be used to study dislocation pipe diffusion and dislocation resistivity in Si. Furthermore, such regrown layers can be used as substrates for further VPE layers that contain very high defect density (i.e., $> 10^6$ dislocation lines per cm^2). Campbell et al. (13) have used such samples to study pipe diffusion of As in Si.

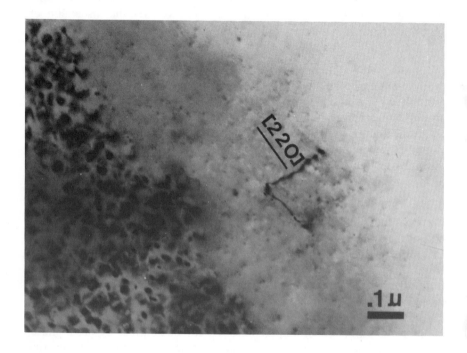

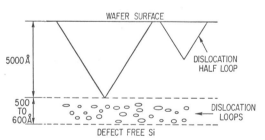

Figure 5.1. *Top*: Bright-field transmission electron micrograph of dislocation structure in Si implanted with 10^{16} Si per cm^2 at 80 keV at room temperature and subsequently annealed at 800°C for 1 h. There are $\sim 2 \times 10^8$ dislocation lines per cm^2. They are perfect, pure edge type and have a Burgers vector of $a/2\langle110\rangle$. *Bottom*: Spatial distribution of the dislocations and loops is shown in the schematic diagram. From reference 12.

Metallic single-crystal films are being used extensively for the investigation of defects in films. These films generally contain defects similar in nature to those in Si described previously. Several applications of these single-crystal metallic films are exemplified as follows:

1. The Study of Surface Atomic Steps. Atomic surface steps are expected to be active sites for surface reactions. The distribution of atomic steps on surfaces may, for this reason, have a close correlation to the kinetics of surface reaction. Investigation of this kind can be performed readily with single-crystal film of heavy metals as shown by Cherns (14). In this case, single-crystal Au films, $\langle 111 \rangle$ oriented and ~ 400 Å thick, were grown on mica substrates and examined in a high-resolution electron microscope. A set of extra diffraction spots due to surface steps on the film surface was observed. Dark-field image from these extra spots revealed the pattern of the steps on the surface. It seems that further work in this area may prove useful in understanding surface defect-related properties.

2. Self-diffusion Along Dislocations in Metallic Single-Crystal Films. Enhanced diffusion along dislocations has been analyzed by Whipple's (15) and Pavlov's models (16). Details of these analyses are presented in Chapter 7 of this volume. Based on these models, the dislocation diffusion coefficient can be determined. Diffusion measurements from bulk specimens have been found to be unsatisfactory for determining the dislocation diffusion coefficient because lattice diffusion is the overriding diffusion mechanism in bulk samples containing low density of dislocations. To overcome this problem, Gupta (17) has performed a self-diffusion study in Au using Au single-crystal films which were grown epitaxially on (100) MgO and contained dislocations and stacking faults in the range of 10^{10} to 10^{11} defects per cm^2. Depth penetration profiles of tracer in 2 μm-thick films over the temperature range of 247 to 382°C were determined by a sectioning method using the Ar sputtering technique. More of this diffusion work is presented in Chapter 7.

3. Single-Crystal Film as a Diffusion Barrier. Fast interdiffusion in layered thin film structure (see Chapter 9) is well known and has been attributed to grain boundary diffusion. From the standpoint of reliability of multilayered thin film in electronic devices (see Chapter 2), it is desirable to reduce the interdiffusion. The significance of grain boundary diffusion in layered thin films and the advantage of using a single-crystal film as a barrier for the diffusion have been demonstrated (18). In the Pb–Ag–Au film structure it was shown that 3000 Å polycrystalline Ag film cannot prevent Au from reacting with Pb at 200°C producing a Pb$_2$Au compound. However, the reaction is stopped by interposing a 3000 Å single-crystal Ag film between the Pb and the Au films.

4. Interfacial Misfit Dislocation and Their Motions Due to Interdiffusion. This subject has been extensively reviewed by Matthews (19, 20), and only a brief discussion will be given here. Misfit dislocations between an overgrowth and a single-crystal substrate or between two consecutively evaporated epitaxial thin films are expected to form when the misfit strain due to lattice parameter difference exceeds the elastic limit. Dislocations of this kind are relatively well defined and equally spaced. The generation and propagation processes of misfit dislocations as a function of overgrowth thickness and shape can be investigated by in situ observations in an electron microscope. When interdiffusion takes place between the two metallic single-crystal layers, vacancies are generated in the vicinity of the interface. Misfit dislocations would therefore experience a driving force for climb motions due to the nonequilibrium concentration of vacancies (20). As a result, misfit dislocations are distributed in the entire volume of the specimen and interdiffusion can be further enhanced. Investigation of this type using single-crystal metallic films can lead to a better understanding of the interaction and may prove useful in predicting (or help solve) practical problems such as impurity diffusion into Si wafers for device applications; for example, misfit dislocations generated by drive-in diffusion may cause carrier lifetime problems of the device.

5.2.2   Bicrystal Metallic Films

Bicrystal metallic films of Au, Cu, or Al have been used for the study of grain boundary properties such as structure and energy (21). Dislocation networks in grain boundaries of twist, tilt, and mixed type and near the coincidence site relation have been extensively studied using transmission electron microscopy (22, 23). In pure twist type, the grain boundary plane is normal to the axis of rotation; and in pure tilt type, the plane is parallel to the axis. When the angle of rotation falls on a set of specific angles, for example, 36.9° around a (100) cubic axis, a sublattice that consists of lattice points common to both crystals is generated; the bicrystal is said to be in coincidence site relation and the grain boundary to be a singular grain boundary if it locates at a symmetric or a mirror position. Ideally, a rigid or unrelaxed singular boundary contains a high density of coincidence lattice sites. However, computer model calculation shows that atomic relaxation tends to destroy coincidence lattice sites in the boundary but that a periodic structure of the boundary with a period closely related to that of coincidence lattice sites in the unrelaxed singular boundary still remains (24).

When the angle of rotation, the so-called misorientation, is slightly off the coincidence angle, a set or a network of dislocations becomes an integral part of the equilibrium structure of the grain boundary. The spacing between these dislocations and their interactions with point defects and lattice dislocations have been extensively investigated and reviewed. The spacing d of dislocation networks has been plotted as a function of twist angle and found to agree very well with the following equation:

$$d = \frac{|b|}{\Delta\theta'} \tag{1}$$

where $\Delta\theta'$ is the rotational deviation from the nearest coincidence angle and b is the base vector in the DSC lattice or "complete pattern shift lattice" (21). This agreement indicates that singular grain boundaries indeed exist in the twist-type bicrystals, which in turn indicates that there will be cusps in the plot of grain boundary energy versus twist angle. Near the exact coincidence angle, dislocation networks of rather large spacing have frequently been observed, but no network was ever found in the twist bicrystals with a misorientation far from the coincidence angles, that is, the cusps.

Bicrystal films of Au with twist boundaries have been made by welding face to face two single-crystal films grown epitaxially on Ag grown on the cleavage plane of rock salt. After removing the Ag and rock salt by dissolution, the bicrystal Au films are ready for transmission electron microscopy. Dislocations that are of the screw type in the twist grain boundaries can be studied.

Tilt-type bicrystal Au films can be made similarly by first growing two epitaxial Au films on any $(hk0)$ plane of rock salt and then welding them face to face to form a symmetric tilt-type grain boundary in between. Or, an asymmetric tilt-type grain boundary can be made by welding a $(hk0)$ oriented film to one that has been grown on a cleavage plane. In these bicrystals, the plane of grain boundary is parallel to the film surface, so it offers a large grain boundary area for microscopic observation. However, Burger's vectors of edge dislocations in these tilt-type grain boundaries are normal or nearly normal to the film surface, which means that $g \cdot b = 0$ in a dark field, so that it is difficult to obtain an image of these dislocations without tilting. Furthermore, such bicrystals are unsuitable for grain boundary diffusion studies. Grain boundary diffusion can be better carried out by using a tilt bicrystal with its boundary normal to the film surface. There are now two methods that can produce such kinds of tilt bicrystal films. The first one is to transform the twist type as discussed in the last paragraph by annealing (25). The annealing at 200 to 400°C leads to grain boundary migration in Au films, forming columnar grains with only two crystal orientations. The grain boundary in the columnar grain structure is of tilt type, with a tilt angle equal to the twist angle. The second method is to grow an epitaxial Au film on a prefabricated rock salt substrate that had a tilt-type grain boundary (26). Figure 5.2 shows the micrograph of a bicrystal Au film of 10° tilt around the [100] axis made by using the second method. The insert in the figure is a diffraction pattern showing the angle of 10° tilt of the bicrystal.

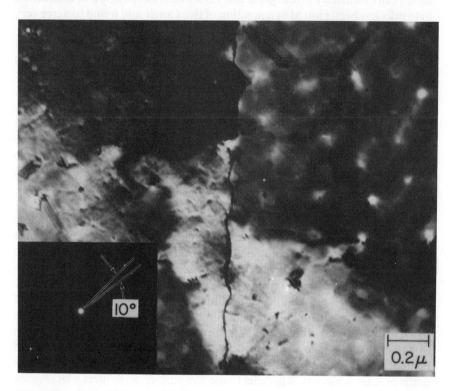

Figure 5.2. Bright-field transmission electron micrograph of a 10° tilt-type grain boundary in a thin film of Au grown on a rock salt bicrystal substrate. Insert is the diffraction pattern showing the 10° tilt. The rock salt substrate was obtained from Prof. C. L. Bauer, Carnegie-Mellon University.

5.2.3 Uniaxial Textured Films

Most polycrystalline thin films, metallic or nonmetallic, on amorphous substrates tend to show a preferred orientation. Since the (111) surface has the lowest surface free energy of *f.c.c.* structure, we expect a polycrystalline thin film of *f.c.c.* structure to show [111] preferred orientation on amorphous substrates. This is true for Al and Pb films on glass. Similarly, one expects a polycrystalline thin film of *h.c.p.* (with small a/c ratio) to show [0001] and *b.c.c.* to show [110] preferred orientation on amorphous substrates; and these are also true, for example, for Co films on glass and Cr films on fused quartz. They are called uniaxial textured films. Since low-cost display, recording, and solar energy devices require large areas of inexpensive substrates, uniaxial textured films on glass or oxidized metal surfaces become potentially technologically important.

Ideally, in a uniaxial textured film, all the columnar grains have a common axis normal to the film surface. The distribution of the angle of tilt between any two neighboring grains in the film may not be random; rather, the distribution at the coincidence angles may be denser because of the lower energy of the singular grain boundaries. In practice, however, it is not easy to obtain an ideal uniaxial textured film. The axis of preferred orientation in most textured films shows a spatial distribution. In fact, the axis may even make an inclination angle with the normal of the film surface if the film was not deposited by normal incidence but rather at inclination. The spatial distribution can be determined using the same principle as that used to determine the pole figure of fiber texture in a wire by x-ray diffraction (27). If the spatial distribution has a rotational symmetry around the preferred axis, a one-dimensional radial distribution is then sufficient to characterize the texture and can be obtained following Harris' method:

$$p_{hkl} = \frac{\dfrac{I'_{hkl}A}{I_{hkl}A'}}{\dfrac{1}{n} \sum\limits_{hkl}^{n} \dfrac{I'_{hkl}A}{I_{hkl}A'}} \tag{2}$$

where p_{hkl} is the texture factor of the hkl reflection of the film; I_{hkl} and I'_{hkl} are integrated intensities for the hkl reflection in a randomly oriented sample and a textured film, respectively; similarly, A and A' are the absorption factors, and A' depends on the film thickness; and $n \gg 1$ is the number of reflections measured. Figure 5.3 shows the radial distribution of the $\langle 111 \rangle$ axis in a 2000 Å Au film deposited on a fused quartz substrate at 400°C. Eleven reflections were measured and compared to those of a Au powder sample. The peak at 70° in the curve shows up because the $\langle 111 \rangle$ axes in Au make an angle of 70.53° with each other.

On the other hand, in the case of no rotational symmetry, a two-dimensional pattern is often needed to represent the texture. The additional dimension can be obtained by fixing the counter at a Bragg reflection and then rotating the film around its normal a complete 360°. A fast display of this kind of texture can be obtained by using negative films in an x-ray camera. Both a cylindrical texture camera (29) and a Read camera (30) can show the texture readily. Quantitative information can also be obtained by reading the density from the negative with a densitometer.

It is obvious that many physical quantities, based on the assumption of randomly oriented grains, require corrections when the properties are

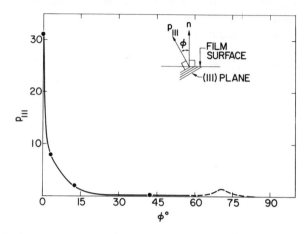

Figure 5.3. Radial distribution of $\langle 111 \rangle$ axes in a Au film of 2000 Å deposited on a fused quartz substrate at 400°C. From reference 32.

measured using textured films. This was the case in applying the x-ray reflection intensity to monitor the rate of interaction in Cu–Au bimetallic films (31).

When a textured film is analyzed by ion backscattering technique, channeling of ions could occur along the uniaxial direction and cause a reduction of backscattered yield. Preliminary results of textured noble metal films on glass substrates show that channeling does occur and that the yield of backscattered ions is reduced (32).

Textured films not only form on amorphous substrates but also on single-crystal substrates, where the misfit is too large to enable heteroepitaxial growth, or on polycrystalline substrates such as a textured stainless steel sheet. However, these subjects are just too broad to be covered here.

5.2.4 Randomly Oriented Fine-Grain Thin Films

In studies where orientation dependence is of great importance, the randomly oriented fine-grain thin film can be used as a reference to show the isotropic properties of the bulk of the thin film, for example, in many studies of the magnetic properties of thin films. The film can also be used to isolate the contribution from grain boundaries by comparing its properties to those of a single-crystal thin film, for example, in the study of electrical conductivity (33). In thick film applications, the concern of texture and orientation dependence is not as critical as that in thin films. It is fair to say that in a film where grain size is comparable to thickness, texture is expected; but in a film where thickness is several times larger than grain size, the microstructure tends to be less oriented.

X-Ray intensity measurement is still the simplest method for monitoring texture in polycrystalline thin films. Often, the pattern of a set of continuous rings in electron diffraction patterns is used to indicate the randomness of the film. This could be misleading because relative intensities among the rings must also be measured and compared to the calculated value. Using x-ray diffraction, one can compare the thin film intensities to those of a powder sample or to the values listed in ASTM files provided that the absorption correction due to thickness difference is made and the diffraction geometry is similar.

5.2.5 Amorphous Thin Films

Amorphous thin films that are stable at ambient temperature have either a random network structure or a random dense packed structure. The former includes oxides, elemental semiconductors, and chalcogenides, and the latter are mainly found in alloyed phases. Random network structures can be regarded as being two interpenetrating random dense packed structures (34). These random structures are characterized by being continuous, rigid, and lacking in long-range order. When examined by diffraction, these structures show broad diffuse peaks in x-ray spectra and halos in electron diffraction patterns. Their images obtained in high-resolution electron microscopy, when subjected to diffractogram analysis, will again show concentric halos. The structure of amorphous elemental semiconductor thin films has been extensively studied (35-38). After careful comparisons using constructed models, computer simulations, diffraction data, and high-resolution images, the prevailing structural picture of these amorphous materials is one of random network rather than microcrystalline, although there are coherent regions of scattering in the structure (37) when viewed in a dark-field transmission electron micrograph. These regions are not inconsistent with a random network structure (38).

Up to now, it has not been shown that any element in the periodic table can be made into an amorphous state and kept stable at ambient temperature. For amorphous alloys, it has not been shown either that a combination of any two elements can be made amorphous. The available data of amorphous alloys formed by splat-cooling show a strong dependence on atomic size and valence electrons, and these alloys can only be made over a very narrow concentration region. From the viewpoint of dense packing and size difference, the narrow concentration ratio can be regarded as dictated by the filling of B atoms in tetrahedral and octahedral voids in the random dense packed structures of A atoms (39). This concept seems to be true in the case of making amorphous alloys by choosing a B element from metalloids and an A element from noble and near-noble metals. From the viewpoint of stabilizing the atomic bonds by charge transfer, the most stable amorphous alloy of Pd–Si is expected and also found to be near the concentration of Pd_4Si (40). In order to suppress the nucleation of crystalline phases during rapid quenching from a melt to produce amorphous phases, the concentration near a deep eutectic point is favored. This seems to be true for the Au–Si system at a concentration near Au_3Si (41). However, vapor deposition onto cold substrates has been found to be a more effective way of obtaining amorphous alloy thin films over a much wider concentration range (42).

Amorphous thin films are also ideal for studying defects and diffusion in an amorphous structure (43). The film in general contains much less impurities than those prepared by quenching bulk liquid droplets. Also the surface of thin film is smooth enough for radioactive tracer plating. Depth penetration of the tracer can be easily determined by using Ar sputtering to erode the film layer by layer. The only concern here is that the temperature of diffusion annealings should be below the glass transition temperature and the crystallization temperature. Crystallization of amorphous phase has been widely studied (44), in particular the enhanced crystallization of amorphous silicon films in contact with metals. This is due to the concern of making Schottky contact to amorphous Si.

Great interest and expectation on the potential use of amorphous thin films have been stimulated by the applications first in fast switching devices (45), in magnetic bubble devices (46), and in the doping of amorphous Si film formed by glow discharge deposition (46a). Recently, it has been demonstrated that amorphous Si films can be used as a source for solid phase epitaxial growth of active layers on Si (47, 49). The search for amorphous superconductor thin films has also attracted much attention. However, the critical transition temperature observed in amorphous superconductor films is not as high as that of most crystalline superconductors.

5.2.6   Man-Made Superlattice Thin Films

Recently, the technique of computer-controlled molecular beam epitaxy (MBE) was developed to achieve superfine epitaxial structures with compositional variations in a small thickness dimension of the order 10 to 100 Å (9, 50, 51). Such a multilayer structure consisting of alternating epitaxial layers of GaAs and $Ga_{1-x}Al_xAs$ exhibit unusual electron transport and optical properties due to quantum mechanical effects (9, 52, 53). A one-dimensional periodic potential well is created by the periodic heterojunctions in the superlattice. The width of the allowed energy bands in the wells is determined by the well or barrier width. With increase in the width, the allowed bands are narrowed and approach discrete energy levels for quantum states in a single square well potential. A superlattice period of 50 to 100 Å is large compared to the crystal lattice constant. Such a large period in real space gives rise to minizones in wave vector space, dividing the $E-k$ relation along the superlattice direction into a series of minibands separated by forbidden gaps. If the energy of incident electrons coincides with the quantized energies in the wells, the electrons can then tunnel through the barrier owing to resonance. Such tunneling behavior would lead to negative differential resistance (9). This unusual transport property was first predicted by quantum mechanical calculations and has now been observed in man-made superlattice thin film systems.

The composition of a superlattice can be analyzed by Auger electron spectroscopy combined with sputtering (9). The compositional profile of a periodic structure consisting of 50 Å layers of GaAs and $Ga_{0.75}Al_{0.25}As$ is shown in Figure 5.4. The steady decrease in the peak-to-valley ratio with increasing depth of removed material is attributed to spatially varying sputtering rates on the scale of the area of the focused primary electron beam (typically 100 μm in diameter).

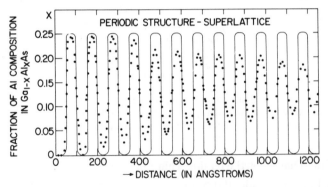

Figure 5.4.   Composition profile of a periodic structure measured by a combination of ion sputter-etching and Auger electron spectroscopy. From reference 49.

The structure of a superlattice can be analyzed by large- and small-angle x-ray scattering measurements. For large-angle x-ray diffraction, diffraction peaks from both the host crystal and the superlattice can be observed. The small-angle x-ray interference technique is best suited for an accurate determination of the periodicity of a superlattice (54). In this technique, a monochromatic x-ray beam is incident on the sample surface at very small glancing angle ($\geqslant 1°$). By varying the glancing angle in the vicinity of the critical angle for total reflection, interference maxima and minima in the specularly reflected x-ray beam can be observed due to different indices of refraction in the alternating layers. The interference pattern is then compared with calculated patterns for various assumed periodicities. Accuracy of the order of a few angstroms (~ 2 Å) can be achieved in the determination of periodicity, as illustrated in Figure 5.5. It should be noted that small-angle x-ray scattering technique is also applicable to other thin film systems for accurate film thickness measurements (55).

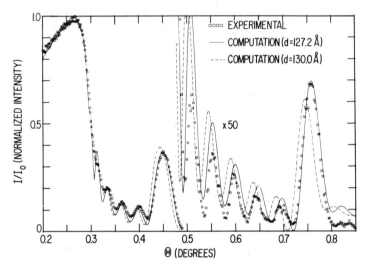

Figure 5.5. X-Ray interference pattern for a periodic structure of GaAs–AlAs with six periods fo 127.2 Å. In comparison with the data points, computer results with different period thicknesses are plotted to show agreement (full curve) and discrepancy (broken curve). From reference 49.

5.2.7 One-Dimensional Metallic Thin Film Lines

Metallic lines 80 Å wide, 100 Å thick, and ~ 10 μm long have been fabricated using a novel electron beam technique (55a). These lines show a fivefold reduction in linewidth compared to those produced by a conventional technique using electron beam and photoresist, where the linewidth is limited to about 1000 Å due to electron scattering within the resist and electron backscattering from the substrate. In the new technique, instead of using photoresist, the linewidth is defined by scanning a 5 Å-wide electron beam (55b) across a metallic film (Au–Pd alloy) to polymerize a fine line of carbon film on the metallic film surface. The source of carbon is from the contamination on the metallic film and from the vacuum pump. Then the metallic line is obtained by using Ar ion beams to sputter off the rest of the film that is not protected by the carbon layer. Thin film lines 80 to 300 Å wide have been repeatedly made. At present, the linewidth is limited by the grain size in the metallic film. It is expected that finer lines of amorphous materials and of finer spacings can be obtained. Obviously, applications of these lines in transport studies and interference of soft x-rays should be of great interest.

5.3 VARIATION OF THE COMPOSITION OF THIN FILMS

Many alloys and compounds of unique compositions have been fabricated in thin film form. This is because thin films can be manufactured under non-equilibrium conditions, resulting in metastable phases and compositions that are relatively stable at room temperature and above. In the case of bulk materials, chemical equilibra (such as solubility) usually dictate the presence of phases and compositions in the material. Many novel deposition techniques of thin films have been developed, allowing for extra dimensions in precision control of thin film compositions. By controlling the composition of thin films, many unique properties not achievable in bulk form can be realized. For example, by varying the nitrogen partial pressure during reactive sputtering of Ta, thin films in the form of beta Ta, body-centered cubic Ta, Ta_2N, or TaN can be obtained. Each of these tantalum-based films has its characteristic resistivity, crystallographic structure, and temperature coefficient of resistance (1). Depending on the application of the Ta films, desirable properties can be tailored and readily obtained. There is an abundant wealth of information on deposition methods of controlling and varying thin film compositions (1–6, 56). We summarize only the salient features in the following sections.

5.3.1 Fixed Composition of Alloy Films by Sputtering

Sputtering is a relatively energetic process of thin film deposition. Atoms from the target surface are ejected by bombardment with energetic particles and condense on the substrate with or without a bias field (56, 57). Typically, the energetic particles are Ar ions with a few keV energies. There are many variants of sputtering modes, such as dc, rf, triode, and reactive sputtering, and ion plating, which is a combination of evaporation and sputtering. Sputtering, with its many advantageous features, is best suited for the deposition of refractory materials and insulators.

One important advantage of sputtering is that the composition of the sputtered film can be the same as the multicomponent or compound target. For this condition to hold, it is necessary that the target material does not decompose and that there is no preferential sputtering of the components. In the case of Bi_2Te_3, PbTe, InSb, and other alloy films, it has been found that little or no compositional change occurs by sputter deposition (2). If preferential sputtering occurs, the target surface composition is altered. However, it is likely that the composition of the altered layer is maintained after establishing equilibrium, and the deposited film will have the same composition as the altered layer, as in the case of $AuCu_3$ alloy (2).

Stoichiometric or relatively well-defined composition of thin films can also be accomplished by reactive sputtering. Oxides and nitrides such as SiO_2, Ta_2O_5, HfO_2, Si_3N_4, and TaN can be readily deposited (1–3). These films are commonly used as capacitors, resistors, and insulators in thin film integrated circuits. One other interesting application of sputtering is the deposition of metallic films of Gd–Co, Gd–Fe, Gd–Co–Au, and Gd–Co–Mo (56, 58). These thin films are amorphous in structure, yet they are magnetically anisotropic and potentially useful for magnetic bubble domain applications.

In the field of superconductivity, there has been intensive research for materials with high transition temperatures. The highest recorded transition temperature superconductor is Nb_3Ge in thin film form (7, 8). A superconducting transition temperature of $> 22°K$ has been obtained in crystalline Nb_3Ge thin films deposited by getter-sputtering, as compared to a T_c of $\sim 5°K$ in bulk samples.

5.3.2   Control of Composition by Using Two Evaporation Sources

Deposition of alloy thin films by a single-source evaporation often runs into the difficulty of maintaining or precisely controlling a constant composition throughout the entire layer or from one evaporation run to the other. This is because fractionation (partial separation of the components) may occur which may lead to the depletion of one component in the evaporation charge, such as in the case of evaporation of Nichrome (80% Ni–20% Cr by weight) (1).

A satisfactory method of fabricating alloy and compound in thin films with controlled compositions is to evaporate each component from a separate source. The substrate temperature is adjusted such that reactions and homogenization between the two components can take place at the substrate. This is known as the "two source" method (2, 59–61). By using this technique, continuously varying or stoichiometric compositions of alloy thin films can be deposited by changing the evaporation power in each source, the substrate temperature, the spacing between two sources, and the relative position of the substrate with respect to the two sources. For example, epitaxial films of $PbTe_xSe_{1-x}$ can be deposited on NaCl crystals by the two-source method, with the value of x ranging from 0 to 1 (59). Stoichiometric GaAs thin films (62), superconducting Nb_3Sn thin films (63), as well as binary alloy films of varying composition with constituents of Cu, Ag, Au, Mg, Sn, Fe, and Ni (64) have all been fabricated by this technique.

Controlled variation of the film composition can also be achieved by producing a wedge-shaped layer using a rotational shutter during evaporation. Hot substrate or a subsequent heat treatment can be used to promote interactions between the initially deposited layer and the wedge-shaped layer to obtain controlled compositional variation across the thin film.

5.3.3 Supersaturation by Vapor Quenching

Alloy thin films with nonequilibrium compositions and metastable phases can be produced by simultaneous evaporation of two components onto a cold substrate. This process is referred to as "vapor quenching" (65). Evaporant atoms arriving from two sources at a substrate at sufficiently low temperature cannot move appreciably. They are "frozen in" at the place of impingement before equilibrium is established. Thin films deposited under such conditions may have compositions exceeding equilibrium solubility limits or exhibit amorphous structure. Investigations on a number of binary systems revealed that extended solubility is obtained in systems where the ratio of atomic radii is close to 1, and amorphous structure results when the ratio exceeds 1.10 (65). For example, supersaturated solid solutions of Co–Cu (65) and Fe–Cu (66), amorphous metallic alloys of Cu–Ag (65) and Cu–Mg (65), and vitreous semiconductor of Bi–Se (66) have been produced this way. In addition, many amorphous films of binary metallic alloys obtained by vapor quenching have shown a much wider range of composition variation than those alloys obtained by splatt-quenching from their melts (42). Thus, by using the technique of vapor quenching, the composition of an amorphous alloy can be varied substantially. Figure 5.6 shows the extent of composition variation of several amorphous alloys prepared by vapor quenching.

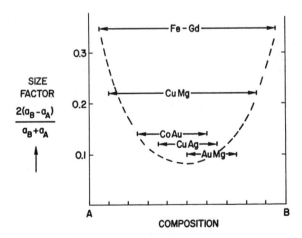

Figure 5.6. Composition of amorphous alloy films with close-packed components. From reference 42.

5.3.4   Adjustment of Composition by Ion Implantation

Ion implantation has long been used to introduce dopant atoms into semiconductors (67). The depths of the implanted atoms are usually confined to the surface layers extending from a few hundred to a few thousand angstroms. In this technique, materials can be implanted with any element for which an ion beam can be generated. The advantages of using ion implantation lie in the precision control of ion species and the concentration level of implanted atoms as well as the reproducibility of the implantation process. As a result, much better manufacturing yield of semiconductor devices can be realized.

In recent years, ion implantation techniques have been applied to metals and other materials to achieve desirable modification of near-surface properties (68–70). For example, the effects of ion implantation on the oxidation behavior of metals, friction and wear properties, corrosion behavior, and catalysis have been investigated. A detailed account of ion-implanted metal layers is given in Chapter 14. Since the range of implanted atoms is of the same order of magnitude as the thickness of commonly used thin films, it seems natural to use ion implantation techniques to adjust the compositions of thin films. Ion implantation offers much better control of the concentration and better uniformity of the altered layer.

The application of ion implantation to thin films has met great interest in the study of superconducting thin films. The introduction of impurities can cause an increase in the superconducting transition temperature (T_c) in some materials, and the introduction of defects can cause decreases in T_c in other materials. For example, implantation of Ne and Xe in oxygen-containing Mo thin films and implantation of B, C, N, P, As, and S in pure Mo thin films causes enhancement of T_c in Mo or Mo-based superconductors due to chemical effects. On the other hand, implantation of He^+ in Nb_3Sn, Nb, and Ta film causes decrease in T_c due to radiation damage. The subject of superconducting thin films has been under intensive research, and ion implantation serves as a convenient means to study the structural and chemical effects.

The investigation of impurity effects on the regrowth behavior of amorphous Si has also been facilitated by ion implantation (71, 72). Amorphous Si layers can be obtained by implanting Si into single-crystal Si substrates near liquid nitrogen temperatures. Selected impurity atoms such as P, As, B, O, C, N, Ar, and Ne are then implanted into the amorphous Si layers separately. The regrowth behavior (epitaxial recrystallization of the amorphous Si layer) is investigated at $\sim 550°C$ as a function of impurity and dose. It is found that certain impurity atoms such as P, B, and As accelerate the regrowth rate, while other impurities such as N, O, C, Ar, and Ne retard the regrowth rate (see Chapter 13). This type of investigation has practical significance on device fabrication.

5.3.5 Formation of Equilibrium and Metastable Phases by Thin Film Reactions

The determination of equilibrium state of solid phase reaction in bulk sample is often a lengthy process, since it takes a long time for bulk samples to equilibrate. In thin film systems, the diffusion distance Dt is often comparable to the film thicknesses. The diffusion coefficients D are often many times larger than those in bulk samples due to high defect concentrations in thin films. These two factors would accelerate the mixing process in thin film samples; therefore equilibrium state at low temperatures can be reached in a much shorter time. For example, in the Au–Al bimetallic thin film system, the equilibrium states as a function of composition can be reached in the order of 60 min at 200°C (73). In the case of metal silicide formation, the equilibrium state can be reached in a few hours at somewhat higher temperatures (see Chapter 10). Because of the fast kinetics of interactions, it is expected that thin film systems can greatly facilitate the determination of equilibrium phase diagrams. This is particularly important for those ternary systems where the equilibrium phase diagram is not available (74).

Phases that are at a metastable state can occur in thin film reactions. For example, Pd_2Si can be formed next to Si over a very large temperature range and used as a contact layer to Si (75); yet according to the Pd–Si binary phase diagram, the phase that is stable with Si should be PdSi.

5.4 FILM DEPOSITION

5.4.1 Introduction

The formation of thin films on a substrate by deposition is basically a phase change phenomenon involving nucleation and growth with the constraint of the substrate. In principle, by controlling the nucleation and growth on the substrate, any of the thin films as described in Sections 5.2 and 5.3 can be obtained and the film structure and composition can be confirmed by the characterization techniques to be discussed in the next section. In practice, because there are too many variables in thin film deposition that can affect the nucleation and growth, we must know how to control them during the deposition. Since the complications of film deposition have been widely covered by many review articles, there is no need to repeat them here. Instead, a rather simple relationship between deposition variables and thin film nucleation and growth is presented in the following to serve as a rough guide in the preparation of films with particular properties.

In general, the important variables that can be controlled in a deposition are vacuum pressure, deposition rate, substrate temperature and substrate structure, and they must be specified in film deposition. A proper choice of these variables is crucial to the film to be deposited, yet the answer to the question of how to specify them must come from an understanding of the effect of these variables on film nucleation and growth. Also, we have to select the deposition technique most suitable for making the film, for example, whether it is by electron beam heating or by rf sputtering. Furthermore, if the film is susceptible to an external field, we should consider whether the presence of such an external field is helpful or not. In the following, we shall first discuss the deposition variables and the effect of external fields and then the various kinds of deposition techniques.

5.4.2 Deposition Variables

If we take the simple relationship as shown in Table 5.1, which indicates that the deposition variables in the left-hand column has a one-to-one correspondence to the parameters of nucleation and growth in the right-hand column,

TABLE 5.1. DEPOSITION VARIABLES

Deposition variables	Parameters of nucleation and growth
Vacuum pressure	impurity content
Deposition rate	supersaturation
Substrate temperature	undercooling
Substrate structure	interfacial energy

the effect of these variables on film nucleation and growth can be understood by the same effect of the parameters on film nucleation and growth (76, 77). A lowering of vacuum pressure corresponds to an increase of purity. Supersaturation goes up with deposition rate, but undercooling increases with decreasing substrate temperature. The selection of a substrate that allows epitaxial film growth means that the film–substrate interfacial energy has the lowest value at the epitaxial condition and any deviation from this condition tends to increase the interfacial energy and the barrier of nucleation. Whether the epitaxial film forms a three-dimensional nucleus or a two-dimensional uniform layer on the substrate depends on the wetting condition, which is governed by the interfacial energies involved. On the other hand, a film on a glass substrate will show much less dependence of interface energy on the film orientation.

It is well known that an increase in impurity content, supersaturation, and undercooling and a decrease in interfacial energy will increase the rate of heterogeneous nucleation (78). Thus, the combination of a poor vacuum, an increase in deposition rate, and a decrease in substrate temperature will lead to the formation of fine-grained thin films (or amorphous films) on a glass substrate. On the contrary, the deposition of an epitaxial film requires high vacuum, slow deposition rate, high substrate temperature, and single-crystal substrate with a low Miller indice surface that has small lattice mismatch with the film. What in reality leads to the formation of an epitaxial film is not to reduce the rate of nucleation to that of one nucleus but rather to increase the tendency of nucleating the oriented nucleus so that all the nuclei on the substrate have the same orientation. To further illustrate the simple relationship discussed in the above, we list in Table 5.2 the conventional deposition conditions for obtaining various kinds of thin films. It should be pointed out that in the deposition of an alloy film, the partial vapor pressure becomes an important variable.

TABLE 5.2.   CONVENTIONAL THIN FILM DEPOSITION CONDITIONS

Deposition variables	Epitaxial film, single crystal $(Ag-NaCl)^a$	Textured film $(Pb-SiO_2)^a$	Randomly oriented film, fine grain $(Cr-SiO_2)^a$	Amorphous film $(\alpha\text{-}Si\text{-}Si)^a$
Vacuum pressure, torr	10^{-9}	10^{-7} to 10^{-9}	10^{-5}	10^{-5} to 10^{-9}
Deposition rate, Å sec^{-1}	1 to 10 (slow)	10 to 100 (medium)	100 (fast)	>100 (fast)
Substrate temperature °C	350 (limited by the decomposition of NaCl)	above the half-melting point of the metal	room temperature or liquid nitrogen temperature	room temperature or liquid nitrogen temperature
Substrate structure	single-crystal surface with good lattice match with the film	amorphous	amorphous or or poly-crystalline metal sheet	amorphous or crystalline

^{a}Composition of film in parentheses given as example.

5.4.3   Effect of External Field on Film Deposition

The application of an external electric field during deposition often causes changes in the structure of thin films. For example, an electric field applied parallel to the substrate plane during the evaporation of Ag on glass substrates has the effect of reducing the critical thickness for forming a continuous film (2). Chopra explained the field effect on the basis of interaction of the field and the electrically charged islands in the initial stage of the deposition. Electrostatic effects cause the islands to coalesce, thus forming a continuous film at a lower critical thickness (2).

The orientation of thin films is often affected by an externally applied field. Chopra (2) discovered that a dc electric field of $\sim 100 \text{ V cm}^{-1}$ applied laterally promotes epitaxial growth of Ag on NaCl substrates. The preferred orientation of thin films can also be controlled by an externally applied electric or magnetic field, as observed in sputtered Ni–Co–Fe films (79). Dendritic growth of evaporated Ag_2Te has been observed (80) under the influence of an electric field. The direction of growth is parallel to the dc field, and the grains are pointing in the direction of cathode to anode. The electrical resistance of thin films may also be reduced because of field-induced coalescence.

A field applied normal to the substrate is also known to enhance the growth rate of epitaxial Ge, Si, and GaAs film formed by chemical vapor deposition techniques (81).

5.4.4 Deposition Techniques

Thin film deposition techniques have been described in detail in the literature (1–3, 5, 81a, 81b). Generally speaking, the conventional techniques are evaporation, sputtering, and chemical methods. Recently, a novel technique for thin film deposition by means of ion beam has been developed (82–84). In this technique, a selected ion current with energies ranging between 20 and 100 V is extracted from an ion source and directed to the substrate for deposition. Ion beam techniques have the following advantageous features: (i) the beam energy, direction, current, and isotopic purity can be precisely controlled; (ii) ion beam deposition can take place during evaporation or in a chemically active environment. These features may lead to better film adhesion, improved epitaxy, cleaner interfaces, and sometimes unique film structures. For example, man-made polycrystalline diamond-like carbon films have been deposited with this technique (83, 84). Table 5.3 gives a brief comparison on common deposition techniques.

5.5 FILM CHARACTERIZATION

5.5.1 Introduction to Characterization Techniques by Scattering Processes

Most modern characterization techniques for thin film structural and compositional analyses are scattering techniques. Scattering methods are particularly well suited for thin film analyses since they tend to have high sensitivity, so that the amount of material needed for analysis is small and both structural and compositional information can be obtained.

In scattering techniques, usually a beam of photons, x-rays, electrons, or charged particles is incident on the specimen surface. The incident beam interacts with the solid sample and gives rise to scattered beams. The scattered

TABLE 5.3. DEPOSITION TECHNIQUES

Deposition technique	Advantages	Disadvantages
Resistance heating evaporation	easy setup for low melting materials	alloy with filament
Electron gun evaporation	apply to most metallic and semiconductor elements	refractory metals; carbon and oxide difficult to evaporate
	produces amorphous elemental semiconductors	
Sputtering	deposit conducting as well as insulating materials; film composition is related to target composition	Ar or any other sputtering gas atoms and molecules incorporated in the film
	produce amorphous metallic and semiconductor	usually high substrate temperatures, intermixing of film and substrate
	easy to apply a biased field	cause damage to substrate surface
Chemical vapor deposition	device quality epitaxial layer with electrical activity; can also deposit polycrystalline layers	more elaborate set up gas flow rate adjustment critical; high substrate temperature
Molecular beam epitaxy	high-quality compound films	elaborate setup
Electrodeposition	wide range of film; large area of uniform thickness	useful mostly to metallic films; impurity problems
Liquid phase epitaxy	good-quality compound films	concentration control and reproducibility problems
Ion-beam techniques	precision control of deposition parameters	slow deposition rate and elaborate setup

beams can also be in the form of photons, x-rays, electrons, and charged particles. By analyzing the scattered beams, structural as well as compositional information of the thin film samples can be obtained. The most common is the x-ray diffraction technique, where the incident and the scattered beams are both x-rays. Table 5.4 gives a brief summary on the scattering techniques commonly used for thin film analysis.

Some of these techniques for compositional analysis are discussed in Chapter 6 in this volume. We shall emphasize in the following the techniques for characterizing surface morphology and internal microstructure of thin films. Since many of the techniques are complementary to each other, it is desirable to use more than one technique in characterizing a thin film.

TABLE 5.4. SCATTERING TECHNIQUES[a]

Incident beam	X-Rays[b]	Electrons[b]	Optical photons[b]	Charged particles[b]
X-Rays	x-ray diffraction and fluorescence	electron spectroscopy for chemical analysis (ESCA)		
Electrons	electron micro-probe, energy dispersive analysis	TEM, SEM, STEM, LEED, HEED, energy loss measurements	Cathodoluminescence	
Optical photons		photoemission	optical microscopy	
Charged particles	ion-induced x-ray analysis	ion-induced secondary electron emission	ion-induced photon analysis	Rutherford back-scattering, SIMS, nuclear reaction analysis

[a]The authors are grateful to Dr. J. F. Ziegler for helpful discussions about this table.
[b]Scattered or emitted beam for analysis.

5.5.2 Surface Morphology

Lateral uniformity of the surface is an important issue in thin film analysis. The accuracy of most of the ion beam techniques depends on the lateral uniformity of sample over dimensions of at least the size of the incident beam. To ensure the applicability of these ion beam techniques, surface uniformity has to be examined. Surface morphological investigation also gives an estimate of the grain size and other indications of surface features, such as cone formation after sputter etching, hillocks and voids due to electromigration, surface crazing marks due to stress relaxation, and the presence of surface defects such as dislocations, pits, and faults. These observations often lead to useful insights into the properties of the entire thin film.

5.5.2.1 *Surface Replica and Scanning Electron Microscopy.*

These techniques have been extensively reviewed in the literature (85). In replication electron microscopy, the surface morphology is revealed by the formation of an intensity contrast by electron transmission through a replica film of the sample surface. Generally speaking, replication microscopy offers better lateral resolution (~ 20 Å) and serves as rather accurate means for surface grain size determination and crystallographic fault observations.

Surface oxide or other impurity films (or particles) can also be stripped for direct observation. However, caution must be exercised when the adhesion between the thin film and the substrate is weak; stripping of replication often causes peeling off of the film and attachment of the film onto the replica layer. Scanning electron microscopy, on the other hand, provides three-dimensional information of the surface to a certain extent. The image results from topographical as well as compositional contrasts. The depth of focus of SEM is superior to that of replica microscopy, although the lateral resolution power (~ 50 to 250 Å in commercial units) is somewhat inferior. Figure 5.7a shows a replica micrograph of the surface of a Sn film deposited on a glass substrate. Grain structure in the Sn film is very clear. Figure 5.7b shows the SEM picture of a Sn whisker grown on a stressed Sn film. Because of its many unique operating modes, SEM is a very powerful instrument when equipped with x-ray analyzing capability. It can be used for the study of surface topography and compositional variations in width as well as in depth on a cleaved cross section of samples, for p–n junction definition.

5.5.2.2 *Stylus Instruments.*

Stylus instruments operate under the principles of dragging a fine-pointed stylus on the sample surface. The surface roughness as well as macroscopic surface steps are traced out, amplified, and recorded on a running chart. Commercially available Taly step instruments are capable of measuring height differences of about 20 Å. Because of its design and construction, the Taly step instrument is suitable for large-area surface roughness investigation and film thickness determination. An example of a Taly step trace together with an optical micrograph showing a depression on a glass sample bombarded by a He ion beam is illustrated in Figure 5.8. Ion-induced surface depression due to volume compaction such as the example shown in Figure 5.8 or volume expansion due to a high-dose ion implantation of silicon single crystals can be readily investigated by Taly step measurements (86).

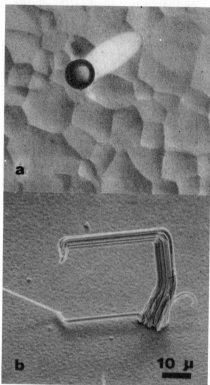

Figure 5.7. (*a*) Replica electron micrograph of the free surface of a Sn film on a fused quartz substrate deposited and kept at room temperature. The plastic sphere has a diameter of 0.5 μm. (*b*) Scanning electron micrograph of a Sn whisker on the free surface of Sn in bimetallic Cu–Sn thin film stored at room temperature.

5.5.3 Internal Microstructure

Internal microstructure of thin films generally means the grain size, stress and strain, texture, defects, epitaxial relationship, ordering structure, and various phases present in thin film samples. It can often be extended to include impurity contents, grain boundary segregation, and compositional gradients. The kinetics of thin film interdiffusions and reactions are extremely sensitive to the microstructure, and to a certain extent the microstructure also affects the driving force and direction of the reaction. It has been shown that the rate of interdiffusion between two epitaxial single-crystal films is quite different from that between two similar but fine-grained thin films (87, 87*a*). Therefore, it is clear that in order to understand thin film interaction, the microstructure should be well characterized before and during the reaction. Any interpretation of thin film kinetics without a careful microstructure study could be

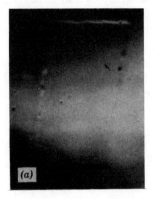

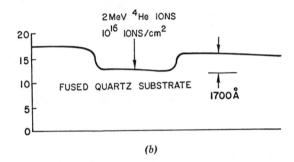

(b)

Figure 5.8. (a) Optical micrograph of the depression on a fused quartz surface produced by 2 MeV He beam during nuclear backscattering analysis. The size of the depression is about 2 mm × 1 mm. (b) Taly step trace of the cross section of the depression shown in Figure 5.8a. Courtesy of J. E. E. Baglin.

misleading. Often, a surprised finding of thin film reaction kinetics is due to the lack of understanding of the microstructure.

Thin films with thicknesses up to 2000 Å can be examined by transmission electron microscopy. For film thicknesses of about and more than 1 μm, conventional x-ray diffraction techniques can be used. Glancing-angle x-ray diffraction techniques are best suited for characterizing the microstructure of thin films with thicknesses ranging from a few hundred angstroms to 1 μm.

5.5.3.1   Glancing-Angle X-Ray Diffraction.   By using a glancing angle of incidence, a large area and hence a large volume of the thin film sample is analyzed by the x-ray beam. The arrangement increases the in-depth resolving power of x-ray diffraction; thin films with a thickness of 100 to 200 Å can be analyzed (88). Except for the difference in diffraction geometry, the principles and procedures of structural analyses and data reduction for the measurements of peak position, profile, and intensity are basically the same as those of the conventional x-ray techniques which can be found in many textbooks (27, 89, 89a).

The common configurations for the glancing-angle technique is the Seemann-Bohlin geometry for diffractometer (88) and the Read geometry for camera (30) as shown in Figure 5.9. In both, a monochromatic x-ray beam is

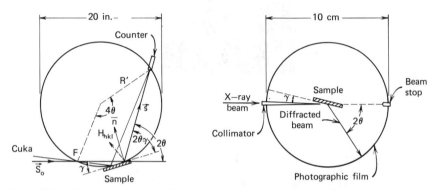

Figure 5.9.   Glancing-angle x-ray diffraction geometries: (*left*) Seemann-Bohlin diffractometer; (*right*) Read camera. From reference 90.

used. The beam is focused in the Seemann-Bohlin geometry so that the foci of the incident and diffracted beam and the sample surface lie on the circumference of the diffraction circle. In the Read camera, the incident beam is collimated through two pinholes. A summary of the comparison of structure analyses by using the two glancing-angle x-ray techniques has been given (90).

5.5.3.2   X-Ray Topography.   X-Ray topographic techniques can be applied to structural investigation of thin films, especially when the films of interest are relatively thick and single-crystal in nature, or when thin films are deposited on single-crystal substrates. These applications have been reviewed recently by several investigators (91, 92). For defect studies, the technique is limited to single-crystal samples with a low defect density, for example, less than 10^6 dislocation lines per cm^2.

There are generally three modes of operation in x-ray topography: (i) reflection mode (Berg-Barrett technique), (ii) transmission mode (Lang technique), and (iii) anomalous transmission mode (Borrmann technique). Applications of these modes of operation to thin films are briefly summarized as follows: crystallinity and perfection investigation of epitaxial layers; generation of interfacial dislocations at heterojunctions; compositional variation in heterostructure; adhesion of thin films on single-crystal substrates; the presence and sign of stress of deposited thin films on single-crystal substrates; and volume changes in single-crystal substrates due to ion implantation and the annealing behavior of the damage introduced by implantation.

5.5.3.3   Reflection Electron Diffraction.

Reflection electron diffraction (RED) is a glancing-angle diffraction technique. Electrons with energies in the range of 10 to 100 keV are reflection diffracted from the sample surface with a very small angle of incidence (of the order of $1°$). Because of the small incident angle of electrons, the actual penetration depth into the sample is small; therefore it is a surface-sensitive analytic tool (93). The penetration depth can be increased by increasing the angle of incidence. For example, Campisano et al. (94) studied the annealing behavior of ion-implanted amorphous Si, using RED and MeV He^+ channeling techniques (see Chapter 6 of this volume). They found that the penetration depth of 60 keV electrons into Si is ~ 40 Å with an incident angle of $\sim 0.4°$, and increases to between 40 and 340 Å when the angle of incidence is increased to $1°$. This variation of penetration depth enables reflection electron diffraction techniques to examine thin film structure as well as interactions between thin films and substrates.

The diffraction patterns from RED generally yield information on the surface structure and cleanliness as well as on the texture and crystallinity of surface layers. This technique has been used to investigate problems of epitaxy and texture formation of thin films on various substrates, of oxidation, and of corrosion of thin films and surfaces. These results have been discussed by Bauer (93) and other investigators. Recently, RED has been used in combination with other analytic techniques such as MeV He^+ channeling to study the reordering behavior of ion-implanted layers and the epitaxial growth of Au on Ge (95) and $NiSi_2$ on Si (96).

5.5.3.4   Transmission Electron Diffraction and Microscopy.

Transmission electron microscopy combined with electron diffraction is a versatile technique for studying structural defects in thin crystals (97, 98). Both the diffraction contrast and crystallographic nature of the defects can be analyzed. To use the technique the sample must be thin enough, ~ 1000 Å, to transmit electrons of 100 keV in most cases. Since the thickness of the film sample can be controlled during the deposition, no thinning of the sample is needed; so the question whether or not the characteristics and the distribution of defects in the sample might have been altered by the thinning process does not exist. However, since most thin films are deposited on thick substrates, the stripping of thin films off their substrates must be considered. In the following, we shall discuss first the stripping of thin films and other methods of preparing thin film samples for transmission electron microscopy study, then briefly the diffraction contrast of defects, and finally the observation of defect formation in the early stage of growth of thin films.

Sample Preparation.   Typically, there are three approaches in preparing thin film samples for transmission electron microscopy study. The first is to select a proper substrate from which the deposited thin film can be stripped easily. For this purpose, the surfaces of cleaved rock salt, mica, MoS_2, and pyrolytic graphite are often used. An extension of this approach is to deposit a thick cushion layer such as Ag or collodion between the film and substrate and to separate them by a preferential etching of the cushion layer. This allows the use of hard-to-cleave substrates. For example, one can use Ag as a cushion layer for Cr on glass and remove the Ag by etching in a dilute solution of nitric acid. The second approach is to use a thin enough substrate so that no stripping will be needed. Since it is crucial that the substrate should cause as little interference as possible with the contrast of the thin film, the most widely used are grids coated with a layer of amorphous carbon. The third approach, which is the most sophisticated, is to use a piece of Si wafer with a window covered by a thin oxide. The observation will be made through the window. A schematic diagram and a SEM micrograph of such a sample is shown in Figure 5.10. We note that if the window has to be opened before the thin film deposition, the oxide film must be grown first, otherwise the oxide may not be required. The oxide is typically of a thickness of 200 Å. It can be thermally grown SiO_2 which forms on both sides of the wafer, or it can be SiO and Si_3N_4 and deposited only on one side of the wafer. To open the window, the pattern of the window can first be defined by a conventional photolithographic technique used in the integrated circuit technology. In the pattern, the edges of the window must be aligned along the [110] directions of a (100) Si wafer. The etching of the window is usually conducted in a solution of KOH, which is known to produce a window frame defined by the {111} planes. The main advantage of this approach is that both the size and position of the window can be strictly controlled. The method has been applied to the microscopic study of thin film electromigration and weak-link superconductors (99, 100).

   It is obvious from the above that only a limited number of substrates are suitable for TEM study. In reality, one often faces the difficulty of removing a thin film from a particular substrate, and one may be tempted to use a substitute for easier thin film stripping. In such a case, it is crucial to be sure that the substitute and the original substrate produce the same kind of thin films. For this reason, it is found acceptable to substitute a bulk glass substrate by an oxidized Si wafer.

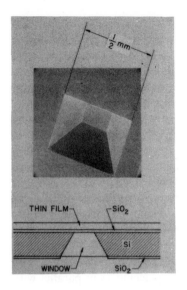

THIN FILM   SiO₂

Si

WINDOW   SiO₂

Figure 5.10.   Schematic diagram and scanning electron micrograph of a square opening in a piece of (100) Si chip. The opening has a dimension of $\frac{1}{2} \times \frac{1}{2}$ mm².

Diffraction Contrast of Defects.   The contrast in a transmission electron micrograph is produced by the variation of intensity of electron beam at the exit surface of the sample. Because the variation is caused by electron diffraction, it is called the diffraction contrast. When a defect is present in a crystal, it may distort the lattice around it to produce a dislocation, or it may cause a discontinuity in the stacking sequence of the lattice to produce a stacking fault. The distortion and discontinuity in general increase the electron diffraction contrast of the defects. According to the kinematic theory of electron diffraction, the amplitude of the diffracted beam through a column containing the defect is given by (97)

$$\phi g = \frac{\pi i}{\xi g} \int_0^t \exp(-2\pi i g \cdot R) \exp(-2\pi i s z) \, dz \qquad (3)$$

where **g** is the reciprocal lattice vector of the reflection; **R** is the displacement vector of the defect; s is the amount of deviation of the reflection from its exact Bragg position ($s \perp$ **g**); t is the sample thickness; and ζg has been shown by the dynamic diffraction theory to be the extinction distance of the sample. The equation holds well for large values of s, but in the limiting case of $s \to 0$, it has been shown by the dynamic theory of contrast that the following substitution should be made:

$$s \rightarrow s + d/dz\,(\mathbf{g}\cdot\mathbf{R}) \tag{4}$$

The application of Eq. 3 to defects with a well-defined $\mathbf{R}$ such as aggregates of point defects, dislocations, stacking faults, twins, and coherent precipitates has been very successful. In these applications, the defect contrast can be analyzed in dark-field micrographs where $Ig = \phi g^* \phi g$, as well as in bright-field micrographs where $I_0 = 1 - I_g$ under the two beam condition. Since the contrast can be varied by varying the reciprocal lattice vector of the reflection, $\mathbf{g}\cdot\mathbf{R}$ can be made equal to, greater, or less than zero for an unambiguous determination of the crystallographic nature of $\mathbf{R}$.

For defects with an ill-defined $\mathbf{R}$ such as an arbitrary large-angle grain boundary, there is no simple way available to analyze their contrast and to determine the $\mathbf{R}$ value. In general, a planar grain boundary needs eight parameters to define it; three are needed to define the orientation of one of crystal with respect to the other, two to define the plane of the boundary, and the rest of the eight to define the amount of displacements in parallel with and normal to the plane of the boundary. It is obvious that these parameters, at least some of them, must be put under control in order to simplify the analysis. For this reason only a few special types of grain boundaries have been studied (101, 102).

Defect Formation in Thin Films.   One area of defect study in thin films that is of particular interest is the observation of defect formation during the early stage of film growth. This is usually done in situ in a microscope; observation is made while the film is being deposited onto a thin substrate. Not only the initial growth mode of thin films in three-dimensional nuclei or in two-dimensional uniform layer can be studied; it is also possible to study at which stage of growth and under what kind of driving force defects are introduced into the film. In some cases, a direct observation of the nucleation of a defect can be made. By understanding the driving force and the mechanism of defect formation, it seems that a controlled deposition of a film with few defects may eventually be achieved.

At present the prevailing model of defect formation in a thin film during its growth is the island-misfit model (20, 103). It has been observed in many cases of thin film epitaxial growth that when two or three misaligned islands come into contact, defects such as stacking faults are formed between the islands to accommodate the misalignment. If the misalignment is rotational, a set of dislocations may form if the angle of the rotation is small, or a twin may form if the angle is near that of the twin orientation. However, the majority of dislocations seems to form at the final stage when the film closes up its holes. Often the residue of a hole is a bunch of dislocations.

In Figure 5.11*a*, a bright-field transmission electron micrograph of a Au film with large openings is shown. The film was grown on the (100) cleavage plane of a piece of rock salt, and the evaporation was arrested at the stage just after most of the Au islands had joined up. It can be seen that there are faults at the joints and that there are not many other defects in the film except one or two dislocations. However, when the evaporation was arrested at a later time when the growth of the islands had nearly closed up all the holes among them, a high density of dislocations was found upon examination of the film, as shown in Figure 11*b*. These micrographs illustrate the island-misfit model as discussed above.

The formation of misfit dislocations along the film–substrate interface has been extensively studied. The common mode of forming a misfit dislocation is by the bending and stretching of a threading dislocation. On the other hand, the nucleation of a misfit dislocation at the edges of epitaxial three-dimensional nuclei has been observed. For a more detailed review of defect formation in thin films, the interested readers are referred to the articles by Matthews (20) and by Stowell (103).

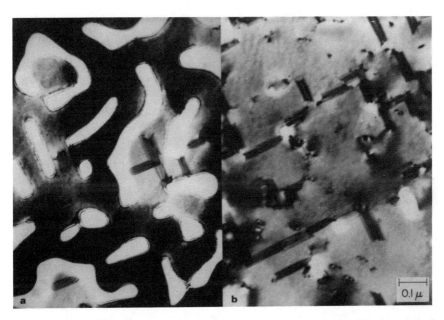

Figure 5.11.   (*a*) Bright field transmission electron micrograph of the early stage of growth of Au film on NaCl. Stacking faults are formed at the joints between Au islands. (*b*) Bright-field transmission electron micrograph of the later stage of growth of Au film on NaCl. Stacking faults, dislocations, and holes are seen.

5.5.3.5 *Microanalysis.*

As the last and concluding section of this chapter, we shall review briefly the present trend of material characterization, which is moving in the direction of microanalysis. Microanalysis means the analysis of the structure and composition of a very small volume of material, in the extreme a single atom. Obviously, it requires high resolution and high sensitivity. There are two typical examples. One is the use of field ion microscopy to observe single atoms on refractory metal surfaces (104, 105). Diffusion of a single atom on the surface can be studied. Combined with an ion probe, impurity atoms can also be studied. Another is the achievement in the high-brightness 5 Å scanning transmission electron microscopy (106) to analyze a minute amount of grain boundary segregation and to image surface single atoms.

One basic requirement of microanalysis is to have a high lateral resolution of the order of atomic dimensions. This is not available in most techniques for thin film and surface characterization. In fact, these techniques often are sacrificed in their lateral resolution in order to gain in-depth resolution. The typical example is the glancing-angle x-ray diffraction. A depth resolution of 200 Å is obtained by a glancing angle of incidence, yet by doing so a sample area of about 0.1 to 1 cm^2 is needed. In the case of ion backscattering, the cross section of an ion beam is typically 1 to 2 mm^2. The primary electron beam in Auger electron spectroscopy is much smaller; still it has a size of $\sim 1 \ \mu m^2$.

To improve the lateral resolution of techniques based on scattering processes, the size of the primary beam must first be reduced. This can be done by the use of a smaller beam source followed by a beam-focusing operation. When the source is small, it must be able to produce a high-intensity primary beam. The compound LaB_6 has been found suitable as a high-intensity electron emitter. The electron beam, in particular, can now be focused into a spot as small as 5 Å (55*b*). With such an electron beam, a high-resolution scanning transmission electron microscope at present is capable of diffracting and chemically analyzing in ultrahigh vacuum a very small volume of material ($\sim 10^7 \ Å^3$) and at the same time imagining a large area of the sample by scanning the beam. The capability of the microdiffraction and chemical analysis is extremely attractive. The diffraction can be used, for example, to determine the structure and orientation of a single precipitate. The chemical analysis, which utilizes either the energy loss spectroscopy of the transmitted electrons or the energy dispersive analysis of the x-rays produced, can be applied to grain boundary segregation (107), interface segregation, or concentration gradient in a thin nonstoichiometric compound. We expect that in the near future much more progress will be made in the microanalysis techniques and their applications in thin films.

ACKNOWLEDGMENTS

The authors are grateful to J. J. Cuomo, J. W. Matthews, J. W. Mayer, W. Reuter, and R. J. Wagner for helpful comments and to C. F. Aliotta for the replica and scanning electron micrographs. One of us (S.S.L.) is also grateful for the partial financial support of the Office of Naval Research (D. Ferry and L. Cooper).

REFERENCES

1. R. W. Berry, P. M. Hall, and M. T. Harris, *Thin Film Technology*, Van Nostrand, Princeton, N.J. (1968), p. 221.
2. K. L. Chopra, *Thin Film Phenomena*, McGraw-Hill, New York (1969).
3. L. I. Maissel and R. Glang, Ed., *Handbook of Thin Film Technology*," McGraw-Hill, New York (1970), Chapter 3.
4. J. C. Anderson, Ed., *The Use of Thin Films in Physical Investigations*, Academic Press, New York (1966).
5. J. W. Matthews, Ed., *Epitaxial Growth*, Vols. 1 and 2, Academic Press, New York (1976).
6. G. Hass and R. E. Thun, Eds., *Physics of Thin Films*, Vol. 1 to Vol. 6, Academic Press, New York (1963 to 1971). G. Hass, M. H. Francombe, and R. W. Hoffman, Eds., *Physics of Thin Films*, Vol. 7 and Vol. 8, Academic Press, New York (1973 and 1975).
7. J. R. Gavaler, *Appl. Phys. Lett.*, **23**, 480 (1975).
8. L. R. Testardi, R. L. Meek, J. M. Poate, W. A. Royer, A. R. Storm, and J. H. Wernick, *Phys. Rev. B*, 4304 (1975).
9. L. Esaki and L. L. Chang, *Thin Solid Films*, **36**, 285 (1976).
10. G. R. Booker and B. A. Unvala, *Phil. Mag.*, **11**, 11 (1965).
11. G. R. Booker and R. Stickler, *J. Appl. Phys.*, **33**, 3281 (1962).
12. L. D. Glowinski, K. N. Tu, and P. S. Ho, *Appl. Phys. Lett.*, **28**, 312 (1976).
13. D. R. Campbell, K. N. Tu, and R. O. Schwenker, *Thin Solid Films*, **25**, 213 (1975).
14. D. Cherns, *Phil. Mag.*, **30**, 549 (1974).
15. R. T. Whipple, *Phil. Mag.*, **45**, 1224 (1954).
16. P. V. Pavlov, V. A. Panteleev, and A. V. Maiorov, *Sov. Phys. Solid State*, **6**, 305 (1964).
17. D. Gupta, *Phys. Rev.*, **B7**, 586 (1973).
18. K. N. Tu, *J. Appl. Phys.*, **43**, 1303 (1972).
19. J. W. Matthews, in *Physics of Thin Films*, G. Hass and R. E. Thun, Eds., Vol. 4, Academic Press, New York (1967), p. 137.
20. J. W. Matthews, *Epitaxial Growth*, Academic Press, New York (1976), Chapter 8.
21. R. W. Balluffi, Y. Komen, and T. Schober, *Surface Sci.*, **31**, 68 (1972).
22. T. Schober and R. W. Balluffi, *Phil. Mag.*, **21**, 109 (1970).
23. T. Schober and R. W. Balluffi, *Phys. Status Solidi (b)*, **44**, 103 (1974).
24. M. J. Weins, *Surface Sci.*, **31**, 138 (1970).
25. T. Y. Tan, J. C. M. Hwang, P. J. Goodhew, and R. W. Balluffi, *Thin Solid Films*, **33**, 1 (1976).
26. F. Cosandey, S. K. Kang, and C. L. Bauer, *J. de Phys.*, **36**, C4-17 (1975).
27. C. S. Barrett and T. B. Massalski, *Structure of Metals*, 3rd ed., McGraw-Hill, New York (1966).
28. G. B. Harris, *Phil. Mag.*, **43**, 113 (1952).
29. C. A. Wallace and R. C. C. Ward, *J. Appl. Cryst.*, **8**, 255 (1975).
30. M. H. Read and D. H. Hansler, *Thin Solid Films*, **10**, 123 (1972).
31. K. N. Tu and B. S. Berry, *J. Appl. Phys.*, **43**, 3283 (1972).
32. H. H. Andersen, K. N. Tu, and J. F. Ziegler, *Nuclear Instruments and Methods*, in press.

33. A. F. Mayadas and M. Shatzkes, *Phys. Rev.*, **1**, 1382 (1970).
34. P. Chaudhari, J. F. Graczyk, D. Henderson, and P. Steinhardt, *Phil. Mag.*, **31**, 727 (1975).
35. S. C. Moss and J. F. Graczyk, *Phys. Rev. Lett.*, **23**, 1167 (1969).
36. D. Turnbull and D. E. Polk, *J. Non-Cryst. Solids*, **8–10**, 19 (1972).
37. A. Howie, D. Krivanek, and M. L. Rudee, *Phil. Mag.*, **27**, 235 (1973).
38. P. Chaudhari, J. F. Graczyk, and S. R. Herd, in *Physics of Structurally Disordered Solids*, S. S. Mitra, Ed., Plenum Press, New York (1976).
39. D. Turnbull, *J. Phys.* (Paris), **35**, C4, 1 (1974).
40. M. H. Cohen, H. Fritzsche, and S. R. Ovshinsky, *Phys. Rev. Lett.*, **22**, 1065 (1969).
41. A. Hiraki, A. Shimizu, M. Iwami, T. Narusawa, and S. Komiya, *Appl. Phys. Lett.*, **26**, 57 (1975).
42. S. Mader, *Thin Solid Films*, **35**, 195 (1976).
43. D. Gupta, K. N. Tu, and K. W. Asai, *Phys. Rev. Lett.*, **35**, 796 (1975).
44. H. S. Chen, *Appl. Phys. Lett.*, **29**, 12 (1976).
45. P. Duwez, *Ann. Rev. Mat. Sci.*, **6**, 83 (1976).
46. P. Chaudhari, J. J. Cuomo, and R. J. Gambino, *IBM J. Res. Develop.*, **17**, 66 (1973).
46a. W. E. Spear, P. G. LeComber, S. Kinmond, and M. H. Brodsky, *Appl. Phys. Lett.*, **18**, 105 (1976).
47. S. S. Lau, C. Canali, Z. L. Liau, K. Nakamura, M-A. Nicolet, J. W. Mayer, R. Blattner, and C. A. Evans, Jr., *Appl. Phys. Lett.*, **28**, 148 (1976).
48. J. O. McCaldin, *J. Vac. Sci. Technol.*, **11**, 990 (1974).
49. Z. L. Liau, S. U. Campisano, C. Canali, S. S. Lau, and J. W. Mayer, *J. Electrochem. Soc.*, **122**, 1696 (1975).
50. L. L. Chang, L. Esaki, W. E. Howard, and R. Ludeke, *J. Vac. Sci. Technol.*, **10**, 11 (1973).
51. A. Y. Cho, *J. Appl. Phys.*, **41**, 782 (1970).
52. R. Tsu and L. Esaki, *Appl. Phys. Lett.*, **22**, 562 (1973).
53. R. Dingle, A. C. Gossard, and W. Wiegmann, *Phys. Rev. Lett.*, **34**, 1327 (1975).
54. A. Segmuller, *Thin Solid Films*, **18**, 287 (1973).
55. A. Segmuller, *Advan. X-Ray Anal.*, **13**, 455 (1970).
55a. A. N. Broers, W. W. Molzen, J. J. Cuomo, and N. D. Wittels, *Appl. Phys. Lett.*, **29**, 596 (1976).
55b. A. N. Broers, *Appl. Phys. Lett.*, **22**, 610 (1973).
56. J. J. Cuomo and R. J. Gambino, *J. Vac. Sci. Technol.*, **12**, 79 (1975).
57. J. J. Cuomo, R. J. Gambino and R. Rosenberg, *J. Vac. Sci. Technol.*, **11**, 34 (1974).
58. J. J. Cuomo, P. Chaudhari, and R. J. Gambino, *J. Electron. Mater.*, **3**, 517 (1974).
59. R. F. Bis, A. S. Rodolakis, and J. N. Zemel, *Rev. Sci. Inst.*, **36**, 1626 (1965).
60. K. G. Gunther, in *The Use of Thin Film in Physical Investigations*, J. C. Anderson, Ed., Academic Press, New York (1966).
61. R. Glang, "Vacuum Evaporation," in *Handbook of Thin Film Technology*, L. I. Marsell and R. Glang, Eds., McGraw-Hill, New York (1970), Chapter 3, pp. 1–90.
62. R. B. Belser, *J. Appl. Phys.*, **31**, 562 (1960).
63. C. A. Neugebauer, *J. Appl. Phys.*, **35**, 3599 (1964).
64. R. F. Steinberg and D. M. Scruggs, *J. Appl. Phys.*, **37**, 4586 (1966).
65. S. Mader, *J. Vac. Sci. Technol.*, **2**, 35 (1965).
66. E. Kneller, *J. Appl. Phys.*, **35**, 2210 (1964).
66a. J. C. Schottmiller, F. Ryan, and T. Taylor, Ext. Abst. 14th AVS Symp., Pittsburg, Pa. (1967).
67. J. W. Mayer, L. Eriksson, and J. A. Davies, *Ion Implantation in Semiconductors*, Academic Press, New York (1970).
68. G. Carter, J. S. Colligan, and W. A. Grant, Eds., *Applications of Ion Beams to Materials*, 1975, *Inst. Phys. Conf. Ser. 28 London and New York* (1976), p. 183.

69. S. T. Picraux, E. P. EerNisse, and F. L. Vook, Eds., *Applications of Ion Beams to Materials*, Plenum Press, New York (1974), p. 15.
70. J. F. Ziegler, Ed., *New Uses of Ion Accelerators*, Plenum Press, New York and London (1975), p. 323.
71. L. Csepregi, W. K. Chu, H. Muller, J. W. Mayer, and T. Sigmon, *Radiation Eff.*, **28**, 227 (1976).
72. E. F. Kennedy, L. Csepregi, J. W. Mayer, and T. Sigmon, to be published.
73. S. U. Campisano, G. Foti, E. Rimini, S. S. Lau, and J. W. Mayer, *Phil. Mag.*, **31**, 903 (1975).
74. K. N. Tu and D. A. Chance, *J. Appl. Phys.*, **46**, 3229 (1975).
75. C. J. Kircher, *Solid State Electron.*, **14**, 507 (1971).
76. J. P. Hirth and G. M. Pound, *Condensation and Evaporation-Nucleation and Growth Kinetics*, Pergamon Press, Oxford (1963).
77. J. A. Venables and G. L. Price, in *Epitaxial Growth*, J. W. Matthews, Ed., Academic Press, New York (1976).
78. J. W. Christian, *The Theory of Transformations in Metals and Alloys*, Pergamon Press, Oxford (1965).
79. H. W. Larson and G. A. Walker, *J. Appl. Phys.*, **38**, 11 (1967).
80. C. J. Paparoditis, in M. H. Francombe, and H. Sato, Eds., *Single Crystal Films*, Pergamon Press, New York (1964), p. 79.
81. Y. Tarui, H. Teshima, K. Okura, and A. Minamiya, *J. Electrochem. Soc.*, **110**, 1167 (1963).
81a. D. C. Larson, in *Methods of Experimental Physics*, Vol. 11, R. V. Coleman, Ed., Academic Press, New York (1974).
81b. J. L. Vossen and W. Kern, Eds., *Thin Film Processes*, to be published by Academic Press, New York (1977).
82. J. S. Colligon, W. A. Grant, J. S. Williams, and R. P. W. Lawson, "Deposition of Thin Films by Retardation of an Isotope Separator Beam," in *Applications of Ion Beam to Materials*, G. Carter, J. S. Colligon, and W. A. Grant, Eds., Inst. Phys. Conf. Ser. 28, Institute of Physics, London (1965).
83. E. G. Spencer, P. H. Schmidt, D. C. Joy, and F. J. Sansalone, *Appl. Phys. Lett.*, **29**, 118 (1976).
84. S. Aisenberg and R. Chabor, *J. Appl. Phys.*, **42**, 2953 (1971).
85. L. E. Murr, *Electron Optical Applications in Materials Science*, McGraw-Hill, New York (1970).
86. K. N. Tu, P. Chaudhari, K. Lal, B. L. Crowder, and S. I. Tan, *J. Appl. Phys.*, **43**, 4262 (1972).
87. H. E. Cook and J. E. Hilliard, *J. Appl. Phys.*, **40**, 2191 (1969).
87a. R. G. Kirsch, J. M. Poate, and M. Eibschuz, *Appl. Phys. Lett.*, **29**, 772 (1976).
88. R. Feder and B. S. Berry, *J. Appl. Cryst.*, **3**, 372 (1970).
89. B. E. Warren, *X-Ray Diffraction*, Addison-Wesley, Reading, Mass. (1969).
89a. A. J. C. Wilson, *Elements of X-Ray Crystallography*, Addison-Wesley, Reading, Mass. (1970).
90. S. S. Lau, W. K. Chu, J. W. Mayer, and K. N. Tu, *Thin Solid Films*, **23**, 205 (1974).
91. G. H. Schwuttke, "X-ray Topography," in *Epitaxial Growth*, Vol. A, J. W. Matthews, Ed., Academic Press, New York (1975), p. 281.
92. G. A. Rozgonyi and D. C. Miller, *Thin Solid Films*, **31**, 185 (1976).
93. E. Bauer, "Reflection Electron Diffraction," in *Techniques of Metals Research*, Vol. 2, R. F. Bunshah, Ed., Wiley-Interscience, New York (1969), p. 501.
94. S. U. Campisano et al., to be published.
95. C. Lorsi, S. U. Campisano, G. Foti, E. Rimini, and G. Vitali, *Thin Solid Films*, **32**, 315 (1976).
96. K. N. Tu, E. I. Alessandrini, W. K. Chu, H. Krautle, and J. W. Mayer, *Japan. J. Appl. Phys.*, Suppl. 2, Part 1, 669 (1974).
97. P. B. Hirch, A. Howie, R. B. Nicholson, D. W. Pashley, and M. J. Whelan, *Electron Microscopy of Thin Crystals*, Butterworth, London (1965).

98. M. J. Whelan, in *The Physics of Metals,*" Vol. 2, P. B. Hirch, Ed., Cambridge Press (1976).
99. R. B. Laiborwitz, S. R. Herd, and P. Chaudhari, *Phil. Mag.,* **28**, 1155 (1973).
100. I. A. Blech and E. S. Meieran, *J. Appl. Phys.,* **40**, 485 (1969).
101. R. C. Pond and D. A. Smith, in *Proceedings of the 6th European Congress on Electron Microscopy,* Vol. 1, D. G. Brandon, Ed., Tal International Publishing (1976), p. 233.
102. P. H. Pumphrey, in *Grain Boundary Structure and Properties,* G. A. Chadwick and D. A. Smith, Eds., Academic Press, New York (1976), Chapter 5.
103. M. J. Stowell, in *Epitaxial Growth,* J. W. Matthews, Ed., Academic Press, New York (1976), Chapter 5.
104. W. R. Graham and G. Ehrlich, *Thin Solid Films,* **25**, 85 (1975).
105. T. T. Tsong, P. Cowan, and G. Kellogg, *Thin Solid Films,* **25**, 97 (1975).
106. A. V. Crewe, J. P. Langmore, and M. S. Isaacson, in *Physical Aspects of Electron Microscopy and Microbeam Analysis,* B. M. Siegel and D. R. Beanan, Eds., Wiley, New York (1975), p. 47.
107. R. G. Faulkner and K. Norrgard, in *Proceedings of the 6th European Congress on Electron Microscopy,* Vol. 1, D. G. Brandon, Ed., Tal International Publishing (1976), p. 465.

6

DEPTH PROFILING TECHNIQUES

J. W. Mayer

California Institute of Technology, Pasadena, California

J. M. Poate

Bell Laboratories, Murray Hill, New Jersey

6.1 INTRODUCTION

Characterization of thin film interdiffusion and interfacial reactions poses a nearly impossible question: how does one analyze a layer 10 to 100 atoms in thickness (or less) that is located some 1000 or 10,000 atom layers below the surface? In the 1960s the question would have been dismissed because it was out of reach of available experimental techniques. In the 1970s the question became painfully relevant as the importance of thin film reactions in technological applications became evident. In fact, one of the major themes of this monograph rests on the premise that one can analyze thin layers and extract information on atomic composition as a function of depth. In this chapter we discuss some of the experimental approaches to the problem of depth profiles and attempt to assess the relative merits and deficiencies inherent in the different approaches.

In characterizing thin films, one generally thinks in terms of laterally uniform situations as sketched in Figure 6.1: interfaces, compound formation, and interdiffusion where the thicknesses of the films are much less than lateral dimensions. Although idealized, the sketches serve to focus attention on the nature of the problem. As we will see, the most difficult situation to character-

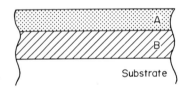

(a) Uniform layers, Sharp interfaces

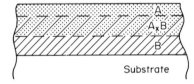

(b) Compound Formation

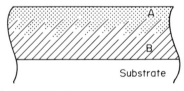

(c) Interdiffusion

Figure 6.1. Idealized sketch (a) of thin film couple A, B deposited on substrate; (b) couple reacts to form compound at interface with definite stoichiometry A_xB and well-defined interfaces; (c) couple interdiffuses with concentration gradient with fastest diffusion along grain boundaries.

ize is an interface located well below the surface. Only in a few cases has it been possible to describe the nature of the interface. Most studies have dealt with the identity and kinetics of compound formation and the profiles of interdiffusing species.

Essentially, the experimental problem is to determine atomic composition as a function of depth. There have been two approaches to this problem (Fig. 6.2). One approach utilizes a beam of energetic particles, typically MeV He ions, and relies upon the energy loss of the particles on their inward and/or outward trajectories to determine depth information. This particle energy loss technique is best illustrated by Rutherford backscattering spectrometry, which is straightforward to carry out but requires an accelerator.

The other approach is to section the sample by eroding (or sputtering) the sample with a beam of particles, typically keV Ar ions. As the sample is sectioned, the newly exposed surface layer can be analyzed. There are two

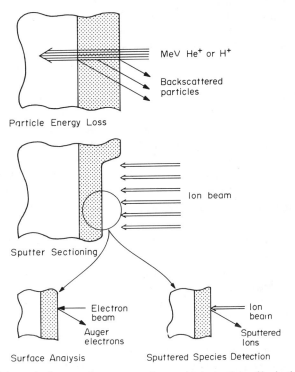

Figure 6.2. Schematic diagram of two approaches to obtain depth profiles in thin films. With particle energy loss techniques, the thickness of the layer is determined from the energy loss of the energetic particles. With sputter sectioning techniques, the surface composition can be directly analyzed either by surface analysis or sputtered species detection.

broad categories for the many commercial instruments now available: surface analysis methods, such as Auger electron spectroscopy or ion scattering spectrometry; and sputtered species detection methods, such as secondary ion mass spectrometry. The first class of methods gives information on the sputtered surface and the second, on the material removed. In a sense, then, these two methods are complementary.

Since our focus is on depth profiles, we do not concentrate at length on the techniques but rather give the salient features. We do, however, attempt to give sufficient background so depth profiles can be interpreted. This is a relatively easy task for the particle energy loss techniques because the physical processes are understood and documented. However, we are faced with certain fundamental problems in quantifying the sputtering techniques when dealing with thin film reactions that involve interfaces and compounds. Sputtering is a collision cascade phenomenon whose details are not understood for the problems of interest here. For this reason we will discuss in Section 6.3 the problems inherent in sputtering compound materials.

It is becoming increasingly apparent that the only way to come to a satisfactory understanding of these thin film phenomena is by the use of more than one depth profiling technique along with other techniques that give information on phase identification as well as lateral uniformity. In this chapter we present profiling techniques. In the preceding chapter diffraction techniques were discussed, which give the phase identification.

The first studies of interdiffusion phenomena in thin films appear to be those of DuMond and Youtz (1) in 1935 who pointed out the applicability of x-ray diffraction for studying low-temperature diffusion in periodic thin layers of Cu and Au. In their report they state, "It would seem likely that we have here an excellent way of studying intimately the diffusion of atoms in the solid state." And in a later work (2) they obtain the diffusion coefficients of Au in Cu at room temperature. Such indirect methods of obtaining interdiffusion parameters have been pursued in the intervening years, but the most active areas have been associated with either the particle energy loss or sputter sectioning techniques.

Nuclear physicists have a lengthy tradition of using particle scattering techniques for detecting surface impurities on their targets. Rubin (3) was the first to review these techniques in the light of surface or thin film analysis, and Sippel (4) in 1959 used the backscattering of 1.88 MeV protons and deuterons to measure the diffusion of Au in Cu from a 20 Å film of Au plated on Cu. Use of the backscattering technique remained somewhat dormant until laboratories became involved with channeling and ion implantation in semiconductors. These activities were also stimulated by the use of solid-state detectors to detect the backscattered particles. The areas of ion implantation (5, 6) and channeling have been reviewed fully elsewhere (7, 8).

Diffraction and particle energy loss techniques are predated by the venerable field of sputtering (9), which has been the subject of study for over a century. It is, however, only in the past five to 10 years that the sputter sectioning techniques have been combined with the surface analysis techniques to give depth profiling information (10). It is undoubtedly the technological imperatives of such areas as the microelectronics industries that have stimulated the development of these depth profiling techniques.

6.2   PARTICLE ENERGY LOSS TECHNIQUES

An energetic ion traversing a solid loses energy in two processes: either to electrons in the solid or to nuclear scattering events. The electronic interactions dominate at high energies. Although being the major source of energy loss, these interactions produce very little deflection of the penetrating particles. To first order for MeV protons or He ions, the energy loss in traversing thin layers is linear with depth. In this case depth refers to the number of atoms per cm^2, not the physical thickness in Å. This distinction arises because the energy loss processes are determined by scattering cross sections which have the units of areal density (at. cm^{-2}). To convert to linear thicknesses, density values have to be assumed or measured. Whether we are considering backscattering, nuclear reactions, or ion-induced x-rays, it is the energy loss of the incident and/or emerging particle which gives a measure of the depth at which the scattering or reaction took place. A compilation of most of the relevant energy loss and cross section data now exists (11).

6.2.1   Rutherford Backscattering Spectrometry

The techniques of Rutherford backscattering come directly from the field of low-energy nuclear physics. Development of particle accelerators has evolved since the 1930s, and most commonly used in backscattering studies is the Van de Graaff electrostatic accelerator. Numerous Van de Graaff accelerators equipped for particle–solid studies exist throughout the world. Typical experimental arrangements have been described in detail elsewhere (12). Energetic ion beams in the range of a few hundred keV to several MeV are produced in the accelerator and analyzed, magnetically or electrostatically, to give an energetically well-defined beam of particles. The ions are then passed along a beam tube through collimating apertures to the target chamber; typical beam spot sizes are of the order of 1 mm in diameter. It should be emphasized that to date most target chambers have been maintained at pressures approximately 10^{-6} to 10^{-7} torr. The energetic ion beams are very insensitive to thin adsorbed layers of hydrocarbons or water vapor on the surface of samples. This means that profiling measurements need not be carried out in UHV chambers, and sample turn around times can be very rapid. If, however, thin film samples are prepared *in situ*, then UHV or good vacuum conditions must be maintained. Several such chambers are in the course of construction (13). Ion beam techniques in conjunction with channeling and blocking are showing great promise for surface structure analysis, and several UHV chambers devoted to this purpose are now in use (14). The use of the channeling technique for determining the lattice site location of impurities in single crystals is reviewed in Chapter 14.

The backscattered particles have to be detected with an energy analyzer. Surface barrier semiconductor detectors (15) are used almost exclusively for routine measurements in the energy range of 250 keV and higher. The energy resolution of a surface barrier detector is typically 15 keV for MeV ^{4}He ions. This energy resolution corresponds to a depth resolution between 150 and 300 Å for most solids. If cooled detectors and associated electronics are utilized, then resolution can be improved to about a value of 10 keV for MeV ^{4}He ions and 7 keV for several hundred keV ^{4}He ions (16). Considerably better depth resolution can be achieved with the use of magnetic and electrostatic energy analysis. These are not in routine use because of the significantly larger times involved in data acquisition due to the small acceptance solid angles and to the sequential accumulation of data in small energy increments. The ultimate limitation, however, to resolution is the energy straggling of the beam traversing matter. This is one of the reasons why only moderate energy resolution systems employing solid-state detectors are so widely used. For example, for films several thousand Å thick, straggling and energy resolution of surface barrier detectors can become comparable. Moreover, the widespread use of these solid-state detectors in nuclear physics has led to the development of very compact, stable, linear, and reliable electronics and data acquisition systems. The energy spectra of the output charge pulses are typically displayed in multichannel analyzers where the channel numbers are linearly related to the particle energy.

We are primarily interested in depth profiles, so we shall now give spectra of a thin film reaction on Si. The upper section in Figure 6.3 shows a 1000 Å Ni film on Si (17). Nearly all the incident ^{4}He beam penetrates microns into the target before it is stopped. We are interested in the relatively small fraction of particles that are backscattered and detected. It is these particles that give mass and depth information. Particles scattered from the front surface of the Ni have an energy given by the following kinematic equation:

$$E_1 = E_0 K \tag{1}$$

where the kinematic factor K for 2 MeV ^{4}He backscattered at a laboratory angle of $170°$ is 0.76 for Ni and 0.57 for Si. The kinematic factor is derived simply from conservation of energy and momentum for projectile of mass M_1 and scattering from target M_2 at a laboratory angle θ, as follows:

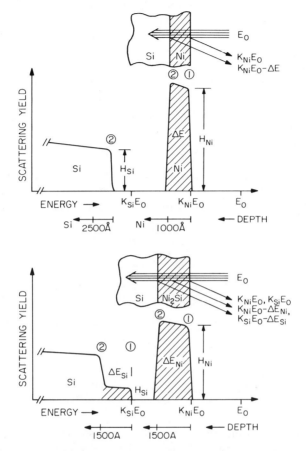

Figure 6.3. Schematic backscattering spectra for MeV ^{4}He ions incident on 1000 Å Ni film on Si (*top*) and after reaction to form Ni$_2$Si (*bottom*). Depth scales are indicated below the energy axes. Circled numbers 1 and 2 indicate the outer surface and the interface, respectively.

$$K = \left[\frac{M_1 \cos\theta + \sqrt{M_2^2 - M_1^2 \sin^2\theta}}{M_1 + M_2} \right]^2 \qquad (2)$$

Different masses can thus be easily identified by the different backscattered energies.

As particles traverse the solid, they lose energy along their incident path at a rate of about 64 eV per Å (assuming a bulk density for Ni of 8.9 g cm^{-3}) In thin film analysis, to a good approximation, energy loss is linear with thickness. Thus, a 2 MeV particle will lose 64 keV penetrating to the Ni–Si interface (marked 2 in the inserts of Fig. 6.3). Immediately after scattering from the interface, particles scattered from Ni will have an energy of 1477 keV derived from $K_{Ni} \times (E_0 - 64)$. On their outward path, particles will have slightly different energy loss due to the energy dependence of the energy loss processes, in this case 69 eV $\mathring{A}^{-1}$. On emerging from the surface, the ^{4}He ions scattered from Ni at the interface will have an energy of 1408 keV.

The total energy difference ΔE between particles scattered at the surface and near the interface is 118 keV, a value which can be derived from the following equation (11, 12):

$$\Delta E = \Delta x[S] = N \, \Delta x[\varepsilon] \tag{3}$$

where $[S]$ is the energy loss factor, $[\varepsilon]$ is the stopping power factor, and N is the atomic density of Ni ($N = 9.13 \times 10^{22}$ at. cm^{-3}); and hence $N \, \Delta x$ is the number of Ni atoms per cm^2 in the film. The stopping cross section ε is defined as

$$\varepsilon = \frac{1}{N} \frac{dE}{dx} \tag{4}$$

We define the stopping cross section factor in the surface energy approximation for particles incident normal to the surface as:

$$[\varepsilon] = K\varepsilon(E_0) + \frac{1}{|\cos \theta|} \varepsilon(E_1) \tag{5a}$$

and the energy loss factor as

$$[S] = K \frac{dE}{dx}(E_0) + \frac{1}{|\cos \theta|} \frac{dE}{dx}(E_1) \tag{5b}$$

where the values of ε and dE/dx are evaluated at the incident energy E_0 and emergent energy E_1, respectively, in the surface energy approximation. This expression just allows for the energy loss for the incident and emergent trajectories. Figure 6.4 shows a plot of the energy loss factor in eV $\mathring{A}^{-1}$ of ^{4}He ions traversing Si, Ni, and Pt as a function of incident energy for $\theta = 170°$. The energy loss processes of ions in solids are sufficiently well understood and measured for tabulations of energy loss (11, 18–20) to give light ion energy losses to an absolute accuracy of 10%. Most uncertainties in measurements of energy loss come from target thickness determinations; the relative accuracies of the tabulations as a function of energy are much better than 10%.

From Eq. 3 we can calculate the depth resolution Δx of the technique which is determined by energy uncertainties ΔE. For example, the depth resolution for a 2 MeV ^{4}He beam in Si, assuming a solid-state detector resolution of 15 keV, is 200 $\mathring{A}$. Dramatic improvements in depth resolution can be obtained by improving the resolution of the detection system (the resolution of MeV

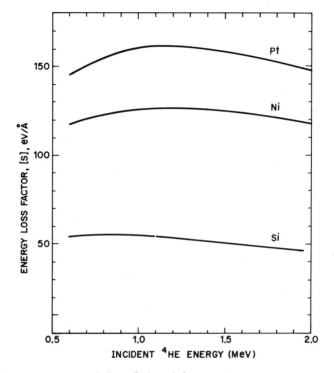

Figure 6.4.   Energy loss factor, [S], (eV Å^{-1}) for ^{4}He ions incident on Si, Ni, and Pt for $\theta = 170°$. Data taken from reference 20.

electrostatic accelerators are usually better than 1 keV). Detector resolution can be improved substantially by using magnetic and electrostatic analysis, and work is starting in this area for thin film analysis. An increase in depth resolution can also be obtained in the near surface region by tilting the target with respect to the beam.

There is, however, a fundamental limit to the depth resolution obtainable with ion beam energy loss techniques due to energy loss straggling in matter. Figure 6.5 shows a calculation (21) of target thickness that will produce an energy straggling of 15 keV in transmission or 21 keV in backscattering. This curve is calculated assuming Bohr's theory (22) of straggling where the energy distribution is Gaussian, with a mean square deviation of

$$\Omega^2 = 4\pi Z_1^2 e^4 Z_2 N \, dx \tag{6}$$

where Z_1 and Z_2 are the atomic numbers of projectile and target, respectively; and $e^2 = 1.44 \times 10^{-13}$ MeV-cm. It can be seen from the figure that in Si, for

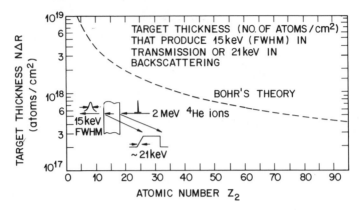

Figure 6.5.   Target thickness that will produce an energy straggling of 21 keV for the back-scattering of 2 MeV ^{4}He ions. From reference 21.

example ($Z_1 = 14$ and $N = 5 \times 10^{22}$ at. cm^{-3}), films approximately 5000 Å thick can be analyzed before straggling becomes comparable to detector resolution. However, if the detector resolution was 1 keV, straggling would be important for films only 10 Å thick. The Bohr theory is not applicable to such thin films, as the energy distributions become asymmetric (23), but the same conclusions are still valid.

As mentioned above, it is the number of atoms per cm^2, $N\,\Delta x$, the areal density, which is determined in these energy loss measurements. For convenience in quick interpretation of thin film spectra, one usually assumes bulk atomic densities and relates energy loss ΔE to linear thicknesses. Of course, the thin film densities can be determined rather simply from a direct thickness measurement of Δx combined with the backscattering measurement of $N\,\Delta x$. The use of the linear depth scale, Eq. 3, introduces absolute errors of less than 5% for film thicknesses of several thousand Å. Rather straightforward calculations can be made to correct this 5% deviation and to determine a depth scale in much thicker films.

To first approximations, the total number A of backscattered and detected particles is given by

$$A = \frac{d\sigma}{d\Omega}\, N\, \Delta x\, Q\, \Delta\Omega \tag{7}$$

where $d\sigma/d\Omega$ is the differential Rutherford scattering cross section, Q is the number of incident particles per cm^2, and $\Delta\Omega$ the acceptance angle of the particle detector. The Rutherford scattering cross section, in laboratory coordinates, is given by

$$\frac{d\sigma}{d\Omega} = 1.296 \left[\frac{Z_1 Z_2}{E_0}\right]^2 \left[\csc^4 \frac{\theta}{2} - 2\left(\frac{M_1}{M_2}\right)^2 + \cdots\right] \times 10^{-27} \text{ cm}^2 \qquad (8)$$

The next term in this expansion is of order $(M_1/M_2)^4$ and can be ignored (E is in MeV). The only approximation in Eq. 7 stems from the fact that the Rutherford scattering cross section is proportional to E^{-2}, and hence there is an increase in σ as the particles penetrate the sample and decrease in energy. In thin film samples this correction is generally less than 10%; for example, the ratio of σ values at the interface to the front surface is 1.06 in the case of a 1000 Å film of Ni. Again, simple analytic corrections can be made.

The heights H_{Ni} and H_{Si} at the surface in counts of backscattering yield are simply related to the stopping powers and cross sections:

$$H = Q \frac{d\sigma}{d\Omega} \Delta\Omega \frac{\delta E_1}{[\varepsilon]} \qquad (9)$$

where δE_1 is the energy per channel. Again to first approximation, $A = H \times \Delta E/\delta E_1$. These relations indicate that the number of atoms per cm^2 in the films can be accurately determined. In general, however, we are interested in reaction products or interdiffusion profiles, and the lower portion of Figure 6.3 shows schematically that the Ni film reacted to form Ni_2Si.

After reaction, the Ni signal ΔE_{Ni} has spread slightly, owing to the presence of Si atoms contributing to the energy loss. The Si signal exhibits a step corresponding to Si in the Ni_2Si. The width of the Si step, ΔE_{Si}, is somewhat narrower than ΔE_{Ni} because of kinematic factors, due principally to K in Eq. 5.

It should be noted that the ratio of the heights H_{Ni}/H_{Si} of Ni to Si in the silicide layer gives the composition of the layer. One must take into account the differences in scattering cross sections and slight differences in stopping cross section along the outward path for particles scattered from Ni and Si atoms. The expression of the concentration ratio is given by

$$\frac{N_{Ni}}{N_{Si}} = \frac{H_{Ni}}{H_{Si}} \frac{\left(\dfrac{d\sigma}{d\Omega}\right)_{Si}}{\left(\dfrac{d\sigma}{d\Omega}\right)_{Ni}} \frac{[\varepsilon_{Ni}]}{[\varepsilon_{Si}]} \qquad (10)$$

and to within an accuracy of 5 to 10% in most cases for MeV ^{4}He ions by

$$\frac{N_{Ni}}{N_{Si}} = \frac{H_{Ni}}{H_{Si}} \left\{\frac{Z_{Si}}{Z_{Ni}}\right\}^2 \qquad (11)$$

as $[\varepsilon_{Ni}] \approx [\varepsilon_{Si}]$ and $\left(\dfrac{d\sigma}{d\Omega}\right)_{Si} \propto (Z_{Si})^2$, $\left(\dfrac{d\sigma}{d\Omega}\right)_{Ni} \propto (Z_{Ni})^2$. Consequently one can determine rather quickly the concentration ratio of the reaction products in the layer. To determine the phases that are present, one generally uses glancing-angle x-ray diffraction as described in the preceding chapter.

The ratio of the Ni heights before and after reaction can also be used to determine the concentration ratio of the reaction products by the following expression:

$$\frac{H_{\text{Ni}}^{\text{Ni}}}{H_{\text{Ni}-\text{Si}}^{\text{Ni}}} = \frac{N_{\text{Ni}} + N_{\text{Si}}}{N_{\text{Ni}}} \frac{[\varepsilon_{\text{Ni}-\text{Si}}^{\text{Ni}}]}{[\varepsilon_{\text{Ni}}^{\text{Ni}}]} = 1 + \frac{N_{\text{Si}}}{N_{\text{Ni}}} \frac{[\varepsilon_{\text{Si}}^{\text{Ni}}]}{[\varepsilon_{\text{Ni}}^{\text{Ni}}]} \tag{12}$$

where the superscripts refer to the scattering atoms and the subscripts, to the alloy the beam traverses. This expression assumes Bragg's rule (24) that atomic stopping cross sections can be added to give the compound stopping cross section, namely,

$$\varepsilon_{\text{Ni}-\text{Si}} = \frac{N_{\text{Ni}}}{N_{\text{Ni}} + N_{\text{Si}}} \varepsilon_{\text{Ni}} + \frac{N_{\text{Si}}}{N_{\text{Ni}} + N_{\text{Si}}} \varepsilon_{\text{Si}} \tag{13}$$

where N_{Ni} and N_{Si} are the atomic densities in the alloy (at. cm^{-3}).

In Eq. 12 the scattering cross sections cancel; however, the stopping cross sections cannot be neglected because we are now comparing energy losses in two different materials. This expression is useful, for example, in determining the concentration ratios of heavy metal oxides, where it is difficult to measure the heights of the oxygen signal directly. The accuracy of this expression is determined primarily by the accuracy of stopping power measurements in the different materials.

In the example we have given, there is no overlap in the backscattering energy spectra of the signals from the different elements in the thin film target. Diffusion profiles or thickness of reaction products can therefore be extracted easily in the example we have shown. There are many cases, however, where overlap in the signals from the different components in the target can occur. This problem can often be overcome in sample preparation by changing the sequence of deposition. The signals from the different constituents can then be directly observed. A schematic example (25) of the use of this technique in studying interdiffusion in Pd–Au is shown in Figure 6.6. In this example the diffusion profiles of both Au diffusing into Pd (and Pd into Au) can be determined from both (a) and (b). However, in the sequence (b), no signal subtraction techniques are required to extract the diffusion profiles.

There are situations, however, where one is not allowed this flexibility in sample preparation, inasmuch as changing the deposition sequence on a substrate may change some physical parameters, for example, grain size. In those cases where overlap of the signals does occur, the diffusion profiles

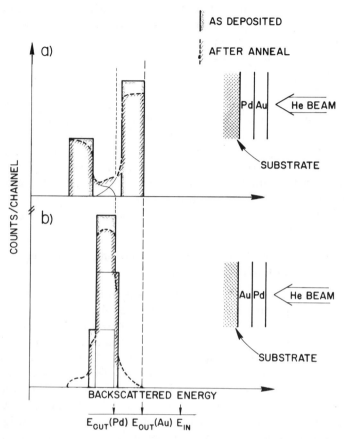

Figure 6.6.   Backscattering schematic showing how the concentration profiles can be extracted directly by reversing the order of the deposited films. As indicated in the cross-sectional view of the sample, the Au film is deposited on the Pd film in (*a*) and the Pd on Au in (*b*). From reference 25.

must be extracted by spectral stripping techniques. These can be carried out in a straightforward fashion as scattering and stopping cross sections are known (26). Computer programs for unscrambling these spectra now exist (27).

Of course, there is another simple experimental technique that provides additional information on depth profiles—that is tilting the target with respect to the incident beam to change the effective thickness of the layers. Not only does this give added information on the identity of surface species (because the energy position of signals from surface species does not shift with tilt angle), but it can also dramatically increase the depth resolution of backscattering. By use of grazing-angle incidence (28), improvements in depth resolution have been reported of nearly an order of magnitude over that obtained for normal incidence. Extremely flat surfaces are required at such grazing angles of incidence.

The power of backscattering lies then in the ability to determine compositional depth profiles quantitatively. There are two obvious limitations to the technique: (a) it is difficult to determine the identity of high-mass species since the mass resolution degrades with increasing mass and the scattering and stopping cross section are nearly identical; and (b) it is difficult to detect low levels of light-mass impurities in heavier-mass substrates because of the unfavorable ratio of scattering cross sections. In the following section we will briefly discuss other nuclear techniques that have been successfully used in conjunction with backscattering to overcome these limitations.

6.2.2   Ion-Induced X-Rays, Nuclear Reactions, and Resonant or Non-Rutherford Backscattering

The identifications of high-mass constituents in targets can be carried out by means of ion-induced x-rays. Experimentally, this only requires the addition of an x-ray detector, typically a Si–Li detector, to the backscattering chamber. The x-ray spectra then provide elemental identification. This approach is generally not required in studies of thin film reactions because the constituents in the films are known from the deposition sequence. Moreover, there is the greater problem that the x-ray technique does not have sufficient depth resolution to permit extraction of depth profiles (29). Some depth information can nevertheless be obtained from a knowledge of the energy dependence of the x-ray excitation cross section. This approach was used in studies of oxides on GaAs (30) where depth information was obtained by tilting the target so that x-ray generation was confined to surface layers.

Ions can also be used to induce nuclear reactions. This is a subject with a history dating back to the origin of accelerators (31). In material analysis the advantage of the use of nuclear reactions lies in their ability to detect light-mass impurities in heavy material. Reactions can be chosen to isolate most of the low-mass nuclei with elemental specificity (32, 33). Contrary to the case of ion-induced x-rays, depth information can be obtained from those reactions where charged particles are emitted and hence energy is lost along the inward and outward paths.

At the present time considerable activity is concentrated on the detection of the isotopes of hydrogen and helium in matter (34). This interest is generated over concern regarding, for example, hydrogen incorporation in the walls of fusion reactors. One approach (35) to the detection of these isotopes is to use nuclear reactions such as $^1H(^{19}F, \alpha\gamma)^{16}O$. In this case, 15 to 20 MeV fluorine ions are used to bombard the target containing hydrogen, and one then detects the emitted gamma ray. A technique which is accessible for accelerators in the 1 to 3 MeV region is to use the enhanced elastic scattering cross section available, for example, for protons on low-Z targets (36). In the previous discussion on Rutherford backscattering techniques, the cross section depends on Coulomb scattering and varies essentially as $(Z_1Z_2/E)^2$. The physical reason that the cross sections are Rutherford for 1 to 2 MeV ^4He particles on most targets is that the incident ^4He particle does not penetrate the Coulomb barrier and hence contributions from elastic nuclear scattering, due to the nuclear force, can be ignored. The Coulomb barrier height is given by

$$E_c = Z_1 Z_2 e^2 / R$$

where R is the nuclear interaction radius. For example, $E_c \sim 5$ MeV for ^4He on Si, but $E_c \sim 0.5$ Mev for protons on ^4He. At 2.8 MeV, for example, the elastic cross section for protons on ^4He is enhanced by a factor of approximately 300 over the Rutherford cross section; for protons on deuterium, at this energy the enhancement is a factor of approximately 260.

Figure 6.7 shows a 2.8 MeV proton backscattering spectrum (36) from an

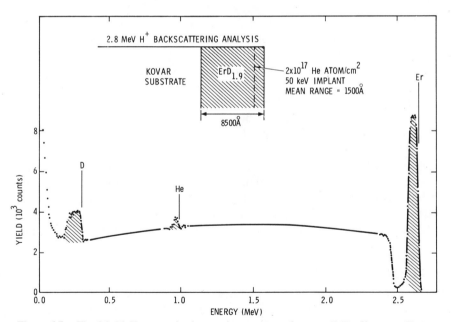

Figure 6.7. The 2.8 MeV proton backscattering spectrum from an ErD_2 film on a Kovar substrate. From reference 36.

ErD_2 film on a Kovar substrate. The film has also been implanted with 2×10^{17} He atoms per cm^2 at 50 KeV. In the spectrum the contribution from the deuterium and helium are clearly visible, and the depth profiles of both can be extracted. When the measurement is performed in an energy range where the scattering cross section is Rutherford, these signals are not observed.

There are also resonances in the nuclear reaction and elastic scattering cross sections, and these can be used not only for signal enhancement but also for depth profiling if the resonance is sufficiently narrow (37). For depth profiling in this case, the energy of the incident beam has to be varied thus sweeping the resonance through the solid. The resonance in the $^1H(^{19}F, \alpha\gamma)^{16}O$ reaction at 16.44 MeV, for example, was used to measure hydrogen depth profiles in lunar samples (35).

6.3   SPUTTER SECTIONING

Sputtering is the process by which an incident beam of energetic particles causes the removal of surface atoms. For certain combinations of particle and target, this is a very powerful technique for sectioning surface layers. It is, however, a destructive technique in that the sample is eroded and there is considerable atomic rearrangement of the remaining surface layer.

The technique is attractive in that keV ion beams can be utilized for the sputter sectioning and a variety of commercial surface analysis instruments exist for characterizing the sectioned surface. It is a natural adjunct for existing thin film technologies and is used extensively.

In order to understand the capabilities of sputtering as a sectioning technique, it is necessary to discuss the various physical and chemical processes involved. We shall start with a discussion of the sputter sectioning of monoelemental materials, where there has been extensive theoretical and experimental work. The step, however, to sputter sectioning of thin films, which naturally involves alloys and interfaces, is not trivial. The fact that there is deposition of energy from the sputtering particles leads to changes in the interface and atomic composition of the systems being studied. This can give rise to misleading depth profile data because changes produced by the energetic ions can overwhelm the changes in the depth profile which one is attempting to study. There has been a voluminous literature (10) on sputtering and sputter sectioning, and we shall only cite a few references pertinent to the discussion.

6.3.1   Monoelemental Layers

Measurements of the sputtering yields of monoelemental targets under carefully controlled conditions where dose effects are minimal have shown good agreement with theoretical predictions. The agreement between theory and experiment is illustrated in Figure 6.8, which shows the sputtering yield

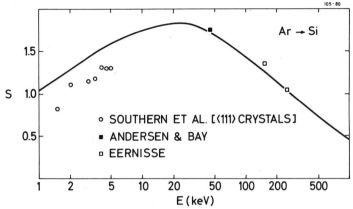

Figure 6.8. Energy dependence of the Ar sputtering yield (S = atoms per ion) of Si. The solid line represents the calculations of Sigmund (41) and the data points are from references 38, 39, and 49.

of Si as a function of Ar ion energy (38–40). The calculations by Sigmund (41) (solid line) are based upon the collision cascade concept of sputtering. The incident ion makes multiple collisions with target atoms which in turn collide with other target atoms, generating a collision cascade. Momentum can thus be returned to the surface atoms; as a result, those atoms which obtain sufficient energy to overcome the surface binding will be sputtered. Consequently, the sputtering yield is proportional to the amount of energy deposited in atomic motion and is inversely proportional to the surface binding energy.

The collision cascade provides good predictions as long as the energy density in the cascade is not too high. The influence of energy density was shown directly by Andersen and Bay (42) who measured sputtering yields for atomic and molecular ions. Markedly higher yields per incident ion are observed for molecular sputtering. The increased yield can be explained qualitatively on the basis of the thermal spike model in which all the atoms within the collision cascade are moved during some time interval within the lifetime of the spike.

These collision cascade theories do not include the incorporation of the sputtering species within the host lattice (i.e., ion implantation). Such dose effects are unavoidable in sputter sectioning and are particularly accentuated when there is a chemical reaction. The violent fluctuations in the sputtering yield as a function of the atomic number of the incident ion, observed by Almen and Bruce (43), have been ascribed to dose effects (44). Figure 6.9, for example, shows the data of Andersen and Bay (42) for the sputtering of Si for 15 different 45 keV ions. Care was taken to avoid dose effects, and it can

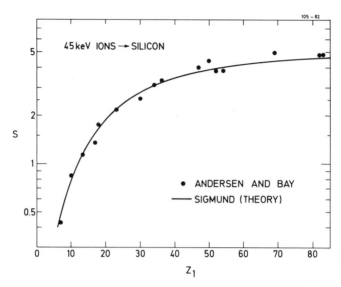

Figure 6.9.   The Z_1 dependence of the sputtering of Si. From reference 42.

be seen that there is good agreement with Sigmund's theory; no fluctuations with Z_1 are observed. Sputtering yields can deviate from theoretical predictions by more than an order of magnitude for some ion–substrate combinations due to the influence of incorporation of the sputtering species. We should note that dose effects appear to be minimal for the inert gas ions Ne, Ar, Kr, and Xe. For these, one need only consider the dilution in the effective atomic density of substrate atoms caused by the concentration of implanted projectiles. As a rough estimate, the surface concentration of implanted projectiles is proportional to the inverse of the sputtering yield (S^{-1}).

For these reasons it is always desirable to measure the sputtering yield for a given ion–target combination. An example of such a measurement is shown in Figure 6.10 for 900 eV Ar sputtering of Pt (45). Figure 6.10a shows the backscattering spectra for 1.9 MeV ^{4}He incident on a 1500 Å Pt film on Si, before and after sputtering. The decrease in film thickness for increasing Ar ion dose is shown in Figure 6.10b. These results were obtained by integrating the yield in the Pt signal of the backscattering spectrum. From the linear slope we determine the sputtering yield to be 1.7. Tables 6.1, 6.2, and 6.3 show sputtering yields from a recent compilation (10) at 500 eV, 1 keV, and 10 keV, respectively, for He, Ne, Ar, Kr, and Xe ions. More detailed data and source references can be found in surveys of sputtering (9, 46–49).

The sputtering yields of MeV light ions for most materials are of the order of 10^{-3}. Rutherford backscattering analysis will therefore cause the sputtering of only a very small fraction of a monolayer during a typical run.

6.3.2   Layered Structures and Interface Effects

One of the concerns in thin film investigations is the nature of the interface between two metal layers, and, metal–semiconductor or metal–oxide layers. Conceptually, sputter sectioning would be an ideal method to evaluate such interfaces since the covering layer could be eroded in a controlled fashion. However, even a cursory glance at the literature indicates that this is a deceptively difficult task for sputter sectioning. The reasons for this are due to a variety of causes, a few of which we shall isolate now. Within the collision cascade, for example, atoms are displaced from their lattice position. This results in considerable mixing at interfaces. One estimate (50) is that

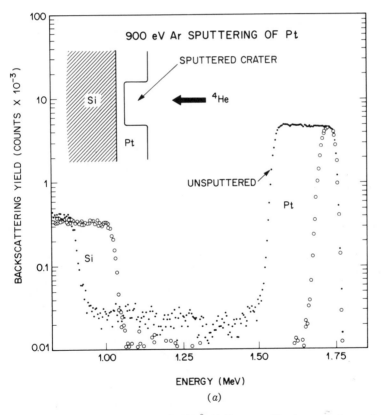

Figure 6.10.   (a) Backscattering spectra of 1500 Å Pt films on a Si substrate before and after sputtering with 900 eV Ar.

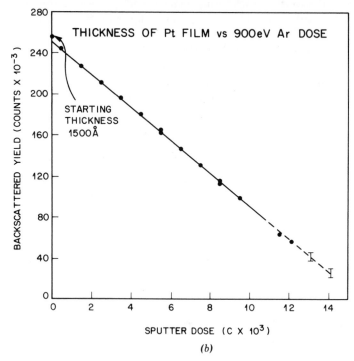

Figure 6.10. (*b*) Thickness of Pt film remaining after 900 eV Ar bombardment. The two error bars do not represent statistical uncertainties but rounding in the crater bottom. From reference 45.

degradation of depth resolution is about the projected range of the incident sputtering projectile. This can be 130 Å for 10 keV Ar in Si.

Another closely associated phenomenon is the triggering of phase formation causes by intermixing or enhanced reaction rates at interfaces. This has been observed for Pd or Pt films on Si (51, 52) where silicides were formed at room temperature by bombarding the films with Ar ions. The observed formation of phases was not a temperature effect but occurred when the projected range of the incident ion coincided with the depth of the interface beneath the surface. These results unequivocally indicate that one cannot probe an interface by sputter sectioning without seriously perturbing the very interface under investigation. This is the Catch-22 of sputtering.

There are other effects, more geometric in nature, associated with the preferential etching of certain surfaces. One of these is the preferential etching at grain boundaries of polycrystalline metal films. The degradation of depth

TABLE 6.1. SPUTTERING YIELDS FOR 500 eV ION BOMBARDMENT[a]

	Sputtering yield, atoms per ion				
	He	Ne	Ar	Kr	Xe
Be	0.2	0.4	0.5	0.5	0.4
C	0.07		0.12	0.13	0.17
Al	0.16	0.7	1	0.8	0.6
Si	0.13	0.5	0.5	0.5	0.4
Ti	0.07	0.4	0.65	0.6	0.4
V	0.06	0.5	0.65	0.6	0.6
Cr	0.17	1	1.2	1.4	1.5
Fe	0.15	0.9	1.1	1.1	1.0
Co	0.13	0.9	1.2	1.1	1.1
Ni	0.16	1.1	1.45	1.3	1.2
Cu	0.24	1.8	2.4	2.4	2.1
Ge	0.08	0.7	1.1	1.1	1.0
Y	0.05	0.5	0.7	0.7	0.5
Zr	0.02	0.4	0.7	0.6	0.6
Nb	0.03	0.3	0.6	0.6	0.5
Mo	0.03	0.5	0.8	0.9	0.9
Rh	0.06	0.7	1.3	1.4	1.4
Pd	0.13	1.2	2.1	2.2	2.2
Ag	0.2	1.8	3.1	3.3	3.3
Gd	0.03	0.5	0.8	1.1	1.2
Hf	0.01	0.3	0.7	0.8	
Ta	0.01	0.3	0.6	0.9	0.9
W	0.01	0.3	0.6	0.9	1.0
Os	0.01	0.4	0.9	1.3	1.3
Pt	0.03	0.6	1.4	1.8	1.9
Au	0.07	1.1	2.4	3.1	3.0
Th		0.3	0.6	1.0	1.1
U		0.5	0.9	1.3	

[a]From reference 10.

profiles due to this phenomenon can be quite pronounced. Another phenomenon is cone formation, which is typically found when sputtering layers with differing sputtering yields (10, 49). A beautiful example is the depth profile of Cu–Ni thin film layers obtained by secondary ion mass spectrometry. Figure 6.11a shows this work of Hofer and Liebl (53) for sputtering with either Ar^+ or N_2^+ projectiles. With Ar ions the interfaces are degraded

TABLE 6.2.   SPUTTERING YIELDS FOR
1 keV ION BOMBARDMENT[a]

| | Sputtering yield, atoms per ion | | | |
	Ne	Ar	Kr	Xe
Fe	1.1	1.3	1.4	1.8
Ni	2	2.2	2.1	2.2
Cu	2.7	3.6	3.6	3.2
Mo	0.6	1.1	1.3	1.5
Ag	2.5	3.8	4.5	
Au		3.6		
Si		0.6		

[a]From reference 10.

TABLE 6.3.   SPUTTERING YIELDS FOR
10 keV ION BOMBARDMENT[a]

| | Sputtering yield, atoms per ion | | | |
	Ne	Ar	Kr	Xe
Cu	3.2	6.6	8	10
Ag		8.8	15	16
Au	3.7	8.4	15	20
Fe		1.0		
Mo		2.2		
Ti		2.1		

[a]From reference 10.

due to cone formation (shown in Fig. 6.11b) starting at the transition from the first Ni layer to the Cu layer. On the other hand, the measured Ni depth profile is more sharply defined, and cone formation is suppressed when the N_2^+ beam is employed. The authors suggest that the chemical activity of the implanted N atoms is responsible for the increased uniformity of erosion. They cite similar examples for Ar and O beams.

6.3.3   Compound and Alloy Layers

We are primarily interested in studying the reaction or interdiffusion of films. The sputtering of multielemental material, therefore, has to be understood. We shall discuss the two main aspects of this problem—first the composition of the sputtered species and second the composition of the sputtered layer.

On simple conservation of matter arguments, the composition of the sputtered species should mirror that of the bulk host. This is often called "stoichiometric sputtering." This simple condition rests on the assumptions that equilibrium has been attained, that there is no long-range bulk diffusion, and that the concentration of implanted sputtering ions does not markedly perturb the system. Surprisingly, only a few experiments (45, 54) have been addressed to this problem; however, systematic studies are underway. An example of such an experiment (45) is shown in the backscattering spectra of Figure 6.12 for the 900 eV Ar sputtering of PtSi. By monitoring the change in the Pt and Si signals one could deduce the sputtering yields for the individual species. The object of this experiment was to measure these alloy yields and compare them with measurements of elemental Pt (Fig. 6.10) and Si sputtering yields. Measurements were also carried out on the Ni silicide system. There are two significant points that are raised by the results shown in Table 6.4. First, the compound yields are not related in any simple fashion to the elemental sputtering yields. Second, we find stoichiometric sputtering. Although the ratios of Si to Pt and Si to Ni partial sputtering yields in the silicides are 1.1 and 1.05, respectively, these numbers in fact reflect the actual measured composition of the thin film silicides. The phases are indeed PtSi and NiSi, but there appears to be a 5 to 10% excess of Si.

The most crucial aspect of these results, however, is that they indicate that there is no *a priori* justification for using elemental sputtering yields. One cannot use the aphorism that the sputtering rate is determined by the value of the lowest elemental sputtering yield. Consequently, depth calculations based on elemental sputtering yields are suspect unless there is other experimental evidence to the contrary. Haff and Switkowski (55) have presented a physical model for sputtering of compound targets. A critical parameter is the surface binding energy. They found that the choice of one value of the binding energy would lead to a reasonable fit between theory and experiment.

We now return to the case of the composition of the eroded surface. It has been well established (56) that the outermost surface layers are altered in composition to maintain stoichiometric sputtering. The dynamics of this process are not understood, involving as they do diffusion and other related processes. We are equally concerned with chemical and phase changes that can occur well below the surface of the sample due to energy deposited by the energetic ions. Kelly and Sanders (57) have reviewed the numerous examples of changes in oxide composition that can occur under sputtering conditions. We illustrate such below-surface composition changes with a backscattering

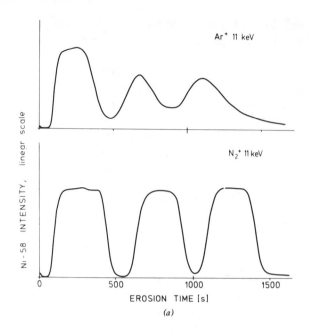

Figure 6.11. (*a*) Ni-58 profiles obtained on a Cu–Ni multilayer sandwich consisting of 510 Å Ni, 500 Å Cu, 500 Å Ni, 500 Å Cu, and 500 Å Ni between 300 Å Cu on top and Si substrate at bottom. Profiling with Ar$^+$ primary ions causes cone formation as shown below. (*b*) SEM view of an Ar$^+$ bombarded area where the irradiation was stopped just before the Si substrate was reached. From reference 53.

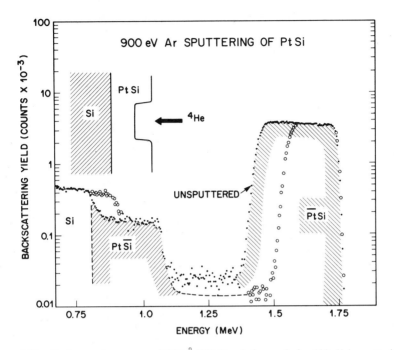

Figure 6.12.   Backscattering spectra of 2900 Å PtSi films before and after 900 eV Ar sputtering. From reference 45.

spectra (45) of a 20 keV Ar sputtered PtSi film (Fig. 6.13). The shape of the Pt signal indicates that an enhancement of the Pt-to-Si ratio has taken place over a range of some 200 Å, the Ar projected range.

The composition ratio in fact approaches that of Pt_2Si. Definite phase identification is difficult as the Ar beam disorders the silicide. Murti and Kelly (58) have made positive identification of phase changes in oxides under bombardment conditions using electron diffraction techniques. Indeed, it was the pioneering work of Gillam (59), using electron diffraction techniques, which first demonstrated that surface compositional changes could occur in bombarded Cu–Au alloys.

All of the effects detailed above make it imperative that sputter sectioning of alloys be approached with several different experimental techniques. Whenever possible, independent measurements should be utilized in determining such depth profiles.

TABLE 6.4. SPUTTERING YIELDS FOR Ar BOMBARDMENT OF Pt, Si, Ni, PtSi, AND NiSi[a]

	Sputtering yield, atoms per ion	
	900 eV	20 keV
S_{Pt}	1.7	4.1
$S_{\overline{Pt}Si}$	0.8_5	2.2
$S_{Pt\overline{Si}}$	0.9_2	2.5
S_{Si}	0.5_2	1.5
S_{Ni}		4.9
$S_{\overline{Ni}Si}$		2.2
$S_{Ni\overline{Si}}$		2.4

[a]From reference 45. The bar over the element indicates the partial sputtering yield of that element in the silicide. The total sputtering yield of the silicide, S_{PtSi}, for example, is given by $S_{\overline{Pt}Si} + S_{Pt\overline{Si}}$.

6.4   SURFACE ANALYSIS COMBINED WITH SPUTTER SECTIONING

The wide variety of laboratory and commercial instruments capable of providing depth profile information based on sputter sectioning has led to a large increase in the number of studies in this field. The aim of this section is to review the salient features of the two general approaches, surface analysis and sputtered species detection, without presenting a detailed description of the methods.

This area has been covered in recent excellent reviews and books where many of the references are listed (56, 60–62). One can also cite recent conferences which contain numerous papers highlighting these techniques (63, 64).

6.4.1   Surface Analysis Methods—AES and ISS

As discussed previously, the concept underlying these techniques is to use a surface-sensitive probe. Two techniques have received most attention:

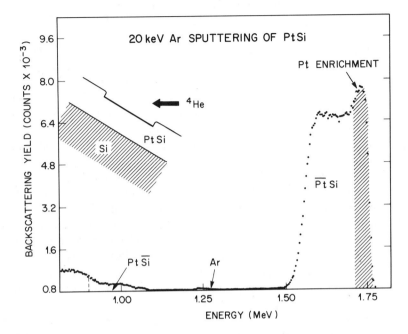

Figure 6.13. Backscattering spectra of PtSi films after 20 keV Ar sputtering. Starting thickness, 1700 Å; final thickness, 1100 Å. The film has been tilted 60° to increase depth resolution. From reference 45.

Auger electron spectroscopy (AES) and ion scattering spectrometry (ISS). In AES, the atomic core levels are ionized by an incident electron beam at keV energies and the subsequent radiationless Auger transition and escape of the Auger electron provide elemental identification through energy analysis of the emitted electron. This technique is surface sensitive because of the limited mean free path of the Auger electrons, typically in the range of 4 to 20 Å.

A similar method is x-ray photoelectron spectroscopy (XPS), which is also referred to as electron spectroscopy for chemical analysis (ESCA). In XPS, the sample is bombarded with a beam of monoenergetic keV x-rays, and photoelectrons are emitted. Elemental identification is provided through measurement of the energy of the emitted photoelectron, which is characteristic of the energy levels of the excited atom in the target material. As in AES, the technique is surface sensitive because of the limited mean free path from which photoelectrons can escape.

In the other surface technique, ISS, the sample is bombarded with keV ions, and the backscattered ions are energy analyzed. As in Rutherford backscattering, elemental analysis is provided through the kinematics of the elastic binary collision. The technique is surface sensitive in that the characteristic sharp peaks (65) in the energy spectra represent scattering from only the first layer or two at the surface. These peaks kinematically coincide with surface scattering, and their existence is thought (66) to be due to two phenomena: (i) the large scattering cross sections at these energies, which lead to attenuation of the ion beam as it penetrates, and (ii) the more complete neutralization of the particles that penetrate more deeply before being scattered.

All of these techniques measure the surface composition after sputtering has taken place. As mentioned in Section 4.3, selective sputtering effects can seriously affect the measurement of composition profiles, especially at interfaces where there are strong concentration gradients. With AES, it appears possible to monitor some of these phenomena by detecting both the low- and the high-energy Auger electrons. These electrons have different escape depths (67). This approach only differentiates compositional changes in the outermost layers (5 to 20 Å) and will not give information on sputter induced composition charges at greater depths.

We shall illustrate the application of these surface analysis methods with two depth profiles. Figure 6.14 shows an AES profile (68) of a multilayer single-crystal periodic thin film structure in which the structure consists of alternate 50 Å layers of GaAs and GaAlAs. Although the depth resolution degrades as a function of depth, it is far superior to the depth resolution that could be obtained from Rutherford backscattering on this type of sample (69). These unique structures are prepared by molecular beam epitaxy techniques which produce interfaces that are extremely sharp and well defined (70). This sample is chosen to represent the remarkably good depth resolution that can be achieved under favorable conditions. Unfortunately, it is not clear yet what constitutes favorable conditions in light of cone formation, atomic mixing, and preferential sputtering effects that are known to occur in other systems.

A problem of considerable interest is that of the Si–SiO$_2$ interface. Figure 6.15 shows the depth profile of a 96 Å oxide on $\langle 100 \rangle$ Si as obtained by ISS and sputter sectioning (71). This profile shows the presence of stoichiometric SiO$_2$ at the outermost layer of the oxide and the presence of excess Si at the Si–SiO$_2$ interface, corresponding to a region about 20 Å in thickness. A similar conclusion was reached in channeling effect measurements using MeV ^{4}He ions (72). However, with the channeling measurements, the depth resolution was not sufficient to ascertain whether the excess Si was present

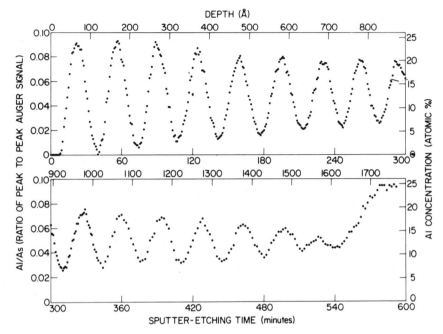

Figure 6.14. Compositional profile of a periodic superlattice, each period consisting of a 50 Å layer of GaAs and $Ga_{0.75}Al_{0.25}As$. Ar sputtering energy is 1.5 keV. From reference 68.

in the oxide or represented disorder in the underlying Si crystal. Again, the example represented in this figure gives evidence for the remarkable depth resolution that can be obtained under certain conditions.

In both these depth profiles, Figures 6.14 and 6.15, obtained using AES and ISS, the depth scales have been given in angstroms rather than in sputtering times. Physical depths could be used because the superlattice was independently characterized using angle sectioning and the oxide thickness was measured by ellipsometry. In the ISS results, it was assumed that the same sputter rate applied to the entire SiO_2 layer and to the interfacial region. The data in Figure 6.15 do indicate the existence of excess Si (hatched region in Fig. 6.15), but the broad interface, approximately 60 Å, seems to be instrumental rather than physical in origin.

The question also remains as to the manner in which quantitative values are assigned to the concentration ordinates in Figures 6.14 and 6.15. In fact, absolute values are not used. In the AES profile, only the relative Al signal is plotted. One of the major efforts in AES is to perform quantitative

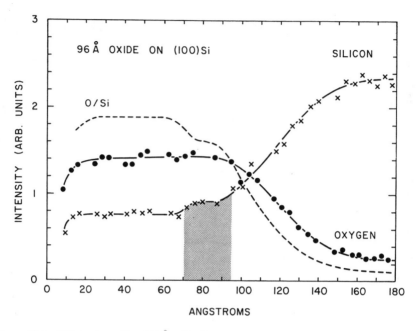

Figure 6.15. ISS depth profile of 96 Å oxide film on (100) Si. Shaded area gives the region of excess Si corresponding to 1.4×10^{14} excess Si atoms per cm^2 per monomolecular layer. From reference 71.

microanalysis as opposed to the semiquantitative approach. Although substantial progress has been made in this direction, there is a lack of quantitative information regarding the fundamental mechanisms in the production and escape of Auger electrons as well as experimental factors involving analyzer sensitivity and surface roughness (73).

Similarly, in ISS there is a lack of information regarding the yield of the backscattered ions since the yields depend not only on low-energy atomic cross sections but also on neutralization phenomena. (Such difficulties do not arise in Rutherford backscattering with MeV ions and solid-state detectors because the cross sections and detection efficiencies for scattered particles are known.) However, the field of low-energy particle–solid interactions is very active, and several groups are trying to resolve these problems associated with ion–surface interactions (66, 74). It is difficult therefore at this stage to assign absolute concentration values in ISS, and generally comparisons are made with standards. For example, in the present case (Fig. 6.15) the signals were calibrated against those obtained from thick thermal oxides.

6.4.2   Sputtered-Species Detection Method—SIMS

The detection of the sputtered species would appear to be the most straight-forward technique for depth profiling since both the sample sectioning and signal generation could be accomplished simultaneously. Moreover, if all the sputtered species are detected, some of the limitations inherent in sputter sectioning should be removed. It is this promise that has provided much of the impetus for development of techniques such as secondary ion mass spectrometry (SIMS) and surface composition analysis by neutral ion impact radiation (SCANIIR) (75).

The mechanisms leading to ejection of sputtered particles have been discussed previously for both these techniques. In SIMS, the charged sputtered species are detected; and in SCANIIR, the light emitted from the deexcitation of sputtered species is detected. Both these methods critically depend on a knowledge of the ionization and deexcitation phenomena of the sputtered species as they leave the surface. This is a problem of considerable scientific complexity because of the many parameters involved, such as energy distribution of the sputtered species and the influence of the surface conditions on the electron exchange between surface and sputtered species. For example secondary ion yields may differ by orders of magnitude for the same element in different materials, whereas the actual sputtered yield certainly does not change by this magnitude.

Several points should be emphasized. The sputtered-species detection techniques have the highest sensitivity for impurity detection, and the components are sputtered at rates proportional to the bulk composition after equilibrium has been attained (i.e., stoichiometric sputtering). They have the drawback that all sputtered species must be detected if the data are to represent bulk concentrations. This is a problem because the sputtered-species detection methods do not involve measurement of all the sputtered material but in reality only a very small fraction that is charged or optically decaying. (Most of the sputtered species are neutral or undergo radiationless deexcitation.)

Under appropriate conditions, generally analysis of amorphous oxide layers, excellent depth profiles can be obtained. We shall illustrate the application of these methods with a SIMS depth profile (Fig. 6.16) of a Ta_2O_5 film containing a layer enriched to 2% in ^{18}O. In this work (76) the $Ta^{18}O^+$ signal was monitored using 8.5 keV N_2^+ as the sputtering probe. The presence of the ^{18}O layer is clearly indicated. The sputtering time-to-depth scale conversion was established on the basis of independent measurements of the thickness of the Ta_2O_5 film. One point that should be noted in connection with SIMS analysis is the strong influence of O in enhancing the secondary ion yield. This has led to the technique (76) of sputtering with O beams instead of the more usual inert gas beams. Alternatively, a high partial pressure of O_2 in the target chambers is employed.

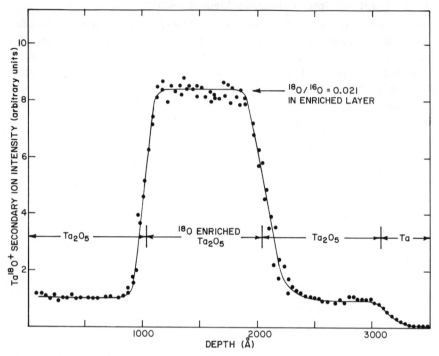

Figure 6.16. Depth concentration profile of ^{18}O in a Ta_2O_5 sample containing a layer enriched in ^{18}O. The ^{18}O is 2% of the oxygen in the enriched layer. The $Ta^{18}O^+$ signal preceding and following the enriched layer peak is due to the ^{18}O level in normal oxygen. From reference 76.

There are numerous other examples that can be cited which also reflect sharp depth profiles. The major problem is to obtain quantitative numbers for the composition of the layers. Although the techniques are extremely sensitive for the detection of trace impurities, absolute calibration relies on comparison with data from standards.

6.5   ANALYSIS BY BOTH ENERGY LOSS AND SPUTTER SECTIONING

From the previous discussions it should be evident that the optimum way to obtain depth information is to employ more than one analytic method. This approach has already been followed in combining surface analysis and sputtered species detection systems, for example, AES plus SIMS (77). These combinations suffer the obvious limitation that they employ sputter sectioning to obtain depth information. This problem can be overcome by employing energy loss techniques to quantitatively determine composition and depth scales along with one of the sputter sectioning techniques which can provide, for example, added information on the depth distributions of light-mass impurities and possibly better depth resolution.

6.5.1 Backscattering and AES

Figure 6.17 shows diffusion profiles of Pd in Au thin films (78). In this work the depth scale was determined by backscattering spectrometry. AES was used to measure the depth profiles where the absolute concentrations were calibrated by backscattering. In this application, AES had the advantage that the relative yields and depths could be measured without the overlap of signals that was found in the backscattering analysis. For example, the Pd surface peak was obscured by the Au signal in backscattering analysis but was readily visible in the Auger data. It should be noted that the interface obtained from backscattering appears sharper than that obtained from AES plus sputtering. This behavior is typically found in comparison of sputter sectioning and backscattering analysis of polycrystalline metal films (79). This worsening of the depth resolution in sputter sectioning is probably due to preferential erosion at grain boundaries.

Another strong advantage in utilizing AES in conjunction with Rutherford backscattering is the sensitivity of AES to oxygen or other light-mass impurities. Oxygen is ubiquitous and must always be suspect when surface or interface behavior is being investigated. Figure 6.18a shows the AES profile, and Figure 6.18b, the Rutherford backscattering profile (80) of a PtSi film formed by annealing a Pt film on Si in an O_2 ambient at $600°C$. In the AES profile, the O signal near the surface is clearly visible; in the backscattering spectrum, the O signal is swamped by scattering from the Si substrate. How-

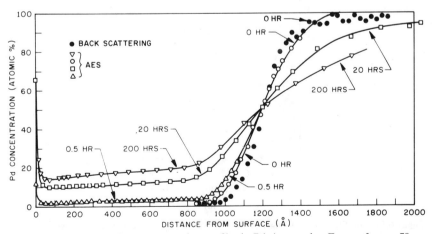

Figure 6.17. Palladium concentration profiles in Pd–Au couples. From reference 78.

ever, the dip in the Pt signal around 1.8 MeV indicates the presence of low-mass impurities, in this case O and Si.

6.5.2   Backscattering and SIMS

There are similar advantages in the utilization of backscattering to determine depth scales and concentrations and of SIMS to determine the presence of either light-mass or low-level impurities. Figure 6.19 shows SIMS and backscattering profiles for a Ta_2O_5 film with a F contaminant at the interface (81). It was clearly impossible to identify this contaminant with backscattering although its presence is manifested by the dip in the Ta yield at the interface. Similarly, the thickness of the Ta_2O_5 film could not be determined directly from sputter sectioning but the identity of the contaminant could be deduced directly from SIMS.

Backscattering, SIMS, and AES have recently been combined to examine the formation of Pt silicide films (82). The object was not only to detect low-level or light-mass impurities but also to evaluate the three techniques. In this example it was found that the concentration of impurities was sufficiently high so that AES was adequate in spite of its lower sensitivity as com-

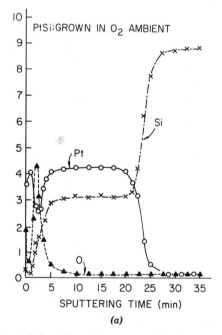

Figure 6.18.   (a) Pt, Si, and O Auger intensities vs. sputtering time from PtSi grown for 30 min in O_2 at 600°C. From reference 80.

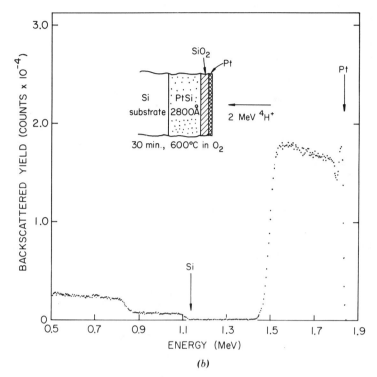

Figure 6.18.   (b) Backscattering energy spectrum for a sample of PtSi grown for 30 min in O_2 at 600°C. From reference 80.

pared to SIMS. Moreover, the ion sputtering yield as measured by SIMS fluctuated violently at the interface.

6.6   LATERAL INHOMOGENEITIES

The underlying assumption in the preceding discussions was that the thin film structures are laterally homogenous. However, in many cases this is an unwarranted assumption. Both sputter sectioning and backscattering data have to be treated with extreme caution if lateral inhomogeneities are expected. In backscattering measurements, the beam spot size is usually of the order of 1 mm, and the compositional information is averaged over a dimension considerably greater than the film thickness. Good progress, however, is being made in producing MeV ion beams with μm-size beam spots (83). In the sputtering techniques, although the analyzing probe diameter can be

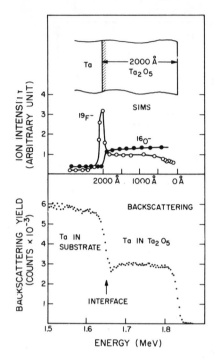

Figure 6.19. *Top*: SIMS depth profile of Ta_2O_5 layer containing fluorine at the interface. *Bottom*: Rutherford backscattering (2 MeV ^4He) from the same Ta_2O_5 layer. From reference 81.

very small, the craters that are eroded have diameters typically in excess of 100 μm, and mm sizes are more common (50, 73, 76).

The specimens must therefore be examined with an imaging technique that has a resolution of at least the film thickness. Scanning electron microscopy (SEM) is ideally suited, and the necessity of characterizing the sample with SEM has been demonstrated in numerous examples (see Chapter 5).

Figure 6.20 shows the backscattering spectra (84) of a 100 Å Cu film deposited on Cr before and after a high-temperature anneal. The simplest interpretation of such spectra is that the Cu and Cr have interdiffused. However, inspection of the SEM micrograph, insert in Figure 6.20, indicates that the Cu has balled up on the surface with a typical ball diameter of 1000 Å. It is the backscattering and penetration through these balls that give rise to the pseudodiffusion profiles of Cu in Cr.

6.7 INDIRECT TECHNIQUES

It would be misleading to suggest that particle energy loss or sputter sectioning techniques are the only ways of gaining information on interdiffusion of

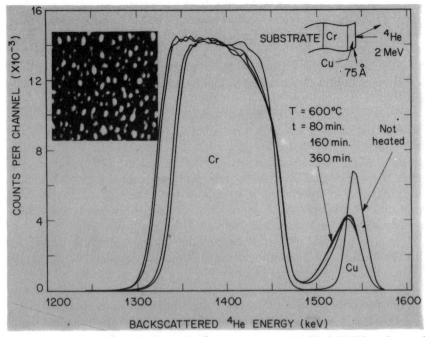

Figure 6.20. Backscattering spectra of 75 Å Cu film on Cr with SEM (26,000 ×). Insert of couple after treatment at 700°C for 2 h. After reference 84.

reactions between thin films. We have, however, concentrated on the techniques that directly give depth profiles in thin films.

Many properties of thin films depend sensitively upon the amount of interdiffusion or reaction. These have been studied by x-ray (1, 2, 85–88) and electron diffraction techniques (89, 90), changes in work functions (91), and optical reflectivity (92, 93) at surfaces as well as changes in resistivity (92). It is somewhat difficult to proceed from such observations to a determination of depth profiles in thin films. As these methods do not involve depth profiling *per se*, we shall not discuss them. The results from such measurements will, however, be discussed elsewhere in this book.

6.8 CONCLUSIONS

We have tried to present both the major strengths and weaknesses of the various methods that have been used to obtain depth profiles in thin film systems. It is fairly clear that experimental techniques are now available to determine composition and impurity concentration as a function of depth. However, there have not been satisfactory demonstrations that a detailed picture of interface regions can be obtained under routine evaluation conditions. Particle energy loss techniques tend to present washed-out profiles owing to the loss of depth resolution caused by either detector resolution or energy straggling. Sputtering techniques tend to alter the profile at the interface because of atomic intermixing or indicate broad interfaces because of artifacts in the erosion process. These problems are receiving increased attention as studies of thin film reactions focus on the properties of the interface region.

Another difficulty in the present profiling techniques is to distinguish between the various diffusion mechanisms. The lateral resolution of the techniques discussed here are not adequate, for example, to distinguish diffusion along individual grain boundaries in fine-grained polycrystalline films, but rather give an average over many grain boundaries. One must often make an interpretation of grain boundary effects based on the shape and the time–temperature dependence of the diffusion profile. Such approaches will be discussed in Chapters 7 and 9.

Numerous review articles have compared depth profiling techniques (94–98). Rather than repeating these treatments, we shall list the major features discussed in previous sections.

1. Particle Energy Loss, Primarily Rutherford Backscattering. Its power lies in the fact that quantitative numbers for the composition and depth of major constituents can be obtained without requirements for calibration standards. Moreover, the technique is nondestructive. Although it has good sensitivity ($\leqslant 0.1$ at. %) for heavy elements in light substrates, it is relatively insensitive to the light elements—oxygen, carbon, and nitrogen —that tend to be universal contaminants in thin film systems. One can overcome this handicap by use of nuclear reaction techniques, but there has not been extensive use of this approach. Instead, it is more common to find AES techniques used to provide the added information on the distribution of light elements. The particle energy loss techniques provide rapid analysis (typically 10 to 15 min per sample) but require an accelerator and the concomitant requirement of laboratory space.

2. Sputter Sectioning. Under optimum conditions, sputtering provides nearly layer-by-layer removal—an atomic microtome. The closest approach to ideal targets for controlled sectioning are amorphous layers or materials rendered amorphous by the sputtering process. In other cases, nonuniform erosion as well as preferential sputtering and atomic intermixing can occur. At present, the major problem is to obtain a clear picture of the processes involved in sputtering of multicomponent films.

    a. Surface Layer Analysis. Auger electron spectroscopy is the commonest of all the techniques used with sputter sectioning to provide quantitative analysis of surface layers. Its sensitivity is in the range of 0.1 at. %, comparable to that of Rutherford backscattering but not as sensitive as SIMS. Similar comments can be made in regard to ion scattering spectrometry (ISS). Perhaps the only differences between the two techniques, from a surface analysis point of view, is that ISS detects primarily the outermost layer and can be more difficult to calibrate. The major drawbacks to these techniques are associated with the sputtering process such as changes in the surface composition.

b. Sputtered-Species Detection. Conceptually, SIMS and SCANIIR are the ideal surface analytic tools since they measure the species that are removed and have high sensitivity. However, the charge exchange and excitation processes at the surface make these techniques difficult to quantify in terms of absolute values. Vacuum requirements are stringent. A common use of these techniques is to profile samples where high sensitivity is required (such as boron in silicon) and matrix effects have been determined from standards of a similar composition. Under these conditions, impurity concentrations can be determined to parts per million.

The natural approach is to use a combination of analytic techniques such as Rutherford backscattering plus AES or AES plus SIMS. As shown by examples cited in earlier sections, the use of complementary techniques provides information not obtainable by any one depth profiling technique.

In conclusion, we stress again the importance of employing some technique, whether optical microscope, scanning electron microscope, or electron microprobe, to examine the lateral characteristics of the thin film system. As a rule of thumb, one should employ techniques with a lateral resolution comparable to the thickness of the film. However, even visual inspection sometimes provides adequate warning about nonuniformities in the film structure.

REFERENCES

1. J. W. M. DuMond and J. P. Youtz, *Phys. Rev.*, **48**, 703 (1935).
2. J. W. M. DuMond and J. P. Youtz, *J. Appl. Phys.*, **11**, 357 (1940).
3. S. Rubin, in *Treatise on Analytical Chemistry*, I. M. Kolthoff and P. J. Elving, Eds., Interscience, New York (1959), p. 2075.
4. R. F. Sippel, *Phys. Rev.*, **115**, 1441 (1959).
5. J. W. Mayer, L. Eriksson, and J. A. Davies, *Ion Implantation in Semiconductors*, Academic Press, New York (1970).
6. G. Dearnaley, J. H. Freeman, R. S. Nelson, and J. Stephen, *Ion Implantation*, North Holland, Amsterdam (1973).
7. D. V. Morgan, Ed., *Channeling, Theory, Observation and Applications*, Wiley, London (1973).
8. D. E. Gemmell, *Rev. Mod. Phys.*, **46**, 129 (1974).
9. G. Carter and J. S. Colligon, *Ion Bombardment of Solids*, Heinemann, London (1968).
10. G. K. Wehner, in *Methods of Surface Analysis*, A. W. Czanderna, Ed., Elsevier, Amsterdam (1975).
11. J. W. Mayer and E. Rimini, eds., *Ion Beam Handbook for Material Analysis*, Academic Press, New York (1977).

12. W. K. Chu, J. W. Mayer, M-A. Nicolet, *Backscattering Spectrometry*, Academic Press, New York (1977).

13. J. E. Baglin and W. N. Hammer, in *Ion Beam Surface Layer Analysis*, O. Meyer, G. Linker, and F. Käppeler, Eds., Plenum Press, New York (1976), p. 447.

14. W. C. Turkenburg, W. Soszka, F. W. Saris, H. H. Kersten, and B. G. Colenbrander, *Nucl. Instr. Methods*, **132**, 587 (1976).

15. G. Dearnaley and D. C. Northrop, *Semiconductor Counters for Nuclear Radiations*, E. and F. N. Spon, London (1966).

16. R. R. Hart, H. L. Dunlap, A. J. Mohr, and O. J. Marsh, *Thin Solid Films*, **19**, 137 (1973).

17. K. N. Tu, W. K. Chu, and J. W. Mayer, *Thin Solid Films*, **25**, 403 (1975).

18. F. H. Eisen, G. J. Clark, J. Bøttiger, and J. M. Poate, *Radiation Effects*, **15**, 31 (1972).

19. L. C. Northcliffe and R. F. Schilling, *Nuclear Data Tables*, **A7**, 233 (1970).

20. J. F. Ziegler and W. K. Chu, *Atomic Data and Nuclear Data Tables*, **13**, 463 (1974).

21. W. K. Chu, in *Ion Beam Handbook for Material Analysis*, J. W. Mayer and E. Rimini, Eds., Academic Press, New York (1977), Chapter 1.

22. N. Bohr, *Mat. Fys. Medd. Dan Vid. Selsk.*, **18**, 8 (1948).

23. P. V. Vavilov, *Zh. Exper. Teor. Fiz.*, **32**, 320 (1957); Transl. *JETP.*, **5**, 749 (1957).

24. W. H. Bragg and R. Kleeman, *Phil. Mag.*, **10**, S318 (1905).

25. J. M. Poate, P. A. Turner, W. J. DeBonte, and J. Yahalom, *J. Appl. Phys.*, **46**, 4275 (1975).

26. D. K. Brice, *Thin Solid Films*, **19**, 121 (1973).

27. J. F. Ziegler, R. F. Lever, and J. K. Hirvonen, in *Ion Beam Surface Layer Analysis*, O. Meyer, G. Linker, and F. Käppeler, Eds., Plenum Press, New York (1976), p. 163.

28. J. S. Williams, *ibid.*, p. 223.

29. J. A. Cairns and L. C. Feldman, in *New Uses of Ion Accelerators*, J. F. Ziegler, Ed., Plenum Press, New York (1975), p. 431.

30. L. C. Feldman, J. M. Poate, F. Ermanis, and B. Schwartz, *Thin Solid Films*, **19**, 81 (1973).

31. J. D. Cockcroft and E. T. S. Walton, *Proc. Roy. Soc.* (London), **A129**, 477 (1930).

32. G. Amsel, J. P. Nadai, E. D'Artemare, D. David, E. Girard, and J. Moulin, *Nucl. Instr. Methods*, **92**, 481 (1971).

33. E. A. Wolicki, in *New Uses of Ion Accelerators*, J. F. Ziegler, Ed., Plenum Press, New York 1975), p. 159.

34. J. Bøttiger, S. T. Picraux, and N. Rud, in *Ion Beam Surface Layer Analysis*, O. Meyer, G. Linker, and F. Kappeler, Eds., Plenum Press, New York (1976), p. 811.

35. D. D. Leich and T. A. Tombrello, *Nucl. Instr. Methods*, **108**, 67 (1973).

36. R. A. Langley, in *Radiation Effects and Tritium Technology for Fusion Reactors*, Vol. IV, (1976), p. 158. US Dept. of Commerce, Publication No. CONF-750989.

37. K. L. Dunning, G. K. Hubler, J. Gomas, W. H. Lucke, and H. L. Hughes, *Thin Solid Films*, **19**, 145 (1973).

38. H. H. Andersen and H. L. Bay, *J. Appl. Phys.*, **46**, 1919 (1975).

39. A. L. Southern, W. R. Willis, and M. T. Robinson, *J. Appl. Phys.*, **34**, 153 (1963).

40. E. P. EerNisse, *J. Appl. Phys.*, **42**, 480 (1971).

41. P. Sigmund, *Phys. Rev.*, **184**, 383 (1969).

42. H. H. Andersen and H. L. Bay, *Radiat. Eff.*, **19**, 139 (1973).

43. O. Almen and G. Bruce, *Nucl. Instr. Methods*, **11**, 279 (1961).

44. H. H. Andersen and H. L. Bay, *Radiat. Eff.*, **13**, 67 (1972).

45. J. M. Poate, W. L. Brown, R. Homer, W. M. Augustyniak, J. W. Mayer, K. N. Tu, and W. F. van der Weg, *Nucl. Instr. Methods*, **132**, 345 (1976).

46. G. K. Wehner and G. S. Andersen, in *Handbook of Thin Film Technology*, L. I. Maissel and R. Glang, Eds., McGraw-Hill, New York (1970), Chapter 3.

47. H. H. Andersen, *Proc. 7 Yugoslav Symposium on Physics of Ionized Gases*, V. Vujnovic, Ed. (1975), published by Institute of Physics, University of Zagreb, Yugoslavia.

48. G. McCracken, *Rep. Progr. Phys.*, **38**, 241 (1975).
49. P. D. Townsend, J. C. Kelly, and N. E. W. Hartley, *Ion Implantation, Sputtering and Their Application*, Academic Press, London (1976).
50. H. Liebl, *J. Vac. Sci. Technol.*, **12**, 385 (1975).
51. W. F. van der Weg, D. Sigurd, and J. W. Mayer, in *Applications of Ion Beams to Metals*, S. T. Picraux, E. P. EerNisse, and F. L. Vook Eds., Plenum Press, New York (1974), p. 209.
52. J. M. Poate and T. C. Tisone, *Appl. Phys. Lett.*, **24**, 391 (1974).
53. W. O. Hofer and H. Liebl, *Appl. Phys.*, **8**, 359 (1975).
54. W. T. Ogar, N. T. Olson, and H. P. Smith, *J. Appl. Phys.*, **40**, 4997 (1969).
55. P. K. Haff and Z. E. Switkowski, *Appl. Phys. Lett.*, **29**, 549 (1976).
56. J. W. Coburn, *J. Vac. Sci. Technol.*, **13**, 1037 (1976).
57. R. Kelly and J. B. Sanders, *Nucl. Instr. Methods*, **132**, 335 (1976).
58. D. K. Murti and R. Kelly, *Thin Solid Films*, **33**, 149 (1976).
59. E. Gillam, *J. Phys. Chem. Sol.*, **11**, 55 (1959).
60. P. F. Kane and G. B. Larrabee, Eds., *Characterization of Solid Surfaces*, Plenum Press, New York (1974).
61. A. W. Czanderna, Ed., *Methods of Surface Analysis*, Elsevier, Amsterdam (1975).
62. R. B. Anderson and P. T. Dawson, Eds., *Characterization of Surfaces and Adsorbed Species*, Academic Press, New York (1976).
63. J. W. Mayer and J. F. Ziegler, Eds., Conf. on Ion Beam surface Layer Analysis, *Thin Solid Films*, **19**, (1973).
64. 22nd National Symposium of the American Vacuum Society, Oct. 1975, Philadelphia, *J. Vac. Sci. Techol.*, **13** (No. 1), (1976).
65. D. P. Smith, *J. Appl. Phys.*, **38**, 340 (1967).
66. T. M. Buck, Y.-S. Chen, G. H. Wheatley, and W. F. van der Weg, *Surface Sci.*, **47**, 244 (1975).
67. J. M. McDavid and S. C. Fain, Jr., *Surface Sci.*, **52**, 161 (1975).
68. R. Ludeke, L. Esaki, and L. L. Chang, *Appl. Phys. Lett.*, **24**, 417 (1974).
69. J. W. Mayer, J. F. Ziegler, L. L. Chang, R. Tsu, and L. Esaki, *J. Appl. Phys.*, **44**, 2322 (1973).
70. R. Dingle, in *Festkörperprobleme* (*Advances in Solid State Physics*), Vol. 15, H. J. Queisser, Ed. Pergamon/Vieweg, Braunschweig (1975), p. 21.
71. W. L. Harrington, R. E. Honig, A. M. Goodman, and R. Williams, *Appl. Phys. Lett.*, **27**, 644 (1975).
72. T. W. Sigmon, W. K. Chu, E. Lugujjo, and J. W. Mayer, *Appl. Phys. Lett.*, **24**, 105 (1974).
73. A. Joshi, L. E. Davis, and P. W. Palmberg, in *Methods of Surface Analysis*, A. W. Czanderna, Ed., Elsevier, Amsterdam (1975), p. 218.
74. E. Taglauer and W. Heiland, *Appl. Phys.*, **9**, 261 (1976).
75. C. W. White, D. L. Simms, and N. H. Tolk, in *Characterization of Solid Surfaces*, P. F. Kane and G. B. Larrabee, Eds., Plenum Press, New York (1974), p. 641.
76. J. A. McHugh, in *Methods of Surface Analysis*, A. W. Czanderna, Ed., Elsevier, Amsterdam (1975), p. 273.
77. S. Komiya, T. Narusawa, and T. Satake, *J. Vac. Sci. Technol.*, **12**, 361 (1975).
78. P. M. Hall, J. M. Morabito, and J. M. Poate, *Thin Solid Films*, **33**, 107 (1976).
79. K. Nakamura, M.-A. Nicolet, J. W. Mayer, R. J. Blattner, and C. A. Evans, Jr., *J. Appl. Phys.*, **11**, 4678 (1975).
80. R. J. Blattner, C. A. Evans, S. S. Lau, J. W. Mayer, and B. M. Ullrich, *J. Electrochem. Soc.*, **122**, 1733 (1975).
81. W. K. Chu, M.-A. Nicolet, J. W. Mayer, and C. A. Evans, Jr., *Anal. Chem.*, **46**, 2137 (1974).
82. J. B. Bindell, J. W. Colby, D. R. Wonsidler, J. M. Poate, D. K. Conley, and T. C. Tisone, *Thin Solid Films*, **37**, 441 (1976).
83. J. A. Cookson and F. D. Pilling, *Thin Solid Films*, **19**, 381 (1973).

84. J. E. E. Baglin and F. M. d'Heurle, in *Ion Beam Surface Layer Analysis*, O. Meyer, G. Linker, and F. Käppeler, Eds., Plenum Press, New York (1976), p. 385.
85. H. E. Cook and J. E. Hillard, *J. Appl. Phys.*, **40**, 2191 (1969).
86. K. N. Tu and B. S. Berry, *J. Appl. Phys.*, **43**, 3283 (1972).
87. C. R. Houska, *Thin Solid Films*, **25**, 451 (1975).
88. M. Murakami, D. DeFontaine, and J. Fodor, *J. Appl. Phys.*, **47**, 2850 (1976).
89. T. C. Tisone and J. Drobeck, *J. Vac. Sci. Technol.*, **9**, 271 (1971).
90. E. M. Horl and K. H. Rieder, *J. Vac. Sci. Technol.*, **9**, 276 (1971).
91. R. E. Thomas and G. A. Haas, *J. Appl. Phys.*, **43**, 4900 (1972).
92. C. Weaver, in *Physics of Thin Films*, Vol. 6, (1971), p. 301.
93. W. B. Nowak and R. N. Dyer, *J. Vac. Sci. Technol.*, **9**, 279 (1972).
94. J. W. Mayer and A. Turos, *Thin Solid Films*, **19**, 1 (1973).
95. R. E. Honig, in *Advances in Mass Spectrometry*, Vol. 6, A. R. West, Ed., Applied Science Publishers, London, (1974), p. 337.
96. J. W. Coburn and E. Kay, *CRC Crit. Rev. Solid State Sci.*, 562 (1974).
97. H. W. Werner, in *Science of Ceramics*, British Ceramic Society, London (1976), p. 55.
98. C. A. Evans, Jr., *J. Vac. Sci. Technol.*, **12**, 144 (1975).

7

GRAIN BOUNDARY DIFFUSION

D. Gupta, D. R. Campbell, and P. S. Ho

IBM Thomas J. Watson Research Center, Yorktown Heights, New York

7.1 INTRODUCTION

It is now well established that diffusion along grain boundaries (GB) in polycrystalline materials is orders of magnitude more rapid than bulk lattice diffusion. The phenomenon of GB diffusion was first formally treated by Fisher (1) and Hoffman and Turnbull (2) in 1951 although references can be found in the literature (3) as far back as 1936. Through the years carefully designed GB diffusion experiments have advanced our understanding of the structure and chemistry of the grain boundaries. The subject has continued to draw the attention of the scientist as evidenced by a number of review articles, notably those by Turnbull (4), Shewmon (5), Gjostein (6, 7), Gleiter and Chalmer (8), and Martin and Perraillon (9). In the past decade the phenomenon has assumed increased significance because of the importance of thin film technology for fabrication of planar microelectronic devices where a quantitative understanding of the mass transport at low temperature is vital. In addition, a number of recent conferences have been devoted to low-temperature diffusion and applications to thin films (10, 11).

While writing this chapter, we have focused our attention on the grain boundary diffusion in thin films rather than on reviewing the large body of information in the bulk polycrystalline materials afresh. It is intended to serve as a comprehensive review of thin film diffusion experiments and interpretation of the results. The mathematical analyses of the GB diffusion experiments are generally more complicated than those of diffusion in the lattice owing to the multiplicity of the diffusion paths. As GB diffusion can seldom be decoupled from the lattice, the diffusing species is likely to leak into the adjoining lattice. The extent of this leakage determines the types of kinetics referred to by Harrison (12) as A, B, and C. These three types of kinetics define the temperature regimes where various experimental techniques are applicable and also the mathematical analysis that would be appropriate for extraction of the diffusion coefficients. Commonly used boundary conditions such as the semi-infinite depth of the specimen compared to the diffusion distance in the GB, and large enough grain size are often inapplicable to thin films for obvious reasons. The film thickness and the grain size in some cases are comparable to the effective diffusion distance. The state of the source of the diffusing species—depletable or constant—is another consideration to be taken into account for development of a rigorous mathematical analysis. In the particular case of the surface accumulation method, the state of diffusant becomes important on the entrance as well as the exit surfaces. We have attempted to cover these aspects while reviewing the mathematical analytic models.

A principal direction the measuring techniques for thin film diffusion have taken is the use of ion backsputtering for in-depth profiling of the diffusant utilizing either radioactive tracers or surface analytic techniques that have recently become available. Another direction taken is the measurement of surface accumulation of the diffusant on the exit surface of the specimen using primarily surface analytic techniques and in a few instances radioactive tracer techniques. The surface analytic techniques are typified by Auger electron spectroscopy (AES). These techniques have generated much of the diffusion data available in thin films. In addition, we discuss some attempts to measure diffusion laterally in the film plane using radioactive tracer scanning, auto-radiography, or electron microprobe analysis. Relatively speaking, lateral diffusion measurements appear to be in their early stages of development, and available data are rather sparse.

Some critical data in thin films and in the bulk polycrystalline materials have been discussed in the context of the present understanding of the GB diffusion phenomenon. However, we have not covered the structure depend-ence of GB diffusion owing to review articles already available (9) on the bulk studies and relative lack of pertinent data in thin films. The effect of solutes on the GB diffusion, particularly for the solvent species, has a special importance in thin films as it has been found beneficial in retarding GB diffusion in many situations. A thermodynamic model is presented to explain some of the observations of the solute effect on the GB diffusion.

7.2 REVIEW OF ANALYSES

7.2.1 Grain Boundary Diffusion Kinetics

In the most literal sense, the term grain boundary diffusion refers to atomic movements occurring in the interfacial region separating two grains. For diffusion purposes, many investigators regard the region of enhanced dif-fusivity as essentially two-dimensional with a thickness comparable to the atomic spacing. However, in a less restrictive sense, the term grain boundary diffusion implies a process in which both boundary and lattice diffusion occur. In fact, some lattice diffusion is essential to the sort of profiling experi-ments in which grain boundary diffusion is measured. This is because the amount of diffusing species which reside within the grain boundary regions is ordinarily too small to be detected, and it is the concentrations created by material diffusing out from the boundary and into the interior of adjoining grains that is measured.

The extent to which lattice diffusion influences the material transport along boundaries determines the kinetic regime which prevails. With refer-ence to Figure 7.1 and similar to Harrison (12), we designate three types of kinetics, called A, B, and C. The distinguishing feature of A-kinetics is the extensive lattice diffusion that causes the diffusion fields from adjoining grains to overlap. The boundaries are shown here as parallel slabs with spacing $2L$. In the course of the mathematical analysis, the lateral diffusion flux J_x is required to be zero at the midpoint between grains, $x = L$. This

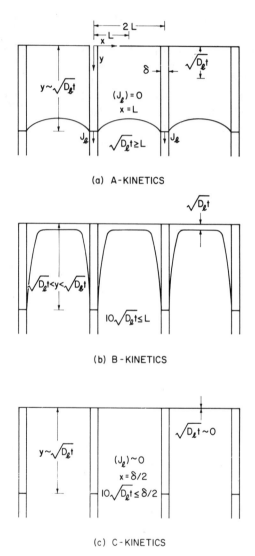

Figure 7.1. Schematic representation of A-, B-, and C-kinetics. Vertical lines indicate grain boundaries, and curved lines are isoconcentration contours. The diffusion source coincides with the top horizontal lines.

situation is distinctly different from that occurring in B-kinetics, where each boundary is assumed to be isolated and the flux at large distances in the x-direction approaches zero. In C-kinetics, lattice diffusion is considered negligible and significant atomic transport occurs only within the boundaries. Grain boundary diffusion measurements in bulk materials invariably involve B-kinetics. Little evidence has been obtained from profiling experiments on bulk materials that clearly supports the existence of A- or C-kinetics. In this respect, thin films are both unique and challenging as candidates for diffusion studies because the very high density of structural defects makes it possible to observe any of the three types of kinetics given the appropriate annealing conditions (13–15).

7.2.2   Analysis for Bulk Samples

Quantitative studies of grain boundary diffusion historically began with bulk samples and utilized either serial sectioning techniques or autoradiography. The kinetic regime was B-type, as only isolated boundaries had to be considered. The first analysis of combined lattice and grain boundary (GB) diffusion for the geometry shown in Figure 7.2 was performed by

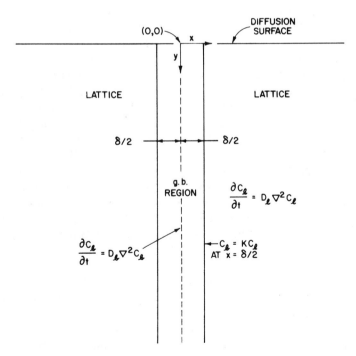

Figure 7.2.   Geometric model for grain boundary diffusion in the B-kinetics regime.

Fisher (1). He solved the coupled lattice and GB diffusion equations in an approximate fashion by making some simplifying assumptions. From the analytic point of view, the most serious approximation was to assume that the concentration C_b within the boundary changes so slowly with time t that the term $\partial C_b/\partial t$ could be set equal to zero. This greatly simplified the method of solution, and, in particular, transform techniques were not needed. Subsequently, Whipple (16) solved the problem exactly by a method of Fourier-Laplace transforms. The usual difficulty in performing the complex inversion integral was overcome by an innovative series of transformations resulting in a final solution for concentration as a function of distance and time, $C(x,t)$, expressed as a real integral. The integral can be evaluated numerically to high precision. Whipple assumed the diffusion surface was in contact with an infinite source. Suzuoka (17) later solved the case for instantaneous source conditions utilizing Whipple's transformations to evaluate the complex inversion integral.

Although in principle the Whipple and Suzuoka analysis provided exact continuum model solutions, in practice many investigators continued to use Fisher's analysis because it was simpler to apply. Later, interpretive papers by several authors including LeClaire (18), Cannon and Stark (19), and Suzuoka (20) clearly established the superiority of the Whipple (16) and Suzuoka (17) solutions and also demonstrated the use of simple and accurate techniques for extracting the GB diffusion coefficient from experimental data. Cannon and Stark (19) presented a graphic method based on infinite source conditions (Whipple's solution) in which the fractional change in concentration from the surface to some penetration depth could be related to the GB diffusion coefficient through a set of normalized curves. LeClaire (18) and Suzuoka (20) gave formulas that related the slope of the penetration profile to the GB diffusion coefficient for infinite and instantaneous source conditions, respectively. In any of these techniques, the GB diffusivity cannot be resolved separately but is determined as the product δD_b, where δ is the boundary width and D_b is the GB diffusion coefficient. All of the solutions discussed so far were originally cast for self-diffusion, and therefore impurity effects were not taken into account. Gibbs (21) incorporated the solute segregation factor K into the analysis and showed that profile measurements would yield values of $K\delta D_b$ for impurity diffusion instead of δD_b as in the self-diffusion case.

As of this writing, the analysis of penetration profiles is based almost exclusively on continuum solutions to the coupled lattice and GB diffusion equations. A significant development involving solutions based on an atomic model that takes into account discrete jumps in both the GB and the lattice has been reported by Benoist and Martin (22). Contrary to the continuum case, their model does not assume that equilibrium is instantly established between the distribution of diffusing species in the boundary and in the interior of the grain adjacent to the boundary. As a consequence, penetration profiles calculated using the atomic model show small but noticeable differences with continuum profiles at large penetration depths in the B-kinetics regime. These differences become more significant for small diffusion lengths in the lattice, particularly as the regime of C-kinetics is approached (23). The significance of the atomic model lies not so much in its ability to predict deviations from the continuum model but in the insight it affords into the role of the atomic jumps or exchanges between the grain boundary sites and sites in the adjoining lattice. It has already been demonstrated that the δ appearing in the continuum model has no conceptual counterpart in the atomistic model. In the context of a simple bicrystal model, it is replaced by the jump distance in the lattice. In a more generalized version (23) of the model, the effects of the GB structural details can be incorporated into the solutions and the resulting influence on measurable diffusion parameters can be displayed.

7.2.3 Analysis for Thin Films

The first attempt to address the complex problem of thin film diffusion in a realistic fashion was reported by Unnam, Carpenter, and Houska (24). Using a finite difference approach, these authors modeled diffusion from a single crystal substrate into a columnar grained film. Their model took account of separate diffusivities in the lattice and grain boundaries and included a concentration-dependent factor in both. The construction of the model also made it possible to include the effects of finite film thickness, closely spaced boundaries, and top surface diffusion. A numerical analysis was performed using a finite difference technique to calculate composition as a function of position and time. The composition profiles were used to calculate x-ray line intensities, and diffusion and structural parameters could be determined by adjusting their values until the calculations agreed with experiment.

Analytical solutions for thin film diffusion are necessarily more restrictive than numerical evaluations. For example, the concentration dependence of diffusivity is not incorporated, and only a few idealized surface boundary conditions can be considered. Such solutions are still quite useful since they make it possible to directly assess the influence of the film thickness, grain size, and surface boundary conditions on the penetration profile. Gilmer and Farrell (25, 26) have reported analytical solutions for coupled lattice and grain boundary diffusion that included finite thickness effects. Solutions were provided for both isolated boundaries (25) and an array of parallel boundaries (26). Surface diffusion along the backside of the film was also included using numerical techniques (25). For profile evaluation, an important result of this work was the strong curvature evident in plots of concentration versus (depth)$^{6/5}$. These same plots are quite linear over a wide range of conditions for the infinitely thick case. For certain conditions the authors demonstrated that estimates of diffusivity that involved taking the slope of finite-thickness profiles could easily be orders of magnitude too large and suggested two alternative means of estimating diffusivity. One involved measuring the curvature, that is, the second derivative of profile at the backside of the films, and the other involved measuring the level of the concentration in the GB portion of the profile as a function of time and temperature. The backside of the film has to be a diffusion barrier to apply either of the above techniques.

Campbell (27) has also analyzed the problem of parallel grain boundaries using Fourier-Laplace transforms instead of the Fourier analysis approach taken by Gilmer and Farrell (26). The form of this solution makes it useful for interpreting the behavior of profiles as a function of grain boundary spacing. In particular, the solution predicts that as grain boundary spacing is reduced, a point will be reached at which the profile will no longer possess a composite structure, that is, with distinct lattice and boundary portions, and will correspond to a one-dimensional planar solution governed only by D_b. This result provides additional insight into the significance of A-kinetics for both bulk and thin film samples and also lends itself to simple methods for determining D_b directly, as will be discussed in a subsequent section.

A relatively new and promising technique for measuring GB diffusion is permeation through a thin film combined with Auger electron spectroscopy to detect the accumulation of material on the back or exit side surface. The first permeation analysis that took into account the effects of coupled diffusion fluxes along structural defects and in the lattice was performed by Wuttig and Birnbaum (28) for dislocations in a thin foil. Because of the analytic complexity inherent in the cylindrical symmetry and the B-kinetics formalism, the flux leakage away from the dislocation and into the lattice was treated in an approximate fashion. This allowed relatively simple analytic solutions to be devised for the finite-thickness geometry corresponding to several idealized source and sink boundary conditions. While the approximate nature of the solutions did not appear to restrict their usefulness in Wuttig and Birnbaum's application, Holloway, Amos, and Nelson (29) have cited difficulties in obtaining agreement with experiment for grain boundary permeation and favor a completely numerical approach. Perhaps this is the most practical way to handle the problem because numerical methods can be flexible with regard to surface and sink boundary conditions. For example, Holloway, Amos, and Nelson were able to include saturating sink conditions, namely,

$$\left(\frac{\partial C}{\partial y}\right)_{y=l} = k(C' - C) \tag{1}$$

where l is the film thickness, k is a constant, and C' is the concentration corresponding to a saturated surface condition. As a point in contrast, a purely analytical solution would be difficult to derive with this boundary condition.

If diffusion lengths in the lattice are kept sufficiently low, the lattice diffusion processes can be neglected altogether. Hwang and Balluffi (30) have adopted this simplification in their analysis of permeation through a thin film with subsequent surface spreading. In this case the kinetic regime is of the C-type. The authors have considered both infinite and depletable source conditions on the diffusion surface. In their treatment the migration of diffusant over the exit surface, also over the source surface for depletable source conditions, is included without approximation.

7.3 ANALYTICAL MODELS

The discussion of analytical models will be divided into four major parts: (i) isolated boundaries corresponding to B-kinetics, (ii) arrays of parallel boundaries with interacting diffusion fields corresponding to A-kinetics, (iii) GB diffusion in the limit of vanishing diffusion length in the lattice corresponding to C-kinetics, and (iv) GB diffusion under a driving force. The first three cases will be further divided according to whether the solutions assume infinite or finite sample thicknesses. Solutions for infinite thickness are suitable for diffusion in bulk samples and for diffusion within the plane of a thin film. Solution for finite thickness are suitable for through-the-thickness diffusion in thin films. In any of the analyses considered here, the lattice and GB diffusion coefficients are assumed concentration independent, and, strictly speaking, the results can only apply to dilute impurity or radiotracer self-diffusion. This restriction has a certain advantage for the interpretation of the mathematical results since the distinctly different effects due to infinite-versus-finite thickness assumptions or isolated-versus-closely spaced boundaries assumptions can be clearly resolved in the context of concentration-independent diffusivities. Concentration-dependent diffusivities have been included in treatments which are more numerical as discussed in Section 7.2.3. Also, it is implicitly assumed throughout that the boundaries are stationary, *i.e.*, no grain growth occurs during the diffusion.

7.3.1 Isolated Boundaries

7.3.1.1 Infinite Thickness. An exact solution to the problem of coupled lattice and grain boundary diffusion treated as a continuum was accomplished by Whipple (16) using a method of Fourier-Laplace transforms. The geometry is indicated in Figure 7.2 where diffusion is occurring from a planar source into a material having a grain boundary perpendicular to the diffusion surface. There are two regions to consider in constructing the solution: (a) the region exterior to the boundary where the lattice diffusivity D_l prevails;

(b) the region within the boundary characterized by width δ and GB diffusivity D_b. The diffusion equations may be written for the region exterior to the boundary as

$$D_l \nabla^2 C_l = \frac{\partial C_l}{\partial t} \tag{2}$$

and inside the slab,

$$D_b \nabla^2 C_b = \frac{\partial C_b}{\partial t} \tag{3}$$

where C_l and C_b indicate lattice grain boundary concentrations, respectively. At the interface between the boundary and the lattice, $x = \pm \delta/2$, and we have

$$C_b = K C_l \tag{4}$$

$$D_b \frac{\partial C_b}{\partial x} = D_l \frac{\partial C_l}{\partial x} \tag{5}$$

The boundary condition expressed by Eq. 4 states simply that the concentrations distributed between the boundary and the lattice are always in equilibrium at the surface that separates the two regions. The parameter K is the solute segregation factor (21) and is included here to take account of impurity effects. The flux leaving perpendicular to the boundary has to equal that entering the interior of an adjoining grain, and this condition is indicated by Eq. 5.

In Whipple's (16) treatment, the concentration at the surface is assumed to stay constant throughout the duration of the diffusion:

$$C(x, 0, t) = C_0 \qquad t \geqslant 0 \tag{6a}$$

$$= 0 \qquad t < 0 \tag{6b}$$

Suzuoka (17) considered the instantaneous source condition:

$$C(x, y, t) = M \, \delta(y) \qquad x = 0, t = 0 \tag{7}$$

and

$$\frac{\partial C}{\partial y} = 0 \qquad y = 0, t > 0 \tag{8}$$

where M is the surface density of the planar source at $t = 0$ and $\delta(y)$ indicates the Dirac delta function. Whipple approximated the concentration within the boundary by the even function

$$C_b(x, y, t) = C_b'(y, t) + [x^2/2 \, C_b''(y, t)] \tag{9}$$

Combining Eq. 9 with Eqs. 3, 4, and 5 and ignoring terms of the order of δ^2

allows one to write

$$D_b \frac{\partial^2 C_l}{\partial y^2} - \frac{2D_l}{K\delta} \frac{\partial C_l}{\partial x} = \left(\frac{D_b}{D_l} - 1\right) \frac{\partial C_l}{\partial t} \qquad (10)$$

which gives a boundary condition for the solution exterior to the grain boundary at $x = \delta/2$.

Whipple's (16) exact solution for C_l is valid for the region exterior to the boundary $|x| > \delta$ and for infinite source conditions, Eq. 6. The particular and homogeneous solutions are designated by subscripts 1 and 2, respectively, where the total concentration C_l is given by

$$C_l = C_1 + C_2 \qquad (11)$$

Here, C_1 is the lattice concentration which is due to the ordinary planar diffusion in the lattice, that is, $C_1 = C_0 \, \text{erfc} \, (\eta/2)$, and C_2 is the lattice concentration which arises because of diffusion out from grain boundaries:

$$C_2 = \frac{\eta C_0}{2\sqrt{\pi}} \int_1^\Lambda \frac{d\sigma}{\sigma^{3/2}} \exp\left(-\eta^2/4\sigma\right) \times \text{erfc}\left[\frac{1}{2}\sqrt{\frac{\Lambda-1}{\Lambda-\sigma}}\left(\varepsilon + \frac{\sigma-1}{K\beta}\right)\right] \qquad (12)$$

The solution is written in terms of reduced variables which are defined as follows:

$$\eta = \frac{y}{\sqrt{D_l t}}, \quad \varepsilon = \frac{x-(\delta/2)}{\sqrt{D_l t}}, \quad \alpha = \frac{\delta}{2\sqrt{D_l t}}, \quad \Lambda = \frac{D_b}{D_l}, \quad \beta = (\Lambda - 1)\alpha \qquad (13)$$

For instantaneous source conditions, the solution provided by Suzuoka (17) gives the lattice contribution as $C_1 = M/\sqrt{\pi D_l t} \, \exp(-\eta^2/4)$ and the boundary portion as

$$C_2 = \frac{M}{4\sqrt{\pi D_l t}} \int_1^\Lambda \left(\frac{\eta^2}{\sigma} - 2\right) \exp\left(-\eta^2/4\sigma\right) \times \text{erfc}\left[\frac{1}{2}\sqrt{\frac{\Lambda-1}{\Lambda-\sigma}}\left(\varepsilon + \frac{\sigma-1}{K\beta}\right)\right] \frac{d\sigma}{\sigma^{3/2}}$$

$$(14)$$

Profiling experiments generally measure a spatial average of C_l symbolized by $\overline{C}_l$. The average concentration is given by

$$\overline{C}_l = C_1 + \int_{-\infty}^{+\infty} C_2 \, dx = C_1 + \overline{C}_2 \qquad (15)$$

Suzuoka gives these as follows:

Infinite Source:

$$\overline{C}_2 = \frac{2\eta C_0}{\sqrt{\pi D_l t}} \int_1 \exp\left(-\eta^2/4\sigma\right) \sqrt{\frac{\Lambda-\sigma}{\Lambda-1}} \times [\exp\left(-X^2\right)/\sqrt{\pi} - X \, \text{erfc} \, X] \frac{d\sigma}{\sigma^{3/2}}$$

$$(16)$$

Instantaneous Source:

$$\bar{C}_2 = \frac{M}{\sqrt{\pi}} \int_1^\Delta \left(\frac{\eta^2}{\sigma} - 2\right) \exp\left(-\eta^2/4\sigma\right) \sqrt{\frac{\Delta - \sigma}{\Delta - 1}}$$

$$\times \left[\exp\left(-X^2\right)/\sqrt{\pi} - X \text{ erfc } X\right] \frac{d\sigma}{\sigma^{3/2}} \tag{17}$$

The quality X is defined by

$$X = \sqrt{\frac{\Delta - 1}{\Delta - \sigma}} \frac{\sigma - 1}{2K\beta} \tag{18}$$

The integrals for $\bar{C}_2$ represent the average concentration arising from a unit length of grain boundary. For a polycrystalline specimen, Suzuoka (17) shows that the profile is given by

$$C_l = C_1 + \frac{1}{d}\bar{C}_2 \tag{19}$$

where d represents the spacing between parallel grains. In other geometries, $2d$ can represent either one side of a columnar grain which has a square cross section with the diffusion surface or the average diameter assuming a circular cross section. Levine and McCallum (31) considered more complex geometries. Equation 19 is based upon the assumption that the diffusion fields from adjoining grain boundaries are not overlapping.

The form in which the integrals C_2 are given makes them suitable for numerical evaluation. Although numerous authors have calculated them or some approximated version, this approach is not ordinarily part of the procedure used to extract δD_b from a penetration profile. This is particularly so for diffusion experiments carried out using bulk samples, where methods involving measuring the slope of the penetration profile can be used to advantage. In thin film specimens, direct integration may have more utility since the grain size and film thickness are built into the analysis and these parameters have considerable influence on the profile.

According to LeClaire (18), the boundary diffusivity can be evaluated from the expression

$$KD_b\delta = \left(\frac{\partial \ln \bar{C}_2}{\partial y^{6/5}}\right)\left(\frac{4D_l}{t}\right)^{1/2}\left(\frac{\partial \ln \bar{C}_2}{\partial[\eta(K\beta)^{-1/2}]^{6/5}}\right)^{5/3} \tag{20}$$

which is an exact expression derived using the definitions of β and η, Eq. 13. This form takes advantage of the findings of Levine and MacCallum (31) who first demonstrated by numerical calculation that the quantity

$$\frac{\partial \ln \bar{C}_2}{\partial[\eta(K\beta)^{-1/2}]^{6/5}} \tag{21}$$

is very nearly independent of $\eta(K\beta)^{-1/2}$. This situation prevails over the range of most GB diffusion experiments in bulk samples, namely, $2 \leqslant \eta(K\beta)^{-1/2} \leqslant 10$. The value of the expression 21 above is not significantly altered by the source conditions. For an infinite source,

$$\frac{\partial \ln \overline{C}_2}{\partial [\eta(K\beta)^{-1/2}]^{6/5}} = 0.78 \qquad \text{for } K\beta \gtrsim 10 \tag{22}$$

and for instantaneous source conditions,

$$\frac{\partial \ln \overline{C}_2}{\partial [\eta(K\beta)^{-1/2}]^{6/5}} = 0.72(K\beta)^{0.008} \tag{23}$$

according to LeClaire (18) and Suzuoka (20), respectively. Equations 22 and 23 differ only by a few percent of each other. Their near equivalence indicates that the slopes of the penetration profiles are not particularly sensitive to the initial boundary conditions, a distinct advantage for this method of determining GB diffusivity. According to Eq. 20, the profile must be linear if the values D_l and D_b are independent of composition or position, and experiments generally confirm this.

7.3.1.2 Finite Thickness.

It has long been recognized that diffusion kinetics in thin films are more rapid than in bulk. In principle, the accelerated kinetics occurring in the films has two obvious causes: (i) shorter diffusion distances and (ii) higher densities of short-circuiting paths such as grain boundaries and dislocations. In practice, it can be difficult to distinguish the separate contributions of these two influences without the aid of extensive mathematical analysis. In this section we present analytical solutions for diffusion in thin films where the boundaries are assumed isolated. This is B-kinetics as discussed in Section 7.2.1, and the particular solutions to be given describe the diffusion of dilute impurities into columnar grained thin films.

Solutions for finite thickness have been derived by Gilmer and Farrell (25) and Campbell (32) using methods of Fourier-Laplace transforms and a series of transformations similar to Whipple's (16). The effects of finite thickness are incorporated by employing finite Fourier transforms and by introducing appropriate boundary conditions at the film surfaces. These analyses have employed a totally reflecting condition on the backside of the film:

$$\left(\frac{\partial C_l}{\partial y}\right)_{y=l} = 0 \tag{24}$$

where l is the film thickness. This boundary condition is idealized since it does not take account of any flux into the substrate or along the interface between the film and the substrate. However, this is plausible enough in some typical experimental situations such as diffusion in a metal film on an inert, amorphous substrate. Furthermore, the profiles calculated under this condition are amenable to some simplified methods of diffusion analysis, as will be discussed later in this section.

The analytic solutions for infinite and instantaneous source conditions, based on the previous analysis of Gilmer and Farrell (25) and Campbell (32), respectively, are given below together with the appropriate section integrals.

1. Constant Source, Reflecting Boundary:

$$C(x, 0, t) = C_0 \qquad \frac{\partial C(x, l, t)}{\partial y} = 0$$

$$C_1 = 1 - \frac{4}{\pi} \sum_{m=1}^{\infty} \frac{1}{(2m-1)} \sin (\mu_{2m-1}\eta) \exp (-\mu_{2m-1}^2) \qquad (25)$$

$$C_2 = \frac{2C_0}{\eta_0} \int_1^\Delta d\sigma \sum_{m=1}^{\infty} \sin (\mu_{2m-1}\eta) \exp (-\mu_{2m-1}^2\sigma)\mu_{2m-1}$$

$$\times \operatorname{erfc}\left[\frac{1}{2}\sqrt{\frac{\Delta-1}{\Delta-\sigma}}\left(\varepsilon+\frac{\sigma-1}{K\beta}\right)\right] \qquad (26)$$

The quantity η_0 is given by

$$\eta_0 = \frac{l}{\sqrt{D_l t}} \qquad (27)$$

and μ_{2m-1} by

$$\mu_{2m-1} = \frac{(2m-1)\pi}{2\eta_0} \qquad (28)$$

The section integral is

$$\overline{C}_2 = \frac{8\sqrt{D_l t}C_0}{d\eta_0} \int_1^\Delta d\sigma \sum_{m=1}^{\infty} \sin (\mu_{2m-1}\eta) \exp (-\mu_{2m-1}^2\sigma)\mu_{2m-1} \sqrt{\frac{\Delta-\sigma}{\Delta-1}}$$

$$\times [\exp (-X^2)/\sqrt{\pi} - X \operatorname{erfc} X] \qquad (29)$$

2. Instantaneous Source, Reflecting Boundary:

$$C(y, t) = M \, \delta(y) \, \delta(t); \qquad \frac{\partial C(0, t)}{\partial y} = \frac{\partial C(l, t)}{\partial y} = 0$$

$$C_1 = \frac{2M}{\eta_0\sqrt{D_l t}} \sum_{m=1}^{\infty} \cos (\mu_{2m}\eta) \exp (-\mu_{2m}^2) \qquad (30)$$

$$C_2 = \frac{-2M}{\eta_0 \sqrt{D_l t}} \int_1^\Delta d\sigma \sum_{m=1}^\infty \cos{(\mu_{2m}\eta)} \exp{(-\mu_{2m}^2 \sigma)}\mu_{2m}^2$$

$$\times \text{erfc}\left[\frac{1}{2}\sqrt{\frac{\Delta -}{\Delta - \iota}} \quad + \frac{\sigma - 1}{K\beta}\right)\right] \tag{31}$$

$$\bar{C}_2 = \frac{8M}{\eta_0} \int_1^\Delta d\sigma \sum_{m=1}^\infty \cos{(\mu_{2m}\eta)} \exp{(-\mu_{2m}^2 \sigma)}\mu_{2m}^2 \sqrt{\frac{\Delta - \sigma}{\Delta - 1}}$$

$$\times \left[\exp{(-X^2)}/\sqrt{\pi} - X \text{ erfc } X\right] \tag{32}$$

The experimentally measured profile will correspond to a sum of C_1 and $\bar{C}_2$ as given by a previous equation, $C_l = C_1 + [(1/d)\bar{C}_2]$, except that now C_1 refers to Eq. 25 or 30 and $\bar{C}_2$, to Eq. 29 or 32.

An important influence of finite thickness on concentration profiles is the introduction of considerable curvature in the grain boundary portion of the profile, whereas the profiles calculated using Whipple's semiinfinite model are invariably straight when plotted as log C versus $y^{6/5}$. This can be seen quite clearly in Figure 7.3, where calculated profiles are plotted using both the finite, Eqs. 25 and 29, and the semiinfinite models from the previous section. The error involved in calculating D_b using an average slope taken from the region $y/l > \frac{1}{2}$ and LeClaire's relationship, Eqs. 20 and 22, which are valid only for the semiinfinite case, can yield "effective" values of boundary diffusivity D_b, which are wrong by orders of magnitude. The situation for constant concentration source conditions and reduced times T, less than 10^{-2}, where $T \equiv \eta_0^{-2}$, is shown graphically in Figure 7.4 (adapted from Fig. 5 in reference 25). In view of this difficulty it is imperative that the slope technique be used only at early times, when $K\beta\eta_0^{-2} \lesssim 10^{-1}$. Gilmer and Farrell have suggested two alternate methods for determining D_b which apply to profiles occurring under constant source conditions.

One method involves determining the curvature κ in the region adjoining the reflecting boundary at $y = l$. The curvature is related to $\bar{C}_2$ through the expression

$$\kappa = \frac{1}{2}\left(\frac{\partial^2 \bar{C}_2}{\partial y^2}\right)_{y=l} \tag{33}$$

Gilmer and Farrell (25) use a value of $\bar{C}_2$ in Eq. 33 which they obtain by a section integral taken over finite limits:

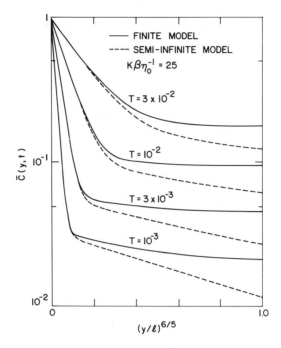

Figure 7.3. Penetration profiles calculated on the basis of semiinfinite and finite thickness boundary conditions. Adapted from Figure 2 in reference 25.

$$\overline{C}_2 \simeq \frac{1}{L} \int_0^L C_2 \, dx \tag{34}$$

where $2L$ is the spacing between parallel grains. For C_2, they use an expression equivalent to Eq. 26 but modified consistent with the assumption that Δ approaches infinity while $K\beta$ remains finite. The expression they derive is

$$\kappa \approx (2L/K\beta\eta_0^{-1})^{-1} \tag{35}$$

which is valid provided that $K\beta\eta_0^{-1} \geqslant 25$ and $\Lambda \gg 1$. The quantity Λ is defined by $L/(D_l t)^{1/2}$ which, when it is about 10 or larger, implies that the diffusion fields of adjoining grains are effectively isolated from each other. The diffusivity is given by

$$K\delta D_b = D_l \, (L\kappa)^{-1} \tag{36}$$

The range of validity of the approximate expression (Eq. 33) can be seen clearly in Figure 7.5 (Fig. 6 in reference 15). According to Eq. 35, a plot of $\kappa L l$ versus $K\beta\eta_0^{-1}$ should be linear as indicated by the solid line. A more exact

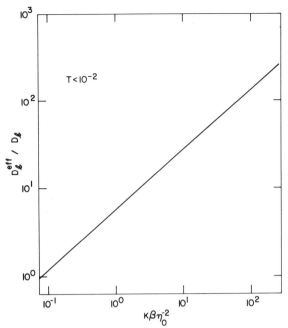

Figure 7.4. Ratio of D_b^{eff} to D_b for various values of $K\beta\eta_0^{-2}$. Adapted from Figure 5 in reference 25.

expression for κ (Eq. 31 in reference 21) gives the dotted lines for several values of T.

Gilmer and Farrell speculate that while Eq. 36 can in principle be used to evaluate D_b, in practice it may be difficult to establish sufficiently precise values of the second derivative at an interface considering the extent of experimental uncertainty usually accompanying profile determinations. To avoid this problem, they suggest another method which involves measuring the concentration level in the vicinity of the interface at $y = l$. For large values of $K\beta\eta_0^{-1}$, the profiles tend to be rather flat over the range $\frac{1}{2} \leqslant y/l \leqslant 1$. A concentration determination somewhere within this region can be adequate and may even be preferable to measurements at the interface because the boundary may not be perfectly reflecting as assumed. The expression for the concentration at $y = l$ is given by

$$\Lambda \bar{C}_2(l, t) = \Lambda + \frac{4}{\pi} \sum_{n-1}^{\infty} \frac{(-1)^{n+1}}{(2n-1)} \exp\left(-\mu_{2n-1}^2\right)$$

$$\times \left\{ \frac{2}{\pi^{1/2}} - \Lambda - \left(\frac{1}{K\beta\mu_{2n-1}^2} \right) \left[1 - \exp\left(K^2\beta^2\mu_{2n-1}^4\right) \operatorname{erfc}\left(K\beta\mu_{2n-1}^2\right) \right] \right\} \qquad (37)$$

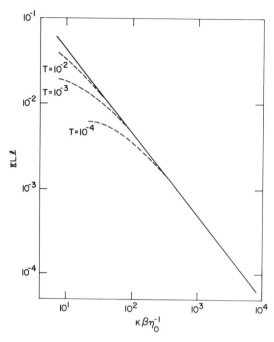

Figure 7.5. Curvature of the profile at the reflecting interface plotted as κLl vs. $K\beta\eta_0^{-1}$. Adapted from Figure 6 in reference 25.

and is based on an expression for $C_2(y, t)$ which was in turn derived with the same assumptions discussed in connection with Eq. 34. The proportionality between $K\beta\mu_{2n-1}^2$ and $K\beta\eta_0^{-2}$ suggests that a plot of $\Lambda C(l, t)$ versus $K\beta\eta_0^{-2}$ will define a "universal curve" relating the concentration to time. The curve is shown in Figure 7.6 (Fig. 7 from reference 25), and it is clear from the figure that the curve is universal for values of $K\beta\eta_0^{-1}$ exceeding 2.5. Therefore, by measuring $\bar{C}_2$ at a particular time and knowing D_l, the grain size and the film thickness, one can establish $K\beta$ and subsequently $K\delta D_b$ using Eq. 13. The quantity Λ appearing in Eq. 37 incorporates the term L, where $2L$ is the spacing between parallel boundaries. For columnar grains making a circular cross section with the diffusion surface, one may substitute one fourth of the average grain diameter for L provided $\Lambda \gg 1$.

In employing Figure 7.6 for estimating D_b, it may be useful to keep the following points in mind: Establish several values of $\bar{C}_2(l, t)$ at early times, namely, $K\beta\eta_0^{-2} \lesssim 1$ because $\Lambda\bar{C}_2(l, t)$ later becomes insensitive to $K\beta$ and likewise D_b. In fact, on the horizontal portion of the curve, the kinetics are dominated by lattice diffusion Values of $K\beta$ should be reasonably large,

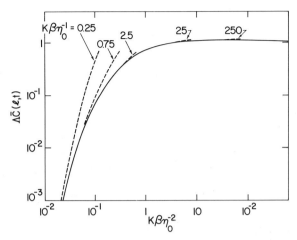

Figure 7.6.   Universal curve relating concentration in the grain boundary portion of the profile to the combined parameters, $K\beta\eta_0^{-2}$. Adapted from Figure 7 in reference 25.

$K\beta \gtrsim 8$, a condition which usually attains in thin film experiments. In order to avoid the possible influence of the bulk portion of the profile, measured values of concentration relative to C_0 should not exceed 0.20.

7.3.1.3 Method of Superposition.   The solutions for diffusion in films with finite thickness have been derived so far on the basis of Fourier transforms. Such boundary problems can also be solved by superimposing solutions of finite thickness. Consider the case of the reflecting boundary with

$$\left(\frac{\partial C(y, t)}{\partial y}\right)_{y=l} = 0$$

If $C(y, t)$ is the integrated concentration for films with infinite thickness, then the concentration distribution for thickness l can be generated by repeatedly folding $C(y, t)$ at $y=0$ and $y=l$ (33). The solution can be written as

$$C'(y, t) = \sum_n [C(2nl + y, t) + C[2(n+1)l - y, t]] \tag{38}$$

$C'(y, t)$ can be readily shown to satisfy the reflecting boundary conditions. Since each term in the summation is a solution to the diffusion problem, it can be shown by simple change of variables that the summation C' is a solution to Eq. 2 subject to the boundary condition of Eq. 10.

In a similar manner, one can derive the solution for an absorbing boundary, that is, $C(l, t) = 0$, to be (33)

$$C''(y, t) = \sum_n [C(2nl + y, t) - C(2(n+1)l - y, t)] \tag{39}$$

The superposition method can also be used to derive the solution for diffusion from a source of finite thickness, a condition which is intermediate between the Suzuoka and the Whipple solutions. Let h be the source thickness; the solution is simply

$$C'''(y, t) = C(y, t) - C(y + h, t) \tag{40}$$

where $C(y, t)$ is the Whipple solution for infinite thickness. Substituting the expressions of $C'(y, t)$ and $C''(y, t)$ as $C(y, t)$ into Eq. 40, one can derive the solutions for diffusion from a finite source into a film with finite thickness and subject to reflecting and absorbing boundary conditions. Such solutions may be of some interest in actual experiments since measurements are often carried out in finite films with finite source layers.

The solutions derived from the superposition method are convenient to use once the basic solutions for infinite thickness have been computed. However, their usefulness depends on the speed of the convergence in the series sums. This can be estimated from the y dependence of $C(y, t)$. To a good approximation, $C(y, t)$ for infinite thickness decreases about exponentially with $\eta\beta^{-1/2}$, which is related to y as

$$\eta\beta^{-1/2} = \left(\frac{D_l}{\delta D_b}\right)^{1/2} \frac{y}{(D_l t)^{1/4}}$$

To assume a suitable convergence, $\eta\beta^{-1/2}$ should be at least about unity when $y = l$. Clearly, this favors diffusion experiments with relatively short annealing times. The requirement is actually not too stringent; an estimate based on the self-diffusion of Au tracers (13) indicates that it can be satisfied for a film thickness of 1000 Å with an annealing time of 10^3 s and temperatures between 200 and 400°C.

7.3.2 Parallel Boundaries

In certain types of thin film diffusion processes, such as the homogenization of bimetallic film couples, the diffusion fields from adjoining grains can overlap so extensively that solutions for isolated grains cannot be expected to account for the kinetic behavior, even though the grain size is incorporated through the appropriate scaling factor (Eq. 19) or a finite limit on the section integral (Eq. 34). The analysis has to be approached as a periodic boundary value problem, and the simplest problem in this category is an array of parallel and equally spaced grain boundaries. The problem is structured similarly to the isolated boundary as there are two regions to consider, the interior of the GB slab and the exterior region of adjoining lattice. The same differential equations, Eqs. 2 and 10, and boundary conditions, Eq 6 or Eqs. 7, 8, and 4, apply as in the isolated boundary cases, with one exception. The boundary condition on the concentration in the direction perpendicular to the grain boundary plane (x-direction) is given by

$$\left(\frac{\partial C}{\partial x}\right)_{x=L} = 0 \tag{41}$$

instead of $C(\infty, y, t) = 0$, as in the isolated boundary case. Due to the symmetry of the problem, it is only necessary to find a solution in the region extending between the boundary $x = \pm\delta/2$ and the midpoint between two grains, $x = \pm L$.

7.3.2.1 Infinite Thickness.   A solution for the problem of multiple parallel boundaries has been given by Campbell (27) who used Fourier-Laplace transforms. As in the isolated boundary case (Section 7.3.1.1), this procedure transforms the partial differential equations into simpler forms which have elementary solutions. However, the complex inversion integral necessary to convert the solutions of the transformed diffusion equations into concentrations as a function of position and time is considerably more difficult because of the appearance of an infinite number of singularities along the imaginary axis. Although the problem of the infinite poles has not been solved exactly, useful approximate solutions have been obtained. Potential applications of these infinite thickness solutions include lateral spreading of a solute stripe in the plane of a fine-grained thin film due to diffusion or electromigration and possibly diffusion through-the-thickness in fine-grained materials such as films that are made sufficiently thick to ignore finite thickness considerations. The kinetic regime is of the A-type, indicating that the diffusion fields from adjoining grains interact.

For infinite source conditions, the solution for parallel boundaries in an infinitely thick specimen gives the customary bulk portion,

$$C_1 = C_0 \text{ erfc } (\eta/2)$$

and for the boundary portion,

$$C_2 = -\frac{iC_0\eta}{2\pi^{3/2}} \int_1^\Delta \frac{d\sigma}{\sigma^{3/2}} \exp\left(-\eta^2/4\sigma\right) \int_{-i\infty}^{+i\infty} I(v)\, dv \tag{42}$$

For an instantaneous source,

$$C_1 = M/\sqrt{\pi D_l t} \, \exp\left(-\eta^2/4\right)$$

and

$$C_2 = -\frac{iM}{4\pi^{3/2}\sqrt{D_l t}} \int_1^\Delta \frac{d\sigma}{\sigma^{3/2}} \left(\frac{\eta^2}{\sigma} - 2\right) \exp\left(-\eta^2/4\sigma\right) \int_{-i\infty}^{+i\infty} I(v)\, dv \tag{43}$$

The quantity $I(v)$ is given by

$$I(v) = \exp\left[v^2 \left(\frac{\Delta-\sigma}{\Delta-1}\right) - v\frac{(\sigma-1)}{K\,\Delta\alpha}\tanh(v\Lambda) \right] \frac{\cosh(v(\Lambda-\varepsilon))}{\cosh(\Lambda v)}\frac{1}{v} \tag{44}$$

and has poles on the imaginary axis at

$$v = 0, \ \pm\frac{(2n-1)}{2\Lambda}\pi i \tag{45}$$

The contribution to the contour integral over $I(v)$ comes entirely from small semicircular regions constructed to avoid the poles along the imaginary axis, Figure 7.7. These semicircular regions of radius r subtend an angle θ which varies from $-\pi/2$ to $\pi/2$ and therefore extend from the imaginary axis into the right-hand plane where the real values of v are positive. The contour integral along the imaginary axis is then replaced by a summation of integrals

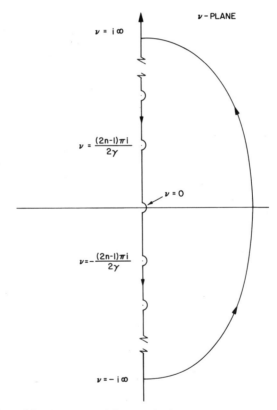

Figure 7.7.   Contour path for complex inversion integral in v-plane.

over the semicircular regions.

$$\int_{-i\infty}^{+i\infty} I(v)\, dv = \lim_{r\to 0} \int_{-\pi/2}^{+\pi/2} I(\rho)\, d\rho + \lim_{r\to 0} \sum_{n=1}^{\infty} \int_{-\pi/2}^{+\pi/2} I(\pm y_n + \rho)\, d\rho \quad (46)$$

where

$$\rho = re^{i\theta} \quad \text{and} \quad y_n = (2n-1)\pi/2\Lambda \tag{47}$$

The contribution from the pole at the origin gives

$$\lim_{r\to 0} \int_{-\pi/2}^{+\pi/2} I(\rho)\, d\rho = i\pi \tag{48}$$

and performing the integration over σ from 1 to Δ as indicated in Eq. 42 yields the following contribution to C_2: C_0 erfc $(\eta/2\sqrt{\Delta}) - C_0$ erfc $(\eta/2)$. The terms partially cancel with C_1 so that

$$C_l = C_1 + C_2 = C_0 \text{ erfc } (\eta/2\sqrt{\Delta}) + \text{contributions from the remaining poles} \tag{49}$$

The equivalent expression for the instantaneous source has a bulk-like term of the form $(M/\sqrt{\pi D_b t}) \exp(-\eta^2/4\Delta)$. Performing a section integral (Eq. 34) over Eq. 49 or the equivalent expression for instantaneous source conditions gives the average concentration as determined by profiling.

Because of the complexity of the complex inversion integral, any analytic evaluation of the integral over the semicircular contours requires approximation. Analytic approximations presented elsewhere (27) were derived assuming that the product $\Lambda\rho$ may be taken sufficiently small that sinh $(\Lambda\rho) \simeq \Lambda\rho$, but this has the effect of limiting the applicability of the solution to those cases where Λ is not too large, say, $\Lambda < 10$. Solutions based on this approximation cannot be expected to yield close numerical agreement with the isolated boundary solution, Eq. 16, since solutions valid for large values of Λ are needed to isolate the emerging diffusion fields from adjoining boundaries. However, the approximate solution gives a qualitatively clear account of the change in the nature of the penetration profile with decreasing boundary spacing, that is, decreasing Λ. This is demonstrated in Figure 7.8 for instantaneous source conditions. The diffusion parameters are those for Au at 444°C (13).

At relatively wide spacing, $\Lambda = 10$, the profile shows the usual composite structure consisting of a high-concentration region with shallow penetration corresponding to lattice diffusion and a low-concentration region that is deeply penetrating arising from grain boundary diffusion. As Λ decreases from 10 to 2, the grain boundary portion of the profile rises proportionately reflecting the greater grain boundary length per unit area of diffusion surface. At the same time the lattice portion of the profile diminishes since a larger

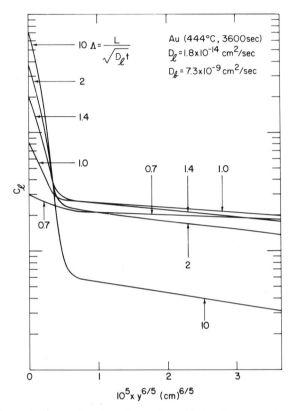

Figure 7.8.　Concentration profiles for various values of Λ showing progressive changes from a composite profile with distinct lattice and boundary portions to a single boundary portion.

fraction of the depletable source material is now distributed to greater depths *via* grain boundaries. The lattice portion also has a more shallow slope, and it is not difficult to see how this effect could easily be mistaken as an enhanced lattice diffusivity. The effect substantially increases as Λ progressively decreases. Hart (35) has predicted that in the regime of Λ-kinetics the profile can be described by a simple one-dimensional function in which the lattice diffusivity is augmented by the term fD_b, where f is the fraction of sites in the diffusion plane associated with defect structures, in this case, grain boundaries. In the expression for C_l, Eq. 49, there are no terms appearing in which D_l and fD_b appear as a sum, which suggests that Hart's analysis is not appropriate for the problem under consideration.

As Λ continues to decrease to about 0.1 (not shown), the lateral spreading away from grain boundaries is so extensive that the lattice diffusion no longer poses any limitation in penetration to further depths. When this situation occurs, the terms arising from poles other than the one at the origin are vanishingly small and the penetration front becomes planar. Then the concentration profile is simply given by

$$C = \frac{M}{\sqrt{\pi D_b t}} \exp(-\eta^2/4\Delta) = \frac{M}{\sqrt{\pi D_b t}} \exp(-x^2/4D_b t) \tag{50}$$

Since only D_b appears in the argument of the exponential and is not a product with the boundary width, it should be possible to measure D_b directly by the analysis of profiles where $\Lambda \lesssim 0.1$. This condition should be obtainable for diffusion experiments such as lateral spreading or electromigration of a cross stripe in the plane of a thin film, for in thin film materials a small grain size can often be retained even after fairly strenuous annealing treatments.

7.3.2.2   Finite Thickness.

A solution for coupled lattice and grain boundary diffusion that includes both parallel grain boundary and finite thickness effects has been derived by Gilmer and Farrell (26) using Fourier analysis techniques. They construct a solution from functions F, where F is given by

$$F(x, y, t) = \sum_{n,m=1}^{\infty} A_{nm} X_{nm}(x) \, Y_n(y) \, T_{nm}(t) \tag{51}$$

The terms X_{nm}, Y_n, and T_{nm} are chosen for reasons of symmetry as

$$X_{nm} = \cos \alpha_{nm} x \tag{52}$$

$$Y_n = \sin \beta_n y \tag{53}$$

and

$$T_{nm} = \exp\{-D_l t(\alpha_{nm}^2 + \beta_n^2)\} \tag{54}$$

Substituting $F(x, y, t)$ as a trial solution into Eq. 10, the boundary condition at the $x = \delta/2$ interface becomes, after some manipulation,

$$(\Delta - 1)\beta_n^2 - \alpha_n^2 = (2\alpha_{nm}/\delta)\tan(\alpha_{nm}L) \tag{55}$$

Values of β_n are determined by the boundary condition at $y = l$. For a reflecting boundary, the condition $(\partial C/\partial y)_{y=l} = 0$ requires that $\beta_n = (2n-1)\pi/2l$; for an absorbing boundary, $C(x, l, t) = 0$, and $\beta_n = n\pi/l$. Values of α_{nm} can be determined from the transcendental relationship Eq. 55. For any given β_n, there exists an infinite set of α_{nm} which will satisfy Eq. 55 due to the periodicity of the tangent function.

The solution satisfying all the necessary boundary conditions is given by

$$C(x, y, t) = 1 - \sum_{n,m} A_{nm} \, X_{nm}(x) \, Y_n(y) \, T_{nm}(t) \tag{56}$$

where

$$A_{nm} = [(8/\alpha_{nm}L) \sin (\alpha_{nm}L) + (4\delta/L) \cos (\alpha_{nm}L)]$$
$$\times \{[1 + (2\alpha_{nm}L)^{-1} \sin (2\alpha_{nm}L) + (\delta/L) \cos^2 (\alpha_{nm}L)]\pi(2n-1)\}^{-1} \qquad (57)$$

The average concentration or section integral is given by

$$\bar{C}(y, t) = 1 - \sum_{n,m} \left(\frac{1}{L\alpha_{nm}}\right) A_{nm} \sin (\alpha_{nm}L) \, Y_n(y) \, T_{nm}(t) \qquad (58)$$

When the extent of spreading of diffusion fields from neighboring parallel boundaries is small relative to the boundary spacing, then Eq. 58 above gives results essentially identical to those of the isolated boundry Eq. 29. This equivalence allows one to apply the method of profile analysis previously described for isolated boundaries and, in particular, the universal curve of Figure 7.6.

Values of $C(l, t)$ should vary inversely with L, provided that the diffusion fields are isolated. The occurrence of this situation over a considerable range of diffusion conditions is demonstrated in Figure 7.9, (Fig. 4 from reference 26), where the quantity $(L/l)C(l, t)$ is plotted versus L/l. The regions where $(L/l)C(l, t)$ is independent of L/l correspond to isolated diffusion fields. At small values of L/l the relationship fails to hold because the over-

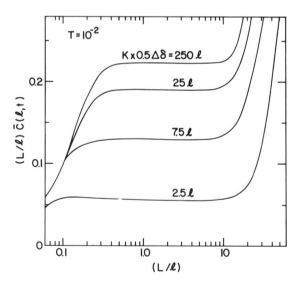

Figure 7.9. Plot of concentration in the GB boundary portion of the profile vs. boundary spacing. The flat portions of the curves indicate regions where the concentration level changes in direct proportion to the boundary spacing. Adapted from Figure 4 in reference 26.

lapping diffusion fields alter the nature of the diffusion profile in a manner unique to A-kinetics (see Fig. 7.1). At large values of L/l, the process is dominated by ordinary lattice diffusion processes since the importance of grain boundaries to $C(l, t)$ is diminished due to their wide spacing.

The utility of the universal curve for calculating grain boundary diffusion coefficients is dependent on how closely the diffusion process adheres to the reflecting boundary condition at the $y = l$ interface. Conditions which may arise and could substantially change the values of $C(l, t)$ from those in the universal curve are absorbing boundary conditions and fast surface diffusion. Absorption (i.e., removal) of a considerable fraction of the flux at the interface will produce a negative curvature and values of $C(l, t)$ that approach zero. Fast surface diffusion will increase $C(l, t)$ above that occurring for reflecting boundary conditions. As the flux exits from a grain boundary at $y = l$, it spreads out along the backside of the film and diffuses into the interior of the grain by ordinary lattice diffusion. This increases $C(l, t)$ to values comparable to C_0, the source concentration at $y = 0$, and creates a positive curvature in the concentration profile. Both the absorbing boundary condition and fast surface diffusion have been treated analytically and numerically, respectively, and typical profiles have been calculated by Gilmer and Farrell (26).

7.3.3   C-Kinetics

7.3.3.1 Infinite Thickness.   Before attempting to apply an analysis of GB diffusion profiles in the C-kinetics regime, it is essential to establish a criterion for safely ignoring lattice diffusion. This can be accomplished by observing the behavior or profiles derived for B-kinetics in the limit of diminishing diffusion length in the lattice: $\sqrt{D_l t} \simeq 0$. It is convenient to examine profiles for infinitely thick samples since these have simple limiting forms.

As $\sqrt{D_l t}$ begins to approach δ, it becomes necessary to consider the amount of material residing in the boundary itself. This quantity is designated as C_b and may be obtained from Eqs. 4 and 12 by replacing ε by zero. We have

$$C_b = \frac{K\eta C_0}{2\sqrt{\pi}} \int_1^\Delta \frac{d\sigma}{\sigma^{3/2}} \exp\left(-\eta^2/4\sigma\right) \operatorname{erfc}\left[\frac{1}{2}\sqrt{\frac{\Delta-1}{\Delta-\sigma}}\left(\frac{\sigma-1}{K\beta}\right)\right] \tag{59}$$

for infinite source conditions, and the analogous expression for instantaneous source conditions can be obtained from Eqs. 4 and 14.

The grain boundary portion of the profile is expressed as the sum of two contributions,

$$C = 1/d(\bar{C}_2 + C_b) \tag{60}$$

where $\bar{C}_2$ is given by Eq. 16. The results of a numerical calculation of Eq. 60 for various values of $\delta/2\sqrt{D_l t}$ is shown in Figure 7.10. As α approaches 10, the profile merges with a complementary error function. This indicates that leakage away from the boundary is negligible since the complementary error function would be obtained for constant source conditions provided that leakage to the lattice is ignored.

It is clear from examining the separate terms in Eq. 60 that C will approach a complementary function as α increases. First, the quantity $\bar{C}_2$ approaches zero as α increases. If Δ and y are fixed, and α and η will change proportionately. At the large values of η which typify profile calculations in C-kinetics, a significant contribution to the integral over σ in C_2, Eq. (16), comes from

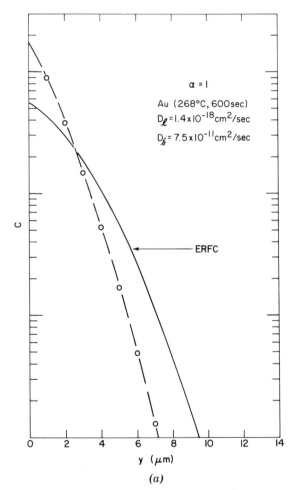

Figure 7.10. Approach of the GB portion of the profile to a complementary error function form with increasing α: (a) 1; (b) 10; (c) 100. Boundary and lattice diffusivities are for pure Au. From reference 14.

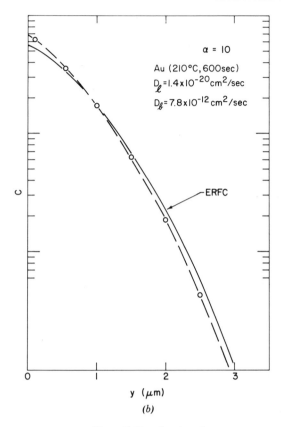

Figure 7.10.   Continued.

the vicinity of the upper limit, Δ, due to the influence of the factor $exp\,(-\eta^2/4\sigma)$. This is increasingly the case as α and η grow larger. Consequently, values of X, Eq. 18, are large over the contributing range of σ. This makes the term in square brackets in Eq. 16 approach zero since for large X, the complementary error function can be expanded to give $X\,\text{erfc}\,X \simeq \exp\,(-X^2)/\sqrt{\pi}$. Secondly, in the integration over σ in C_b, Eq. (59), the term involving erfc may be replaced by unity over most of the contributing range in σ, and integration over the remaining integrand gives a complementary error function:

$$C_b = \frac{K\eta C_0}{2\sqrt{\pi}} \int_1^\Delta \frac{d\sigma}{\sigma^{3/2}} \exp\,(-\eta^2/4\sigma) = KC_0\,\text{erfc}\,(\eta/\Delta) \qquad (61)$$

The analogous treatment for instantaneous source conditions will give a Gaussian form for the GB profile. We will not treat this case further as the

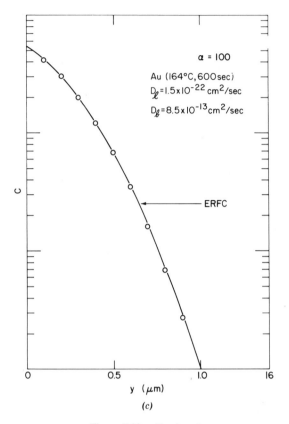

Figure 7.10.   Continued.

Gaussian form is probably irrelevant to most experimental situations. The derivation neglects the motion of the source material over the surface to the grain boundary. The inclusion of this effect is necessary and will lead to a different form for the penetration profile as will be evident by the analysis presented in the next section (7.3.3.2).

The term K appearing in the expression for C_b, Eqs. 59 and 61, implies that the boundary condition where the grain boundary intercepts the entrance surface is given by

$$C(x, 0, t) = KC_0 \qquad -\delta \leqslant x \leqslant \delta$$

This condition is also implicit in the B-kinetics derivations for C_2, Eqs. 12 and 14, for reasons of mathematical convenience. In a more rigorous analysis, the segregation between surface and boundary may involve a different segregation factor, K', as will be assumed in Section 7.3.3.2. An analogous assumption is also made for instantaneous source conditions. This simplification is of little consequence in the B-kinetics regime since the boundary concentration must equilibrate with the surrounding bulk regions where Eq. 4 applies.

Benoist and Martin (23) have also investigated the approach to C-kinetics using their atomic model. These results have particular significance to C-kinetics since the analysis was performed without explicitly requiring that the distribution of diffusing species be equilibrated between the boundary and the immediately adjoining lattice regions. Similarly to the continuum analysis presented here, they show that their exact solution, which incorporates the diffusing species both within and without the boundary, has a simpler limiting form for large values of α. In their case the solution becomes a Gaussian solution consistant with their assumption of instantaneous source conditions. They find the onset of the Gaussian form occurring between α values of 2 to 8. This is comparable but possibly a little less stringent than the criteria set by the continuum analysis model. The continuum model may be expected to give a more conservative condition on α because equilibrium conditions were imposed.

7.3.3.2 Finite Thickness.

This section summarizes the analysis by Hwang and Balluffi (30) of permeation through grain boundaries together with subsequent surface spreading for the C-kinetics regime. With reference to Figure 7.11, the origin of coordinates is located at the intersection of a grain boundary with the sink or exit surface. The source or entrance surface is located at $y = -l$ where l is the film thickness. An array of parallel boundaries is considered, and due to symmetry the solution may be restricted to $0 \leqslant x \leqslant L$, where $2L$ is the spacing between grains. The concentration of material within

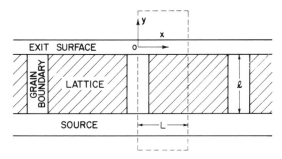

Figure 7.11. Geometric model used for permeation and combined surface spreading analysis. From reference 30.

the boundary is given by C_b and on the exit and entrance surfaces by C_s and C_s', respectively. The width of the surface is designated by δ'.

For C-kinetics, the diffusion equations in the boundary and on the exit surface are given by

$$\frac{\partial C_b}{\partial t} = D_b \frac{\partial^2 C_b(y, t)}{\partial x^2} \qquad -l \leqslant y \leqslant 0 \tag{62}$$

$$\frac{\partial C_s}{\partial t} = D_s \frac{\partial^2 C_s(x, t)}{\partial x^2} \qquad 0 \leqslant x \leqslant L \tag{63}$$

Initially,

$$C_b = C_s = 0 \quad \text{for } t = 0$$

everywhere except at the entrance surface. The boundary conditions at the origin are

$$C_b = K' C_s \tag{64}$$

$$\frac{\delta}{2} D_b \frac{\partial C_b}{\partial y} = \delta' D_s \frac{\partial C_s}{\partial x} \tag{65}$$

At the midpoint between grains on the exit surface, $x = L$, and

$$\frac{\partial C_s}{\partial x} = 0 \tag{66}$$

The source condition for an infinite source is given by

$$C_b = K' C_0 \tag{67}$$

at the point $y = -l$, $x = 0$. For a thin layer of source material, there are the additional relations

$$\frac{\partial C_s'}{\partial t} = D_s' \frac{\partial^2 C_s'}{\partial x^2} \qquad 0 \leqslant x \leqslant L \tag{68}$$

and at the point of $y = -l$, $x = 0$,

$$C_b = K' C_s' \tag{69}$$

$$-\frac{\delta}{2} D_b \frac{\partial C_b}{\partial y} = \delta' D_s' \frac{\partial C_s'}{\partial x} \tag{70}$$

Here, the source condition is

$$C_s' = M \delta(y + l) \, \delta(t) \tag{71}$$

The authors derive the following expressions for boundary and surface concentrations using Laplace transforms. For an infinite source,

$$C_{b0} = 1 - 2 \sum_{n=1}^{\infty} \exp(-\alpha_n^2 \tau_b)$$

$$\times \frac{HF \cos(\alpha_n F) \cos(\alpha_n Y') + \sin(\alpha_n F) \sin(\alpha_n Y')}{\alpha_n \{(1+H)F \sin(\alpha_n) \cos(\alpha_n F) + (1+HF^2) \cos(\alpha_n) \sin(\alpha_n F)\}} \tag{72}$$

$$C_{s0} = 1 - 2 \sum_{n=1}^{\infty} \exp(-\alpha_n^2 \tau_b)$$

$$\times \frac{HF \cos(\alpha_n F)(1-X')}{\alpha_n \{(1+H)F \sin(\alpha_n) \cos(\alpha_n F) + (1+HF^2) \cos(\alpha_n) \sin(\alpha_n F)\}} \tag{73}$$

where α_n is the nth root of the transcendental equation

$$HF \cos\alpha \cos(\alpha F) - \sin\alpha \sin(\alpha F) = 0 \tag{74}$$

The dimensionless quantities used in the above expressions are defined as follows:

$$C_{b0} = \frac{C_b}{K'C_0} \qquad Y' = \frac{y}{l} \qquad \tau_b = \frac{D_b t}{l^2}$$

$$C_{s0} = \frac{C_s}{C_0} \qquad X' = \frac{x}{L} \qquad \tau_s = \frac{D_s t}{L^2} \tag{75}$$

$$H = \frac{K'\delta}{2\delta'} \frac{l}{L} \qquad F^2 = \frac{\tau_b}{\tau_s} \qquad G = HF^2$$

Here, H is the ratio of grain boundary capacity, namely, available sites, to surface capacity. The quantity F^2 is the ratio of the effective time scales of diffusion in the boundary and surface regions.

The concentration on the exit surface averaged over X' is $\overline{C}_{s0}$ and is obtained by the same section integral indicated in Eq. 34:

$$\overline{C}_{s0} = 1 - 2 \sum_{n=1}^{\infty} \exp(-\alpha_n^2 \tau_b)$$

$$\times \frac{H \sin(\alpha_n F)}{\alpha_n^2 \{(1+H)F \sin(\alpha_n) \cos(\alpha_n F) + (1+HF^2) \cos(\alpha_n) \sin(\alpha_n F)\}} \tag{76}$$

The exact solutions presented by Eqs. 72, 74, and 76 can be evaluated by computer once the α_n values are obtained by the solution of Eq. 74. However, Hwang and Balluffi (30) have developed approximate solutions which are considerably easier for analyzing experimental data.

In one approximation, the surface diffusion is considered sufficiently more rapid than boundary diffusion so that the concentration on the exit surface may be approximated by an average value which is a function of time only, namely, $C_s = \bar{C}_s(t)$. The flux at the point $Y' = 0$ is given by

$$-H\frac{\partial C_{b0}}{\partial Y'} = \frac{\partial C_{b0}}{\partial \tau_b} \tag{77}$$

The solution for this flux condition and infinite source condition, Eqs. 62, 64, and 67, is given by

$$C_{b0} = 1 - 2\sum_{n=1}^{\infty} \exp\left(-\gamma_n^2\tau_b\right)\frac{(\gamma_n^2 + H^2)(\sin\gamma_n)(1 + Y')}{\gamma_n(\gamma_n^2 + H^2 + H)} \tag{78}$$

and

$$\bar{C}_{s0} = 1 - 2\sum_{n=1}^{\infty} \exp\left(-\gamma_n^2\tau_b\right)\frac{(\gamma_n^2 + H^2)(\sin\gamma_n)}{(\gamma_n^2 + H^2 + H)\gamma_n} \tag{79}$$

where γ_n is the nth root of

$$\gamma \tan\gamma = H \tag{80}$$

Equations 78 and 79 above can be extracted from Eqs. 72 and 75 by letting $G \ll 1$ and $\tau_s \gg 1$ but keeping H finite.

A further degree of simplification can be obtained if the grain boundary capacity is small compared to that of the exit surface. Then $H \ll 1$ and a quasi-steady state will occur in the boundary region. This implies that the grain boundary profile can readjust to the accumulation of material on the exit surface sufficiently rapidly that a linear profile is always maintained. This can occur if only a small amount of material is needed to establish the profile. Under these conditions the concentration in the profile is given by

$$C_{b0} = -Y' + (1 + Y')C_{s0}(0, t) \tag{81}$$

where $C_{s0}(0, t)$ is the concentration of material at the origin (Fig. 7.11). If this condition is combined with rapid surface diffusion, namely, $G \ll 1$, the combined solution of Eqs. 77 and 81 for infinite source conditions gives

$$\bar{C}_{s0} = 1 - \exp\left(-H\tau_b\right) \tag{82}$$

This expression can also be obtained from Eq. 79 by letting $H \ll 1$, $G \ll 1$, and $\tau_b \gg 1$. Numerical calculations show that Eq. 82 is a good approximation to Eq. 72 for less stringent conditions, namely, $H \simeq 1$ and $G \simeq 1$.

Hwang and Balluffi employed Eq. 82 to determine GB diffusivity after observing the rate of accumulation of Ag on the surface of Au film through which it had permeated. Since they were able to fit data over a wide range in the inverse of temperature, $1/T$, it would appear that the conditions on H and G are not particularly restrictive. Measurements of $\bar{C}_s$ as a function of time enable one to determine the quantity $H\tau_b$ and therefore $(K'\delta/\delta')D_b$. The quantity D_b can be determined to within the uncertainty in the ratio $K'\delta/\delta'$.

By a similar but more complex analysis, Hwang and Balluffi have also obtained the analogous solutions for instantaneous source conditions. These may be given by the following expression for the entrance surface:

$$\bar{C}'_{s0} = \tfrac{1}{2}[1 + \exp(-2H\tau_b)] \qquad (83)$$

and for the exit surface

$$\bar{C}_{s0} = \tfrac{1}{2}[1 - \exp(-2H\tau_b)] \qquad (84)$$

Note that $\bar{C}'_{s0} + \bar{C}_{s0} = 1$, which indicates that GB capacity is assumed to be negligible.

7.3.4 Diffusion Under a Driving Force

The analysis presented so far has been focused on diffusion under a concentration gradient. In many thin film applications, diffusion often occurs under a driving force caused by stress gradients or current passage. The latter is related to the phenomenon of electromigration which is the subject of discussion in Chapter 8. In studying such diffusion phenomena in thin films, it is of basic interest to measure the driving force at the grain boundary. This can be accomplished by measuring the diffusion profile under a driving force provided that the profile can be properly analyzed.

The analysis for electromigration with coupling between the lattice and the grain boundary was first carried out by Martin (35) using a Fisher-type approximation. Later, Ho (36) reported that a general solution of the Whipple type has been obtained. Recently, Tai and Ohring (37) extended the Fisher-type analysis by removing the quasi-steady-state assumption, namely $\partial C_b/\partial t$ in Eq. 5 is not assumed to be zero. Below we outline the general solution based on the Whipple-type analysis.

In Figure 7.12a we show a typical sample geometry used to measure diffusion under a diriving force. The geometry used to derive the solution is given in Figure 7.12b. The diffusion equations with the driving forces acting parallel to the boundary can be written as

$$D_l\left(\frac{\partial^2 C_l}{\partial x^2} + \frac{\partial^2 C_l}{\partial y^2}\right) + \frac{D_l F_l}{kT}\frac{\partial C_l}{\partial y} = \frac{\partial C_l}{\partial t} \qquad \text{for } |x| \geqslant \delta/2 \qquad (85)$$

$$D_b\left(\frac{\partial^2 C_b}{\partial x^2} + \frac{\partial^2 C_b}{\partial y^2}\right) + \frac{D_b F_b}{kT}\frac{\partial C_b}{\partial y} = \frac{\partial C_b}{\partial t} \qquad \text{for } |x| \leqslant \delta/2 \qquad (86)$$

where F and F_b are the driving forces in the lattice and at the grain boundary, respectively. Imposing the boundary conditions of Eqs. 4 and 5, one can deduce from Eq. 86 the following boundary equation for C_l at $x = \delta/2$:

$$\left(\frac{D_b}{D_l} - 1\right)\frac{\partial C_l}{\partial t} = D_b \frac{\partial^2 C_l}{\partial t^2} - \frac{D_l}{K\delta}\frac{\partial C_l}{\partial x} - \frac{D_b}{kT}(F_b - F_l)\frac{\partial C_l}{\partial y} \qquad (87)$$

The above equation is identical to Eq. 10 except for the last term containing the driving forces.

Equation 85 subjected to the condition of Eq. 87 can be solved by the method of Laplace-Fourier transform as used in the Whipple analysis. With the sample geometry as indicated in Figure 7.12a, the source condition is that of an infinite source. The particular and the homogeneous solutions are given respectively as

$$C_1 = C_0 \operatorname{erfc}\left(\frac{\eta - \gamma_l}{2}\right) \qquad (88)$$

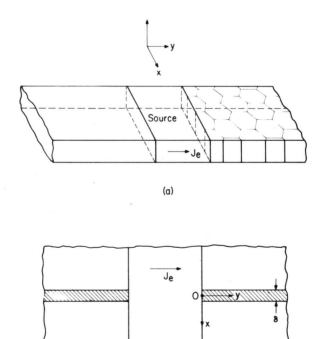

(a)

(b)

Figure 7.12. (a) Cross-stripe sample configuration for measuring grain boundary diffusion under a driving force. (b) Grain geometry for obtaining solution for grain boundary diffusion under a driving force J_e.

and

$$C_2 = \frac{C_0}{2\sqrt{\pi}} \int_1 \frac{d\sigma}{\sigma^{3/2}} \exp - \left[\eta + \gamma_l - (\gamma_b - \gamma_l) \frac{\sigma - 1}{\Delta - 1} \right]^2 \bigg/ 4\sigma$$
$$\times \left[\eta + \gamma_l + (\gamma_b - \gamma_l) \frac{\sigma + 1}{\Delta - 1} \right] \operatorname{erfc} \left[\frac{1}{2} \left(\frac{\Delta - 1}{\Delta - \sigma} \right)^{1/2} \left(\xi + \frac{\sigma - 1}{K\beta} \right) \right] \quad (89)$$

where γ_l and γ_b are defined as

$$\gamma_l \equiv F_1 (D_l t)^{1/2} / kT \equiv v_l \bigg/ \left(\frac{D_l}{t} \right)^{1/2} \quad (90)$$

$$\gamma_b \equiv F_b (D_b t)^{1/2} / kT \equiv v_b \bigg/ \left(\frac{D_l}{t} \right)^{1/2} \quad (91)$$

and all other parameters have been defined in Eq. 13.

The effects of the driving force on the diffusion profile can be visualized by comparing C_1 and C_2 to their counterparts in Eqs. 11 and 12. The ordinary lattice concentration C_1 is shifted along the direction of the driving force by a displacement of $v_l t$:

$$(\eta - \gamma_l) = (y - v_l t)/(D_l t)^{1/2}$$

This means that the lattice reference frame is moving with a speed of v_l. In C_2, the ξ dependence remains unchanged, indicating that a driving force parallel to the grain boundary has no effect on diffusion leaking out from the boundary in a normal direction. However, the overall profile of C_2 is modified in the y-direction due to the two additional terms containing $\gamma_b - \gamma_l$. These two terms are incorporated in such a way as to increase the composition penetration along the direction of the force. The amount of the increase is given in terms of $\gamma_b - \gamma_l$ which is essentially proportional to v_b (since $v_b \gg v_l$). Combining C_1 and C_2, the effect of the driving force is to shift the overall profile by a speed of v_l and to increase the profile penetration along the direction of the force as a result of enhanced grain boundary mobility.

The situation for reversing the driving force can be solved simply by changing the signs of γ_b and γ_l in Eq. 89. The resultant profile would be retarded by the force acting on diffusion at the grain boundary. For a sample configuration as shown in Figure 7.12a, the profiles measured near the two edges of the source region will reflect the effects of opposing driving forces. The combination of the two profiles gives a unique determination of the driving force and the diffusivity at the grain boundary. This approach has been used to investigate electromigration in thin films (49) (see Chapter 8).

Diffusion for multiple boundaries with a parallel driving force has also been solved. The problem consists of solving Eq. 85 subject to the multiple boundary condition of Eq. 41. Using the method of Laplace-Fourier transform as in Section 7.3.2, the solutions for an infinite source are found to be

$$C_1 = C_0 \text{ erfc} \left(\frac{\eta - \gamma_l}{2} \right) \tag{92}$$

and

$$C_2 = -\frac{iC_0}{2\pi^{3/2}} \int_1^\Delta \frac{d\sigma}{\sigma^{3/2}} \exp -\left[\eta + \gamma_l - (\gamma_b - \gamma_l) \frac{\sigma - 1}{\Delta - 1} \right]^2 \Big/ 4\sigma$$

$$\times \left[\eta + \gamma_l - (\gamma_b - \gamma_l) \frac{\sigma + 1}{\Delta - 1} \right] \cdot \int_{-i\infty}^{+i\infty} I(v) \, dv \tag{93}$$

where the function $I(v)$ has been defined in Eq. 44. Again, the driving force has no effect on diffusion normal to the boundary as one would expect. It is worth noting that the effect on profile penetration along the force direction for multiple boundaries is identical to the case of the isolated boundary. The effect found here on the diffusion normal to the driving force can be shown to hold for general boundaries.

There is an interesting case for multiple boundaries. This is the case when γ_b is small compared to unity; then only the leading term in the series form of the integral over $I(v)$ would be important (see discussion in Section 7.3.2). In this case, one can carry out the integration over σ explicitly. Using the approximation that $\gamma_b - \gamma_l \simeq \gamma_b(\Delta - 1/\Delta)$, the results give a combined profile of

$$C_1 + C_2 = C_0 \text{ erfc} \left(\frac{\eta - \gamma_b}{2\Delta^{1/2}} \right) = C_0 \text{ erfc} \left[\frac{y - v_b t}{2(D_b t)^{1/2}} \right] \tag{94}$$

So in this case, similar to the type A diffusion kinetics without a driving force, the lateral spreading from the grain boundaries is so extensive that the overall profile has a uniform front with a composition penetration depending only on the grain boundary parameters of v_b and D_b. Thus, situations exist for carrying out diffusion experiments under A-kinetics to measure directly the grain boundary driving force.

7.4 EXPERIMENTAL TECHNIQUES

One of the objectives in measuring diffusion in thin films is to determine the grain boundary diffusivity. There are two basic types of measurements: one measures the diffusant profile directly, and the other determines the diffusion flux through the film by measuring the material accumulation on the exit surface. A number of techniques have been developed for carrying out these two types of measurements in thin films. For profile measurements, the limited thickness of the film (typically a few microns) requires the capability of removing thin layers of materials of thickness down to a few tens of angstroms. To achieve this, ion sputtering has emerged as the principal method. For measuring surface accumulation, the trend appears to be the use of various surface analytic techniques, such as AES and ESCA (electron spectroscopy for chemical analysis). In Sections 7.4.1 and 7.4.2, we shall discuss the use of radioactive tracers in combination with ion sputtering for measuring diffusion profiles in the depth of the films and diffusion measurements by the lateral spreading method on the film surfaces. In Section 7.4.3, applications of surface analytic techniques for measuring surface accumulation and sputter profiling will be discussed. These discussions will be limited mainly to measurements of self-diffusion or diffusion from a very thin source where the compositional identity of the host is maintained. Other methods, such as He ion backscattering and x-ray diffraction, which have often been used to study interdiffusion in layered films having comparable thickness, will not be discussed. The problem of interdiffusion is the subject of Chapter 9 in this monograph.

Before discussing the specific experimental techniques, it is useful to mention some general considerations for carrying out diffusion measurements in thin films. Of particular interest are the effect of the grain structure and the annealing condition on the diffusion kinetics. The combination of these factors may yield a particular type of diffusion kinetics which can dictate the choice of the experimental techniques. The mathematical analysis presented in the previous Sections 7.2 and 7.3 have elucidated the variation of the diffusion profile due to variable coupling of the lattice and grain boundary diffusions. The results specify the conditions for observing the A, B, and C types of diffusion kinetics. For A-kinetics, when the lattice diffusion distance $(D_l t)^{1/2}$ is comparable to or larger than L, where $2L$ is the intergranular spacing, the diffusing atoms sample the lattice and the GB sites alike and the diffusion front becomes almost planar. In this situation, surface accumulation methods would be difficult to apply because of the difficulty in separating the contributions to mass transport through grain boundaries from that in the lattice. Instead, profile measurements using lateral diffusion from a long stripe source would be more suitable. In the case of B-kinetics, several techniques may be applicable, and the selection of measuring techniques can reflect the nature of the diffusing materials and other experimental limitations. On the other hand, for C-kinetics, most of the atomic transport is confined to the grain boundaries. With the limited quantity of diffusant at the boundaries, profiling would become rather difficult because of the limiting sensitivity of the measuring techniques, typically 0.1 to 1%. In this case, surface accumulation would be more expedient. However, the higher sensitivity of the radioactive tracer technique can often extend the profiling measurement into the type C regime. The situation can also be improved by reducing the grain diameter in the sample thereby increasing the flux of the diffusing atoms.

Among the factors controlling diffusion kinetics, the grain structure appears to have been ignored in many thin film diffusion studies. The grain structure, comprising the grain size and relative orientations, can influence not only the extent of the coupling between lattice and boundary diffusion but, perhaps more importantly, the intrinsic nature of grain boundary diffusion as well. The influence of grain boundary structure on diffusion and defect interactions has attracted considerable interest recently (9), but no experimental studies in thin film diffusion in this context have been reported. We found that in most cases the structure information reported is limited to the grain size and only occasionally to the preferred orientation. Such information is useful in specifying the type of diffusion kinetics but is still insufficient for discussing the effect of atomic structure on the diffusivity.

7.4.1   Radiotracer Profiling

Among the grain boundary diffusion measuring techniques in thin metallic films or polycrystalline solids, the radioactive tracer profiling technique offers a precise and direct method where the actual distribution of the diffusing species normal to the diffusion plane is measured. The amount of the radioactive material used in such studies is typically less than a microgram, and consequently the chemical composition and the base purity of the host remain unaltered even in the parts-per-million range. The radioactive tracer studies, however, require special laboratory areas and are generally time consuming; therefore, they are considered primarily in the context of self-diffusion, that is, diffusion occurring within the host itself without the benefit of any chemical composition gradients, or of trace impurity diffusion in a single-phase host.

Measurements of tracer profiles in films require submicron sectioning since the films themselves are typically only microns thick and diffusion depths are usually maintained shallower at the temperatures of interest to preserve the semiinfinite boundary conditions for the specimens. Obviously, the conventional mechanical methods of material removal are not suitable. The alternatives are (a) electrochemical serial sectioning involving oxidation of the film surface and then stripping the oxide separately, and (b) removing thin layers of material from the specimen surface using rf backsputtering and collecting the sputtered-off material for nuclear counting. These two techniques are discussed below in some detail. Sometimes, self-absorption of the diffusing radioactive tracer in the specimen is employed as a vehicle to eliminate the need for sectioning (38). The level of radioactivity is monitored on the front and/or back surfaces of specimens continually during diffusion annealing at a constant temperature. The technique requires either a pure β-emitting source or an x-ray emitter that is totally absorbed in the specimen. The technique is generally not applicable to thin films since mass absorption coefficients in excess of 10^3 cm^{-1} are required for x-ray emitting diffusant.

7.4.1.1 Electrochemical Serial Sectioning. Davies and co-workers (39) first developed a sensitive electrochemical serial sectioning technique for profiling ranges of energetic ions in Al, Si, and W. It is, in fact, a two-step electrolytic polishing process consisting of anodic oxidation at constant voltage followed by dissolution of the oxide in a solvent. Since the micro-profiling of the diffused tracer is an analogous problem, the Davies technique quickly caught on with diffusion workers, notably Pawel and co-workers (40), who achieved a high degree of precision and reproducibility and were able to measure diffusivities as low as $2.5 \times 10^{-19} \, \text{cm}^2 \, \text{s}^{-1}$. Recently, an improved version of this technique has been used by Campbell et al. (41) to simultaneously measure the diffusion of As in the lattice and along dis-loctions in epitaxially grown Si.

The technique works well when the oxide film formed is protective in nature and its thickness is voltage limited. It has been possible to remove 10 atomic-layer-thick Si in a reliable manner with the material removal preserv-ing the surface topology. In the case of noble metal films such as Au, Ag, and Cu, however, the oxide film formed during anodizing is nonprotective, and its final thickness therefore is not controlled by the applied voltage. The oxide film may actually "blister" during its growth making the surface uneven. The high density of structural defects in thin films further complicates the matter as oxidation may take place perferentially at these sites, ultim-ately developing holes in the film and finally lifting it from the substrate. The preferential oxidation of different atomic species is an added consideration in alloys and particularly in impurity diffusion studies since it alters the composition of the removed layer. The electrochemical sectioning technique has therefore been used only in bulk single crystal of pure metals (42, 43). The electrochemical microsectioning of Au films has been unsuccessful for the same reasons.

7.4.1.2 Microsectioning by Ar Ion Backsputtering. A microsectioning technique utilizing radiofrequency (rf) Ar ion backsputtering was first developed in the IBM laboratory (44) and has since been successfully used in a variety of situations (13–15, 45, 46). Figure 7.13 shows the backsputtering apparatus. The chamber containing the diffused specimen is first prepumped to 10^{-7} torr vacuum. Serial sectioning of the diffused specimen parallel to the diffusion plane involves material removal by sputtering in an Ar glow discharge with a peak-to-peak voltage of 1000 V at 13.56 MHz applied at the cathode (specimen). To minimize the orientation-dependent sputtering or etching of the GBs, the effective power density is kept at $\sim 0.4 \, \text{W cm}^{-2}$ of the cathode area. The sputtered-off material is collected onto aluminum disk planchets. The multianode assembly in the carousel form helps considerably to speed up the sectioning operation, as six consecutive sections can be obtained in a single pumpdown. The principal requirements for reproducible material removal are the stability of the rf power supply and of the flow of Ar gas. The former is controlled to within $\pm 1\%$ of the rf voltage by providing a feedback control circuit, and the latter is maintained at $20\,(\pm 1) \times 10^{-4}$ torr by a system of precision variable leak and throttling valves. During sectioning, a precise control on the geometry is also important in order to maintain a constant fraction of collection of the sputtered-off material onto the planchets, which is typically 30%. A quartz guard ring is used for bulk specimens to eliminate sputtering on the specimen edges; no such ring is, however, necessary for the thin film specimens since the edges do not stand high. A sputtering rate of $\sim 1\,\text{Å sec}^{-1}$ is achieved for most metals. Consequently, by varying the duration of sputtering from 30 s to tens of minutes per section, it is possible to profile the tracer penetration in steps of 30 Å or more with precision. The thickness of sections is obtained from successive weight determination using a microbalance. The penetration distance, therefore, is computed from the cumulative weight loss of the specimen between pumpdowns, the diameter, and the density of the metal.

An important feature of the sputtering technique, in addition to the microsectioning, is that it can be used to implant radioactive tracers onto diffusion specimens. It is possible to first clean the specimen surface, particularly of oxides, and then implant the tracer without breaking vacuum and without any significant heating of the specimen. For this purpose, the carousel and shutter are replaced with a second cathode and shutter assembly, shown in Figure 7.13b. A relatively strong radioactive source of the desired tracer is placed in the sputtering chamber and the diffusion specimen is mounted onto the second cathode assembly. After prepumping the chamber, Ar gas is fed into the system, the roller shutter is closed, and the specimen surface sputter cleaned for a few seconds by energizing the second cathode–shutter assembly at about 2 kV. The roller shutter is then opened so that that specimen and the source of the radioactivity face each other, and the radioactivity from the source is transferred onto the specimen surface by energizing the lower cathode and grounding the specimen. During the above procedures the system is never opened to the ambient. The specimen is then taken out of the chamber, encapsulated in vacuo, and annealed for diffusion.

The backsputtering technique can be used in a variety of situations, notably for metals and homogeneous alloys (43, 47). In the multicomponent systems some perturbation in the compositions may be observed in the first 100 Å penetration because of the differences in the sputter yield of the components. Such problems of using the ion sputtering technique for depth profiling are discussed in Section 7.4.3.2 in the context of AES studies, and the comments generally apply to the radioactive tracer profiling as well. In the latter, however, unfocused Ar ions of energy less than 1 keV are used which tend to minimize the preferred sputtering effect at the GBs and dislocations. Use of

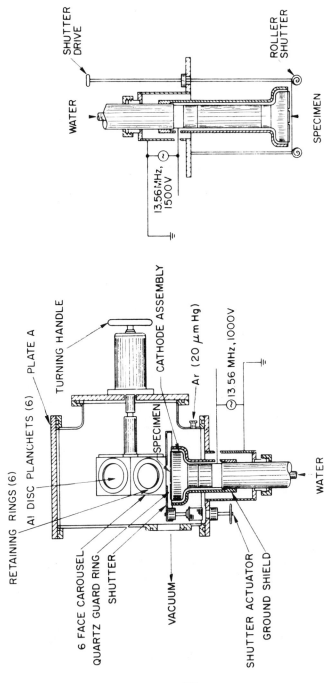

Figure 7.13. Schematic view of (*a*) the Ar backsputtering chamber used in microsectioning of the diffused thin film and bulk specimens, and (*b*) the second cathode assembly used for incorporation of the radioactive tracer prior to diffusion.

lower-power density also minimizes the increase in the surface temperature of the specimen during sputtering, which has been found to level off at $\sim 90°C$. Thus, this is the lower temperature limit for diffusion studies. Also, materials having high vapor pressure cannot be used in this technique.

In Table 7.1 a comparison of three commonly used microprofiling techniques for radioactive tracer penetration is given. It is seen that the rf back-sputtering technique is the most versatile one. It also covers a wide range of measurements for diffusivities varying from 10^{-12} to 10^{-19} cm^2 s^{-1}. Furthermore, we have found the backsputtering sectioning technique to be able to resolve contributions to diffusion from lattice point defects, dislocations, and grain boundaries, which are discussed in Section 7.5.1.

TABLE 7.1. COMPARATIVE MERITS OF LOW-TEMPERATURE DIRECT DIFFUSION TECHNIQUES

Capability	Anodizing and stripping[a]	Surface activity changes	Sputtering[a]
Section μm	0.01–0.1	no sections	0.003–0.1
D, cm^2 s^{-1}[b]	10^{-12}–10^{-16}	10^{-13}–10^{-16}	10^{-12}–10^{-19}
Section form	powder/liquid (NPR)		atoms on planchet
Nuclear absorption	not so serious	based on it	none whatsoever
For dislocations, GB, and thin film work	generally doubtful	measures overall diffusion	excellent resolution among various diffusion modes
Limitations	limited in application and needs development for every new material	extremely limited but non-destructive	should be universal, may not be suitable for high vapor-pressure materials
References	39–43	38	44–46

[a]NPR = No processing required.
[b]On the basis of $(Dt)^{1/2}$ equivalent to 10 sections and $10^3 < t < 10^7$ s.

7.4.2 Lateral Spreading Measurements

Unlike most of the thin film experiments which measure the mass transport normal to the film surface, this method measures the lateral spreading of diffusant along the plane of the film. This type of measurement usually employs samples with a stripe geometry with the source material confined in a narrow region (Fig. 7.12a). The spreading of the source material upon annealing is usually made very large compared to the film thickness, as the profiles can be measured along the length of the stripe without sectioning of the sample. This is the main advantage of this technique. However, the lateral diffusion geometry causes considerable complications to the analysis of the diffusion problem; of particular concern is the additional contribution from the top surface and the bottom interface between the film and its substrate. To our knowledge, there exists no analysis so far which takes into account the complete coupling among surface, interface, grain boundary, and lattice mechanisms. In practice, the relative contribution from the film surfaces as compared to the grain boundaries can be qualitatively estimated by varying the thickness and/or the grain size of the sample.

In spite of the difficulties mentioned above, this technique can be very useful for studying materials with fast diffusion rates. This would be particularly so if the experimental conditions can be designed to yield the type A diffusion kinetics. As discussed in Section 7.3.2, when the diffusion distance in the lattice is large compared to the grain size, that is, when $\Lambda \simeq 0.1$, the intergranular spreading would be so extensive that the diffusion front is essentially planar with a simple exponential profile characterized only by the grain boundary diffusivity (see Fig. 7.8). This condition can be achieved for fast diffusing species using films with fine grain size. Experiments can therefore be designed based on type A kinetics and to measure directly the grain boundary diffusivity instead of the usual product of the diffusivity and the width of the grain boundary.

The techniques used to measure the lateral spreading depend on whether impurity elements or radioactive tracer atoms are used as diffusing species. For impurity atoms, a convenient technique is to use an electron microprobe. This instrument has a lateral resolution of about 1 μm, so a profile with a spreading of about 30 μm can be determined. The concentration calibration for microprobe measurements is well established. The detectability limit for quantitative measurements is generally not better than 0.1 at. %, which may be restrictive for diffusants with low solubility in the film matrix. Such a measurement has been carried out for Cu in Al films (48) and for Ag in Au films (49). For radioactive tracer atoms, two techniques have been developed: one uses direct scanning over the stripe surface, and the other employs an autoradiographic technique. These two methods are described below.

7.4.2.1 Surface Scanning Techniques.   Attempts have been made in recent years to scan the distribution of the radiotracer atoms before and after diffusion annealing in a nondestructive manner using a collimating slit interposed between the nuclear counting probe and a movable stage onto which the specimen is placed. Archbold (50) made an early attempt to scan the GB diffusion profiles of ^{110}Ag in Ag bicrystals, but the experiment had little success since a wide discrepancy was observed between the GB diffusivities obtained by scanning and those determined by the lathe sectioning technique. His difficulties possibly stemmed from the nature of the ^{110}Ag isotope which emits 1.1 MeV γ rays and is hard to collimate because of poor nuclear absorption. Also, the mechanical drive of the stage may have lacked sufficient precision. Gupta (51) used a precision scanning technique for observing lateral self-diffusion of ^{195}Au tracer in thin Au films. Improved versions of the tracer scanning techniques having a positional resettability of a few μm have recently been reported by Ohring and co-workers (52, 53). The apparatus in both investigations essentially consists of a movable stage having a precision stepping motor. Typically, 50 μm wide by 1 cm long radioactive stripes are deposited onto films held on substrates. The specimens are mounted on the stage, aligned, and traversed under a NaI(Tl) scintillation counting probe with an intervening Pb collimating slit 50 μm wide. The ^{195}Au and ^{119}Sn radioisotopes with 81 and 24 keV x-ray emissions are fully absorbed in the 0.5-cm-thick Pb collimating slit. The data consisting of counts for a fixed time interval as a function of position of the radioactive stripe are stored in a 400-channel analyzer. The diffusion kinetics are obtained by profiling the radioactive stripes over distances of hundreds of μm as function of time and temperature of annealing in a nondestructive manner. Although the tracer profiles are typically Gaussian in character, they are distorted and need to be corrected for the finite width of the slit with accompanying edge effects, the natural background counts, and the decay of the radioactivity.

The lateral resolution of the scanning technique is limited by the slit width which introduces uncertainty in the diffusivity expressed by the inequality $4\sqrt{Dt} \geqslant w$, where t (s) is the diffusion annealing time, w (cm) is the detector slit width, and D is the lateral diffusion coefficient. A slit width of 0.022 cm was considered to be the optimum in Sun and Ohring's experiment (53) when the initial tracer stripe width was 0.07 cm. Since practical annealing times are in the range of 10^2 to 10^6 s, the measureable range of diffusion coefficients is 10^{-5} to 10^{-10} cm^2 s^{-1} in thin films by the tracer scanning technique. This limits the application of the technique to systems with relatively large diffusivities. Sun and Ohring were able to carry out diffusion measurements by the scanning technique in Sn films in the temperature range above 0.5 T_m.

7.4.2.2 Autoradiographic Techniques.

Autoradiographic techniques have been widely employed in diffusion studies in bulk materials (54). For GB self-diffusion and GB segregation studies as well, autoradiographic techniques have been found useful and accurate (55, 56). The degree of resolution possible is determined by the nature of nuclear emission from the isotope used. Typically, low-energy (<0.1 MeV) x-rays and beta- or alpha-emitting tracers are convenient, as the Ag grains in the photographic material absorb these emissions well. The spatial resolution available in commercial x-ray films, for example, Kodak type AR, is ~ 3 μm when scanned by an optical microphotodensitometer. Automated densitometers with capability of computer data acquisition are now available which can generate precise contours of the radioactivity over wide areas. Such quantitative autoradiographic studies have recently been made by Renouf (45) for GB studies in bulk Cu bicrystals employing radioactive tracers. A resolution of $\sqrt{D_l t} \geqslant 5 \times 10^{-4}$ cm was obtained, and correspondingly $D_b \geqslant 10^{-13}$ cm^2 s^{-1} was measured. Further improvements on resolution, particularly in small geometries, are possible through the use of extrafine grain emulsions, notably Kodak Nuclear Track Emulsion, instead of films, followed by examining the exposed Ag grains in an electron microscope (57).

The application of the autoradiographic technique to the lateral diffusion on the thin film surfaces is analogous to the scanning techniques described earlier. The comparative merits of two techniques have been discussed by Tai *et al.* (42). Generally, the autoradiographic technique is more convenient for studying small geometries. It does not require as high a level of radioactivity since the film and the specimen can be brought into intimate contact. Corrections due to background radiation and radioactive decay are not required. It is also a nondestructive technique and more expeditious since the specimen can be released for increased diffusion annealing times after an autoradiogram has been made. The equipment required is simple and commercially available, unlike that for the radioactive scanning technique, which has to be specially assembled. Thus, the autoradiographic technique appears to be superior to the radioactive scanning technique. It also has a better ultimate resolution (54), $\sim 10^{-15}$ cm^2 s^{-1}, in comparison to 10^{-10} cm^2 s^{-1} attainable by the scanning technique. The principal difficulty with the use of a microphotodensitometer for analyzing autoradiograms is that the signal measures the transmitted light intensity rather than the Ag density. A one-to-one correspondence between the level of the radioactivity and the observed signal, therefore, breaks down when the former has a large variation on the specimen. Such is indeed the situation in thin film experiments where the intensity of the radioactive source is maintained at a high level to attain an observable flux through the GBs. Recently, Tai and Ohring (58) have measured the Ag density on autoradiograms using Ag-L_α fluorescent x-rays generated by the finely focused electron beam in the scanning electron microscope. The SEM settings are calibrated using a Ag standard. The SEM autoradiograms have considerably superior resolution compared to their optical transmission counterparts.

7.4.3   Surface Techniques

In the last several years, developments of several electron spectroscopy techniques such as AES, ESCA, and ultraviolet photoemission spectroscopy

(UPS) have made available surface-sensitive methods for diffusion measurements in thin films. All these spectroscopy techniques employ energetic particle or photon beams to excite characteristic electrons from surface atoms for chemical analysis. The sampling depth of these techniques is determined by the escape distance of the characteristic electrons, which is usually limited to about 10 atomic layers for energies ranging up to about 2 keV. Such sampling depths make possible surface analysis for most chemical elements heavier than helium. These spectroscopy techniques have been employed for diffusion studies in thin films by measuring surface accumulation or the in-depth composition profile, the latter usually coupled with ion sputter sectioning of the material. The small sampling depth and the capability of detecting small fractions of a monolayer of an element make these techniques particularly attractive for surface accumultion measurements. Principles of these two surface measuring techniques for diffusion studies and some of their limitations are discussed in this section. Our discussion is focused almost exclusively on AES because of the wide scope of applications of the method as compared to the others (59). Nevertheless, the principles discussed for AES can usually be applied equally well to other surface techniques. Not covered here are other techniques such as ion scattering spectroscopy (ISS) (60) and field ion microscopy (FIM) (61), which have been chiefly employed for surface diffusion studies and are therefore beyond the scope of the present discussion.

Surface techniques such as AES when used for diffusion measurements are in principle similar to the tracer techniques discussed previously in Section 7.4.1. There are some interesting distinctions arising from the different methods of signal detection. First, tracer techniques can measure self-diffusion while the Auger method is limited to impurity diffusion. This is clearly an advantage of the tracer method although AES can be applied to a wider choice of material systems that may not have suitable radioactive isotopes. Second, when both methods are employed for profile measurements, the tracer technique can be more sensitive than AES. With proper experimental design, a concentration in the range of 10^{-6} to 10^{-9} can be measured with tracers; in comparison, the detectability limit of AES is usually not better than 10^{-4}. Therefore, the tracer method can measure diffusivities with very dilute concentrations to avoid complications due to concentration dependence. However, when AES is used in the surface accumulation method, since only the concentration on the exit surface is measured, the amount of diffusant can be sufficiently small so that diffusion in dilute concentrations can be measured. Third, the tracer profiling measurement works best when the activity is measured in the material removed from the surface. On the contrary, AES measures directly the concentration on the surface, so it is equally convenient to use with the profiling technique or the surface accumulation method.

7.4.3.1 Surface Accumulation Method.   The surface accumulation method measures the diffusivity by monitoring the amount of diffusant accumulated on the exit surface of the sample. In contrast to the profiling technique, which measures the composition distribution within the film, this method determines the diffusion flux passing through the film. Since surface techniques such as AES can measure a fraction of monolayer of surface coverage, the surface accumulation method requires a relatively small amount of mass transport. Thus, for the same material system, this method can extend diffusion measurements to dilute concentrations and to temperatures below that of the profiling method. Consequently, the method is suitable for experiments in the regime of the type C diffusion kinetics. The use of AES to measure surface accumulation is in principle quite similar to surface segregation studies. The principle of Auger spectroscopy and its applications on surface studies have been reviewed recently (62, 63), so here only its operation is briefly outlined. The essential experimental setup consists of an ultrahigh vacuum chamber equipped with an electron gun and an electron energy analyzer. The electrons from the electron gun, usually operated in the energy range of 3 to 5 keV, generate the Auger electrons upon impact with the surface to be analyzed. The Auger electrons are detected among a large background of secondary electrons by an ac modulation technique, and the signal is measured in its differentiated form. The peak-to-peak height of the differentiated signal is usually used to measure the amount of surface materials, although problems exist associated with the effects of the peak shape factor, backscattering contribution, and electron escape depth for quantitative analysis (64). For diffusion studies, the sample film is usually prepared with a double-layer structure, with the exit surface of the diffusing layer exposed to the electron beam. The sample can be annealed inside or outside the vacuum chamber although the former has two distinct advantages: (i) The surface coverage can be continuously monitored during annealing, and (ii) a clean initial surface can be prepared by ion sputtering after reaching the diffusion temperature. These features are very useful since the surface condition and the initial amount of the surface coverage are important in determining the diffusivity, as is discussed later. Using a multiplexing scheme (65), one can monitor the peak heights of several elements simultaneously so surface contamination or other impurity effects can be detected during diffusion. There are certain complications due to the interaction of the electron beam with the sample surface, such as beam-induced heating (30, 66) and chemical dissociation (67). The beam heating problem is of particular concern to the surface accumulation method since the beam is usually being turned on continuously while monitoring the surface concentration. This increases the effective temperature for diffusion. The effect can be quite significant, particularly for the low-temperature range normally used to study type C kinetics. Recent measurements indicate that even with a beam operating at

10% of the normal intensity, the beam heating on a suspended Au film 1.5 μm thick increases the sample temperature by about 20°C (30).

Besides the instrumentation problems discussed above, there are two important aspects worthy of consideration when carrying out surface accumulation measurements. One is on the diffusion analysis, and the other concerns the quantitative calibration of the Auger signal for obtaining the amount of surface material. These problems are discussed below.

The mathematical analysis for type C diffusion kinetics is discussed in Section 7.3.3. Here, we refer only to the points relevant for experimental consideration. In the type C kinetics, the lattice contribution to the mass transport can usually be ignored in comparison to that in the grain boundary. The mass transport in the boundary can be considered to be essentially decoupled from the lattice, in contrast to the type B kinetics where the leakage from the boundary to the lattice is an essential characteristic. In such a case, the amount of mass transport that can pass through the boundary is very must affected by the disposition of the exit material and the supply of the source material. Therefore, the effects of the exit and source conditions of the diffusant on the permeation rate through the film are important factors to consider in the surface accumulation experiments. This point has been emphasized in the work of Hwang and Balluffi (30) and was demonstrated also in the study by Gibson and Dobson (68) who observed that Ag atoms after diffusing through epitaxial films of Ni and Cu were constrained into a specific crystalline pattern matching that of the exit surface, and the driving force was attributed to minimizing the surface energy of the system.

The diffusion kinetics in the type C regime has been analyzed in detail by Hwang and Balluffi (30), and the results for various combinations of boundary conditions are summarized in Section 7.3.3.2. Of particular interest is the case when the parameters G and H are both small compared to unity; then, the expression for the surface coverage as a function of time is essentially independent of the source condition (see Eq. 75 for the definitions of G and H). This corresponds to the situation where the exit surface has capacity and diffusivity exceeding those of the boundary so that a quasi-steady state of diffusion in the boundary is established most of the time. Under such circumstances, the time variation of the surface coverage can be approximated for the infinite and instantaneous sources by an exponential type of saturation (see Eqs. 82 and 84). Such simple expressions are very convenient to use for data analysis. In Figure 7.14, we show that rate of surface coverage is predicted by the approximate and the exact solutions. For $H < 0.1$, the exact solution is well approximated by Eq. 84. We note also that the coverage rate generally has an S-shaped form which can be seen clearly from the case of $H = 3$. At the present time, the standard procedure for extracting the diffusivity appears to be by fitting the data to curves derived from the diffusion

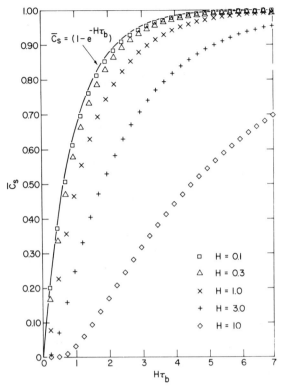

Figure 7.14. Average concentration on the exit surface calculated according to type C kinetics for surface accumulation studies. The data points are numerical solutions calculated as a function of the H parameters, and the solid line is the approximate solution of Eq. 84.

analysis, although a nucleation growth process which is not based on diffusion analysis has also been applied to analyze data obtained by surface accumulation measurements (69).

To convert the Auger data into the amount of surface coverage, quantitative calibration has to be carried out. For homogeneous samples, such a calibration requires considerations of various contributing factors such as backscattering, Auger escape depth, and matrix effects on Auger yields (62, 64, 70). For inhomogeneous samples typically obtained with surface coverage of a second element, there are problems concerning the specific distribution of the surface material. A common practice in calibration of homogeneous samples is to use the elemental sensitivity factor s, which is defined as the proportionality constant between the Auger current I and the atomic concentration X of a particular element (62, 64). The relationship can be expressed in a normalized form as

$$X_i = \frac{I_i/s_i}{\sum\limits_j I_j/s_j} \tag{95}$$

where the subscripts i and j refer to the ith and the jth element and the sum gives the normalization factor with j extending over all the elements on the surface. The sensitivity factor as defined in Eq. 95 would generally be a function of the concentration and the alloy constituents. In principle, it has to be determined by measuring homogeneous standards of varying alloy compositions. To avoid such a laborious procedure, the sensitivity factor is usually determined only with respect to pure elements, namely,

$$s_i = \frac{I_i^0}{I_s^0} \tag{96}$$

where I^0 is the Auger current measured from the surface of the pure elements and the subscript s refers to a standard element. Using this definition, Eq. 95 becomes

$$X_i = \frac{I_i/I_i^0}{\sum\limits_j I_j/I_j^0} \tag{97}$$

This is a simplified relationship commonly used for Auger calibration of homogeneous samples (71).

When Auger calibration is carried out for surface accumulation measurements, one has to consider the specific distribution of the diffusant on the exit surface (72). Here, not only the distribution normal to the exit surface has to be considered but also the uniformity of the lateral spreading on the exit surface. In the Ag/Au study, Hwang and Balluffi (30) considered the problem for two idealized situations: one assumes a uniform coverage of pure Ag on Au, and the other assumes a homogeneous mixing of Ag and Au within the first few atomic layers at the surface. Following Eq. 97, they defined two parameters:

$$p = \frac{I_{69}/I_{69}^0}{I_{2024}/I_{2024}^0} \quad \text{and} \quad q = \frac{I_{356}/I_{356}^0}{I_{2024}/I_{2024}^0} \tag{98}$$

where the subscripts indicate the Auger energies with 69 and 2024 eV for Au and 356 eV for Ag. Using a layer model (72), they showed that the variation between p and q is linear for the homogeneous surface layer but of some exponential form for a pure Ag layer. Therefore, by comparing the predicted and the observed p and q dependences, the composition distribution on the surface can be inferred. The results given in Figure 7.15 are consistent with the homogeneous situation. This establishes the calibration procedure for

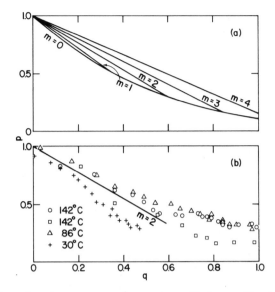

Figure 7.15. Dependence between parameters p and q used in Auger calibration of Ag coverage on Au films. In (*a*), the predicted dependence is given for the pure Ag coverage case ($m=0$) and the homogeneous mixing of Ag and Au ($m \neq 0$). In (*b*), the measured values of p and g from diffusion measurements are given with the straight line indicating the theoretical results for $m=2$.

thin film study of Ag/Au diffusion. In Figure 7.16, some of the surface accumulation data obtained from Ag and Au are shown.

Diffusion measurements employing the Auger surface accumulation method have appeared in the literature (69, 73) since 1973. Due to the sensitivity of this method, most of the experiments were carried out at relatively low temperatures. For example, the Ag/Au work mentioned above was performed between 30 and 270°C, a range of about 0.3 T_m of Au. A majority of the diffusion measurements have used various simplified procedures to analyze the diffusion kinetics or to calibrate the Auger signal, which inevitably introduces errors in the results. The results of the surface accumulation experiments are summarized in Section 7.5.

7.4.3.2 Auger Sputter Profiling. AES can be combined with ion sputtering to measure diffusion profiles. For this type of measurement, an ion sputter gun with energy up to a few keV is incorporated into the Auger vacuum chamber. During measurement, the ion gun is operated simultaneously with the electron gun, so the Auger signals can be continuously recorded as a function of the sputter time. To derive the diffusion profile, the time variation of the Auger signal has to be converted into the depth variation of the

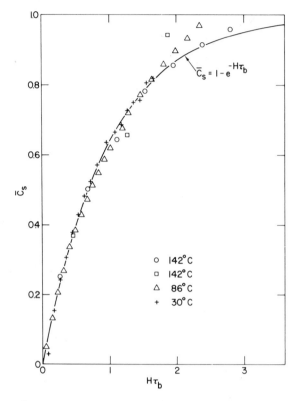

Figure 7.16.   Diffusion results of Ag in Au films at 30, 86, and 142° C showing the measured surface accumulation of Ag atoms as a function of normalized time.

composition. This requires the conversion of the sputter time to the sample thickness and that of the Auger signal to the surface composition. Both these conversions have been found to be affected by the sputtering process (74, 75), so here lie some of the difficulties at the present time of using the Auger sputtering technique for profile measurements. There are three such effects: surface roughness, preferred sputtering, and knock-in effects. The existence of these sputter damage effects has been observed in earlier investigations of the sputtering phenomenon (75, 76). For diffusion studies in thin film couples, since the main concern is to measure the composition profile which may have a rather steep gradient, the situation is usually different from that of homogeneous alloy surfaces. Below, we discuss the effects on profile measurement due to such sputter damages and describe methods which can be used to extract the composition profile.

Surface roughness describes the development of rough morphology on the surface during sputtering. It is caused by the nonuniform removal of materials from certain crystallographic planes or defect regions such as grain boundaries or precipitate particles. In special cases, cone-like structures have been observed on the surface (77). However, for sputter profiling with 1 or 2 keV ions, the rough morphology developed on most thin film surfaces is more uniform. Even for such surface morphology the depth resolution can be considerably reduced since compositions from different depths within the surface roughness can now contribute to the Auger signal. The problem can be formulated by expressing the measured profile $h(x)$ in terms of the actual profile via the following integral (78):

$$h(x) = \int_{-\infty}^{+\infty} c(x-y)g(x-y, y)\,dy \tag{99}$$

where $g(x, y)$ is the resolution function characterizing the sputter broadening effect due to surface roughness. The integral is in a convolution form. The procedure of inverting the integral for determining $c(x)$ is called the deconvolution method, which requires the knowledge of the resolution function.

A set of resolution functions measured for detecting Cu in Ni films as a function of depth is given in Figure 7.17. The half-width of the resolution function was found to increase linearly with the sputter distances (about 7% of the depth). Numerical methods were used for the deconvolution with the

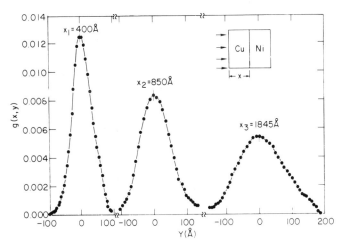

Figure 7.17. Normalized resolution functions obtained for Cu in Cu/Ni couples at different thicknesses. The measured signals are the differential peak-to-peak heights of the Cu 920 eV Auger line.

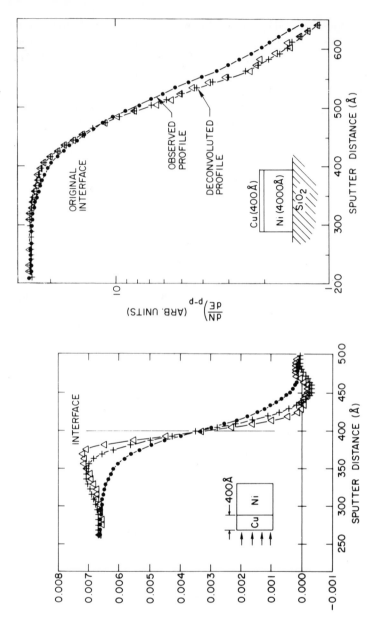

Figure 7.18. (*Left*) Results of deconvolution on the normalized Cu Auger peak measured at a Cu/Ni interface before annealing. (*Right*) Semilog plots of the results obtained by deconvoluting the same interface after annealing for 4 h at 300°C: (●) is observed profile; (+) deconvoluted profile with noise subtraction; (Δ) deconvoluted profile with no noise subtraction.

criterion of obtaining the smoothest profile or the most probable slope of the profile. (The latter is aimed at optimizing the results for diffusion studies.) In Figure 7.18, we show the profiles for a Cu/Ni thin film diffusion couple before and after annealing at $300°$C for 4 hr obtained by deconvolution based on the resolution functions of Figure 7.17. Using the deconvolution method, the broadening of the initial interface is reduced from 73 to 42 Å, and the slope of the semilog plot of the diffusion profile is increased by about 25%.

Judging from the example given above, surface roughness appears to be an important factor to be considered in sputter profiling measurements. It is useful to consider the extent of such an effect for the usual measurement of diffusion profiles. Based on Eq. 99, the effect can be seen to depend on the width of the profile relative to that of the resolution function; the broadening effect is important when the profile is not much broader than the resolution function. In fact, one can show (79) that if both $c(x)$ and $g(x, y)$ are of the Gaussian form and have half-widths Δ_c and Δ_g, then the half-width of $h(x)$ equals $(\Delta_c^2 + \Delta_g^2)^{1/2}$. Therefore, we have an interesting situation, namely, that the broader the original profile, the less would be the effect induced by sputter broadening. For thin film measurements, this means that the effect would be more on the initial part of the profile with steep gradients than in the tail part with shallow gradients. For type B diffusion kinetics, these two parts can usually be attributed separately to lattice and grain boundary diffusivities, so the apparent increase for bulk diffusivity as measured by sputter profiling techniques is expected to be considerably greater than for the grain boundary diffusivity. The measurement errors of these two parameters can be evaluated on the basis of Eq. 99.

Closely related to the surface roughness effect is the knock-in effect which describes the impinging of the surface atoms into the material underneath upon impacts of the sputter ions. This effect is expected to increase the profile broadening and to cause an asymmetry in the profile skewing along the direction of the ion beam. Its effect on profile broadening is usually measured in combination with surface roughness although by varying the ion energy one can qualitatively observe the extent of the knock-in effect (80). Such a measurement has been carried out in an unannealed Au/Ag/Au sandwich structure, and the measured half-widths of the Ag profile are plotted in Figure 7.19 as a function of the ion energy where the average roughness of the sputtered surfaces measured by an electron replica technique is also included for comparison. These results indicate clearly the importance of the knock-in effect above 1 keV of the ion energy for the particular system studied. For most measurements, this effect can be minimized by using low-energy ions, and we found 1 keV to be about the upper limit for Ar ions in most applications.

Preferred sputtering is caused by differences in sputter yields of the constituent elements on the surface. The effect is limited mainly to a thin surface

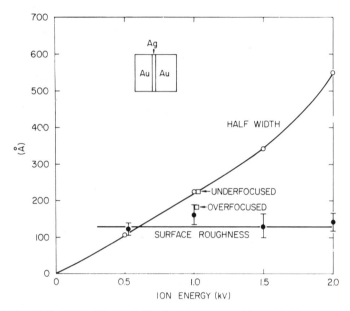

Figure 7.19. Half-widths of the resolution function measured in Au/Ag/Au sandwich films as a function of Ar ion energy. The surface roughness is measured by electron replica techniques on the sputtered surface. The difference gives a measure of the ion knock-in effect. The focusing condition of the ion gun appears to have minimal effect as indicated by the overfocused and underfocused data points at 1 keV.

layer (called the altered layer) where atomic mixing occurs as a result of sputtering (75, 76, 81). The behavior of preferred sputtering for homogeneous binary alloy surfaces has been analyzed by kinetic models (82–85). These models predict the rate of change in the average composition within the altered layer by considering the mass balance on the sputtered surface. The composition is expected to vary exponentially with time in the initial period and then reaches a quasi-steady state with the following composition ratio:

$$\frac{X_A(\infty)}{X_B(\infty)} = \frac{S_B N_A}{S_A N_B} \frac{X_A(0)}{X_B(0)} \tag{100}$$

where $X(\infty)$ and $X(0)$ are the final and the initial atomic concentrations, respectively; S is the individual sputter yield; and N is the atomic density of the pure element. The subscripts A and B denote the elements of the alloy.

Thus, the effect of preferred sputtering for profile measurements has to be corrected in a different way for the initial transient period and the quasi-steady state. The influence is not only in changing the surface composition but also in the sputter rate since the total sputter yield depends on the surface composition. Concerning the sputter rate, there is an additional complication due to the matrix effect on the individual sputter yields, that is, the sputter yield of individual elements in an alloy may vary according to the alloy composition. The time constant for the transient period depends on the sputter yields and the ion flux. When translated into a sputter depth, we found that for many thin film materials sputtered with 1 keV Ar ions, the extent of the transient period corresponds to about 20 to 30 atomic layers (82, 86). For diffusion measurements, this implies that within such distances from the source region, the correction for preferred sputtering is most complicated, and the observed profile is least reliable. However, even the effect in the transient period can be corrected on the basis of the kinetic analysis provided that the sputter yields and the time constant are known. During the steady state, if one assumes that a quasi-equilibrium condition always exists in the presence of the diffusion composition gradient, Eq. 100 can be used to deduce the actual concentration. To do this based on data obtained by surface measuring techniques such as AES, it is important to compare the altered layer thickness with the electron escape distance. They must be at least comparable or with the former being larger in order to assure the measured signal originated completely from the sputtered layer.

Therefore, in principle, the effect of preferred sputtering on profile measurements can be corrected on the basis of information obtained from sputtering studies on homogeneous alloy surfaces as a function of composition. Such quantitative studies have been made by using Ar ions of energy from 0.5 to 2 KeV on Cu/Ni (82) and Ag/Au (86) over the whole range of composition and on Au/Cr up to 15 at. % of Cr (87) and on Au/Cu. Ag/Cu, Ni/Cr, and Fe/Cr (88). For the first two systems, the steady state was found to be established after removal of about 30 Å of the surface material, and the thickness of the altered layer varied between 10 and 30 Å. The individual sputter yields and their ratio were found, within the error limits, to be independent of the ion energy and the alloy composition. The results on the sputter yields for the Au/Ag system are shown in Figure 7.20 as a function of composition.

Converting the observed Auger signal of the sputtered surface to the actual composition requires quantitative calibrations. Without sputtering effects, one can show from Eq. 95 that for a homogeneous binary alloy AB,

$$\frac{X_B}{X_A} = s_A \frac{I_B}{I_A} \tag{101}$$

where s_A is the sensitivity factor for the A element, with the B element used as the standard. This relationship indicates that the observed Auger signal ratio I_B/I_A can be converted directly to the atomic ratio X_B/X_A if s_A is given. Since s_A may vary according to the composition, it should be determined by

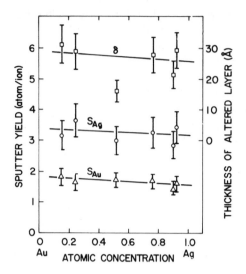

Figure 7.20.   Individual sputter yields of Ag and Au atoms in their alloys and thickness of the altered layer as a function of composition. Sputtering was done using 1 keV Ar ions.

calibrating a series of standards with varying alloy compositions. When preferred sputtering is taken into account, the surface composition is affected. According to Eq. 100, Eq. 101 can be rewritten for the quasi-steady state as

$$\frac{I_B(\infty)}{I_A(\infty)} = \frac{1}{s_A} \cdot \frac{S_A N_B}{S_B N_A} \cdot \frac{X_B(0)}{X_A(0)} \tag{102}$$

This provides a relationship to convert the Auger signal ratio observed on the sputtered surface to the original composition. Calibration measurements have been carried out on Ag/Au homogeneous films subject to Ar sputtering (86). Results are given in Figure 7.21 as a function of composition where R_0, the Auger signal ratio on the unsputtered surface, was determined by an extrapolation procedure. The slopes of the log–log plots for both R_∞ and R_0 were found to be unity within experimental errors indicating a constant sensitivity factor and a constant sputter yield ratio independent of the composition. Similar results have been obtained also for the Cu/Ni system (82).

As seen from the results in Figures 7.20 and 7.21, the correction of the preferred sputtering effects for profile measurements in Ag/Au is relatively straightforward due to the invariance of the sensitivity factor and the sputter yield ratio. Such simple behavior was not observed for the Au/Cr system (87),

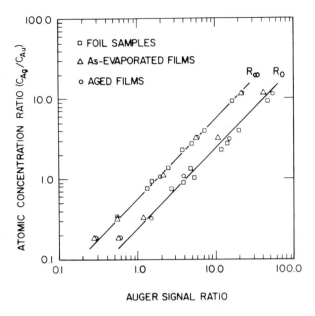

Figure 7.21.   Log–log plots of the Auger calibration curves for Ag/Au alloys; R_∞ is the Auger signal ratio measured at the steady state, and R_0 is the extrapolated ratio for the initial surface. The Auger signal ratio is that of the Ag 348 to 354 eV peak over the Au 69 eV peak.

where the sputter yield of Au was found to depend on Cr concentration. In a related study of Pt/Si films (89), the problem was further complicated by the possibility of forming compound layers on the sputtered surface.

Because of the difficulty in correcting the sputtering effects discussed above, few quantitative measurements on thin film diffusion using Auger sputter profiling technique have been reported, in spite of the common use of this technique for qualitative observation of thin film reactions. It appears now that an analytical basis exists for measuring the sputtering effects and to correct them in the measured profiles. Thus, one can expect that the Auger profiling techniques will have increasing use for quantitative diffusion measurements. In Figure 7.22 we give, as an example, a diffusion profile for Cu in an Al film at 175°C (90) where the grain boundary portion of the profile can be clearly delineated from the initial portion of lattice diffusion; and the solid line represents the profile predicted by the Whipple solution based on the diffusivities given in the figure. Included also are the dashed lines indicating the estimated uncertainty due to surface roughness.

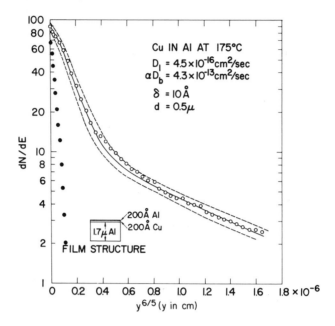

Figure 7.22. Diffusion profile measured for Cu in Al after annealing at 175°C for 85 min by the Auger sputtering technique. Solid curve is the profile calculated according to the Whipple solution with parameters given herein. Dashed lines give estimated errors due to surface roughness. Solid circles show the unannealed profile.

7.5 EXPERIMENTAL RESULTS

In this section we survey only those experiments which measure directly the mass transport of the diffusing species in thin films. We have not included results obtained from indirect methods such as from changes in resistance, contact potential, and surface reflectivity because of the difficulty in data interpretation. Also, the results of interdiffusion such as those obtained by He ion backscattering and x-ray diffraction will not be reviewed here since they are discussed in Chapter 9 of this monograph. For the same reason, results of electromigration are not covered. This section is divided into two parts: one summarizes results obtained by the tracer profiling technique and the other, those obtained by surface techniques and lateral spreading measurements. In the first part, the emphasis is placed on the Au results since systematic studies have been carried out on measuring Au tracer diffusion in thin films and bulk specimens along with the effect of impurity additions on grain boundary diffusion. In the second part, a wide variety of data are reviewed in metallic systems obtained chiefly by surface techniques but mostly not as detailed as the Au tracer work. At this relatively early stage of diffusion study in thin films, the results in the second part appear to be insufficient to infer grain boundary diffusion behavior on a broad basis. It will be our endeavor to point out the problems which have confronted the grain boundary diffusion studies in thin films.

7.5.1   Tracer Profiling Results

7.5.1.1 Nature of the Tracer Profiles.   Typical tracer penetration profiles in a polycrystalline bulk Au specimen (14), epitaxial Au films on (001) MgO (13), and polycrystalline Au films on fused SiO_2 (15) are shown respectively in Figures 7.23, 7.24, and 7.25. The bulk specimens in general contain small-

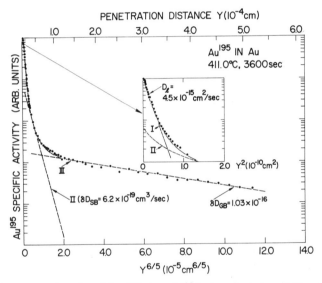

Figure 7.23.   Penetration plot of the diffusion of ^{195}Au in a Au polycrystalline specimen. A three-term exponential fitting procedure marked I, II, and III yielded diffusion coefficients in the lattice, along subgrains, and along large-angle grain boundaries respectively.

angle subgrains ~ 5 μm in diameter within the large-angle GBs. Consequently, the tracer penetration profiles are expected to relate consecutively to diffusion in lattice, along subgrains, and in the GBs as marked by regions I, II, and III in Figure 7.23. If the lattice diffusion distance is maintained sufficiently small in the experiment in comparison to the grain or subgrain diameter and the annealing condition is controlled to yield typical type B kinetics, it is possible to compute the individual diffusion coefficients involved

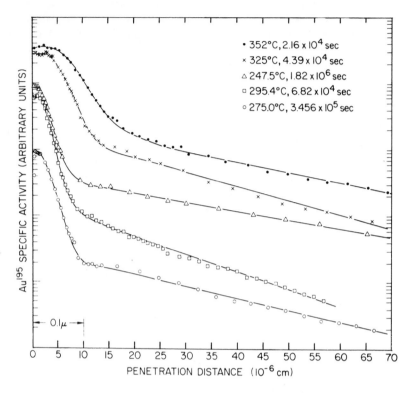

Figure 7.24. ^{195}Au tracer penetration profiles in (001) epitaxial Au films on MgO substrates. Note the sharp transition from lattice to dislocation diffusion in the lowest temperature profile.

in the three processes. The lattice diffusion distances in these experiments have been maintained in the 100 to 1000 Å range, which is $\ll 5$ μm the subgrain diameter. Also, the diffusion distances have been controlled by annealing conditions to be sufficiently different in the subgrains and GBs with respect to that in the lattice. The penetration profiles can thus be decomposed into three segments by curve fitting, using 2, $\frac{6}{5}$, and $\frac{6}{5}$ powers of the penetration distance. The choice of the powers is motivated by the Gaussian behavior of tracer diffusion in the lattice and Whipple-Suzuoka-type (22, 23) behavior in the high-diffusivity paths. Thus, the profiles in Au-polycrystalline bulk specimens are described by

$$C = p \exp\left(-PY^2\right) + m \exp\left(-MY^{6/5}\right) + n \exp\left(-NY^{6/5}\right) \tag{103}$$

where C is the specific activity of the tracer at penetration distance Y, and p, m, n and P, M, N, are the coefficients of the exponentials terms in the

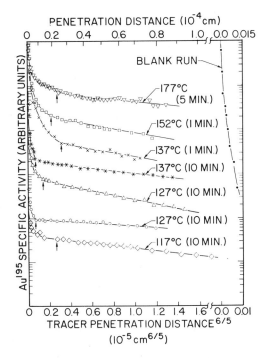

Figure 7.25. ¹⁹⁵Au tracer penetration profiles in polycrystalline Au films on fused quartz substrates with a thin Mo adhesive layer. Note the change of scale on the right for the "blank" run. Arrows indicate the points beyond which data were used for analysis.

respective powers of Y. The coefficient P yields $D_l(P = 1/4D_lt)$, and M or N leads to the combined GB or subgrain diffusion coefficients according to the Whipple-Suzuoka solution for instantaneous source condition (Eq. 23), which for large β and tracer diffusion in pure host reduces to

$$\delta D_b = 0.661 \left(\frac{\partial \ln C_b}{\partial y^{6/5}}\right)^{-5/3} \left(\frac{4D_l}{t}\right)^{1/2} \tag{104}$$

In the above, the subscript b refers to either subgrain or GB diffusion coefficient according to the second term on the right-hand side of Eq. 103 identified either by M or N. The validity and limitation of using Eq. 104 for analyzing diffusion profiles in the type B regime are discussed in Section 7.3.1.1.

In Figure 7.24, Au (195) tracer profiles in (001) epitaxial Au films are shown (13). These films did not contain high-angle GBs but instead a high density of dissociated dislocations in the range of 10^{10} to 10^{11} lines per cm² lying in {111} planes with Burgers vector $\mathbf{a}/6 \langle 211 \rangle$, where $\mathbf{a}$ is the lattice parameter, such that they threaded the film's thickness. In all probability the high density of dislocations is conducive for them to arrange in networks analogous to subgrains in bulk specimens. In the absence of high-angle GBs, the tracer penetration profiles would contain only two regions comprising of lattice diffusion and diffusion along the dissociated dislocations. The penetration profiles, therefore, were analyzed for these paths. The resulting diffusion coefficients along the dissociated dislocations were indeed similar to those found later in the subgrains (region II, Fig. 7.23) in polycrystalline bulk specimens (14). In the case of polycrystalline Au films (15) with large-angle grains of less than 1 μm in diameter, the subgrain contribution is expected to be absent since any substructure would be difficult to be accommodated therein. Thus, the tracer profile would be largely related to the final term in Eq. 103. Since GB diffusion measurements in these films were possible only at temperatures $< 177° C$, the diffusion kinetics are expected to be in type C regime. The data analysis based on Eq. 104 was less rigorous than the exact solutions which have since been developed as discussed in Section 7.3.3.1. The errors introduced, however, appear to be small. Also, the error intro- duced on account of the film thickness is small since value of $\beta \eta_0^{-2}$ is < 0.1 where the ratio of the effective and actual GB diffusivities becomes unity, as seen in Figure 7.4.

It is interesting to compare the contributions of the lattice, the average high-angle GBs, and dissociated dislocations to the penetration profiles as a function of the grain size or the nature of the dislocations and their density present in the films and the temperature of diffusion. These effects have recently been discussed by Balluffi and Blakely (91). Their conclusions are shown in Figure 7.26 where the dominant mass transport processes for steady-state diffusion are shown within the broken rectangles as a function of the grain size in thin films, the dissociated dislocation densities, and the melting point normalized temperature (T/T_m). Despite the fact that the tracer studies are not steady-state experiments, their observation still appear to be valid. The dominant diffusion paths in the epitaxial films according to Figure 7.26 with the condition log $1/g.s. = 0$ are expected to be the dissociated dis- locations and lattice at temperature $< 0.5 \, T_m$ with densities in excess of 10^{10} lines per cm^2. This, indeed, is the situation in Figure 7.24 where the temperatures are $< 0.5 \, T_m$, and the lattice diffusion distance $2\sqrt{D_l t}$ is < 500 Å compared to the dislocation diffusion distance in excess of 0.7 μm. In the polycrystalline films, having log $1/g.s. \sim -4$, on the other hand, the large- angle GBs only are expected to be the diffusion paths at temperatures $< 0.3 \, T_m$ since the lattice diffusion becomes vanishingly small, which is also clearly seen in Figure 7.25. In the preceding case the maximum temperature is 0.34 T_m and the grain size is ~ 0.5 μm. The bulk specimens, having a grain size $> 10^{-2}$ cm and dissociated dislocation densities $< 10^8$ lines per cm^2,

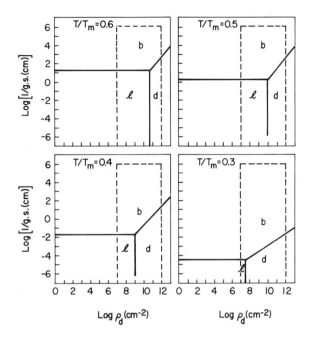

Figure 7.26. Regimes of grain size (g.s.) and dislocation density ρ_d over which lattice diffusion (*l*), grain boundary diffusion (*b*), or dislocation diffusion (*d*) is dominant during steady-state diffusion through a thin film specimen of an fcc metal as function of the homologous temperature (T/T_m). After reference 91.

should be near the triple point as all three processes would be observed at T is $\sim 0.6\ T_m$ which may be the case if the departure from the steady state can be accounted for.

7.5.1.2 Grain Boundary Diffusion Data. We have compiled in Table 7.2 the recent data on GB diffusion measured by the tracer profiling technique in thin films and some bulk specimens including dilute alloys. The data listed in Table 7.2 are displayed in Figure 7.27 together with selected results obtained by some nontracer techniques, which are listed in Table 7.3. A reduced temperature abscissa (T_m/T) is used since diffusion in general correlates well with the absolute melting point (7) of the host and the data in various systems can be displayed with ease. While comparing the diffusivities in various systems obtained by differing techniques, two factors must be considered: All diffusivities measured by profiling or surface spreading include a correlation factor which was estimated to be about 0.54 for self-diffusion at the grain boundary (92); but for impurity tracers this factor is not

TABLE 7.2.   RADIOACTIVE TRACER GRAIN BOUNDARY DIFFUSION DATA IN SELECTED SYSTEMS

Diffusant / matrix[a]	Method of Measurement	δD^0, cm^3 s^{-1}	Activation energy, eV	Remarks[b] (references)
^{195}Au / Au(epi. film)	ion sputtering	1.9×10^{-10}	1.16	diss. disl. (13)
^{195}Au / Au (p. bulk)	ion sputtering	3.1×10^{-10}	0.88	high-angle GBs (14)
^{195}Au / Au (p. film)	ion sputtering	9.0×10^{-10}	1.0	high-angle GBs (15)
^{195}Au / Au–1.2 Ta (p. bulk)	ion sputtering	5×10^{-7}	1.26	high-angle GBs (47)
^{195}Au / Au–1.2 Ta (p. bulk)	ion sputtering	1×10^{-9}	1.2	low-angle GBs (47)
^{195}Au / Au–1.2 Ta(p. bulk)	ion sputtering	1×10^{-9}	1.2	low-angle GBs (47)
^{195}Au / Ni–0.5 Co (p. film)	ion sputtering	1.4×10^{-10}	1.6	high-angle GBs (94)
^{110}Ag / Ag (p. bulk)	lathe sectioning	1.3×10^{-9}	0.80	high-angle GBs (92)
Ni / Ni bicrystal	autoradiography	7×10^{-10}	1.08	45° ⟨100⟩ (95)
^{63}Ni / Ni bicrystal	autoradiography	2.2×10^{-8}	1.77	10° ⟨112⟩ tilt (96)

[a]p. = polycrystalline matrix; epi. = epitaxial.
[b]diss. disl. = dissociated dislocations

known (2). For impurity diffusion the GB diffusivity includes a solute segregation factor K (see Section 7.3.1.1) or K' (see Section 7.3.3.2), which is usually greater than unity. The magnitude of K appears to depend on the temperature, the nature of the solute itself, and the solute solid solubility limit in the host. The solid solubility limit in the host, in fact, has been shown to be related to the GB–solute binding energy (93). These factors may cause considerable scatter in the impurity diffusion data.

In Figure 7.27 we observe that the self-diffusion data in pure Au–both in

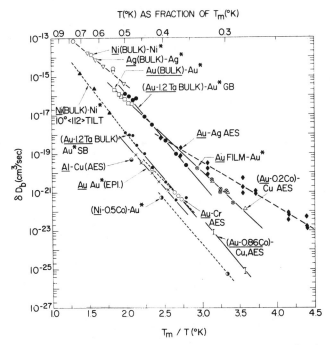

T(°K) AS FRACTION OF T_m(°K)

Figure 7.27. Arrhenius plots of the combined diffusion coefficients δD_b (cm^3 s^{-1}) along grain boundaries and dislocations in some fcc metals and alloys vs. the reciprocal normalized temperature (T_m/T).

the polycrystalline films and the bulk specimens—and several AES data in films selected from Table 7.3 follow about the same Arrhenius dependence. This Arrhenius behavior interestingly extrapolates at T_m to a value of GB diffusivity of about 10^{-5} cm^2 s^{-1} assuming δ of 3 Å. It seemingly correlates with the magnitude of diffusion coefficients in the liquid phase at T_m postulated by Gjostein (7). Diffusivity in the GBs in the solid phase equal to that in the liquid phase at T_m could imply a highly disordered or liquid-like structure of the GBs in the regions of poor fit and of the dislocation cores, and both involve high concentrations of mobile defects providing for the enhanced diffusion. The self-diffusion in the Au–1.2 at. % Ta alloy along the GBs (47) in Figure 7.27 shows a crossover with the data in pure Au, indicating that the effect of solute addition (Ta) is variable on the GB diffusion in Au. This is an important effect and is the subject of detailed discussion in Section 7.6.

Another group of the data in Figure 7.27 consist of (a) ^{195}Au tracer diffusion measurements in the epitaxial Au films (13), in the Au–1.2 at. % Ta in the subgrains (47), and in the Ni–0.5 at. % Co films (94); and (b) several AES

TABLE 7.3. RESULTS OF THIN FILM DIFFUSION MEASURED BY SURFACE TECHNIQUES AND LATERAL SPREADING METHOD

Diffusant matrix	Method of measurement	Temperature, °C	Diffusivity, $cm^2 s^{-1}$	Activation energy, eV	Reference
$\dfrac{\text{Ag}}{\text{Au}}$	AES surface accumulation	30–260	$10^{-5}(D_b^0)$	0.63	30
$\dfrac{\text{Ag}}{\text{Au}}$	AES sputter profiling	30	$\sim 2 \times 10^{-18}$		98
$\dfrac{\text{Ag}}{\text{Au}}$	He ion scattering	30	$\sim 10^{-14}$		99
$\dfrac{\text{Ag}}{\text{Au}}$	AES surface accumulation	150–260	$10^{-6}(D_b^0)$	1.1	100
$\dfrac{\text{Cr}}{\text{Au}}$	AES surface accumulation	210–293	$10^{-3}(D_b^0)$	1.09	29
$\dfrac{\text{Cu}}{\text{Au–0.2 Co}}$	AES surface accumulation	100 125 150	6.4×10^{-15} 3.6×10^{-14} 1.3×10^{-13}		101
$\dfrac{\text{Cu}}{\text{Au}}$	He ion backscattering	25	$\sim 10^{-17}$		102
$\dfrac{\text{Co}}{\text{Au–0.86 Co}}$	AES sputter profiling	100 150	3.3×10^{-18} 2.4×10^{-16}		103
$\dfrac{\text{Cu}}{\text{Al}}$	AES sputter profiling	175	8.6×10^{-13}		90
$\dfrac{\text{Cu}}{\text{Al}}$	lateral spreading	150–300	$10(D_b^0)$	0.81	104
$\dfrac{\text{Au}}{\text{Ag}}$	ESCA surface accumulation	141	4.3×10^{-17}		105
$\dfrac{\text{Pt}}{\text{Cr}}$	AES surface accumulation	400–580	$10^{-2}(D_b^0)$	1.69	106
$\dfrac{\text{Ag}}{\text{Cu}}$	AES surface accumulation	300	$\sim 10^{-14}$		68
$\dfrac{\text{Pb}}{\text{Cu}}$	AES surface accumulation	125–230		0.52	73
$\dfrac{\text{Ag}}{\text{Ni}}$	AES surface accumulation	300	$\sim 10^{-18}$		68
$\dfrac{\text{Au}}{\text{Pt}}$	AES surface accumulation	250–350		0.96	107
$\dfrac{\text{Si}}{\text{W}}$	AES surface accumulation	670–850		2.6–3.2	69
$\dfrac{\text{Sn}}{\text{Sn}}$	tracer scanning	142–213	1.8×10^{-5}	0.46	53

diffusion measurements in thin film couples from Table 7.3, namely, the Cr studies in Au films, Cu studies in Au–0.86 at. % Co, and Cu studies in Al films. The [195]Au tracer data in (a) are in the dissociated dislocations which are expected to display lower diffusivities in view of the tighter core structure compared to those in the single dislocations or average high-angle GBs. The activation energy for diffusion along the dissociated dislocation is about 60 to 80% of lattice diffusion energy, compared with 40 to 50% in the average high-angle GBs. These effects have been known for some time and have been discussed earlier by Balluffi (97). It is, however, surprising that the data in the Au–Cr, Ni–0.5 at. % Co, and Au–0.86 at. % Co films should line up so well with those in the epitaxial films.

7.5.2   Results from Other Techniques

Diffusion data obtained using surface techniques and the lateral spreading method are compiled in Table 7.3. Some of the results given here have been included in Figure 7.27 for a comparison with the tracer results. As a result of a wider applicability of the surface techniques for diffusion studies, the experiments listed in Table 7.3 cover more material combinations than the tracer profiling measurements. The majority of the studies were carried out by using Auger surface accumulation technique. Most of the surface measurements were made in a temperature range of about $T_m/3$, which should yield the diffusion kinetics in the type C regime (see discussion in Section 7.4.3). Results given in Table 7.3 also include data obtained by other measuring techniques in a higher-temperature range, and consequently the diffusion condition cannot be classified as type C kinetics. In fact, the Sn lateral spreading measurement (53) can be classified in the type A regime of diffusion kinetics by virtue of its lower melting points.

In tabulating the results given in Table 7.3, we found that the diffusion kinetics, data analysis, and signal calibration have often been treated in a very simplified manner. In fact, some of the diffusivities have been estimated simply by equating the film thickness to the diffusion distance, and the diffusivity so determined is attributed to that of the grain boundary. As discussed in Section 7.4.3.1, the proper use of the surface accumulation method to measure grain boundary diffusivity requires a detailed analysis for the coupled diffusion between the boundary and the source and exit surfaces. The resultant diffusivity should be derived from the rate of coverage on the exit surface. And if AES is used to measure the surface coverage rate, a quantitative calibration has to be carried out to ensure a proper conversion from the measured Auger peak height to the amount of diffusant on the exit surface. For results obtained by the Auger sputter profiling technique, there are similar questions concerning the effects of sputter damage on the diffusivity, a problem which we have discussed in Section 7.4.3.2. In Table 7.3, we have not attempted to reanalyze the observed data, and the diffusivities given there are as reported.

In addition to the problems related to the measuring technique discussed above, two other points are worth noting concerning the effects of sample structure on the measured diffusivity: the first is about the effect of interfacial compound formation, and the other is about grain structure change. In studying diffusion in thin films, the choice of materials has often been motivated by the practical interest in investigating material reactions in multilayered electronic devices, so the material combination is usually not optimized for diffusion studies. Of particular concern in this aspect is the possible effect of compound formation at the diffusion interface on the mass transport. In some of the systems listed in Table 7.3, for example, Cu/Au (101, 102) and Si/W (69), the formation of an interfacial compound has been detected and appeared to have slowed down the mass transport. In this context, an oxide layer at the diffusion interface has also been observed to be an effective barrier for mass transport. Although little information is available presently on diffusion through compound layers, the effect may be significant and certainly deserves consideration. The effect of grain structure change arises from the fact that the bimetallic diffusion couple is often prepared by sequential evaporations at room temperature. Upon diffusion annealing, grain growth or recrystallization often occurs in such samples which can change the intrinsic structure of the diffusion path and may in addition alter the morphology of the diffusion interface. Such effects have frequently been ignored in diffusion studies. Fortunately, the effect can usually be avoided by stabilizing the grain structure of the matrix film before diffusion annealing.

Among the results given in Table 7.3, it is interesting to compare the results of Ag diffusion in Au films. Of the four measurements, the results of Meinel et al. (100) are below those of the others by several orders of magnitude, and the activation energy is considerably higher than that obtained by Hwang and Balluffi (30). The measurements were made with very thin films with thickness varying between 55 and 200 Å, so the results would be most affected by lattice contribution which could lead to the high activation energy and low diffusivities observed. Even discounting this particular experiment, the other three sets of data differ by a factor of about 10^4 at 30°C (Hwang and Balluffi's results give a diffusivity of 1.5×10^{-16} cm^2 s^{-1} at this temperature). The discrepancy may be due to factors such as different grain structures, the measuring techniques, and the fact that the diffusivities in the two room-temperature experiments were estimated from the diffusion distance. A discrepancy of comparable magnitude also exists in the measurements of Cu diffusion in Al films at 175°C by AES profiling (90) and lateral spreading techniques (104), and in the measurements of Cu in Au films by AES (101) and He backscattering techniques (102). At present, the cause for such discrepancies in thin film diffusion results is not well understood.

In spite of the discrepancies among measurements for some of the studies, some of the Auger results displayed in Figure 7.27 seem to show a trend in the temperature dependence of the diffusivity consistent with that of the tracer profiling results. The overall comparison of the results appears to indicate that at the present stage of thin film diffusion studies, quantitative agreements within an order of magnitude are difficult to achieve. However, improvements are certainly expected with better understanding on the analysis of the diffusion kinetics and more appropriate use of the measuring techniques.

7.6 INFLUENCE OF SOLUTES ON GRAIN BOUNDARY DIFFUSION

The effects of solute additions on the diffusion behavior of the solvent atoms in the grain boundaries have special importance in reliability and degradation of thin film device metallization. The solute effect comes about in devices either through multilevel metallization in order to accomplish objectives such as improved adhesion with the substrate or from barriers to prevent corrosion and interdiffusion of the metallic components. Sometimes solutes are directly incorporated in thin films to improve their specific properties. One finds an excellent example of such additions in the beneficial effect of ~ 4 at. % Cu added to Al thin film conductor (108) which reduces the low-temperature electromigration damage markedly and contrary to the normal behavior of solute atoms in the lattice where the electromigration has been known to get accelerated (109). However, not all solutes reduce GB diffusion; surveying the literature (8, 9) in this regard, one comes across many possible trends—reduction of the GB diffusion by some solutes, increase by others, and no observable effect in some cases. It has been shown by Gupta and Rosenberg (47) that, even the same solute, Ta may exhibit a different effect in altering the GB diffusion of Au, as shown in Fig. 7.27. An explanation was first advanced by Rosenberg (110) for the solute effect based on an interaction between the solute atoms and the GB defects which accounted for the reduction in the GB solvent diffusivities, thereby improving electromigration lifetimes. Recently, Gupta (111) has used a thermodynamic model to explain the solute effect on solvent GB diffusion. The model appears to explain the effect of Ta addition in Au and also of Cr, Ti, Te, and Zr additions in Cu reported earlier by Barreau et al. (112).

In this section the solute effect in altering the GB diffusion of the solvent is discussed, and the alloy data in Figure 7.27 are explained phenomenologically by taking into account the solute interaction with the defects both in the lattice and the GBs and the equilibrium solute adsorption at the GB. The underlying thermodynamic principles are also discussed briefly.

7.6.1 Relationship Between Self-diffusion and Grain Boundary Energy

It is now possible to relate the increased atomic mobility in the GBs compared to that in the lattice with the state of vacancies in the two regions through the

Borisov et al. (113) relationship. The basic postulate in their approach is that the Gibbs free energy ΔG_b for activation of an atom vacancy jump in the GB, to the first approximation, is reduced by the absolute GB energy γ_b from its level ΔG_l in the lattice, as shown in Figure 7.28. Thus,

$$\Delta G_b + \gamma_b \simeq \Delta G_l \tag{105}$$

For vacancy diffusion in metals, the following thermodynamic relationships hold in the lattice:

$$\Delta G_l = \Delta H_l - T\Delta S_l \tag{106}$$

$$\Delta H_l = \Delta H_l^m + \Delta H_l^f = Q_l \tag{107}$$

$$\Delta S_l = \Delta S_l^m + \Delta S_l^f \tag{108}$$

and

$$D_l^0 = ga^2 fv(\Delta S_l/R) \tag{109}$$

where the subscripts m and f (see also Fig. 7.28) refer to the formation and motional components of the Gibbs energy, enthalpy, or the entropy for a vacancy; g is a geometric factor; f is the correlation factor for tracer diffusion

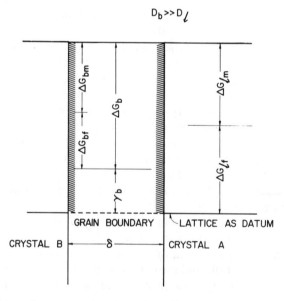

Figure 7.28. Schematic representation of the relationship of the free energies of activation for atom vacancy jumps in the lattice and the grain boundary with the grain boundary absolute energy. After reference 113.

only; a is the lattice parameter; and v is the lattice frequency. Similar relationships can also be written for GB diffusion by replacing the subscript l with b. The geometric factor g is $\frac{1}{12}$ and $\frac{1}{8}$ in lattice and GB, respectively, and the corresponding f values are 0.78146 and ~ 0.54 in the FCC lattice (14, 92). Further, the average jump frequency may be assumed to remain unaltered in the two diffusion paths. The product of the nonexponential terms ga^2fv in Eq. (109) is then $\sim 10^{-3}$ cm^2 s^{-1} both for the lattice and the GBs. By using Eqs. 106 to 109 written for the lattice and the GBs, Eq. 103 can be reduced to the useful form of the Borisov's relationship given below:

$$\gamma_b \, (\text{erg cm}^{-2}) = \left(\frac{4.18 \times 10^7}{2a^2 N_0} \right) [RT \ln (D_b^0/D_l^0) + (Q_l - Q_b)] \qquad (110)$$

where N_0 is the Avogadro number and $\delta \sim a$ is assumed.

Physically, Eqs. 105 and 110 imply that the GBs by virtue of their energies are the preferred sites for mobile defects, primarily the vacancies, and their total activation energy for diffusion $Q_b < Q_l$. Incidentally, the formation and motional components of self-diffusion energy in the GB are expected to be reduced unequally. Experimentally, the ratio of $Q_b/Q_l \simeq 0.5 \ (\pm 0.05)$ is considered appropriate for the average high-angle GBs in random polycrystalline specimens, and $\sim 0.7 \ (\pm 0.1)$ is considered appropriate for special boundaries including the cases of the epitaxial and textured films (97). The validity of the relationship of Borisov et al. has been checked several times (114, 115) by using the self-diffusion data in the lattice and GBs available in some metals for calculating the GB energies as a function of temperature and comparing them with those obtained by direct means.

In Figures 7.29 and 7.30, the GB energies thus computed are shown for the cases of (a) Au and Au–1.2 at. % Ta alloy (47) and (b) Cu and Cu alloys from the Ag tracer diffusion of Barreau et al. (112). The observed temperature dependences of γ_b for the above cases are listed in Table 7.4, from which γ_b can be linearly extrapolated to 850°C for the purpose of making comparisons with those obtained by direct means by Hilliard and co-workers (117). The extrapolated values of γ_b for Au and Cu are respectively 378 and 638 ergs cm^{-2} and the corresponding values from the direct measurements are 370 and 630 ergs cm^{-2}, showing an agreement of a few percent in both cases. It appears that the computation of GB energies from the self-diffusion data can be reliable when the data are available with good precision and at low enough temperatures. The small negative temperature coefficients of γ_b for pure Au and Cu listed in Table 7.4 can be interpreted to within a magnitude as entropies of the GBs. Furthermore, the GB energies obtained for the Au and Cu alloys are lower than those in the pure metals and have large positive temperature coefficients. Both the observations are related to the solute effect in altering the solvent GB diffusion. It is also noteworthy in Figures 7.29 and 7.30 that

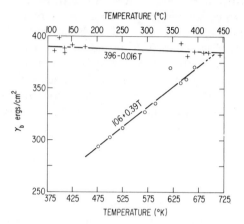

Figure 7.29. Temperature dependence of the grain boundary energies in Au (+) and Au–1.2 at. % Ta alloy (○) determined from the self-diffusion (^{195}Au) measurements in the lattice and the grain boundaries. Note the change of slope in the latter.

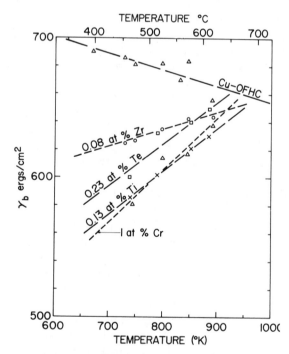

Figure 7.30. Temperature dependence of grain boundary energies in Cu and Cu alloys determined from the diffusion of ^{110}Ag tracer. From reference 112.

236

TABLE 7.4.   GB ENERGIES FROM SELF-DIFFUSION AND THE GB ENRICHMENT OF THE SOLUTE IN SOME ALLOYS

Material, at. %	Temp. range, °C	γ_b, ergs cm^{-2} (T in °K)	GB enrichment C_b/C_0	ΔE_a, kcal mole^{-1}	ΔS_a R units	Solubility, at.%
Au[a]	367–444	$396 - 0.016T$	—			
Au–1.2 Ta[a]	204–394	$106 + 0.39T$	2000–100	7.6	1.3	~0.5[c]
Cu[b]	398–618	$776 - 0.123T$	—			
Cu–1.0 Cr[b]	475–618	$314 + 0.364T$	256–21	20.0	4.0	~0.06[c]
Cu–0.13 Ti[b]	465–611	$353 + 0.315T$	2000–360	15.0	1.3	~0.5[c]
Cu–0.08 Zr[b]	456–618	$536 + 0.124T$	1500–170	15.6	1.6	~0.01[c]
Cu–0.23 Te[b]	465–611	$376 + 0.311T$	~2×10^5	—	—	trace[d]

[a]^{195}Au diffusion. From D. Gupta, *Phil. Mag.*, **33** 189 (1976).
[b]^{110}Ag diffusion. From G. Barreau, G. Brunel, G. Cizeron, and P. Lacombe, *Mem. Sci. Rev. Metall.*, **68**, 357 (1971).
[c]Solubilities are obtained from *Constitution of Binary Alloys* by M. Hansen, II, Ed. (1958) and its supplements.
[d]GB precipitates possible.

the γ_b values in alloys tend to approach the values in the pure metals as temperatures increase because of the solute dispersion effect into the adjoining lattice, and any crossover between the GB energy plots is not permissible.

7.6.2   Effect of Solutes on Grain Boundary Energies Derived from Self-diffusion

The effect of solute in lowering the GB energies in solids has been known for sometime, and the underlying thermodynamic bases have been discussed by Murr (118) and Hondros (93, 119). In essence, the Gibbs adsorption theorem is considered to be applicable to the GBs as well, analogous to the lowering of surface tension in liquids, but only in situations where the solute segregation at GBs is reversible and of equilibrium nature. The dilute homogeneous binary alloys at constant temperature are consequently relevant in this context which are also of interest in the present discussion. The GB energy lowering obtained in Figures 7.28 and 7.29 from the diffusion data following the solute additions have been analyzed to determine the solute excess at the GBs.

McLean's (120) has derived the following expression for the equilibrium GB segregation in solids:

$$C_b = C_0 \exp(\Delta G_a/RT)/[1 + C_0 \exp(\Delta Ga/RT)] \qquad (111)$$

where C_b is the solute concentration at the GB, C_0 is the solute concentration in the lattice, and ΔG_a is the associated solute-binding free-energy difference between the GB and the lattice sites. Hondros (119) combined the concept of the Gibbsian energy lowering with Eq. 111 as follows:

$$\gamma_{bp} - \gamma_b = RT \phi \ln[1 + C_0 \exp(\Delta G_a/RT)] \qquad (112)$$

where γ_{bp} is the GB energy of the pure solvent metal and ϕ depends on the level of the GB saturation to be taken equal to unity. The GB energy lowering derived from the self-diffusion measurements is assumed to be valid for the evaluation of ΔG_a and hence the GB solute excess C_b. It is further assumed that $\Delta G_a = \Delta E_a - T\,\Delta S_a$, where ΔE_a is the energy and ΔS_a is the entropy change associated with the GB–solute interaction, and all these quantities can be determined individually from the temperature dependence of the lattice and GB diffusivities in the pure solvent and the alloy. We note that all the diffusivities in both the lattice and the GBs have to be for the *solvent species* as the solute diffusivities do not enter in the above formalism. In Figure 7.31 the Arrhenius behaviors of the GB solute enrichment factors (C_b/C_0) determined according to Eq. 112 are displayed from a recent publication (111); the values of the solute–GB interaction parameters ΔE_a and ΔS_a are listed in Table 7.4. It is seen that the GB–solute interaction parameters obtained from the self-diffusion measurements in general have the same magnitudes as those obtained by other techniques (93) and are commensurate with the prevailing strain energy models. Incidentally, the GB–solute enrichment factors obtained from diffusion measurements also show the general correlation with solid solubility similar to those obtained by other techniques, notably the AES (111).

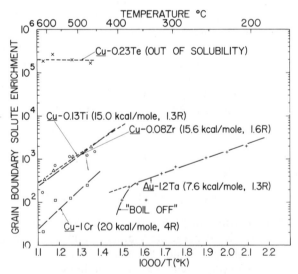

Figure 7.31. Arrhenius dependence of the GB solute enrichment factors in Au–1.2 at. % Ta and Cu alloys.

7.6.3.   Variable Effects of Solutes on Grain Boundary Diffusion

For explaining the reported variable effects of the solute additions on the GB diffusion, Eqs. 110 and 112 are combined, taking ϕ, the GB saturation factor, equal to unity and the terms rearranged as follows:

$$\left[\left(\frac{D_b^p}{D_b^a}\right)\left(\frac{D_l^a}{D_l^p}\right)\right]^{1/2} - 1 = C_0 \exp\left(\Delta G_a/RT\right) \tag{113}$$

where the superscripts a and p refer to the alloy and the pure solvent, respectively. It is immediately clear from Eq. 113 how the changes in the GB diffusion in the solvent are brought about by the solute addition. The solute GB interaction free energy only partially explains the effect, and in addition the changes in the lattice diffusion also need to be accounted for. All the three possibilities for solute effect on the GB diffusivities—enhancement, supression, or no effect alluded to earlier—can be explained according to Eq. 113. These cases are discussed below individually.

7.6.3.1 Solute Increasing the GB Diffusion at High Temperatures.   Since for dilute alloys we always have $C_0 \ll 1$, at some high temperature the product $C_0 \exp\left(\Delta G_a/RT\right)$ in Eq. 113 is sufficiently small so that it can be neglected. Thus, Eq. 113 reduces to

$$\left(\frac{D_b^p}{D_b^a}\right)\left(\frac{D_l^a}{D_l^p}\right) \sim 1 \tag{114}$$

As D_l^a is always greater than D_l^p it will be required that D_b^p be the less than D_b^a, which is the case of the GB diffusion enhancement in the Au–1.2 at. % Ta at high temperatures compared to pure Au (see Fig. 7.27). The effect, however, would only extend to a limited temperature region since the ratio D_l^a/D_l^p itself declines and approaches unity with increasing temperatures.

7.6.3.2 Solute Decreasing the GB Diffusion at Low Temperatures.   At lower temperatures, despite $C_0 \ll 1$ in dilute alloys, the product $C_0 \exp\left(\Delta G_a/RT\right) \gg 1$; and even though $D_l^a/D_l^p \geqslant 1$, it may not be sufficiently large to satisfy the relationship (113). Consequently, $D_b^p/D_b^a > 1$ would be required, thus resulting in *decreased* GB diffusion of the solvent due to the solute addition.

Between the effects discussed above, of course, there would be a region of no effect because of the crossover.

In general, the contributing factors for the variable solute effects on the GB diffusion are expected to be the solute–GB interaction parameters (ΔE_a and ΔS_a) and the self-diffusion parameters both in the lattice and the GBs, namely D_l^0, D_b^0, Q_l, and Q_b for pure and alloyed solvent. It is therefore not possible generally to predict the solute effect with any certainty in situations where these informations are not fully available. Also, the temperature regions where these effects may be observed are expected to be influenced by contaminants and the thin film growth conditions.

REFERENCES

1. J. C. Fisher, *J. Appl. Phys.*, **22**, 74 (1951).
2. R. E. Hoffman and D. Turnbull, *J. Appl. Phys.*, **23**, 634 (1951).
3. R. F. Mehl, *Trans. Amer. Inst. Min. Metall. Eng.*, **122**, 11 (1936).
4. D. Turnbull, *Atom Movements*, Amer. Soc. Metals, Cleveland (1951), p. 129.
5. P. G. Shewmon, *Diffusion in Solids*, McGraw-Hill, New York (1963), p. 164.
6. N. A. Gjostein, in *Techniques of Metals Research*, Vol. IV, part 2, R. A. Rapp, Ed., Interscience, New York (1970), p. 405.
7. N. A. Gjostein, *Diffusion*, Amer. Soc. Metals, Metals Park, Ohio (1973), p. 241.
8. H. Gleiter and B. Chalmers, *Progr. Mater. Sci.*, **16**, 77 (1972).
9. G. Martin and B. Peraillon, Proceedings of International Conf. on Grain Boundaries in Metals, *J. de Phys.*, **36**, *Coll. C4*-165 (1975).
10. A. Gangulee, P. S. Ho, and K. N. Tu, Eds., *Low Temperature Diffusion and Applications to Thin Films*, Elsevier, Lausanne (1975).
11. J. L. Walter, J. H. Westbrook, and D. A. Woodford, Eds., *Grain Boundaries in Engineering Materials*, Claitor's Pub. Div., Baton Rouge, La. (1975), p. 181.
12. L. G. Harrison, *Trans. Faraday Soc.*, **57**, 1191 (1961).
13. D. Gupta, *Phys. Rev.*, **7**, 586 (1973).
14. D. Gupta, *J. Appl. Phys.*, **44**, 4455 (1973).
15. D. Gupta and K. W. Asai, *Thin Solid Films*, **22**, 121 (1974).
16. R. T. P. Whipple, *Phil. Mag.*, **45**, 1225 (1954).
17. T. Suzuoka, *Trans. Jap. Inst. Met.*, **2**, 25 (1961).
18. A. D. LeClaire, *Brit. J. Appl. Phys.*, **14**, 351 (1963).
19. R. F. Cannon and J. P. Stark, *J. Appl. Phys.*, **40**, 4361 (1969).
20. T. Suzuoka, *J. Phys. Soc. Jap.*, **19**, 839 (1964).
21. G. B. Gibbs, *Phys. Stat. Solidi*, **16**, K27 (1966).
22. P. Benoist and G. Martin, *Thin Solid Films*, **25**, 181 (1975).
23. P. Benoist and G. Martin, *J. de Phys.*, **36**, C4, p. 213 (1975).
24. J. Unnam, J. A. Carpenter, and C. R. Houska, *J. Appl. Phys.*, **44**, 1957 (1973).
25. G. H. Gilmer and H. H. Farrell, *J. Appl. Phys.*, **47**, 3792 (1976).
26. G. H. Gilmer and H. H. Farrell, *J. Appl. Phys.*, **47**, 4373 (1976).
27. D. R. Campbell, *Bull. Amer. Phys. Soc.*, **19**, 347 (1974); detailed results to be published.
28. M. Wuttig and H. K. Birnbaum, *Phys. Rev.*, **147**, 495 (1966).
29. P. H. Holloway, D. E. Amos, and G. C. Nelson, *J. Appl. Phys.*, **47**, 3769 (1976); G. C. Nelson and P. H. Holloway, ASTM Special Publ. 596, Philadelphia (1976), p. 68.
30. J. C. M. Hwang, Ph.D. Thesis, Cornell University, Ithaca, New York (1977); J. C. M. Hwang and R. W. Balluffi, to be published.
31. H. S. Levine and C. J. Mac Callum, *J. Appl. Phys.*, **31**, 595 (1960).
32. D. R. Campbell, to be published.
33. P. S. Ho, to be published.
34. E. W. Hart, *Acta Met.*, **5**, 597 (1957).
35. G. Martin, *Phys. Stat. Solidi*, **A14**, 183 (1972).
36. P. S. Ho, J. E. Lewis and J. K. Howard, *Thin Solid Films*, **25**, 301 (1975); detailed results to be published by P. S. Ho.
37. K. L. Tai and M. Ohring, *J. Appl. Phys.*, **48**, 28 (1977).
38. A. Gainotti and L. Zecchina, *Nuovo Cimento*, **40B**, 295 (1965).
39. J. A. Davies, J. Friesen, and J. D. McIntyre, *Can. J. Chem.*, **38**, 1526 (1960).
40. R. E. Pawel and T. S. Lundy, *J. Phys. Chem. Solids*, **26**, 937 (1965).
41. D. R. Campbell, K. N. Tu, and R. O. Schwenker, *Thin Solid Films*, **25**, 213 (1975).
42. W. Rupp, U. Ermert, and R. Sizmann, *Phys. Stat. Solidi*, **33**, 509 (1969).

43. N. Q. Lam, S. J. Rothman, H. Mehrer, and L. J. Nowicki, *Phys. Stat. Solidi*, **57**, 225 (1973).
44. D. Gupta and R. T. C. Tsui, *Appl. Phys. Lett.*, **17**, 294 (1970).
45. K. Maier, H. Mehrer, E. Lessmann, and W. Schüle, *Phys. Stat. Solidi*, **78**, 689 (1976).
46. D. Gupta, *Thin Solid Films*, **25**, 231 (1975).
47. D. Gupta and R. Rosenberg, *Thin Solid Films*, **25**, 171 (1975).
48. P. S. Ho and J. K. Howard, *J. Appl. Phys.*, **45**, 3229 (1974).
49. P. S. Ho, J. E. Lewis, and J. K. Howard, *Thin Solid Films*, **25**, 301 (1975).
50. T. F. Archbold, Ph.D. Thesis, Purdue University (1961).
51. D. Gupta, *Abs. Bull. Metall. Soc. AIME*, **4**, 117 (1970).
52. K. L. Tai, P. H. Sun, and M. Ohring, *Thin Solid Films*, **25**, 343 (1975).
53. P. H. Sun and M. Ohring, *J. Appl. Phys.*, **47**, 478 (1976).
54. T. J. Renouf, *Phil. Mag.*, **9**, 781 (1964).
55. T. J. Renouf, *Phil. Mag.*, **22**, 359 (1970).
56. J. P. Stark and W. R. Upthegrove, *Trans. Amer. Soc. Met.*, **59**, 479 (1966).
57. A. W. Rogers, *Techniques in Autoradiography*, Elsevier, New York (1967).
58. K. L. Tai and M. Ohring, *J. Appl. Phys.*, **48**, 36 (1977).
59. For a compilation of AES work, see *Auger Electron Spectroscopy*, P. F. Kane and G. B. Larrabee, Eds., Plenum Press, New York (1977).
60. T. M. Bulk, in *Methods of Surface Analysis*, Vol. I, A. W. Czanderna, Ed., Elsevier, Amsterdam and New York (1975), Chapter 3.
61. W. R. Graham and G. Ehrlich, *Thin Solid Films*, **25**, 85 (1975); T. T. Tsong, P. Cowan, and G. Kellog, *ibid.*, p. 97.
62. C. C. Chang, in *Characterization of Solid Surfaces*, P. F. Kane and G. B. Larrabee, Eds., Plenum Press, New York (1974), p. 509.
63. A. Joshi, L. E. Davis, and P. W. Palmberg, in *Methods of Surface Analysis*, A. W. Czanderna, Ed., Elsevier, Amsterdam and New York (1975), Chapter 5.
64. P. W. Palmberg, *Anal. Chem.*, **45**, 549 (1973); *J. Vac. Sci. Technol.*, **13**, 214 (1976).
65. P. W. Palmberg, *J. Vac. Sci. Technol.*, **9**, 160 (1972).
66. P. S. Ho and J. E. Lewis, to be published.
67. J. Ahn, C. R. Perleberg, D. L. Wilcox, J. W. Coburn, and H. F. Singers, *J. Appl. Phys.*, **46**, 4581 (1975).
68. M. J. Gibson and P. J. Dobson, *J. Phys. F: Met. Phys.*, **5**, 1828 (1975).
69. C. C. Chang and G. Quintana, *J. Electron. Spectrom. Related Phenom.*, **2**, 363 (1973).
70. J. M. Morabito, *Surf. Sci.*, **49**, 318 (1975).
71. C. C. Chang, *Surf. Sci.*, **48**, 9 (1975); see also references 64 and 65.
72. S. H. Overbury and G. A. Somorjai, *Surf. Sci.*, **55**, 209 (1976).
73. J. W. Wilson, *Phil. Mag.*, **27**, 1467 (1973).
74. J. W. Coburn and E. Kay, *Crit. Rev. Solid State Sci.*, **4**, 561 (1974).
75. G. K. Wehner, in *Methods of Surface Analysis*, A. W. Czanderna, Ed., Elsevier, Amsterdam and New York (1975), Chapter 1.
76. H. H. Anderson, *Proc. 7th Symp. on Physics of Ionized Gases*, Roving, Yugoslavia (1974).
77. R. S. Nelson and D. J. Mazey, *Rad. Eff.*, **18**, 199 (1973).
78. P. S. Ho and J. E. Lewis, *Surf. Sci.*, **55**, 335 (1976).
79. W. K. Chu, private communication.
80. P. S. Ho and J. E. Lewis, unpublished results; see also G. K. Wehner, in *Scanning Electron Microscopy*, Part I, I.I.T. Research Institute, Illinois (1975), p. 133.
81. P. Sigmund, *Phys. Rev.*, **184**, 383 (1969); see also references 63 and 74.
82. P. S. Ho, J. E. Lewis, H. S. Wildman, and J. K. Howard, *Surf. Sci.*, **57**, 393 (1976).
83. W. L. Patterson and G. A. Shirn, *J. Vac. Sci. Technol.*, **4**, 343 (1967).
84. H. Shimizu, M. Ono and K. Nakayama, *Surf. Sci.*, **36**, 817 (1973).

85. H. F. Winters and J. W. Coburn, *Appl. Phys. Lett.*, **28**, 176 (1976).
86. P. S. Ho, J. E. Lewis, and J. K. Howard, *J. Vac. Sci. Technol.*, **14**, 322 (1977).
87. P. H. Holloway, *Surf. Sci.* **66**, 479 (1977).
88. P. M. Hall and J. M. Morabito, *Surf. Sci.*, **62**, 1 (1977).
89. J. M. Poate, W. L. Brown, R. Homer, W. M. Angnstyniak, J. W. Mayer, K. N. Tu, and W. F. van der Weg, *Nucl. Instr. Methods.*, **132**, 345 (1976).
90. H. S. Wildman, J. K. Howard, and P. S. Ho, *J. Vac. Sci. Technol.*, **12**, 75 (1975).
91. R. W. Balluffi and J. M. Blakely, *Thin Solid Films*, **25**, 363 (1975).
92. J. T. Robinson and N. L. Peterson, *Surf. Sci.*, **31**, 586 (1972).
93. E. D. Hondros, *J. de Phys.*, **36**, Coll 4-117 (1975).
94. D. Gupta and K. W. Asai, *Elect. Chem. Soc. Abstr.*, **75-1**, 255 (1975).
95. W. R. Upthegrove and M. J. Sinnott, *Trans. Amer. Soc. Met.*, **50**, 1031 (1958).
96. R. F. Cannon and J. P. Stark, *J. Appl. Phys.*, **40**, 4366 (1969).
97. R. W. Balluffi, *Phys. Stat. Solidi*, **42**, 11 (1970).
98. A. W. Czanderna and R. Summerwatter, *J. Vac. Sci. Technol.*, **13**, 384 (1976).
99. R. G. Kirsch, J. M. Poate, and M. Eibschutz, *Appl. Phys. Lett.*, **29**, 772 (1976).
100. K. Meinel, M. Klaua and H. Bethge, *Thin Solid Films*, **34**, 157 (1976).
101. H. G. Tompkins and M. R. Pinnel, *J. Appl. Phys.*, **47**, 3804 (1976).
102. J. A. Borders, *Thin Solid Films*, **19**, 359 (1973).
103. H. G. Tompkins, *J. Electrochem. Soc.*, **122**, 983 (1975).
104. P. S. Ho and J. K. Howard, to be published.
105. J. K. Wood, J. L. Alvarez, and R. Y. Manghan, *Thin Solid Films*, **29**, 359 (1975).
106. S. Danyluk, G. E. McGuire, K. M. Koliwad, and M. G. Yang, *Thin Solid Films*, **25**, 483 (1975).
107. C. C. Chang and G. Quintana, *Appl. Phys. Lett.*, **29**, 453 (1976).
108. I. Ames, F. M. d'Heurle, and R. E. Horstmann, *IBM J. Res. Dev.*, **14**, 461 (1970).
109. P. S. Ho, *Phys. Rev.*, **8**, 4534 (1973).
110. R. Rosenberg, *J. Vac. Sci. Technol.*, **9**, 263 (1972); R. Rosenberg, A. F. Mayadas, and D. Gupta, *Surf. Sci.*, **31**, 566 (1972).
111. D. Gupta, *Metall. Trans. A.*, 8A, 1431 (1977).
112. G. Barreau, G. Brunel, G. Cizeron, and P. LaCombe, *Mem. Sci. Rev. Metall.*, **68**, 357 (1971).
113. V. T. Borisov, V. M. Golikov, and G. V. Scherbedinskiy, *Phys. Met. Metall.*, **17**, 80 (1964).
114. D. Gupta, *Phil. Mag.*, **33**, 189 (1976).
115. J. Pelleg, *Phil. Mag.*, **14**, 594 (1966).
116. M. Hansen, *The Constitution of Binary Alloys II* (1958) and its supplements by R. P. Elliott (1965) and F. A. Shunk (1969); McGraw Hill, New York.
117. J. E. Hilliard, M. Cohen, and B. L. Averbach, *Acta Met.*, **8**, 26 (1960).
118. L. E. Murr, in *Interfacial Phenomena in Metals and Alloys*, Addison Wesley, Reading, Mass. (1975), p. 33.
119. E. D. Hondros, *Proc. Interface Conference*, Australian Inst. of Metals, Butterworth, London (1969), p. 77.
120. D. McLean, *Grain Boundaries in Metals*, Oxford University Press, London (1957), p. 128.

8

ELECTROMIGRATION IN THIN FILMS

F. M. d'Heurle and P. S. Ho

IBM Thomas J. Watson Research Center, Yorktown Heights, New York

8.1 INTRODUCTION

In the general context of interactions between metallic thin films, electromigration occupies a particular place since the phenomena involved do not result from material transport caused by chemical potential gradients. Instead, the transport in electromigration is due to the interaction between the atoms of a conductor and a direct current flowing through this conductor. Although alloy effects can play a significant role in electromigration studies on thin films, the main concerns are with self-transport, the motion of atoms in a pure material, and the motion of the solvent atoms in an alloy. However, electromigration, like many other transport phenomena in thin films, generally occurs via grain boundary diffusion, so that the mechanism of atomic transport is common to electromigration and most thin film interactions of technological interest. In electromigration the driving force is usually considered to be the sum of two effects: the electrostatic interaction between the electric field and the ionic core of the atoms stripped of their valence electrons, and a friction force between these ions and the flowing charge carriers, which is often called the "electron wind" force. In metals, which are good electrical conductors, the electron wind force is usually dominant.

Electromigration effects are not expected under ac conditions because of the cancellation in the mass transport occurring repeatedly in opposite directions. Even under direct currents at the low current densities that normally obtain in most applications, electromigration can be neglected. In the present semiconductor technology, for reasons which are discussed further on, the metallic thin films used as conductors are submitted to current densities which are 100 to 1000 times larger than the "normal" densities. Under such conditions electromigration can cause severe damage as seen in Figure 1, which shows holes formed by electromigration in the negative terminal of an Al thin film conductor (1). Here, the film is a single crystal. In a polycrystalline film under the same conditions of temperature and current density, the transport would have occurred more rapidly and the conductor would have quickly failed through a process whereby the holes seen in Figure 8.1 would be joined together along some grain boundaries, thus forming a crack across the conductor and making it electrically discontinuous.

The interest in studying electromigration in thin films started approximately 10 years ago after Blech and Sello (2) observed that electromigration in Al thin film conductors could be a significant source of failure in planar semiconductor circuits. This discovery ushered a period of intense activity in the investigation of electromigration phenomena in thin films, which in certain respects are quite distinct from the effects that had formerly been studied in "bulk" or liquid samples. A fairly comprehensive review of the ear y work on electromigration in thin films appeared in 1973 (3). In the last

Figure 8.1. Scanning electron microscope picture of voids formed by electromigration in the negative terminal of a single-crystal Al–3wt % Mg film. Test conditions: $176°C$, 4×10^6 A cm^{-2}, 16,000 h. From reference 1.

few years, with the development of new measuring techniques, a considerable amount of quantitative measurements has been carried out on pure as well as alloyed films. In the present paper the experimental results will be analyzed, with emphasis placed on the most commonly studied materials, in the hope that some general trends in the study of electromigration in thin films shall be brought to light. The chapter is essentially divided into three parts. First, the nature of mass transport at grain boundaries and the specific thin film problems of ion flux divergence are discussed. Next, the methods of measurements and the results obtained in pure and alloyed films are summarized.

In the last part the failure models used to predict electromigration lifetimes and various methods used for increasing these lifetimes are evaluated.

Although concern about electromigration in thin films is quite recent, other electromigration phenomena have been studied for a long time. In 1861, Gerardin (4) discovered electromigration in liquid alloys of Pb–Sn, K–Na, Au, and Bi in Hg. At the time he did not distinguish the results obtained with liquid metals from those observed concurrently with molten salts. The first inclination is to think in terms of the electrostatic interactions between the field and metallic ions. Skaupy (5) in 1914 suggested the importance of the interaction between atoms and moving charge carriers. The significance of this latter factor was underlined in experiments conducted by Seith and Wever (6) who showed that in a series of solid alloys the direction of transport varies with the band structure and with the sign of the charge carriers. Essentially similar mathematical formulations of the electromigration forces have been derived by Fiks (7), Huntington and Grone (8), and Bosvieux and Friedel (9). In many of the early studies, electromigration was used to investigate the interactions between point defects and charge carriers.

The subject has been reviewed recently in a series of books and monographs (10–14), including one that refers mostly to liquid alloys (12) and two that deal simultaneously with the related topic of thermomigration (13, 14). Electromigration phenomena in bulk materials continue to elicit interest. Work in this field now includes studies of electromigration in anisotropic crystals Sn, (15), Mg (16), and Be (17), of the transport of fast diffusing impurities in Pb (18), and of the light isotopes H and D in Ag (19). Investigators prompted by the practical goal of obtaining metals of high purity have extended their work on the removal of the interstitial elements C, N, and O from transition metals Mo and U (20), Nb and Ta (21, 22), Sc (23), Fe, and Ti (24). This purification work is now encompassing studies of the rare earths Gd (25), Tb (26), Yb, Lu, and Gd (27). Technically important electrical instabilities may have motivated studies of electromigration in chalcogenide glasses (28–30). Other recently published works (31–35) include two studies on electromigration in Au (31, 33) and one on α-brass (34).

8.2 GENERAL ASPECTS OF ELECTROMIGRATION IN THIN FILMS

In discussing electromigration in thin films, it is useful to start with a description of those salient features which distinguish electromigration in films from its counterpart in bulk materials. But for a few exceptions, for example, a study of the electromigration of Sb in an Ag bicrystal (36), studies of electromigration in bulk samples have been conducted at relatively high temperatures (close to the melting point), at moderate current densities (approximately 10^4 A cm^{-2}), under conditions where the dominating mode of transport is lattice diffusion. The parameters selected for the study of thin films have been largely dictated by the conditions that prevail in the use of thin films as conductors in electronic devices. Thus, electromigration phenomena in thin films have been studied mostly at high current densities (10^5 to 10^7 A cm^{-2}), at relatively low temperatures ($0.3T_m < T < 0.7T_m$, where T_m is the melting temperature of the material), under conditions where lattice diffusion becomes vanishingly small, and where transport occurs mostly via grain boundary diffusion. While the dissipation of Joule heat sets limits to the maximum current density which may be attained in bulk samples, the effect of thermal generation in thin film samples is minimized by the very small cross section of the conductors, and the very effective cooling by the substrates (which is especially true for substrates with a high thermal conductivity like Si). The preponderant role of grain boundary diffusion in thin films results not only from the low experimental temperatures, as already noted, but also from the usually very small grain size. While a significant amount of attention continues to be paid to the electromigration of impurities in bulk samples, the role of alloy additions plays a more important role in the investigations of electromigration in thin films. The starting point was the discovery that Cu additions are quite effective in reducing electromigration damage in Al thin film conductors (37). Because of the technical implications much of the work on alloy additions has been centered on lifetime investigations. Yet, through the use of a specific measurement technique, quantitative data have been obtained on solute grain boundary transport (38). These have yielded important information on the mechanism of electromigration in grain boundaries.

8.2.1   Mass Transport

In a lattice the atomic flux J_l due to electromigration can be expressed as

$$J_l = \frac{N_l D_l}{kT} Z_l^* eE \tag{1}$$

where N_l is the atomic density, D_l is the diffusion coefficient (uncorrelated for pure metals), Z^*e is the effective charge, and kT has its usual meaning. The subscript l denotes parameters of the lattice. From theoretical considerations the effective charge is given by (8)

$$Z_l^* = z\left[1 - \gamma \left(\frac{\rho_d}{N_d}\right)\left(\frac{N_l}{\rho_l}\right)\frac{|m^*|}{m^*}\right] \tag{2}$$

where z is the electron/atom ratio (the number of condution electrons per atom) and the product $(\rho_d/N_d)(N_l/\rho_l)$ is the ratio of the specific resistivity of the moving defects to that of the lattice atoms. The term $|m^*|/m^*$ indicates the change in the force direction with the sign of the charge carriers; γ is an averaging term which is introduced to take into account the variation of the force along the length of an elemental atomic jump (it is usually assumed to be 0.5). In the right-hand side of Eq. 2, the first term, equal to z, is the electrostatic force between the ions and the field, while the second term, which is negative, is the "electron wind" force originating from the interaction between the ions and the moving charge carriers. The value of Z^* for Au (31) was determined to be about -7 around $1200°$ K, which indicates that the wind force is considerably larger (by a factor of 8) than the electrostatic force. In general, this is true for all metals which are good electrical conductors. While Eq. 2 gives a relatively straightforward formulation of the effective charge, the detailed nature of the electromigration forces is quite complex. A realistic approach requires considerations of the band structure of the metals, of screening effects, of the coupling between the moving defects and the lattice, both at the equilibrium positions and at the saddle points. Moreover, for matrix atoms and for substitutional impurities in close-packed metals, the situation is further complicated by the presence of the vacancies required for any diffusive motion. The considerable interest in these matters is reflected by a large number of publications (39–57). Among these, one may single out the use of the pseudopotential method to calculate electromigration forces (41, 57) and considerations about modulation of the charge carriers in the vicinity of a moving ion (43, 54). The discussion of these questions, which at best deal only with the lattice and ignore the more complex problem of the forces at play in grain boundaries, is beyond the scope of the present paper.

In polycrystalline films, it is necessary to take into account the transport which occurs via short circuit paths, mostly grain boundaries. For this purpose it is convenient to write a new equation for the atomic flux which is similar to Eq. 2. For a film with an ideally textured grain structure, as shown in Figure 8.2, the flux can be expressed as

$$J_b = N_b \frac{\delta}{d} \frac{D_b}{kT} Z_b^* e E \tag{3}$$

where δ is the effective width of the boundary, d is the average grain size, and the subscript b refers to boundary parameters. Equation 3 is based on the assumption that in all grain boundaries the transport occurs in channels with the same characteristics. For real films, Eq. 3 should be modified to take into account variations in the transport parameters of individual grain boundaries both as a function of the intrinsic structural properties of the boundaries and as a function of their geometrically random orientation. This latter effect has been analyzed for diffusion in polycrystalline bulk samples (58). Assuming that the respective grain boundary diffusivities can be adequately approximated by a mean value, in a sample with a perfectly random orientation the geometric correction is about 0.6. It is approximately equal to the average of the square of the cosine of the tilt angles between the boundaries and the

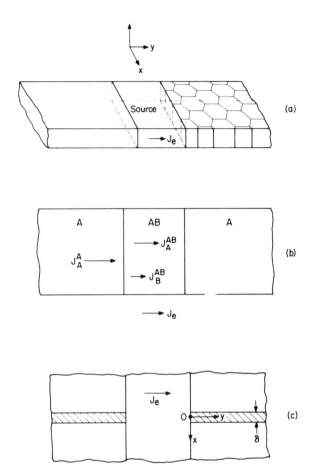

Figure 8.2. (*a*) Cross-stripe sample configuration for electromigration. The equiaxed grain structure with perfect texture normal to the surface is shown on the anode side. (*b*) Atomic fluxes at the cross-stripe edge due to the current J_e. (*c*) Grain geometry for obtaining a solution for grain boundary electromigration. From reference 87.

direction of mass flow. The same correction factor would be valid only for relatively thick films with a grain size smaller than the film thickness. For simplicity, it will be assumed that the grain boundary parameters in Eq. 3 are always suitably averaged so that the expression of the equation need not be changed.

The grain boundary flux given by Eq. 3 is not as well defined as the lattice flux given by Eq. 1 because of the imprecision in the available knowledge of grain boundary parameters, even if one assumes that the diffusivity and the effective charge for the grain boundaries are known. Precise values cannot be ascribed to N_b in the absence of a proper understanding of the grain boundary structures. For impurity atoms, this is further complicated by grain boundary segregation. One usually writes $N_b = \beta N_l$ where β, the enrichment factor, is a measure of the adsorption of solute atoms on grain boundaries. The value of β is expected to vary as $\exp(H_a/kT)$, with the heat of adsorption H_a usually being a positive quantity. The exact magnitude of β is generally unknown although a measure of it can be obtained from the empirical observation that for a number of alloy systems β is approximately proportional to the inverse of the solubility limit (59). When required, δ is usually assumed to be 10 Å. For lattice electromigration, the flux of atoms is directly proportional to the flux of charge carriers. For grain boundary transport, the problem is considerably complicated by uncertainties relating to the different resistivities in the boundaries and in the bulk, and the way these factors affect the local current density in (or at) the grain boundaries. As a result the local value of E at a boundary may differ from the overall average. Thus, even though one may experimentally measure J_b, and perhaps also D_b, there remains a considerable uncertainty about values of Z_b^* calculated from Eq. 3.

For usual polycrystalline materials such as most thin film conductors, the total atomic flux is the sum of the terms given by Eqs. 1 and 2 for lattice and grain boundary transport. One might have to consider other transport paths, such as along dislocations or along surfaces. However, in most thin films, transport along grain boundaries usually dominates. At 175°C, the ratio J_b/J_l has been estimated to be 10^6 for Al electromigration in Al films (60). In the case of the electromigration of Cu in Al films (61), a condition for which the various parameters are known with better precision, at 225°C that ratio is about 10^4. Hence, for practical purposes in considering electromigration phenomena in thin films, one may generally neglect the lattice mechanism of diffusion and pay attention to grain boundary transport exclusively.

8.2.2   Flux Divergence

Failure remains a matter of central importance in the study of electromigration in thin films. The principal mode of failure is the formation of open circuits due to ruptures in the continuity of the conductors in areas where atomic transport results in material depletion. Although being the most dramatic and the easiest to study, this is not the only damage caused by electromigration. The converse phenomenon of mass accumulation, in the form of hillocks or whiskers, may cause short circuits between adjacent or superimposed conductor lines. Alternatively, mass accumulation can cause breakage through insulating or protective layers and lead to subsequent corrosion. Electromigration is not the only cause of failure in metallizations on semiconductor devices; an overall view of the limitations of metallizations for such purposes can be obtained from recent articles (62–64) and in Chapter 2 of this book.

In the ideal case where temperature gradients could be eliminated, electromigration by itself in an homogeneous conductor cannot cause mass accumulation or depletion even if there are variations in the conductor cross section. The formation of electromigration damage requires the existence of specific sources of divergence in the flux of migrationg ions. To examine this problem, consider the following continuity equation where for simplicity only the transport along grain boundaries is considered:

$$\frac{dN_b}{dt} = -\nabla \cdot J_b + \frac{N_b - N_b^0}{\tau} \tag{4}$$

where τ is the average lifetime of the moving defect at the boundaries and N_b and N_b^0 are respectively the instantaneous and equilibrium defect concentrations at the boundaries. Equation 4 indicates that the local defect concentration depends both on the flux divergence and the lifetime of the defects. A steady-state condition, namely, $dN_b/dt = 0$, obtains when the defect supersaturation is proportional to the flux divergence. Including the diffusion term due to concentration gradients, Eq. 3 can be rewritten in the following form:

$$J_b = \frac{\delta D_b}{d}\left(-\nabla N_b + \frac{N_b}{kT}Z_b^* eE\right) \tag{5}$$

Its divergence becomes:

$$\nabla \cdot J_b = \frac{\delta D_b}{d}\left[-\nabla^2 N_b + \frac{N_b}{kT}Z_b^* eE\left(\frac{\nabla N_b}{N_b} + \frac{\nabla Z_b^*}{Z_b^*} - \frac{\nabla T}{T}\right)\right] + \left(\frac{\nabla D_b}{D_b} - \frac{\nabla d}{d}\right)J_b \tag{6}$$

Discontinuities in the value of the atomic flux J_b can be caused by a variety of factors: gradients in the concentration of defects, in the temperature, in the effective charge, in the diffusion coefficient, and in the grain size. Several of these factors, particularly the last two, are specific to polycrystalline materials; usually they could be ignored in considering bulk electromigration phenomena. In thin films a vanishing flux divergence requires the proper balance between the various contributions in Eq. 6, a condition that may be expected to prevail only under a very restricted set of circumstances.

Most of the factors contributing to the inhomogeneity of the atomic flux are related to the microstructure of the films. This is not true, however, of the thermal gradients. These have been put to good use by many an investigator of electromigration in bulk samples (8, 11). In thin films the importance of thermal effects tends to be minimized by the heat conductivity of the films themselves combined with cooling by the substrates, through the interface. Temperature gradients are almost non-existent along the length of the conductors, although they may play a role at the end of the conductors at the terminals with a large cross section. Thus, in tests with thin films it is usually found that failures are randomly distributed over the length of the conductors. However, failures caused by temperature gradients, generally localized at the negative end of the conductors, can be observed whenever the heat generated by the Joule effect cannot be properly dissipated. Under a given set of experimental conditions, this may occur in conductors with relatively large cross sections, or with a relatively high resistivity (especially with alloys), or in tests conducted at very high current densities. In practical applications, where the conductors may be used with active devices (with heat generated at the junctions) with resistors and under particular modes of heat sinking, the role of temperature gradients can be much more complicated. In experiments with unsupported films, such as for transmission electron microscopic studies, the separate formation of hillocks and voids at the two terminals is usually observed (65). The influence of thermal effects on the lifetimes of thin film conductors is analyzed in Section 8.6 on failure.

Other sources of flux divergence are related to the microstructure of the films. Many inhomogeneities occur as a result of local changes in microscopic configuration or the atomic structure of the grain boundaries. Two well-known inhomogeneities are those due to a localized change in grain size (66) and those found at the junction of three grain boundaries (the so-called triple point) (33). At a triple point, flux divergences can occur as a result of a nonequilibrium geometry or different intrinsic transport properties along the three boundaries. Visualizing the geometric effect is an easy matter; understanding the transport properties is less straightforward since these originate from the atomic interactions between the migrating defects and the boundaries. Because it still requires much clarification, this question has attracted a good deal of interest recently. The values of δD_b are known to vary extensively according to the nature of the boundaries. For example, the diffusion of Zn in the boundary of Al bicrystals (with a common [001] or [011] axis) can change by two or three orders of magnitude with changes in the tilt angle between the two adjoining crystals (67). The density of moving defects N_b also depends on the structure of the boundaries. There is no direct experimental information on the structural dependence of N_b. However, a large number of considerations, namely, about the internal atom distribution in grain boundaries (68), about variations in the energy of boundaries (69) and the adsorption of defects (70), and about the density of maximum energy sites (71), clearly indicate that N_b should be expected to vary from boundaries to boundaries. Variations in the effective charge Z_b^* as related to grain structures are unknown at present. Observations of failures

due to irregularities in grain size have been made in Al films (66). The importance of electromigration damage at triple points is greatly reduced in films
with a marked preferred orientation (72). This is presumed to result from a
greater uniformity in grain size as well as a narrower variation in grain
boundary diffusion than in films with a random grain orientation.

Because of the general tendency to use conductors of increasingly narrow
widths, another source of divergence has become significant: this one results
from geometry. Assuming a film with a perfectly regular array of grains, and
conductors made therefrom with a perfectly uniform width, the number of
paths (grain boundaries) for the atomic flux will vary regularly by plus or
minus one unit along the length of the conductors. The resulting divergence
is not negligible with grain diameters generally of the order of 1 μm and
conductor widths of the order of 10 μm or less.

8.2.3 Temperature Distribution and Current Crowding

The study of electromigration phenomena in thin films requires that the
temperature distribution along the length of the conductors be known. As
already mentioned, the considerable current-carrying capability of such
conductors results from the very effective heat transfer through the film–
substrate interface. Until recently the effect of interfacial cooling on the
temperature distribution has not been thoroughly discussed in the open
literature. The related problem of current crowding, which refers to the local
increase in current density caused by changes in the conductor geometry,
has also been relatively ignored. However, current crowding is important
when considering thermal effects in failure analysis, since it can cause an
increase in the rate of failure propagation through the combined effect of the
increase in current density and the subsequent increase in temperature.

The heat flow problem was first formulated for a conductor with localized
changes in the dimensions of the cross section (73). Because of the varying
sample geometry the results were obtained in an integral form and had to be
evaluated numerically. Recently, detailed solutions, including the time variations of the temperature, have been obtained for a uniform conductor (74).
Since a steady state was shown to be established in a short period of 10^{-3} s,
for simplicity only the steady state shall be discussed here.

Consider the ideal case of a uniform conductor with thickness L and length
l. The heat flow equation to be solved for the steady state has the form

$$K \frac{d^2 T}{dx^2} - j\mu \frac{dT}{dx} + j^2 \rho - \frac{H}{L}(T - T_0) = 0 \qquad (7)$$

where K is the thermal conductivity of the film and μ is the Thomson coefficient. The term $H(T - T_0)/L$ represents the heat loss through the film–
substrate interface, which is simply assumed to be proportional to the differ-

ence between the temperature of the film and that of the substrate T_0, with H being the heat transfer coefficient. The linear form of the heat transfer enables one to obtain an analytic solution. For convenience, the term due to the Thomson effect is neglected. For $T = T_0$ at $x = \pm l/2$, the temperature distribution is given by the following solution:

$$T - T_0 = \frac{j^2 \rho_0}{\dfrac{H}{L} - j^2 \rho_0 \alpha} \left[1 - \frac{\cosh (ax/2)}{\cosh (al/2)} \right] \tag{8}$$

where

$$a = \left[\left(\frac{H}{L} - j^2 \rho_0 \alpha \right) K \right]^{1/2}$$

and α, the temperature coefficient of electrical resistivity, is defined according to $\rho = \rho_0 [1 + \alpha (T - T_0)]$.

Under normal test conditions, the interfacial cooling term in Eq. 7 is almost equal to the Joule heating term. Were it not so, thin film conductors would not be able to sustain current densities so much greater than those sustained by bulk conductors without burning out. Other things being equal, usual thin film conductors would have an enhanced ability to carry current because of their small cross section and their relatively large external surface area which tends to increase the cooling efficiency. However, this geometric effect alone is small in comparison to the cooling effect resulting from the contact with a heat-conducting substrate. The ability of substrates to carry heat away depends on the selection of materials at the interface as well as on the size of the conductors. Cooling should be increasingly efficient for narrow conductors because of the bidimensional heat flow on both sides of the longitudinal edges of the conductors. However, ultimately the temperature and the heat balance are limited by the ability of the substrates to dissipate heat. Under certain conditions with small Si chips on the usual transitor headers, the main barrier to heat flow has been found at the chip–header interface, with the chips being almost isothermal.

With the last two terms, Joule heat and interfacial heat dissipation, in Eq. 7 being almost equal, the average temperature increase of a thin film conductor can be readily estimated if H is known. Even without knowing H, one can see that the temperature should vary approximately as j^2. It is important to note that the value of H is usually large enough to ensure that the temperature of a thin film conductor should be approximately constant over most of the conductor length, except at the very ends with terminals of increased cross section. Thus, the formation of localized hot spots is likely to occur only at points of poor adhesion or where a hole or a crack develops. Heat effects for the latter situation have been analyzed under the assumption that the local current density remains unchanged (74). It was found that for a good conductor the local temperature increase should be quite small.

A complete analysis of this problem would have to include the thermal effect resulting from current crowding, which contributes an additional term to the heat generated in the conductor. In practice, because of the small dimensions of the usual test conductors, it is convenient to determine H by measuring only the maximum temperature. Such measurements have been carried out by means of an infrared microscope on an Al conductor deposited on an oxidized Si wafer (75). For j in the range of 10^6 A cm^{-2}, H was found to be about 10 joules cm^{-2} s^{-1} °C^{-1}. This value is in agreement with measurements obtained by measuring the average temperature increase in test conductors on Si chips mounted on transitor headers.

The introduction of geometric defects in an otherwise uniform conductor has two effects on the current density. The first effect is an increase in the average current density which is inversely proportional to the remaining cross section of the conductor. The second effect, current crowding, causes the current density to increase in the immediate vicinity of a defect to a value above the average. In the immediate periphery of a spherical hole, at the two lateral positions (on each side of the hole with respect to the direction of current flow), the current density has been shown to be twice the average value. The magnitude of this effect increases as the size of the hole becomes commensurate with the cross-sectional dimensions of the conductor. An analysis has been presented (76) for the case of a rectangular crack developing vertically from the surface of a film toward the substrate, which may be considered as a rather idealized model for failure through grain boundary grooving. The magnitude of the current crowding effect was found to depend sensitively on the crack dimensions, with the increase in current density specifically due to current crowding usually exceeding the increase due to the average geometric effect. For a crack with a width of 40 Å (of the order of the usually assumed grain boundary width of 10 Å), extending halfway through the thickness of a conductor, the maximum current density at the tip of the crack is found to be about 100 times the average value. This is probably an overestimate since in reality one is unlikely to find the combination of such a narrow crack with sharp edges.

A detailed calculation of the change in temperature induced by current crowding around a cylindrical void in a thin film has been made recently (77). The results indicate that the associated thermal and current crowding effects can significantly change the mass transport near the void, so that they must be considered in analyzing the propagation of defects leading to failure. Further details are discussed in Section 8.6 dealing with failure models.

8.3 TECHNIQUES OF MEASUREMENT

In measuring electromigration in thin films, the main purpose is to determine the effective charge and diffusivity of ions migrating at grain boundaries. For very thin films, one must also consider possible contributions from the surface. There are two basic types of measurements for thin film electromigration: (a) direct measurements of the magnitude of the ion flux, or of certain parameters, such as conductor lifetimes and resistance changes, which can be related to the mass transport; and (b) measurements of the change in the composition distribution caused by the passage of a direct current. Flux measurements determine only the product of the diffusivity and the effective charge but not each individually. To separate these two parameters, an additional measurement, usually on the diffusivity, would be required. Unlike bulk studies, measurements on grain boundary diffusivity have been carried out only for a limited number of systems, so a majority of the thin film electromigration data give the product of δD_b and Z_b^*. One exception for this type of measurement is the drift velocity experiments (78) which can be used to determine the electromigration driving force or the effective charge, although an alternate measurement of the film stress is required. On the other hand, a proper design for experiments measuring composition changes can be used to determine δD_b and Z_b^* independently. This can be done with the cross-stripe experiment for example.

Thin film measurements are usually carried out at about $T_m/2$, where the bulk contribution to the mass transport is not important. In this temperature range, mass transport is readily observable in most metal films with a current density of about 10^6 A cm^{-2}. Under these test conditions, the temperature distribution of a thin film stripe can be easily affected by local variations of adhesion and sample geometry. Accurate temperature measurement of the test stripe usually presents a problem, particularly for TEM samples without supporting substrates. In addition, there have been few serious attempts to control the grain structure in test samples. These two factors have caused considerable variations in experimental results.

The commonly used measurement techniques are described below.

8.3.1 Measurements of Mass Accumulation or Depletion

This type of experiment measures directly the net amount of mass transport along the current flow direction through the sample. The sample is usually prepared in a stripe form with large end pads. Upon the passage of a direct current, a temperature profile and an atomic flux are established as shown schematically in Figure 8.3a and 8.3b. In Figure 8.3c the corresponding flux divergence is plotted. The net mass transport is usually manifested by formations of voids and hillocks although it is possible to observe uniform thickness changes if local divergent sites are not readily available. It can be seen that the total amount of mass accumulation or depletion due to flux divergence is equal to the net mass transport at the maximum temperature, which is assumed to be the same as the total volume of voids or hillocks.

The aim of the experiment is to measure the total volume of the voids and/

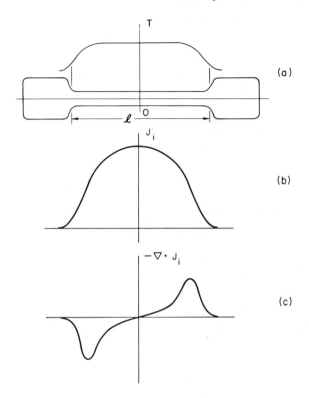

Figure 8.3. Schematic diagrams showing (a) the sample configuration and the temperature distribution for measuring mass accumulation or depletion, (b) the atomic flux induced by electromigration, and (c) the negative of the divergence of the atom flux.

or hillocks. Several techniques have been employed for such measurement, with transmission electron microscopy (TEM) (65) and electron microprobe techniques (79) being the most commonly used. Other methods of using scanning electron microscopy (80) or measuring the intensity change of a scanning transmitted electron beam (81) have also been successfully applied. The TEM technique has the advantage of providing direct observation and with good dimensional accuracy. However, because it usually required suspended films, accurate temperature measurements present a problem due to the existence of large thermal gradients.

This type of measurement determines directly the amount of mass transport, so the results can be interpreted easily. The method can be used to measure impurity as well as self-electromigration. It has the disadvantage that the results give only the magnitude of the ion flux, not the values of the effective charge and the diffusivity separately.

8.3.2   Lifetest of Conductor Lines

The lifetime test of conductor lines is the most commonly used method for evaluating electromigration resistance. In such tests, a group of identical conductor lines are tested at a certain current density and temperature. The choice of the test conditions and the number of conductors used can vary considerably depending on the information (and its statistical accuracy) to be extracted from the results. The results are usually given in terms of a median time to failure (MTF), or t_{50}, which is the time to reach a failure of 50% of the stripes. Extracting information on electromigration transport from lifetest results is indirect and should be done with caution, in spite of its common practice. The difficulty is mainly due to the complicated nature of the physical processes leading from mass transport to stripe failure; mass transport is only one of the factors in the process. Effects of a thermal and structural nature can significantly change the magnitude of t_{50} and its temperature dependence, a problem which is discussed in Section 8.6.

In spite of the basic difficulties, lifetime tests are frequently used for studying electromigration behavior of thin films. The value of the activation energy so determined has often been assumed to be that of electromigration at grain boundaries, an assumption that is not strictly valid (Section 8.6.2). Usually, the activation energy is the most useful information that one can extract from lifetest results. It is difficult to extract information on the effective charge and the diffusivity from lifetest results.

8.3.3   Resistance Measurements

The electrical resistance of a conductor usually changes with time upon the passage of a direct current. Such changes can be due to material accumulation and depletion along the conductor as a result of electromigration. They can also be due to changes in film microstructure or crack formation. However, it has been pointed out that during the early stage of testing, the changes come primarily from mass transport (82). The fractional change in the resistance, $\Delta R/R_0$, for samples without changes in microstructures can be related to the ion velocity at grain boundaries v_b by the following relationship:

$$\frac{\Delta R}{R_0} = \text{constant} \times \frac{v_b t}{l} \tag{9}$$

where t is the test time, l is the stripe length between two measuring electrodes, and the value of the constant is close to unity for films with an average grain size commensurate with the thickness of the film.

Using the above relationship, one can measure the value of v_b averaged over the distance between two electrodes from the resistance change. This technique has been used extensively in studying electromigration in pure films. The standard sample configuration and a typical set of resistance changes measured for Al are shown in Figure 8.4. The sample has several voltage taps for monitoring the resistance change along the stripes, hence the mass transport can be obtained as a function of distance. Such information enables one to infer the direction of mass transport if the structure is uniform along the length of the conductor and if the temperature distribution is determined. In Figure 8.4, the initial resistance changes indeed appear to vary linearly with time although it is not clear that the changes can be attributed completely to mass transport alone.

This technique has the advantage of being simple and that the resistance change can be measured accurately. The difficulty is to prove that the observed resistance change can be attributed to electromigration only. The ion velocity measured from the resistance change is usually within an order of magnitude in agreement with other measurements although the direction of ion migration for Au, Ag, and Cu as deduced from relative resistance changes is toward the cathode (83), a direction opposite to that of lattice and other thin film measurements.

An extension of the resistometric technique can be used to measure the activation energy (84). It is based on measuring the change in the rate of resistance increase due to a sudden change in the temperature. Let us assume that the rate of resistance change is due to the accumulation of a thermally activated defect characterized by an activation energy for diffusion, Q. If the temperature is changed suddenly from T_1 to T_2 at a constant current density,

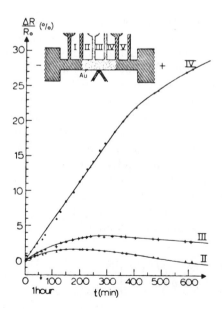

Figure 8.4.   Relative resistance change $\Delta R/R_0$ as a function of time for a gold thin film under electromigration. Test conditions: $T = 356^\circ$ C, $J = 6.3 \times 10^5$ A cm^{-2}, $d = 1.6 \times 10^{-5}$ cm. The sample configuration for measuring resistance changes along the length of the stripe is also illustrated. From reference 102.

the slope of the resistance curve will change in such a way that

$$\frac{(\dot{R})_{T_1}}{(\dot{R})_{T_2}} = \exp\left[-\frac{Q}{k}\left(\frac{1}{T_1} - \frac{1}{T_2}\right)\right] \tag{10}$$

Here, it is assumed that with a constant current density and with a rapid temperature change, the basic process for defect electromigration remains the same.

This technique is simple to use although it yields only the activation energy. It also has the advantage that the effect of microstructure variations can be reduced since only the ratio of the rates of resistance change is required. Nevertheless, it is an indirect method; its validity requires that the relationship between resistance change and electromigration be understood.

8.3.4   The Cross-Stripe Experiment

This experiment was first conceived to study the effects of Cu addition on electromigration in Al films (85). When applied to study solute electromigration, the effective charge and the diffusivity can be determined separately (38). The technique can also be applied to study self-electromigration if radioactive tracer atoms are used (86).

When used to study alloying effects, the experiment is similar in spirit to the Kirkendall study for bulk diffusion couples (87, 88). It employs a cross-stripe sample with a geometry as shown in Figure 8.2a. Corresponding to the Kirkendall interfaces, there are two edges separating the pure metal A from the alloy AB. Depending on the flux balance between the alloy and the pure metal, voids or hillocks will form near each of the edges. For example, in Figure 8.2b, if $J_A > J_A + J_B$, hillocks form on the left edge and voids form on the right edge. By measuring the rate of mass accumulation and depletion, one can determine the change in the mass transport due to alloy additions.

The impurity atoms at the edges of the alloy zone are subjected to driving forces from the current and the concentration gradient. These two driving forces superimpose along the same direction at one of the edges but in opposite directions at the other. The combinations provide two independent measurements for electromigration and diffusion driving forces; thus, the effective charge and the diffusivity can be independently determined. To derive these two parameters, one has to analyze the composition profiles at the stripe edges with grain boundaries acting as fast diffusion paths (see Fig. 8.2c.) The analysis has been carried out according to a Fisher-type approximation with the result that the slope of the logarithm of the concentration against the distance equals (89, 38)

$$A_{\pm} = \frac{Z_b^* eE}{kT} - \left[\left(\frac{Z_b^* eE}{wkT}\right)^2 \mp \frac{2}{\beta D_b \delta}\left(\frac{D_l}{\pi t}\right)^{1/2}\right]^{1/2} \tag{11}$$

where the $+$ and $-$ signs indicate slopes for the anode and the cathode edges, respectively, and D_b is assumed to be much larger than D_l. In this type of experiment, it is very desirable to check the self-consistency of the electromigration results by measuring the diffusivity from the edge profile in a sample without current flow. In that case, the slope gives only the diffusivity (setting Z_b^* to zero in Eq. 11).

The analysis for grain boundary electromigration has also been extended to the more general type of Whipple analysis (90). The details of this analysis are given in Chapter 7 of this book (Section 7.2.4) where the mathematical analysis of grain boundary diffusion is reviewed. Recently, the Fisher-type analysis has been generalized to nonquasi-steady-state electromigration (91). This new analysis is most suitable to treat low-temperature situations when the leakage of atoms into the lattice is essentially normal to the grain boundaries, a condition which simplifies the treatment considerably.

The edge profiles of the inpurity can be measured by using an electron microprobe. For studying matrix ions, measurements of the tracer profiles have been attempted by direct scan counting (92) or by scanning autoradiography (86). So far, these measurements have been somewhat restricted to relatively high temperatures by the lack of spatial resolution. Elaborate schemes in scanning and data processing appear to be promising for improving the resolution.

Another approach, also based on the cross-stripe geometry, can be used to measure Z_b and δD_b^* of impurity atoms. This is to measure the time dependence of the propagation of the void and the hillock fronts (93). The experiment is based on the assumption that voids and hillocks will form when the local solute concentration reaches a certain critical value. During electromigration, the position of this concentration shifts, which is reflected by formations of new voids and hillocks along the direction of mass transport. An analysis similar to that used to derive Eq. 11 can be used to account for the motion of the void and hillock fronts. By combining the migration rates of the void and hillock fronts or by a least-squares fit to the time dependence of the void front displacement, one can determine δD_b and Z_b^* separately. The results obtained by this method for Cu in Al (93) are in good agreement with those derived from profile measurements (38).

8.3.5   Measurement of Drift Velocity

This technique is a recent development (78). The experiment employs samples formed by depositing materials in stripe forms onto a substrate with high resistivity and low migration rate. A typical sample configuration is shown in Figure 8.5a. When a voltage is applied, the current is confined to flow through the test stripe because of its low resistivity. Because of electromigration, the material will be depleted from one end of the stripe and accumulated on the

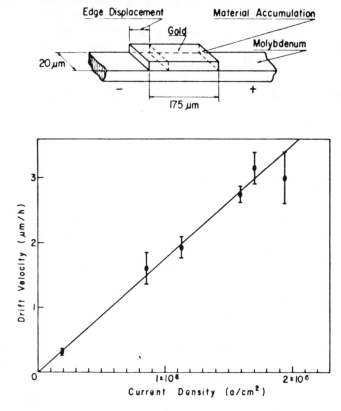

Figure 8.5. *Top*: schematic drawing of the sample configuration used to measure drift velocity in Au films. *Bottom*: drift velocity measured as a function of the current density at 430°C for Au. From reference 78.

other. The drift velocity of the stripe end provides a measure of the mass transport.

This method has been applied to study electromigration in Al (94) and Au (78) films. Results of a typical experiment of Au are shown in Figure 8.5b. It was observed that the stripe end will move only when the current density exceeds a threshold value, and the end will flow backward when the current is removed. These results were attributed to the presence of a stress gradient which is produced as a result of the mass accumulation and mass depletion at the two ends of the sample. The threshold current density corresponds to the minimum driving force required to overcome the effect of the stress gradient. In a steady-state condition, the drift velocity can be expressed as

$$v_d = \frac{ND_i}{kT} \left[eZ_i^* \rho j - \frac{\partial}{\partial x} (\mu_0 + \Omega \sigma_{nn}) \right] \tag{12}$$

where D_i and Z_i are the average ion diffusivity and effective charge, respectively; Ω is the atomic volume; σ_{nn} is the stress normal to the grain boundary; and μ_0 is the chemical potential for zero stress. The second term of v_d gives the backflow velocity associated with the stress gradient. If this term does not depend on the current density, the slope of the v_d-versus-j plot can be used to calculate the product of the effective charge and the diffusivity for the migrating ions.

At the threshold current density j_c, one has

$$eZ_i^* \rho j_c = \Omega \frac{\partial \sigma_{nn}}{\partial x} \tag{13}$$

This provides a unique relationship for determining Z_i^* if the stress gradient can be measured. Such a measurement of Z_i^* has been attempted for Al films (95).

8.4   RESULTS WITH PURE FILMS

In reviewing results for pure films, data obtained from mass transport measurements are summarized first and then results on lifetime tests. With respect to the mass transport results, the data will be expressed in terms of the product of the effective charge times the diffusivity at grain boundaries. This presentation not only gives parameters directly related to grain boundary transport but also factors out the influence of grain size on the magnitude of the ion flux. For the lifetest results, only the activation energies will be tabulated. This parameter is less affected by the test conditions than the magnitude of the lifetime, although it still may not be the same for electromigration since factors (to be discussed in Section 8.6) other than mass transport can influence the temperature dependence of the lifetime.

Most of the mass transport data are expressed in terms of v_i/j or $v_i T/j$, with v_i being the average ion velocity, sometimes called the marker velocity. We use Eq. 3 to deduce the value of $\delta D_b Z_b^*$ from the mass transport data. However, there is some question whether one should use v_i/j or $v_i T/j$ in an Arrhenius plot to determine the activation energy for electromigration. If one assigns the activation energy to the diffusivity, as is usually accepted for bulk studies, the answer will depend on the temperature dependence of Z_b. If Z_b' is constant, the slope of the v_i/j plot would give the activation energy since the ratio of ρ/T is almost independent of T. If Z_b^* varies linearly with $1/T$, as would be the case when the wind force is dominant and its temperature dependence comes only from ρ_1 (see Eq. 2), then plotting $v_i^* T/j$ would be appropriate.

Due to the lack of information on the temperature dependence of Z_b^*, this problem is difficult to resolve. In presenting the results here, we give simply the activation energy as reported.

In the following sections, the results on Al and Au will be summarized first, and then the results for all other metals will be given.

8.4.1   Aluminum

The mass transport in Al films due to electromigration has been measured by TEM (65, 96, 97), electron microprobe (79), SEM (81), resistometric (82, 98), and drift velocity (94) techniques. There are also a number of lifetime measurements (37, 72, 99, 100). In Figure 8.6, the values of $\delta D_b Z_b^*$ derived from data of mass transport measurements are plotted as a function of $1/T$. In converting the data, the average value of the grain size as given was used. The plot covers a temperature range from about 150 to 550°C. As indicated from the change of the slope in two of the results, the lattice contribution appears to become important for temperatures exceeding about 350°C, although such a slope change is not seen in others. Below 350°C, the overall agreement in the data is within two orders of magnitude even though the slopes, hence the activa-

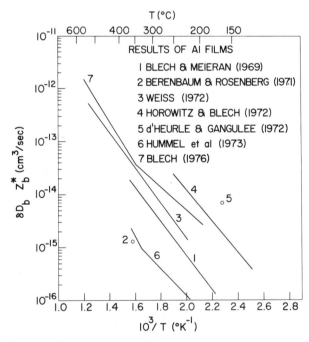

Figure 8.6.   Summary of mass transport results measured for electromigration in pure Al films.

tion energies, are in relative good agreement. This magnitude of the difference in the results is rather typical of electromigration measurements in thin films. This reflects errors due to temperature measurement, calibration of the amount of mass transport, and the difficulty in controlling the film microstructure. In fact, considering the factors that can contribute to measurement errors, the discrepancy in the data is to be expected.

The separation of Z_b^* or δD_b from their product requires an independent measurement of one of these two parameters. To our knowledge, no direct measurement of grain boundary self-diffusion for Al has been reported. A preliminary measurement on Z_b^* based on the value of the threshold current density in a drift velocity experiment has been reported recently (95) (see Section 8.3.5). A value of about unity was considered to be the best estimates for Z_b^*; a higher value in the range of 10 to 100 would require an unusually high film stress (about 10^{10} to 10^{11} dynes cm^{-2}) and a preexponential factor of $D_b\delta$ too low to be physically probable. On the other hand, the preexponential factor of $D_b\delta Z_b^*$ estimated from mass accumulation results (80) has been found to be about two orders of magnitude higher than the value derived from drift velocity measurements. However, there exists the possibility in the drift velocity experiment of having the surface as an effective backstreaming path which, when combined with a better measurement of the stress gradient, would increase the value of Z_b^*. At present, a reasonable estimate of Z_b^* in Al appears to be about 10.

The activation energies determined from the mass transport data in Figure 8.6 are compiled in Table 8.1. Since there is no information on the temperature dependence of Z_b^*, the activation energy may be slightly different from that for grain boundary diffusion. The activation energies obtained from lifetime measurements are also compiled in Table 8.1. Compared to those from mass transport measurements, the values are more spread, with a variation from 0.3 to 1.1 eV. It is worth mentioning that the values of the lifetime vary by more than three orders of magnitude, which is about 100 times worse than the measurements of mass transport. These comparisons indicate the extent of difficulties in using lifetime tests for quantitative studies of electromigration in thin films.

8.4.2   Gold

As was done for Al, the results of mass transport studies in Au are shown in Figure 8.7 by plotting the product $\delta D_b Z_b^*$ as a function of $1/T$. The activation energies from mass transport experiment and lifetime tests are listed in Table 8.2. The data in Figure 8.7 were obtained by different techniques: TEM (101), resistance (102), tracer autoradiography (103), and drift velocity measurements (78). The spreading of the mass transport data in Figure 8.7 is about one order of magnitude better than that of Al. The variation among

TABLE 8.1.   ACTIVATION ENERGIES FOR ELECTROMIGRATION IN THIN Al FILMS

Method of investigation	Temperature range, °C	Method of temperature determination	Activation energy, eV	Grain size, μm	Reference
Void growth (TEM)	180–350	resistometric	0.7 ± 0.2	1/2 to several μ	65
Void growth (TEM)	360 ± 40	resistometric	0.7 ± 0.1	——~4——	—— 96
Void growth (TEM)	175–350	resistometric	0.63	~2	97
Mass depletion (SEM)	240–550	resistometric	0.7	1	81
Drift velocity	250–400	furnace temp.	0.55 (g.b.)^a 1.3 (bulk)	1–1.5	94
Resistance change	220–360	resistometric	0.53 (g.b.) 1.2 (bulk)	~0.3	82
Resistance change	60–150	resistometric	0.5–0.6		84
Resistance	100–200	resistometric	0.5–0.6		98
Lifetest	109–260	power dissipation	0.48–0.8	1.2–8	99
Lifetest	140–300	resistometric	0.51–0.73	2–8	72
Lifetest	75–380	resistometric	0.41	1	100
Lifetest	125–175	resistometric	0.5–0.7		37

^{a}g.b. = Grain boundary.

the activation energies is also less than that of Al, even among the lifetime tests. The different underlayer materials used for adhesion purposes do not seem to influence the activation energy. Self-diffusion along Au grain boundaries for thin film samples has been measured by using radioactive tracers combined with sputter-etching techniques (108). The results give an activation energy of 0.95 eV and a preexponential factor of δD_b of about 3×10^{-10} cm^3 s^{-1}. The values of the activation energy in Table 8.2 seem to be generally in good agreement with the tracer result. The agreement among lifetime measurements is somewhat unexpected in view of the different film thickness, deposition method, and the underlayer material used.

With the tracer data of δD_b one can estimate the value of Z_b^* from the data given in Figure 8.7. At 500°K, Z_b^* is found to vary between -7 and -67, while at 700°K its value ranges from -2 to -40. It seems that the experimental errors do not allow Z_b^* for Au to be specified to better than one order of magnitude although its value appears to be in the range of -5 to -50 between 500 and 700°K. The experimental errors also seem to preclude the determination of the temperature dependence of Z_b^* although Z_b^*, as cal-

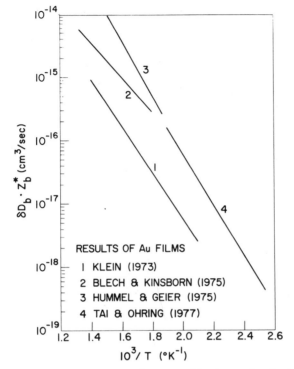

Figure 8.7.   Summary of mass transport results measured for electromigration in pure Au films.

culated from data of the individual experiments, increases with decreasing temperature. (Such a temperature dependence may not be intrinsic to Z_b^* itself since small errors in the activation energy would introduce an apparent temperature dependence to Z_b^*.)

There is an interesting question concerning the direction of mass transport for electromigration in Au thin films. This question has attracted considerable interest recently because of its possible implication on the nature of electromigration driving force in grain boundaries. A group of investigators have carried out several electromigration experiments (82, 102, 109, 110) to elucidate this problem. They concluded that electromigration in Au films is directed toward the cathode, opposite to that of lattice electromigration. The main supporting evidences include the observations of maximum resistance increase at the anode portion of the film (82, 102), the anode–cathode polarity of void–hillock formation (82), and the displacement toward the cathode direction of Au tracer atoms (109). Most of the observations were made on films deposited directly on glass substrates at room temperature. A

TABLE 8.2.   ACTIVATION ENERGIES FOR ELECTROMIGRATION IN THIN Au FILMS

Method of investigation	Temperature range °C	Method of temperature determination	Activation energy, eV	Underlayer material	Reference
Void growth (TEM)	210–425	resistometric	0.8 ± 0.2	Cr or no underlayer	101
Drift velocity	260–500	resistometric	0.6 ± 0.1 0.9 (anomalous sample)	Mo	78
Resistance change	250–390	resistometric	0.98	none	102
Tracer scanning	120–250	ambient temperature	0.80 ± 0.03	Mo	103
Lifetest	200	estimated	∼ 0.9	Ti–W	104
Lifetest	267–452	resistometric and thermal	0.85 ± 0.17	none	105
Lifetest	202–293	resistometric	0.88 ± 0.06	Ni–20% Fe	106
Lifetest	220–500	resistometric	0.42 ± 0.2	Ni–20% Cr	107
Lifetest	536–646	not specified	1.07 ± 0.08	none	77

small number of experiments performed on films deposited on oxidized silicon wafers or with varying grain sizes gave similar results.

Most mass transport measurements for Au do not agree with the cathode direction of mass transport. Of particular interest are the results of the tracer (86, 103) and the drift velocity (78) studies which measured directly the flow of Au atoms. However, all these experiments were carried out on films deposited on some adhesion layer, for example, Ti and Mo. Questions exist on the effect of such a layer on the electromigration direction although at least in one case (103) the direction of the tracer electromigration was found to remain unchanged after removing the Mo underlayer.

The question of substrate material on the electromigration direction in Au has been the object of a recent study (111). The polarity of void (anode) and hillock formation (cathode) was confirmed for films deposited on glass substrates or with underlayers removed. Strong supporting evidence from TEM observations on film microstructure indicates that the observed polarity of defect formation is most probably caused by the structural instability of the film. The instability was attributed to the Joule heating, which was observed to cause large changes of grain size and resistivity in a period of minutes. The resulting divergence caused by the grain size gradient was estimated to be sufficient to overcome the effect of temperature gradients under the test conditions and lead to a reversal of the void and hillock polarity. However, such a structural divergence has been disclaimed (110) to be the cause for observing the reversal polarity of void and hillock growth in films on glass substrates since the void–hillock polarity was found to be independent of the sign of thermal gradient. At present it appears to be a general agreement that the direction of electromigration is along the electron flow in Au films with an adhesive underlayer, but for films deposited on glass substrates the problem remains unresolved.

8.4.3 Other Metals

Electromigration in several other pure films have been investigated and the results are summarized in Table 8.3. Except for Ag and Sn, only the direction of mass transport has been measured for these films. The techniques used were mostly resistance measurement and SEM observations of void and hillock formation.

TABLE 8.3.   RESULTS OF ELECTROMIGRATION IN SOME PURE FILMS

Metal	Method of investigation	Temperature, °C	Transport direction toward	Activation energy, eV	Reference
Ag	resistometric	160–225	cathode	0.3	102
	SEM	225–280		0.95	
Ag	SEM		anode		112
Sn	tracer scanning	142–213	anode	0.43	114
Sn	resistometric SEM	∼ R.T.	anode		83
Cu	resistometric SEM	∼ R.T.	cathode		83
Cu	electron microprobe	290	anode		113
Mg, In, Pb	resistometric SEM	∼ R.T.	anode		83
Co	electron microprobe	540–640	anode	1.5	115

Experiments on Ag were carried out in considerable detail. In addition to the results quoted in Table 8.3, effects of grain size, film stress, and film thickness have been investigated (102). The activation energy of 0.3 eV at low temperatures is presumed to be associated with surface electromigration and the value of 0.95 eV with grain boundary diffusion. The other SEM experiment on Ag (112) has indeed shown the important effect of surface treatment on electromigration in Ag since the coverage of a film with a thin Cr layer was observed to inhibit hillock formation. However, it is not clear that the effect can be exclusively attributed to stopping surface electromigration. Chromium atoms can diffuse into the Ag grain boundaries and alter the diffusion rate. The surface coverage can alter the kinetics of defect annihilation and generation thus affecting void and hillock formation. Finally, it is possible that Cr modifies the interaction of Ag with the atmosphere.

The observations of the electromigration direction toward the cathode in Ag and Cu (83), reported by one group, have not been substantiated by other independent experiments. In fact, direct measurements of mass transport indicate motion toward the anode (112, 113). The evidence presented for cathode motion is similar to that for Au, a problem which was discussed in the previous section. It is not clear whether the structural instability observed for the Au film exists for Ag and Cu films on glass substrates and whether such a change, if it existed, can explain the observed results.

The tracer scanning results on Sn (114) are interesting since the temperature range used was close to the melting point, but the activation energy is still consistent with a grain boundary transport mechanism.

In Co films (115) between 540 and 640°C, the motion was found to be directed toward the anode, in contrast to lattice results obtained at higher temperatures (116). The change in the direction of motion with temperature is believed to be correlated with a change in the sign of the normal Hall coefficient.

8.5 ALLOYING EFFECTS ON MASS TRANSPORT

Starting with the discovery of the beneficial effect of Cu additions on the lifetimes of Al thin film conductors (37), the study of alloys with relatively low solute concentrations has become a significant part of the investigation of electromigration in thin films. Practical considerations dictated that most of this work be concerned with Al and Au films. Some of the alloying studies were aimed at the selection of suitable alloy additions. others were intended to elucidate the specific mechanism at work. The studies relating to mass transport are reviewed first, while results concerning lifetimes are summarized in Section 8.7.

Historically the grain boundary effect due to Cu additions on electromigration in Al thin films was made implicit by early lattice studies in Al. In the lattice it was found that electromigration in impure Al (commercial Al, where Cu is often the main impurity) proceeds at faster rate than in high-purity Al (117), while lifetime studies of Al thin films implied the opposite (37). A later measurement of mass transport in Al–Cu thin films indicated that the increase in lifetimes was more or less commensurate with the decrease in Al grain boundary diffusion (79). To investigate the question, a theoretical study on the effect of alloying additions on the kinetics of the atomic jump process in the lattice was carried out (118). The results indicate that most of the solutes considered for Al, Ag, Au, and Cu, including indeed Cu in Al, would enhance the total (solvent plus solute) electromigration in the lattice. The exceptions, Fe in Cu and Ag in Fe, were attributed to the strong binding energy between Fe impurities and vacancies. Thus, the beneficial effect of alloy impurity in thin films cannot result from a lattice mechanism. A detailed atomistic understanding of the effect of impurities on grain boundary self-diffusion remains missing although an atomic model specifically dealing with the diffusion jumping process has been developed recently (119). This interesting approach has yet to be applied to investigate solute effects on electromigration.

A model has been proposed which treats the solute effect on the basis of the energetics between the impurity atoms and the grain boundary (120). Assuming a reasonable binding energy for the solute atom at the boundary, one can show that the number of boundary sites available for diffusion can be significantly reduced. For example, an energy of 0.2 eV can produce more than an order of magnitude decrease in the number of unbounded sites due to the presence of 1 at. % of solute at 400°K. The smaller number of sites available for diffusion leads to a reduction in the solvent mass transport. The amount of reduction depends on the dissociation energy of the solute from its bounded site. For the example given here, a dissociation energy of 0.1 eV is found to reduce the solvent flux by 90%, and a comparable increase in the lifetime is to be expected.

Unfortunately, it is difficult at present to measure accurately the corresponding binding energies. An empirical observation found that the adsorption of solutes to grain boundaries is almost inversely proportional to their respective solid solubility limits (59, 70). This may reflect the fact that the free-energy change of adding a solute atom from the surface to the lattice is comparable with that of attaching a solute atom in the lattice to the boundary. Some recent theoretical results which take into consideration both elastic and electronic interactions yield binding energies of the order of 0.1 to 0.3 eV for nontransition metal impurities in Al grain boundaries (121). The magnitude of these binding energies indicate that the values used in the model mentioned previously are reasonable for Al films.

The cross-stripe experiments can provide quantitative answers to the following questions concerning solute effects: 1. Does the solute retard or enhance the electromigration of the solvent? This is important since the overall damage comes primarily from the solvent flux. 2. What are the diffusivity and effective charge of solute atoms in grain boundaries?

The principle of this experiment and the corresponding data analysis are described in Section 8.3.4. The technique has been applied to a number of systems: Cu (87), Ni and Co (122) additions in Al, Ag in Au (87), and various elements in Sn (123). The observations on the polarity of void and hillock formation with respect to the direction of current flow indicated that all the impurities studied retard Al electromigration but that Ag has no effect in Au. The results in Sn films indicate that electromigration can be enhanced or retarded depending on changes in the solvent diffusivity.

The cross-stripe experiment can provide another useful information concerning formation of electromigration damage in thin films. The solute concentration C_s^c at the void and hillock fronts represents the concentration at which the flux divergence between solute-poor and solute-rich areas is just enough for damage formation. Within experimental accuracy the values of C_s^c for Cu in Al are the same: 0.5 at.% at 250°C for both fronts. Under the given experimental conditions of temperature and current density, this is the concentration of solute required to reduce the overall flux divergence for void and hillock formation. For practical purposes it may be necessary to utilize solute concentrations greater than C_s^c when the electromigration of the solutes is greater than the electromigration of the solvent. On the other hand, measurable increases in lifetimes have been observed with trace impurities in the ppm range, much below the observed C_s^c, although the lifetests were carried out under different conditions of temperature and current density.

Values of Z_b^* and δD_b have been determined separately for Cu in Al in cross-stripe experiments by measuring the Cu profiles at the two stripe edges (38) and also from the void and hillock propagation rates (93). Results of the profile measurements obtained at 175°C with and without electromigration driving force are shown in Figure 8.8. The slopes of the anode and cathode profiles were used to evaluate Z_b^* and δD_b separately on the basis of Eq. 11. The annealing conditions for these measurements have been chosen in a range where the Fisher-type analysis is a good approximation. The electromigration results have been checked by measuring the diffusivity without the driving force (Figure 8.8b), and the agreement between the diffusivity values was found to be better than 5%. The grain boundary diffusivity for Cu in Al has been measured using the cross-stripe geometry as a function of temperature, and the activation energy was found to be 0.81 ± 0.05 eV (124). A quasi-steady-state method has been used to measure the effective charge of Cu in Al (125). The apparent effective charge which was attributed to that of the grain boundary was found to decrease from −14.9 to −4.1 in the temperature range of 325 to 500°C.

In addition to marker experiments, direct mass transport measurements using TEM or electron microprobe techniques have been carried out extensively in dilute Al alloy films. In the microprobe measurements, mass transport for both the solute and the solvent element can be determined. The TEM experiment measures only the total ion flux which, for dilute alloys,

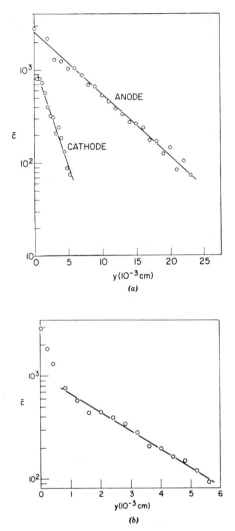

Figure 8.8. (*a*) Copper electromigration profiles in polycrystalline Al films. Test conditions: $T = 250°C$, $J = 1.87 \times 10^5$ A cm^{-2}, $t = 84$ min. For clarity, the anode profile has been shifted upward relatively to the cathode profile. (*b*). Copper profile measured under identical condition but without current passage. From reference 38.

consists primarily of the solvent transport. In Figure 8.9, the values of the product of $\beta D_b \delta \cdot Z_b^*$ for Al–Cu films from various measurements are plotted as a function of temperature. Here, only the product of the diffusivity and the effective charge is given since the mass transport experiments cannot measure these parameters separately. Representative results on $\beta D_b \delta$ and Z_b^* obtained by various studies of Al as well as other alloy films are summarized in Table 8.4.

Comparing the results in Figure 8.9 for Al–Cu and Al films (see also Fig. 8.6 for pure Al results), the total mass transport at grain boundaries is clearly reduced by Cu additions. The reduction is about two orders of magnitude between about 140 and 250°C. This magnitude can be accounted for on the basis of the type of model analysis (120) mentioned above by assuming a binding energy for Cu to a boundary site of about 0.2 eV. However, this model can not explain the difference in the observed activation energies, perhaps as high as 1.1 eV for Al–Cu versus 0.6 eV for Al, since it predicts an increase in the activation energy only about 0.2 eV (coming mainly from the dissociation energy of Cu atoms from grain boundaries). It should be mentioned that such large differences in the activation energy for grain boundary

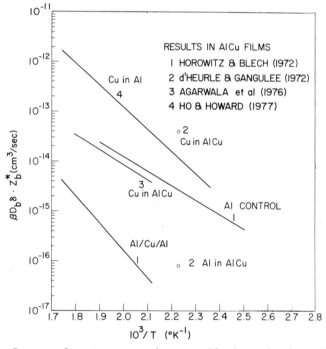

Figure 8.9. Summary of mass transport results measured for electromigration in Al–Cu films.

TABLE 8.4.   SELECTED RESULTS OF SOLUTE ELECTROMIGRATION AT GRAIN BOUNDARIES

Diffusing species	Host metal	$T, °K$	Z_b^*	Z_l^*	βD_b cm² s⁻¹	D_l cm² s⁻¹	Reference
Cu	Al	528	-16.8	-7	1.8×10^{-7}	2.7×10^{-14}	38, 93
Cu	Al+Cu	448	-20^a	-7^a	2.0×10^{-8}	8.1×10^{-17}	79
Mg	Al+Mg	504	-40	-39	6.0×10^{-12}	1.1×10^{13}	127
Ni	Al+Ni	448	-3	?	1.5×10^{-12}	?	128
Al	Al	500	-10	-17	1.5×10^{-9}	3.0×10^{-14}	81, 129
Al	Al+Cu	448	-30^a	-17^b	2.7×10^{-11}	5.3×10^{-16b}	79
Al	Al+Mg	504	-30^a	-17^b	4.7×10^{-11}	2.1×10^{-14b}	127
Al	Al+Ni	448	-30^a	-17^b	5.6×10^{-11}	5.3×10^{-16b}	128
Ag	Au	523	-9.6	-7.4	1.7×10^{-10}	1.1×10^{-18}	87
Au	Au	573	-10	-9.2	1.3×10^{-10}	9.2×10^{-18}	78, 130, 131
Au	Au+Ta	569	—	—	3.2×10^{-11}	10^{-18b}	108
Sb	Ag	750	-230	-100	2.5×10^{-7}	1.2×10^{-12}	89

[a]Values not actually measured but assumed in calculating βD_b.
[b]Values calculated according to the pure metal data.

self-diffusion was not observed when Ta was added to Au, where the increase was found to be only about 0.25 eV (108).

Of particular interest in the results given in Table 8.4 is that the value of the boundary effective charge, in contrast to the diffusivity, is generally not very different from the lattice value. (At present, it is not possible to correct for the vacancy flow and correlation effects (126), which may change the actual value of Z_b^*.) So in comparing the electromigration behavior in the lattice with that in boundaries, the difference appears to be primarily in the atomic mobilities associated with the basic diffusion processes. Therefore, it is possible to reduce electromigration damage by decreasing the grain boundary mobility. Factors such as the film microstructure and the impurity segregation, which are important control factors for grain boundary diffusion, are expected to be important for electromigration as well. When extending these results to use temperatures, one has to consider the temperature effect on the effective charge and the diffusivity. Since the boundary effective charge is not expected to vary much with temperature, the temperature dependence of the electromigration flux would come mainly from the variation of the boundary diffusivity. On the basis of this observation, one can infer that solute additions are effective in reducing electromigration damage at normal operating temperatures because of the reduction in the grain boundary diffusivity and not because of changes in the effective charge.

In Table 8.5 the values of activation energies for electromigration in Al–Cu

TABLE 8.5.   ACTIVATION ENERGIES FOR ELECTROMIGRATION IN Al–Cu FILMS

Method of measurement	Temperature range, °C	Copper concentration wt %	Activation energy, eV	Reference
Cu Migration				
Cross-stripe experiment	140–300	0	0.81	124
Mass accumulation	200–280	1.4	0.55	132
Alloy Mass Transport				
Temperature cycling of resistance change	70–230	0.5 to 8	0.45–0.73	98
Lifetest	105–230	0.3 to 9	0.63–0.83	133
Lifetest	230–280	0 to 12	0.43–0.78	134
Mass depletion TEM	200–330	6	1.1	97

films are summarized. The range of variation in these results is comparable to that of pure Al films. Two of these investigations (132, 133) measured the activation energy for the total (solute plus solvent) mass transport as a function of the Cu concentration. In two of the lifetest experiments (133, 134), the activation energy was found to increase with the Cu concentration with a maximum at about 4 wt. %. However, such variations in lifetimes with Cu concentrations have not been observed in another experiment where the failure time was measured only at 190°C (135).

Several experiments have been performed to investigate the structural morphology of Cu in Al and its change during the electromigration process. A TEM observation was made on single-crystal Al films with 3 wt. % of Cu as a function of the annealing temperature (136). Compared to the precipitation found in bulk alloys, the formation of the coherent clusters and semi-coherent θ-Al$_2$Cu particles were affected by the proximity of the film surfaces. The stable θ-Al$_2$Cu phase was found to precipitate preferentially near the film surface as a result of annealing between 150 and 250°C. In a similar study on polycrystalline Al–Cu films (137), the distribution of the stable θ precipitates was found to be near the surface only when Cu was deposited as a top layer, but throughout the film when Cu was incorporated as a sandwich layer. X-Ray measurements showed that the Cu concentration in the solid solution can vary relatively to the total amount of θ precipitates as a result of annealing. An in situ observation of an Al–Cu stripe during electromigration has been made using the electron beam-induced current mode in SEM (138). The coarsening of the θ particles near the surface and at the anode was found to be at the expenses of the interior θ precipitates. By observing the volume change of the θ precipitates along the stripe, the direction of Cu transport was found toward the anode, and the θ particles acted as Cu sources during electromigration. Failures in the form of crack growth or uniform thinning were mostly observed in areas depleted of Cu. A related study showed that the θ precipitates, when present in sufficient quantity, can modify the grain structure and affect electromigration lifetimes (135). The above experiments demonstrate that the effect of solute additions on electromigration can be relatively complicated and is not always confined only to mass transport.

8.6 FAILURE MODELS

The variety of problems encountered in electromigration testing and in the analysis of failures caused by electromigration has been the object of many studies (139–142). The models to be analyzed presently were established in the hope of providing some degree of understanding to the relatively simple question of failures occurring randomly along the length of homogeneous conductors with a uniform geometry. It is hoped to provide the means of predicting failure at low temperatures and reduced current densities from accelerated test results.

A satisfactory and complete model for failure should be able to duplicate the experimentally observed relations between the lifetimes of thin film conductors and such parameters as material selection, current density, test temperature, the geometric dimensions of the conductors, and the heat conductivity between the conductor and its surroundings. The problem has never been treated as a whole. The literature contains models which propose to explain the relationship between failure times t_{50} and the current density and temperature of the conductors expressed in the simplified form (98)

$$t_{50} = A j^{-n} \exp(\Delta H / kT) \qquad (14a)$$

or slightly modified as

$$t_{50} = A T j^{-n} \exp(\Delta H / kT) \qquad (14b)$$

In the second expression, the extra T term has been added to account for the variations in the electromigration parameters with the temperature of the conductors; see Section 8.4.1. The proportionality term A is characteristic of a given set of conductors: it is a function of the statistical distribution of inhomogeneities (which cause flux divergence), of the structure of the films, and of the geometry of the conductors. Assuming A to be constant, the problem of establishing a relation between t_{50}, n, and ΔH depends on considerations of thermal factors which are analyzed in a first set of models. A second set of models, based on a statistical analysis of such factors as grain size distributions and the length and width of the conductors, relate these factors to the distribution of failure times. It may be said that these latter models are actually models for the quantity A in the above failure expressions.

8.6.1 Thermal Considerations

The main point to be considered in analyzing thermal effects on the lifetimes of conductors is that, as a result of Joule heating, the temperature at which failure propagates increases as the initial porosity, or crack, nucleates and grows. At the beginning of accelerated tests, the temperature of the conductors can be adjusted very precisely by using the conductors themselves as their own resistance thermometers; it is possible to compensate for the initial increment in temperature ΔT_0 caused by the Joule heat at the start of a test. However, once the test actually begins, the resistance of the conductors is almost always found to increase. This, in turn, causes a corresponding increase in temperature which is practically impossible to eliminate, since this would require a continuous readjustment of the ambient temperature. Under normal use conditions, one cannot even adjust the initial temperature of the conductors, which will vary not only as a function of their own resistance but also as a function of fluxtuations in the heat generated by various active, and inactive, devices in any one "chip." Thus, when mentioning the temperature of a conductor, one refers at best to its initial temperature. The test temperature itself is a variable which increases during the duration of the test. The extent of the temperature increase depends on the current density since for the same ΔR, ΔT will be four times as high if the current density is doubled. Therefore, at each stage of the failure process the conductor with the higher current density will fail at a rate that is higher not only in proportion to the relative current densities but, in addition, as a function of the higher increment in temperature. This effect not only influences the measurements of lifetimes as a function of current density (at a nominally constant temperature); it also can create sensible variations in the "measured" activation energies obtained from lifetime tests conducted at constant current density. The basic thermal relation

$$\Delta T = \frac{H'(R_0 + \Delta R)I^2}{1 - H'(R_0 + \Delta R)\alpha I^2} \tag{15}$$

is common to all three models that have been proposed in the literature. It dictates that the temperature shall become infinite whenever $R_0 + \Delta R$ becomes equal to $(H'\alpha I^2)^{-1}$ or when I (the total current through the conductor) is greater than $[H'(R_0 + \Delta R)\alpha]^{0.5}$.

8.6.2 Thermal Models

The models start from the basic transport relationship where the atomic flux, which is proportional to the flux of charge carriers, follows the usual Arrhenius temperature dependence, $\exp(-Q/kT)$. The models specify the exact conductor temperature, where T is taken as the initial conductor temperature T_0 plus the increment ΔT, the excess temperature resulting from the Joule effect and the increase in electrical resistance resulting from failure propagation. The models assume that the divergence in atomic flux that causes failure remains constant during the process of failure propagation. Such an assumption ignores the fact that failure nucleation occurs at the point of highest initial divergence. Model I (144) and Model II (145, 146), which differ only in a minor way, both assume that the current density and hence the electric field vary only with the residual cross section of the conductor, excluding current crowding effects (which are considered in Model III (77). Model I considers the propagation of randomly distributed pores formed at the junctions of three grain boundaries. Model II considers the propagation of a vertical crack across the width of a conductor. Although seemingly different, these two models are quite similar. Indeed, from a thermal point of view Model II becomes identical to Model I if one assumes a crack of increasing width in the extreme where the crack width is commensurate with the length of the conductor, so that as the "crack" propagates the cross section of the conductor is uniformly reduced along its length as in Model I. Thus, the mathematics are approximately the same in either case. The main distinction is that in Model II the current is assumed to be null over some distance on both sides of the crack, which adds an extra term to the resistance increase. Thus, Model II requires an extra equation to specify ΔR as a function of the crack length.

Basically, both models use three simultaneous equations. One differential equation expresses the rate of failure propagation in terms of the atomic transport relation, one equation relates the resistance of the conductor to its instantaneous geometric configuration, and the third equation relates the instantaneous conductor temperature to the instantaneous resistance. The heat dissipation factor H' is assumed to be constant. Since the variable temperature appears as $\exp(-Q/kT)$ in the first equation, an analytic solution may not be available; instead numerical solutions or approximate analytic solutions were used. The initial boundary conditions are given in terms of the initial current density and the initial increase in temperature due to Joule heating (which is a measure of the heat dissipation factor H'). Failure is estimated to occur either when the defects have propagated across the whole width of the conductor or when the temperature has reached the melting point of the material. Realistically, one could choose other criteria for failures, for example, the temperature at which the conductor material will react catastrophically with its environment (eutectic temperatures for Al or Au conductors on Si, oxidation for Cu conductors, etc.).

The results are essentially identical for both Models I and II and for Model III as well. While a first glance at Eq. 1 can lead to the conclusion that the time for failure should be inversely proportional to the current density, the models show that this is true only at low current densities where ΔT remains small during the major part of the failure process. As the current density increases, the time to failure varies as j^{-n}, with n taking progressively large values (above 10). This is shown in Figure 8.10, which is taken from reference 144. The three curves corresponding to three values of the thermal dissipation factor emphasize the importance of this parameter on lifetime measurements. Here, the thermal dissipation is expressed as ΔT_R, which is the initial increase in temperature (before failure nucleation) for a value of the current density

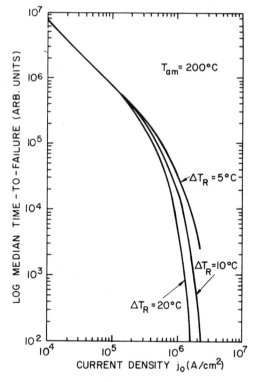

Figure 8.10. The median time to failure as a function of the initial current density J_0 on a log–log plot. The three curves, for three different heat sinking conditions (ΔT_R), correspond to the following test parameters: ambient temperature $T_{am} = 200°C$, initial conductor temperatures $T_0 = 205$, 210, and 220°C. From reference 144.

equal to 1×10^6 A cm^{-2}. These curves can probably explain the rather wide scatter in the reported values of the current density exponent n, even under apparently identical test conditions. From a practical point of view, the results show the difficulty of an extrapolation of lifetimes based on the results of accelerated tests conducted within relatively narrow variations of current density. The use of fixed values of n, often in the range between 2 and 3, to extrapolate lifetimes to normal use conditions (at a relatively low current density), from the results obtained under accelerated test conditions, can obviously lead to greatly overoptimistic projections. This point is made as well in Figure 8.11, which is taken from reference 145. Here the two curves are made to correspond to two sets of conditions. The lower curve belongs to conductors tested in such a way that regardless of the current density, the initial conductor temperature (ambient plus initial Joule effect) is a constant;

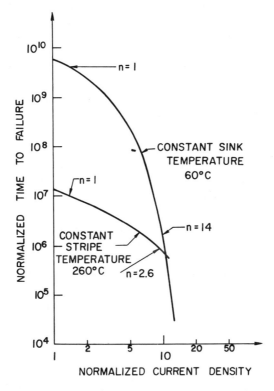

Figure 8.11. Normalized times to failure as a function of normalized current densities on a log–log plot. Upper curve corresponds to a constant ambient temperature ($T_{am} = 60°$C) and continuously increasing initial conductor temperatures T_0. Lower curve corresponds to a constant initial conductor temperature $T_0 = 260°$C. From reference 146.

such conditions may be obtained under accelerated tests. The upper curve belongs to conductors stressed at a constant ambient temperature, with the initial conductor temperature increasing as a function of the Joule thermal generation in the conductors, conditions which can be anticipated to obtain under normal usage.

While in Figure 8.10 it was necessary to plot several curves to demonstrate the effect of different values of the heat dissipation factor, in Figure 8.11 there is only one set of curves. This has been made possible through the normalization of the units on the horizontal coordinate axis. The quantity plotted horizontally is not the logarithm of the current density but the logarithm of the ratio between the current density and the current density required to give an initial Joule effect of $1°C$. Likewise, in Figure 8.11 the lifetimes plotted on the vertical axis have also been normalized: the quantity plotted vertically is not t_{50} but $t_{50}\delta/cw$. This takes into account the fact that the failure time will decrease if the grain boundary width (grain boundary diffusion) increases while it will increase if the width of the propagating crack c or its length (for failure, this is equal to the conductor width w) increases.

One further effect resulting from the inability to define with precision an "average" conductor temperature during the process of failure propagation is that the apparent activation energy ΔH derived on the usual Arrhenius plots from lifetime tests always tends to be smaller than the true activation energy for grain boundary diffusion Q. Essentially, the error arises from plotting the reciprocal of the initial conductor temperature (at best, some data can be found with ambient temperatures) rather than a meaningfully defined "average" temperature to take into account the progressively increasing Joule contribution during the failure of a conductor. This effect will be minimized at low current densities and with high heat dissipation factors, a point that is well underlined in Model III (77). The error should be somewhat proportional to the square of the current density. The variations in apparent activation energy are illustrated in Figure 8.12, which is taken from reference 144. Here, the conductors with an assumed activation energy of 0.70 eV for grain boundary diffusion Q are presumed to be mounted in such a way that, with a current density of 1×10^6 A cm^{-2}, the initial increase in temperature due to Joule heat (ΔT_R) is $10°C$. The apparent activation energy, which is already somewhat too small for a current density of 5×10^5 A cm^{-2}, decreases to a value of 0.65 eV for a current density of 2×10^6 A cm^{-2}. On a sufficiently expanded range of temperatures, each line would be slightly curved, but this is not noticeable over the usually narrow range of experimentally accessible temperatures.

Undoubtedly, such considerations explain in part the scatter in reported activation energies obtained from observations with conductors of different

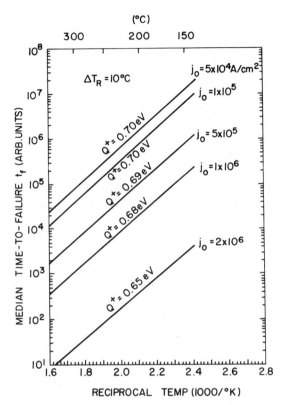

Figure 8.12. Apparent activation energy for failure, ΔH, as a function of the initial current density J_0. The activation energy for grain boundary diffusion Q is 0.7 eV. The heat sinking conditions are such that with a current density of 1×10^6 A cm^{-2} the initial conductor temperature would increase by the quantity $\Delta T_R = 10°$C. From reference 144.

cross sections tested at different current densities, at different temperatures, and certainly with widely varying heat sinking conditions. This does not mean that activation energies derived from failure times are necessarily meaningless. For Au conductors, the activation energy reported from failure times (106), 0.9 ± 0.06 eV, is almost exactly equal to that reported for tracer diffusion measurements in polycrystalline Au (108), 0.95 eV. Such good agreement is probably accidental although, as should be expected, the value derived from electromigration lifetests appears to be somewhat low.

Another source of scattering in activation energy data results from the alternative use of either one of the two failure expressions, Eqs. 14a and b, with and without the T term. For failures in Au conductors tested in the range of 200 to 325°C, the introduction of the T term increases ΔH by 0.04 eV (compare ref. 106, 0.88 eV, and ref. 147, 0.92 eV). The uncertainty with respect to the proper use of either equations can be removed if the variations of Z_b^* as a function of temperature are known.

Neither Model I nor Model II make provision for the exact distribution of electric current in a conductor containing geometric defects. However, the current crowding effect is accurately analyzed in Model III for the special case of a cylindrical hole extending vertically through the thickness of a thin film conductor of Au on an amorphous SiO_2 substrate. In an earlier report (148), it had been pointed out that on both sides of a hole with circular geometry the current density increases to a maximum value that is twice the average value. Model III demonstrates that, because there is no heat generation within the hole itself and little heat generation both in front and in back (with respect to the direction of the flow of charge carriers) of the hole, the temperature and the resistivity of the conductor in the immediate periphery of the hole are lower than average; as a result the maximum current density increases to values that are more than twice the average value. The two points of maximum temperature are slightly displaced outwardly from the two lateral sides of the hole. The maxima in atomic flux are located at some intermediate positions between the maxima in temperature and the maxima in current density at the very edge of the hole. The distributions of the temperature T, of the current density J_e, and of the atomic flux J_a around a cylindrical hole are clearly seen in Figure 8.13, where the three variables are plotted vertically in perspective drawings. Only one quarter-area of the conductor is represented with the hole at the right-hand corner; the arrow indicates the direction of current flow. The results were calculated with the following parameters: Au film thickness, 800 Å; width, 1 cm; hole radius, 0.08 cm; SiO_2 substrate thickness, 0.15 cm; activation energy, 1 eV. On account of the cooling effect of the hole, the maximum atomic flux is not on the periphery of the hole itself but slightly on the outside of it. Unfortunately, the conditions of film dimensions and the choice of substrate are quite distinct from the conditions that obtain for the more usual practical applications. To a certain extent this deficiency has been remedied through the use of a number of adimensional normalization factors.

The calculations for the temperature distribution were found to be in rough agreement with experimental results obtained via infrared microscopy. The analysis of failure times and apparent activation energies is quite similar to that reported for Models I and II. None of the models takes into account the transport of heat along the conductor themselves. While this omission is of no significance for Model I where the pores are presumed to grow randomly and uniformly along the length of the conductors, it is certainly of greater importance for Models II and III, which conform more to the usual observations of localized failures.

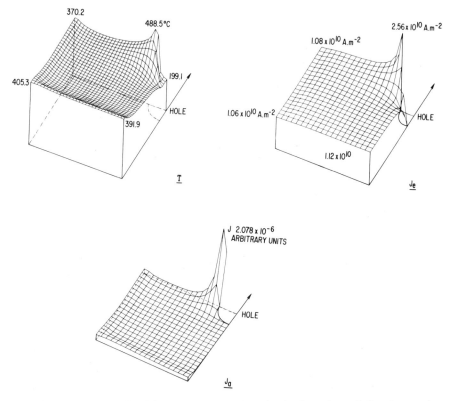

Figure 8.13. Distributions of temperature T, current density J_e, and atomic flux J_a around a cylindrical hole in a Au film on a vitreous SiO_2 substrate. From reference 77.

8.6.3   Statistical Models, Geometry, and Grain Size Effects

Failure times are often found to be distributed as shown in Figure 8.14 (134) more or less according to a lognormal law, whereby a straight line is obtained when the logarithms of the failure times are plotted against the cumulative percentage order (if nine conductors form a test group, the cumulative percentage order will be 10%, 20%, etc.) on a probability scale. The probability of a failure occurring at time t is given by

$$p = \frac{1}{\sqrt{2\pi}\,\sigma t} \exp\left[-\frac{1}{2}\left(\frac{\ln t - \ln t_{50}}{\sigma} \right)^2 \right] \tag{16}$$

where σ, the standard deviation, is a measure of the width of the distribution and t_{50} is the median failure time.

It may be anticipated that in any group of conductors the individual

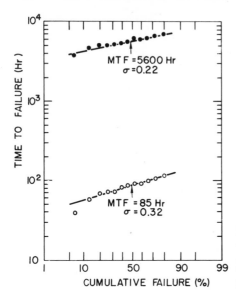

Figure 8.14. Lognormal probability plots of the times to failure for Al (lower curve) and Al–4 wt % Cu–1.7 wt % Si (upper curve). Test conditions: 220°C, 8×10^5 A cm^{-2}. From reference 134.

resistances and the individual heat conduction factors H' will vary according to a Gaussian distribution. For equal currents this should lead to a Gaussian distribution of temperatures (on account of the Joule effect). It has been reported (149) that this effect and the diffusion law $\exp(\Delta H/kT)$ determine the lognormal distribution of failure times. This would indeed be correct if the distribution of $1/T$ were Gaussian. However, because it does not follow from T being Gaussian that $1/T$ should be Gaussian also, the explanation fails. Yet, if conductors are selected with a narrow distribution of resistances and are carefully mounted to obtain equivalent heat sinking for all the conductors, the resulting distribution of failure times should also be narrow, as was observed experimentally. One consequence of these considerations is that with T varying as the square of the current density, the values of σ obtained at high current densities (e.g., in accelerated tests) will be greater (other things being equal) than those which would apply at a low current density under normal use conditions.

For practical applications, where one is often concerned with very small failure rates often at a level which is much below what may be accessible during tests (since the required number of samples would be exceedingly high), it is important not only to know the exact nature of a distribution, its median failure time, and the effects thereon of various processing or design parameters but the spread of the distribution as well. Giving results about median failure times without the corresponding data on the distribution widths can be deceiving since at low failure rates, which are of practical interest, a high t_{50} with a high σ, might be less favorable than a low t_{50} with a low σ. The selection of the correct distribution law is extremely important. In extrapolating test data to low failure rates, an overly pessimistic forecast (by a factor of 10) can result if the test data are fitted to a Weibull rather than a lognormal distribution (150). (For a specific set of data, the magnitude of the error will vary with the size of the sample data, the characteristics of the distribution, and the extent of the extrapolation.)

In a first study (66), mostly of an experimental nature, it was shown that the median time to failure and the standard deviation (obtained from lognormal plots) for Al thin film conductors, either 10 to 15 μm wide, decreased as the length of the conductors increased from 25 to 200 μm. The t_{50} values, which varied approximately as $\exp(1/l)$, where l is the total length, decreased quite precipitously up to about 250 μm and remained approximately constant for longer lengths. The decrease in the standard deviation from a value of about 2 to 0.5 was more gradual, but it, too, remained about constant for lengths greater than 500 μm. The values of t_{50} increased approximately linearly with the width of the conductor—by a factor of 2.5 for conductors from 5 to 15 μm wide. It was shown that the failure distributions, both t_{50} and σ for conductors with length l, could be calculated by numerical methods from the failure distribution of the shortest conductors tested ($\lambda = 25$ μm) by assuming that the long conductors were composed of n ($n = l/\lambda$) short segments randomly chosen. The experimental and calculated values for t_{50} and σ were in quite good agreement. For a simple illustration, take two conductors with a length λ and a median failure time t_m, then place them one after the other to form a new conductor of length 2λ. The probability that this new conductor will have a median failure time smaller than t_m is 0.75, while the probability of the failure time being greater than t_m is only 0.25, hence the decrease in the value of the median failure time with increasing conductor length. A similar argument can be made for the increase in median failure time with increasing conductor width by placing the two conductors of length l side by side rather than end to end. A decrease in lifetime with an increase in the length of the conductors has been experimentally observed with Au films (151). The increase in lifetime resulting from an increase in the width of the conductors has been the object of extensive observations in Al–Cu and Ag films (152). Practical limitations restrict the use of this latter effect in circuit design.

A more complex computer simulation (150) than the one (66) discussed above was based on a structure analysis of real Al thin films. However, there were no attempts to match calculated failure results with experimental tests. The model presumes a conductor to be made of a series of short sections with different lengthwise dimensions, each equal to 1 grain in diameter, randomly chosen according to the grain size distribution. Normal to the length of the conductors each section is made up also of randomly selected grains, the number of them being sufficient to make up the width w of the conductor. The atomic transport within each grain boundary is a function of the randomly chosen angle ϕ between two adjoining grains and is proportional to $\cos\theta$, where θ is the angle between the grain boundary and the direction of current flow. The atomic transport within each section is a constant equal to the sum of the transport in each boundary. The time to failure for a conductor is inversely proportional to the maximum flux divergence between two consecutive sections. In Figure 8.15a and 8.15b, the relationships between the conductor length, the failure time, and the standard deviation are shown (150). In each case, three curves corresponding to films deposited at different temperatures give a measure of the effect of the grain size on both t_{50} and σ. The effect of the conductor width on t_{50} is shown in

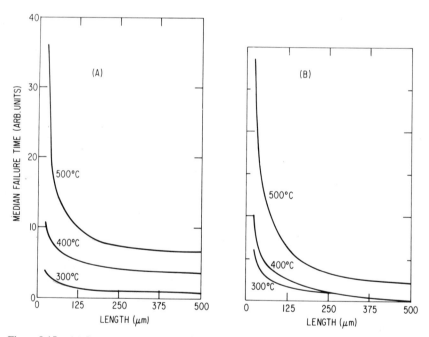

Figure 8.15. (a) Computer-simulated dependence of the lifetime and (b) of the standard deviation σ on length of conductors. The width of the conductors is 10 μm. The three curves correspond to real structures for Al films deposited at 300, 400, and 500°C to a total thickness of approximately 12,000 Å. The film structures were found to fit lognormal distributions of grain diameters with average values of 1.6, 3, and 8 μm and respective standard deviations of 0.6, 0.4, and 0.8. From reference 150.

Figure 8.16; unfortunately, the corresponding curves for σ are not given. Because of the continuous decrease in conductor width made possible by improved lithographic techniques, it is also to be regretted that calculations were not made for widths smaller than 5 μm. One particularly wishes to know how the model would behave as the conductor width varies for values on each side of the average grain size, at a point where one might expect a maximum in divergence and a minimum in t_{50}. Results on Al–Cu conductors (152) display some singularities when the width w of the conductors becomes equal to the average grain size d; the investigators suggest that data on the relations between t_{50}, w and d should be normalized with respect to the ratio w/d.

The distribution of failure times was found to follow closely a lognormal law (150). Quite surprisingly, this result was shown not to depend on the choice of a lognormal distribution for the grain size; when the grain diameters were fitted to a Weibull distribution, the effect on the distribution of failure times was hardly noticeable. The model provides that the median failure times should increase with the average grain size d at a rate that is somewhat less than linear. Considering the inherent scatter of experimental observations, one may say that this result is in good agreement with accelerated test data which show t_{50} increasing linearly with d (see, for example, ref. 153). If studies of the effect of the grain size on σ, either through experimentation or computer simulation, were ever made, they do not seem to have been published, leaving a deplorable lacuna in our understanding. That an increase

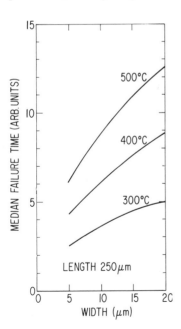

Figure 8.16. Computer-simulated dependence of lifetime t_{50} on width of conductors with a length of 250 μm. The three curves correspond to the same conditions as in Figure 8.15. From reference 150.

in the standard deviation for the grain size distribution should lead to a decrease in t_{50}, accompanied by an increase in σ, requires no elaboration.

8.7   METHODS FOR IMPROVING THIN FILM CONDUCTORS

8.7.1   Grain Size

Increasing the grain size of thin film conductors, thereby decreasing the number of short circuit paths for electrotransport, is a very direct way of improving the median lifetime t_{50} of thin film conductors with respect to electromigration-induced failures. Several instances of the successful use of this technique with Al films have been reported (72, 153, 154). In one case t_{50} increased by a factor of 4 for a grain size increase from 1.2 μm to 8 μm (72), which is almost in agreement with a linear relationship between life-times and grain sizes. Information about the effect of a change in grain size on the distribution of failure times remain scanty. In the case which has just been mentioned, the standard deviation of the lognormal distribution of failure times decreased from about 0.5 for the small-grain (1.2 μm) samples to about 0.3 for the large-grain (8 μm) samples; the width of the conductors, either 10 or 15 μm, was not exactly specified. However, as seen in Figure 8.15b, this decrease in σ cannot be anticipated to be universally observed. In every study of this effect an increase in the grain size has been accompanied by an increase in the apparent activation energy for failure (72, 154). One should be aware that the high activation energy characteristic of large-grain conductors might be due in part to an increased lattice contribution to atomic transport which would cause the activation energy to decrease with the test temperatures. Alternatively, the high activation energy might be due to a structural change in the films: the (111) preferred orientation has been reported to be more dominant in large-grain films than in small-grain films, which have a more random orientation (72).

For all Al films with thickness of about 1 μm, the biggest reported grain size, corresponding to the highest substrate temperature, is of the order of 8 to 10 μm. It is noted that the grain size does not increase much above 2 to 3 μm unless the substrate temperature is increased above 400°C (62). In alloyed films, where the alloying element is added as a layer sandwiched between two layers of Al, heat treatment can cause the growth of extremely large grains (100 μm or more) (155). The resulting failure times are correspond-ingly extremely long. However, with a grain size commensurate with the length of the conductors, the large values of σ minimize the advantage to be gained from the large values of t_{50}. In order to test such effects, Al conductors 8 and 25 μm wide were etched from films with respective thicknesses of 10,000 and 5000 Å. The films had been deposited on oxidized Si wafers at 200°C and subsequently annealed at 560°C. Under accelerated test conditions of 2×10^6 A cm^{-2} and $175°$C the t_{50} for the thick and narrow conductors was found to be 30 times shorter than the t_{50} for thin and wide ones, although as usual the films with the greater thickness can be presumed to have the larger grains (156).

8.7.2 Dielectric Overlayer

The technique of covering thin film conductors with an insulating dielectric layer has also been found beneficial with respect to electromigration failure times. In the earliest implementation, Al conductors were covered with a layer of fused glass (154). Subsequently, improvements in the mean time to failure by a factor of about 10 were reported for Al films covered with a layer of Al oxide obtained by anodization (99, 157). The effect is most marked if care is taken to anodize the sides as well as the top of the conductors. A layer of sputtered SiO$_2$ on Ta–Au–Ta conductors increased the lifetimes of the conductors by a factor of 30 (138). An overlayer of Al$_2$O$_3$ was found to reduce the grain boundary diffusivity of Cu in Al by about 40% (159). It is to be noted that in spite of these observations on dielectric overlayers, the technique is not universally effective; some investigators found no effect for some conductors and dielectric coatings.

In an early study the possibility of compressive stresses preventing the growth of hillocks, and consequently the formation of holes as well, was examined (160). Thermodynamic equilibrium conditions were assumed. It was concluded that the stresses (tensile) in the overlayers would increase to a level too high to be sustained before any effect on the failure times could be observed. More recently, the effect was experimentally observed in Al films while the stresses were measured simultaneously (94, 95). On the basis of these measurements, the forces at work, namely, the value of Z_b^* for Al, were calculated (see Section 8.3.5). The value of Z_b^* thus obtained appears to be suspiciously low, by almost one order of magnitude. Both observation and theory agree that the effect should be maximal for short conductors or under conditions where the sources and drains (hillocks and holes) for the mobile defects are at short distances, such as in small grain conductors. Yet there remains the paradox that if realistic values of electromigration forces are used as the basis for calculations, the resulting stresses are found to be too high, while if one proceeds from the observed stress levels to the value of Z_b^*, the value of Z_b^* is found to be too low. The mechanism at work is not clearly understood.

To the extent that the stresses are an important factor, one can anticipate that the effect of an overcoat will be maximized for thick overlayers and for very thin films, where the small grain size, implying short source and drain distances for vacancies, should also be helpful. The results for Al films 300 to 400 Å thick covered by a sputtered SiO$_2$ layer of 8000 Å are illustrated in

Figure 8.17, which is taken from reference 161. Presumably, the extremely long lifetimes reported for thin (200 Å) Ni–Cr resistors (1000 h at 2.6×10^7 A cm^{-2} and 240°C) are partly due to the presence of an SiO_2 overlayer (162), just as the excellent performance reported for Ni–Fe magnetoresistive elements (no failures in 11 conductors after 1000 h at 1×10^7 A cm^{-2} and 240°C) is partly induced by the sputtered Al_2O_3 coating. Comparison with uncoated elements could not be made because of corrosion effects (163).

8.7.3 Alloy

The addition of a second element is an effective and reliable method for improving electromigration lifetime of thin film conductors which is attracting fairly wide attention (164–166). Although this technique requires some modifications of the deposition process, it is easy and inexpensive to use. Illustrative data of the alloying effect on the median time to failure of Al, Au, and Cu conductors are listed in Table 8.6. For convenience of comparison, the data given were obtained either directly under comparable fabrication and test conditions or by extrapolation using Eq. 13 and the reported values of n and ΔH. In using the data in Table 8.6 to discuss the alloying effects, one

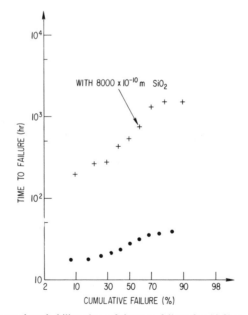

Figure 8.17. Lognormal probability plots of times to failure for Al films with a thickness of 300 to 400 Å, tested at $T_0 = 160°C$ and $J_0 = 3 \times 10^6$ A cm^{-2}. Upper curve corresponds to conductors covered with 8000 Å of sputtered SiO_2; lower curve, to similar conductors without a covering layer of SiO_2. From reference 161.

TABLE 8.6.   MEDIAN FAILURE TIMES FOR Al, Au, AND Cu
CONDUCTORS AS A FUNCTION OF ALLOYING ADDITIONS
($J = 2 \times 10^6$ A cm^{-2})

	Alloy, wt %	MTF, h	Reference
Al at 175°C			
Al		30–45[a]	37, 72
	Si 1.8	100–200	115
	Cu 4	2500	37
	Cu 4, Si 1.7	4000[a]	134
	Ni 1	3000	128
	Cr	8300	79
	Mg 2	1000	127
	Cu 4, Mg 2	10000[a]	168
	Cu 4, Ni 2, Mg 1.5[b]	32000[a]	168
	Au 2	55	169
	Ag 2	45	169
Au at 300°C[c]			
	Ta	4000[a]	158
	Ni–Fe	800	106
	Mo	100[a]	158
	W + Ti	25	104
Cu at 300°C			
Cu		180	113
	Al 1	300	113
	Al 10	6000	161
	Be 1.7[b]	20000[a]	113

[a]Extrapolated by means of appropriate values for n and ΔH.
[b]Deposited from an alloyed source by evaporation; the composition indicated is that of the source.
[c]Gold films on SiO$_2$ need an adhesion layer, which is made of the material indicated in the table. After annealing some degree of alloying takes place.

must bear in mind that the results of lifetests can vary considerably according to the test conditions and the geometry and microstructure of the sample used. Thus, the mean time to failure values as given should be considered only as an or'ler-of-magnitude indication of the alloying effects. Figure 8.14 shows an increase in t_{50} of about two orders of magnitude for Al–1.7 wt % Si–4 wt % Cu as compared to plain Al films (test conditions: 8×10^5 A cm^{-2} and 220°C) (134). The effect of Cu on the lifetimes of Al–Si conductors appears to be quite comparable to its effect on pure Al (37).

8.7.3.1 Aluminum. In the study of the Al–Si–Cu thin films (134), it is confirmed that Si additions have a somewhat beneficial effect on the t_{50} of Al thin film conductors, as had been reported earlier. (167). Both sets of published results indicate that the activation energy for failure of the Al–Si conductors is lower than for pure Al, so that at low temperatures Si additions cease to be advantageous. Additions of Cr (79), Ni (128), and Mg (127) to Al were found to be beneficial. In particular, the effect of Mg was investigated in some depth. However, the usefulness of Mg is limited by its strong reactivity with oxygen (present with SiO_2 in the substrates) and the possible formation of Mg_2Si at Si contacts. Apparently, the effects of some of these alloying elements are additive, perhaps through such mechanism as a mutual reduction of the different diffusion coefficients. In any case, the longest reported lifetimes for Al thin film conductors were obtained with a quaternary alloy of Al with Cu, Mg, and Ni (168). However, the usefulness of such an alloy is restricted by its complexity and by the limitations enforced by the presence of Mg. Attempts at increasing the lifetimes of Al–Cu conductors through the addition of a third element (In, Sn, or Ag) (168) have remained unsuccessful up to now. It is remarkable that additions of Au or Ag to Al, in opposition to additions of Au, were found to be without effect on the lifetimes (169).

8.7.3.2 Gold. Among the results for Au, the addition of Ta to Au thin films is particularly interesting (158). In this case measurements with radioactive tracers showed that the increase in lifetimes is due to a decrease in the grain boundary coefficient for self-diffusion (of Au), accompanied by an increase in activation energy (108). However, at temperatures above $300°C$ the lifetimes of Au films with a Ta adhesion layer continue to be longer than other Au films (with a Mo adhesion layer), although in this range of temperatures Ta does not decrease the grain boundary diffusion of Au. Therefore, there must be factors other than effects on Au diffusion which can influence the conductor lifetime. A similar situation exists for Cr addition to Al where the improvement in lifetimes is not accompanied by a significant reduction in the solvent diffusivity (79).

8.7.3.3 Copper. In Cu thin film conductors, Be additions increase the electromigration lifetimes by about two orders of magnitude (113). Because of the effect of Cu additions to Al thin films, the behavior of Al additions to Cu thin films is especially interesting. A series of tests with Cu (170) conductors containing up to about 20 at. % of Al have been carried out. To ensure structural homogeneity, all of the films were annealed at $500°C$ for 30 min prior to testing. The anneals and the tests were carried in an atmosphere of $85\% N_2$–$15\% H_2$ under conditions which were sufficiently reducing to prevent the oxidation of Cu and sufficiently oxidizing to allow the formation of Al oxide. Low Al concentrations do not improve the lifetimes (113), yet with about 10 at. % or more, the lifetimes are increased by about two orders of magnitude (170). The ineffectiveness of the low concentrations indicates that the mechanism at work in Cu–Al may be different from the mechanism operative in Al–Cu. Since Al additions become beneficial to the electromigration lifetimes of Cu films in a range of concentration where they provide excellent oxidation resistance through the formation of a strong oxide layer, one may hypothesize that the beneficial effect is partly due to the dielectric coating behavior as discussed in the preceding section.

8.7.3.4 Failure in Al–Cu. In the majority of cases, failures of Al–Cu 169) or Al–Si–Cu (134) have been observed to occur in areas of the conductors that are depleted of Cu. This has been confirmed by X-ray fluorescence mapping of an Al–Cu conductor both before and after electromigration (132) and also in experiments conducted with the scanning electron microscope to show massive migration of Cu (135). A failure model has been conceived on the premise that Cu depletion must precede failure, at least in a certain range of Cu concentration (133). Accordingly, the maximum activation energy for failure in Al–Cu films would be approximately equal to the activation energy for Cu depletion. For a saturated solid solution, this is given by the relation

$$\Delta H_{\text{Al-Cu}} = Q_{b\,\text{Al}}^{\text{Cu}} + E_{\text{Al}}^{\text{Cu}} - H_{a\,\text{Al}}^{\text{Cu}} \tag{17}$$

The activation energy for the grain boundary diffusion of Cu in Al $Q_{b\,\text{Al}}^{\text{Cu}}$, is modified by two terms, one additive, $E_{\text{Al}}^{\text{Cu}}$, which is the heat of solution of Cu atoms in Al (or the heat of dissolution of $CuAl_2$ precipitates), the other substractive, $H_{a\,\text{Al}}^{\text{Cu}}$, which is the heat of absorption of Cu atoms on Al grain boundaries. None of these terms can be known from failure experiments alone. The maximum activation energy for failure in Al–Cu was found to be about 0.8 eV for a Cu content of 4 wt % (133). If one estimates $E_{\text{Al}}^{\text{Cu}}$ from the Al–Cu phase diagram to be about 0.4 eV and assumes $H_{a\,\text{Al}}^{\text{Cu}}$ be about 0.2 eV, then $Q_{b\,\text{Al}}^{\text{Cu}}$ would be about 0.6 eV according to the relation above. This is too small in comparison to the value of 0.8 eV found from the cross-stripe measurements (124). However, it may be more correct to assume $E_{\text{Al}}^{\text{Cu}} - H_{a\,\text{Al}}^{\text{Cu}}$ to be approximately equal to zero since the grain boundary adsorption has been found empirically to be roughly proportional to the inverse of the solubility limit (59, 171). If this were the case, $Q_{b\,\text{Al}}^{\text{Cu}}$ would be 0.8 eV from the failure time measurements, in agreement with the cross-stripe measurements. The model cannot hold for all Al films containing Cu; at points of maximum divergence in grain boundary diffusion or where there is excess Cu, one may anticipate to observe failure occurring prior to total Cu depletion. Certainly, this is most likely to be found in the vicinity of large $CuAl_2$ precipitates.

8.7.3.5 Ion Implantation. One form of alloying that has received little attention is that which is obtained by ion implantation. It has been reported that ion implantations of Ne, P, or As in Al, Al–Cu, or Ag films to a level of 1×10^{16} ions per cm^2 have a detrimental effect on the resistance of thin film conductors to electromigration failures (172). Annealing at $450°C$ of a Ne-implanted Al film did not affect the results. Experiments conducted with B and P implants to a level of 4×10^{15} ions per cm^2 in Al films followed by a 40-min anneal at $450°C$ in a N_2–H_2 mixture resulted in a decrease of the median failure times by a factor of 2 or 3. A definite improvement, however, was obtained with O_2 implants. Doses of 2×10^{17} and 1×10^{18} ions per cm^2 were used, and the ion energy was chosen so that the implanted atoms were distributed in the middle third of a 4200 Å sample. Without any anneal after the implant, the t_{50} for the conductors with 10^{18} ions per cm^2 was not significantly different than the t_{50} obtained with pure Al. However, after annealing for 2 h at $350°C$, the t_{50} increased to values of 3300 h and more than 9300 h, respectively, for the 1×10^{18} ions per cm^2 and the 2×10^{17} ions per cm^2 implants (173). For the test conditions of 2×10^6 A cm^{-2} and $175°C$, these failure times are at least ten times longer than the times obtained with pure Al films (under identical conditions.) Analysis of some of the samples by transmission electron microscopy indicated that the implant and heat treatment did not result in grain growth; therefore this can be eliminated as a possible explanation for the observed improvements.

ACKNOWLEDGMENTS

The authors wish to thank co-workers at IBM, N. Ainslie, A. Gangulee, J. K. Howard and R. Landauer, who contributed to their understanding of electromigration phenomena in thin films. They are also grateful to C. Aliotta, A. Ginzberg, J. Lewis, H. Luhn, V. Ranieri, and W. Schug whose work in carrying many tedious experiments cannot be too highly valued. The permission of many colleagues, B. Agarwala, M. Attardo, I. Blech, M. Biscondi, L. Braun, G. van Gurp, L. Herbeuval, R. Hummel, R. Jack, J. Learn, R. Lye, J. C. Rouais, R. Ruttledge, R. Sigsbee, and J. Venables, to reproduce their figures is acknowledged with gratitude.

REFERENCES

1. A. Gangulee and F. M. d'Heurle, Mass transport during electromigration in aluminum-magnesium thin films, *Thin Solid Films*, **25**, 317 (1975).
2. A. Blech and H. Sello, "A Study of Failure Mechanisms in Silicon Planar Epitaxial Transistors," in *Physics of Failure in Electronics*, Vol. 5, T. S. Shilliday and J. Vaccaro, eds., Rome Air Development Center (1966), p. 496.
3. F. d'Heurle and R. Rosenberg, "Electromigration in Thin Films," *Physics of Thin Films*, Vol. 7, G. Hass, M. Francombe, and R. Hoffman, eds. Academic Press, New york (1973), p. 257.
4. M. Gerardin, De l'action des piles sur les sels de potasse et de soude et sur les alliages soumis a la fusion ignée, *C.R. Acad. Sci.*, **53**, 727 (1961).
5. F. Skaupy, Die Elektrizitätsleitung in Metallen, *Verhandl. Deut. Phys. Ges.*, **16**, 156 (1914).

6. W. Seith and H. Wever, Über einen neuen Effekt bei der elektrolytischen Überführung in festen Legierungen, *Z. Elektrochem.*, **59**, 942 (1953).

7. V. B. Fiks, On the mechanism of the mobility of ions in metals, *Sov. Phys. Solid State*, **1**, 14 (1959).

8. H. B. Huntington and A. R. Grone, Current induced marker motion in gold wires, *J. Phys. Chem. Solids*, **20**, 76 (1961).

9. C. Bosvieux and J. Friedel, Sur l'electrolyse des alliages metalliques, *J. Phys. Chem. Solids*, **23**, 123 (1962).

10. J. N. Pratt and R. G. Sellors, *Electrotransport in Metals and Alloys, Diffusion and Defect Monographs*, Trans. Tech. SA, Riehen, Switzerland (1973).

11. H. B. Huntington, "Electromigration in Metals," in *Diffusion in Solids—Recent Developments*, A. S. Nowick and J. J. Burton, eds., Academic Press, New York (1975), p. 303.

12. D. A. Rigney, "Electromigration in Metallic Systems," in *Charge Transfer—Electronic Structure of Alloys*, L. H. Bennett and R. H. Willens, eds., AIME, New York (1974).

13. H. Wever, *Elektro- und Thermotransport in Metallen*, Johann Ambrosius Barth, Leipzig (1973).

14. *Electro- and Thermo-Transport in Metals and Alloys*, R. E. Hummel and H. B. Huntington, eds., AIME, New York (1977).

15. A. Khosla and H. B. Huntington, Electromigration in tin single crystals, *J. Phys. Chem. Solids*, **36**, 395 (1975).

16. J. Wohgemuth, Electromigration in polycrystalline and single crystal magnesium, *J. Phys. Chem. Solids*, **36**, 1025 (1975).

17. V. N. Grinyuk and G. F. Tikhinskii, Electrotransfer in mono- and polycrystalline beryllium, *Fiz. Met. Metall.*, **39**(3), 514 (1975).

18. Ch. Herzig and E. Stracke, Atomic transport of gold and silver in lead in a DC Field, *Phys. Stat. Solidi*, **A27**, 25 (1975).

19. R. E. Einziger and H. B. Huntington, Electromigration and permeation of hydrogen and deuterium in silver, *J. Phys. Chem. Solids*, **35**, 1563 (1974).

20. F. A. Schmidt and O. N. Carlson. Electrotransport of carbon and in molybdenum and uranium, *Metall. Trans.*, **7A**, 127 (1976).

21. R. Kircheim and E. Fromm, Electrotransport of oxygen and nitrogen in niobium and tantalum, *Acta Met.*, **22**, 2397 (1974).

22. E. Fromm, R. Kircheim, and J. Mathuni. Change of electrotransport direction with increasing temperature in the niobium-oxygen system, *J. Less-Common Met.*, **43**, 211 (1975).

23. F. A. Schmidt and O. N. Carlson, Electrotransport of carbon, nitrogen and oxygen in scandium, *J. Less-Common Met.*, **50**, 237 (1976).

24. O. N. Carlson, F. A. Schmidt, and R. R. Lichtenberg, Investigation of reported anomalies in the electrotransport of interstitial solutes in titanium and iron, *Metall. Trans.*, **6A**, 725 (1975).

25. R. G. Jordan and D. W. Jones, The purification of the rare-earth metals. I. Solid state electrolysis of gadolinium, *J. Less-Common Met.*, **31**, 125 (1973).

26. R. G. Jordan, D. W. Jones, and V. J. Hems, The purification of the rare-earth metals. II. Solid state electrotransport processing of terbium, *J. Less-Common Met.*, **42**, 101 (1975).

27. O. N. Carlson, F. A. Schmidt, and D. T. Peterson, Purification of rare-earth metals by electrotransport, *J. Less-Common Met.*, **39**, (1975).

28. E. A. Lebedev, P. Suptiz, and I. Willert, Transport of silver under the influence of an electric field in glassy AsSe, *Phys. Stat. Solidi*, **A28**, 461 (1975).

29. B. I. Boltaks, Z. U. Borisova, U. K. Biktimirova, T. D. Dzhafarov, A. A. Obraztsov, and E. N. Shenyaskaya, Diffusion, electric transport and effect of a silver impurity on the electrical and optical properties of As Se, *Phys. Stat. Solidi*, **A30**, 731 (1975).

30. P. Suptiz, J. Teltov, E. A. Lebedev, and I. Willert., The electromigration of Au in glassy As_2Se_3 *Phys. Stat. Solidi*, **A31**, 31 (1975).
31. Ch. Herzig and D. Cardis, Matter transport in pure gold induced by high direct current densities, *Appl. Phys.*, **5**, 317 (1975).
32. L. J. Gauckler, S. Hofmann, and E. Haessner, The growth of hillocks and voids in gold single crystals by electrotransport, *Acta Met.*, **23**, 1541 (1975).
33. C. H. Herzig and W. Wiemann, Experimental determination of the electrostatic driving force in the electromigration of tin in gold and gold-tin alloys, *Phys. Stat. Solidi*, **A26**, 459 (1974).
34. J. A. Bleay, D. H. Oldham, and D. A. Blackburn, The use of electromigration and thermo-migration to produce non-equilibrium vacancy conditions in α-brass, *Acta Met.* **24**, 81 (1976).
35. A. Yoshikawa, Some analyses of the steady-state method for electrotransport, *Jap. J. Appl. Phys.*, **13**, 599 (1974).
36. G. Martin and P. Truchot, L'electromigration intergranulaire: outil d'etude de la structure de coeur des joints, *Can. Metall. Q.*, **13**, 111 (1974).
37. I. Ames, F. d'Heurle, and R.Horstmann, Reduction of electromigration in aluminum films by copper doping, *IBM J. Res. Dev.*, **14**, 461 (1970).
38. P. S. Ho and J. K. Howard, Grain boundary solute electromigration in polycrystalline films, *J. Appl. Phys.*, **45**, 3229 (1974).
39. G. Frohberg, "Critical remarks on the theory of the 'Direct Field' force," in *Atomic Transport in Solids and Liquids*, A. Loding, ed. Verlag der Zeitschrift fur Naturforschung, Tübingen (1971).
40. M. Gerl, Calculation of the force acting on an impurity in a metal submitted to an electric field or a temperature gradient, *Z. Naturforsch.*, **26a**, 1 (1971).
41. R. S. Sorbello, A pseudopotential based theory of the driving forces for electromigration in metals, *J. Phys. Chem. Solids*, **34**, 937 (1973).
42. A. K. Das and R. Peierls, The force on a moving charge in an electron gas, *J. Phys. C, Solid State Phys.*, **6**, 2811 (1973).
43. R. Landauer and J. W. F. Woo, Driving force in electromigration, *Phys. Rev.*, **B10**, 1266 (1974).
44. P. Kumar and R. S. Sorbello, Linear response theory of the driving forces for electromigration, *Thin Solid Films*, **25**, 25 (1975).
45. R. Landauer, Sources of conduction band polarization in the driving force for electromigration, *Thin Solid Films*, **26**, L1 (1975).
46. H. B. Huntington, On dynamic polarization as a driving force for electromigration, *Thin Solid Films*, **26**, L3 (1975).
47. R. Landauer, Spatial conductivity modulation at metallic point defects, *J. Phys. C, Solid State Phys.*, **8**, 761 (1975).
48. A. K. Das and R. Peierls, The force in electromigration, *J. Phys. C, Solid State Phys.*, **8**, 3348 (1975).
49. R. Landauer, Conductivity modulation at metallic point defects, *Phys. Lett.*, **51A**, 161 (1975).
50. R. Landauer, Electromigration and spatial variations in metallic conductivity, *J. Electron. Mater.*, **4**, 813 (1975).
51. R. Landauer, The Das-Peierls electromigration theorem, *J. Phys. C, Solid State Phys.*, **8**, L389 (1975).
52. L. J. Sham, Microscopic theory of the driving force in electromigration, *Phys. Rev.* **B12**, 3142 (1975).
53. L. Turban and M. Gerl, Driving force in electromigration and the residual resistivity field, *Phys. Rev.* **B13**, 939 (1976).
54. R. Landauer, Driving force electromigration and the residual resistivity field—a reply,

Phys. Rev. **B13**, 942 (1976); R. Landauer, Spurious agreement on the Das-Peierls electromigration theorem, submitted to *Phys. Rev. B.*

55. L. Turban, P. Nozieres, and M. Gerl, Driving force for electromigration of an impurity in a homogeneous metal, *J. Phys.*, **37**, 159 (1976).

56. A. K. Das, Electromigration in a two-band metal and in Hall field: application to an electron-hole drop, *Solid State Comm.*, **18**, 601 (1976).

57. R. S. Sorbello, "Basic Concepts of Electro- and Thermomigration: Driving Forces," in *Electro- and Thermo-Transport in Metals and Alloys*, R. E. Hummel and H. B. Huntington, eds., AIME, New York (1977), p. 2.

58. H. S. Levine and C. J. MacCallum, Grain boundary and lattice diffusion in polycrystalline bodies, *J. Appl. Phys.*, **31**, 595 (1960).

59. D. McLean, Dislocations, vacancies and solutes in grain boundaries, *Can. Metall. Q.*, **13**, 145 (1974).

60. F. d'Heurle and I. Ames, Electromigration in single-crystal aluminum films, *Appl. Phys. Lett.*, **16**, 80 (1970).

61. P. S. Ho, F. M. d'Heurle, and A. Gangulee, "Implications of Electromigration for Device reliability," in *Electro- and Thermo-Transport in Metals and Alloys*, R. E. Hummel and H. B. Huntington, eds., AIME, New York (1977), p. 108.

62. A. J. Learn, Evolution and current status of aluminum metallization, *J. Electrochem. Soc.*, **123**, 894 (1976).

63. E. Philofsky and E. L. Hall, A Review of the Limitations of Aluminum Thin Films on Semiconductor Devices, *IEEE Trans.*, PHP-11, 281 (1975).

64. J. L. Vossen, G. L. Schnable, and W. Kern. Processes for multilevel metallization, *J. Vac. Sci. Technol.*, **11**, 60 (1974).

65. I. A. Blech and E. S. Meieran, Electromigration in thin Al films, *J. Appl. Phys.*, **40**, 485 (1969).

66. B. N. Agarwala, M. J. Attardo, and A. P. Ingraham, Dependence of electromigration-induced failure time on length and width of aluminum thin film conductors, *J. Appl. Phys.*, **41**, 3954 (1970).

67. L. Herbeuval and M. Biscondi, Diffusion du zinc dans les joints de flexion symetriques de l'aluminum, *Can. Metall. Q.*, **13**, 171 (1974); *Grain Boundaries and Interfaces*, P. Chaudhari and J. W. Matthews, eds., North Holland, Amsterdam (1972).

68. G. H. Bishop and B. Chalmers, A coincidence-ledge-dislocation description of grain boundaries, *Scripta Met.*, **2**, 133 (1968).

69. G. Hasson, M. Biscondi, P. Lagarde, J. Levy, and C. Goux, "Structure of Grain Boundaries: Theoretical Determination and Experimental Observations," in *The Nature and Behavior of Grain Boundaries*, H. Hu, ed., Plenum Press, New York (1972), p. 3.

70. E. D. Hondros, Grain boundary segregation. The current situation and future requirements, *J. de Phys.*, **36**, C4-117 (1975).

71. G. Hasson, J. Y. Boos, I. Herbeuval, M. Biscondi, and C. Goux, Theoretical and experimental determinations of grain boundary structures and energies: correlations with various experimental results, *Surf. Sci.*, **31**, 115 (1972).

72. M. J. Attardo and R. Rosenberg, Electromigration damage in aluminum film conductors, *J. Appl. Phys.*, **41**, 2381 (1970).

73. D. W. Jepsen, *Qualitative Description of Temperature Distribution Effects in Aluminum Stripes*, IBM Research Note NC 753, New York (1968).

74. Y.-S. Chaung and H. L. Huang, Temperature distribution on thin film metallizations, *J. Appl. Phys.*, **47**, 1775 (1976).

75. P. S. Ho and L. Glowinski, "Observation of Void Formation Induced by Electromigration in Metallic Films," in *Atomic Transport in Solids and Liquids*, A. Loding, ed., Verlag der Zeitschrift für Naturforschung, Tübingen (1971), p. 122.

76. Y.-S. Chaung and H. L. Huang, Cracked-stripe resistance and current crowding effect, *Jap. J. Appl. Phys.*, **14**, 267 (1975).
77. J. C. Rouais, Role des héterogénéités locales de temperature dans la rupture des couches minces d'or par electromigration, Ph.D. Thesis, Institut National des Sciences Appliqueés, Lyon (1976).
78. I. A. Blech and E. Kinsborn, Electromigration in thin gold films on molybdenum surfaces, *Thin Solid Films*, **25**, 327 (1975).
79. F. M. d'Heurle and A. Gangulee, "Solute Effects on Grain Boundary Electromigration and diffusion," in *Nature and Behavior of Grain Boundaries*, H. Hu, ed., Plenum Press, New York (1972), p. 339.
80. R. Rosenberg, Value of $D_0 Z^*$ for grain boundary electromigration in aluminum films, *J. Appl. Phys.*, **16**, 27 (1970).
81. J. Weiss, Quantitative measurements of the mass distribution in thin films during electrotransport experiments, *Thin Solid Films*, **13**, 169 (1972).
82. R. E. Hummel, R. T. DeHoff, and H. J. Geier, Activation energy for electrotransport in thin aluminum films by resistance measurements, *J. Phys. Chem. Solids*, **37**, 73 (1976).
83. H. M. Breitling and R. E. Hummel, Electromigration in thin silver, copper, gold, indium, tin, lead and magnesium films, *J. Phys. Chem. Solids*, **33**, 845 (1972); R. E. Hummel, Observations on the electromigration in various thin films of group I-IV, *Thin Solid Films.*, **13**, 175 (1972).
84. R. Rosenberg and L. Berenbaum, Resistance monitoring and effects of non-adhesion during electromigration in aluminum films, *Appl. Phys. Lett.*, **12**, 20 (1968).
85. J. K. Howard and R. F. Ross, Hillocks as structural markers for electromigration rate measurements in thin films, *Appl. Phys. Lett.*, **18**, 344 (1971).
86. K. L. Tai, P. H. Sun, and M. Ohring, Lateral self-diffusion and electromigration in thin films, *Thin Solid Films*, **25**, 343 (1975).
87. P. S. Ho, J. E. Lewis, and J. K. Howard, Kirkendall study of electromigration in thin films, *Thin Solid Films*, **25**, 302 (1975).
88. E. O. Kirkendall, Diffusion of zinc in alpha brass, *Trans. AIME*, **147**, 104 (1942).
89. G. Martin, Electromigration intergranuaire de l'antimoine dans l'argent, *Phys. Stat. Solidi*, **A14**, 183 (1972); see also reference 36.
90. P. S. Ho, to be published.
91. K. L. Tai and M. Ohring, Grain-boundary electromigration in thin films. I. Low-temperature theory, *J. Appl. Phys.*, **48**, 28 (1977).
92. K. L. Tai and M. Ohring, A digital tracer scanner for studies of longitudinal self diffusion in thin films, *Rev. Sci. Instr.*, **45**, 9 (1974).
93. P. S. Ho and J. K. Howard, Non linear propagation of void front during electromigration in alloy films, *Appl. Phys. Lett.*, **27**, 261 (1975).
94. I. A. Blech, Electromigration in thin aluminum films on titanium nitride, *J. Appl. Phys.*, **47**, 1203 (1976).
95. I. A. Blech and C. Herring, Stress generation by electromigration, *Appl. Phys. Lett.*, **29**, 131 (1976).
96. L. Berenbaum, Electromigration damage of grain-boundary triple points in Al thin films, *J. Appl. Phys.*, **42**, 880 (1971).
97. S. L. Horowitz and I. A. Blech, Electromigration in Al/Cu/Al films observed by transmission electron microscopy, *Mater. Sci. Eng.*, **10**, 169 (1972).
98. M. C. Shine and F. M. d'Heurle, Activation energy for electromigration in aluminum films alloyed with copper, *IBM J. Res. Dev.*, **15**, 378 (1971).
99. J. R. Black. Electromigration failure modes in aluminum metallizations for semiconductor devices, *Proc. IEEE*, **57**, 1587 (1969).

100. T. Satake, K. Yokoyama, S. Shirakawa, and K. Sawaguchi, Electromigration in aluminum film stripes coated with anodic aluminum oxide films, *Jap. J. Appl. Phys.*, **12**, 518 (1973).

101. B. J. Klein, Electromigration in thin gold films, *J. Phys. F, Met. Phys.*, **3**, 691 (1973).

102. R. E. Hummel and H. J. Geier, Activation energy for electrotransport in thin silver and gold films, *Thin Solid films*, **25**, 335 (1975).

103. K. L. Tai and M. Ohring, Grain-boundary electromigration in thin films. II. Tracer measurements in pure Au, *J. Appl. Phys.*, **48**, 36 (1977).

104. J. C. Blair, C. R. Fuller, P. B. Ghate, and C. T. Haywood. Electromigration-induced failures in, and microstructure and resistivity of, sputtered gold films, *J. Appl. Phys.*, **43**, 307, (1972).

105. J. Rouais, G. Lormand, M. Chevreton, and C. Eyraud, Role des joints de grains dans la rupture des couches minces d'or par electromigration, *C.R. Acad. Sci., Ser.* **B274**, 827 (1972).

106. A. Gangulee and F. M. d'Heurle, The activation energy for electromigration and grain boundary self diffusion in gold, *Scripta Met.*, **7**, 1027 (1973).

107. M. Etzion, I. A. Blech, and Y. Komen. Study of conductive gold film lifetime under high current densities, *J. Appl. Phys.*, **46**, 1455 (1975).

108. D. Gupta and R. Rosenberg, Effect of a solute addition (Ta) on low temperature self-diffusion processes in gold, *Thin Solid Films*, **25**, 171 (1975).

109. R. E. Hummel and R. M. Breitling, On the direction of electromigration in thin silver, gold and copper films, *Appl. Phys. Lett.*, **18**, 373 (1971).

110. R. E. Hummel and R. T. DeHoff, On the controversy about the direction of electrotransport in thin gold films, *Appl. Phys. Lett.*, **27**, 64 (1975).

111. I. A. Blech and R. Rosenberg, On the direction of electromigration in gold thin films, *J. Appl. Phys.*, **46**, 579 (1975).

112. R. Rosenberg and L. Berenbaum, "Electrotransport Damage Mechanisms in Thin Metallic Films," in *Atomic Transport in Solids and Liquids*, Verlag der Zeitschrift für Naturforschung, Tübingen, Germany (1971), p. 113.

113. F. M. d'Heurle and A. Gangulee, Electrotransport in copper alloy films and the defect mechanism in grain boundary diffusion, *Thin Solid Films*, **25**, 531 (1975).

114. P. H. Sun and M. Ohring. Tracer self-diffusion and electromigration in thin tin films, *J. Appl. Phys.*, **47**, 478 (1976).

115. G. J. van Gurp, Electromigration in cobalt films, *Thin Solid Films*, **38**, 295 (1976).

116. P. S. Ho, Electromigration and Soret effect in cobalt, *J. Phys. Chem. Solids*, **27**, 1331 (1966).

117. R. V. Penney, Current-induced mass transport in aluminum, *J. Phys. Chem. Solids*, **25**, 335 (1964).

118. P. S. Ho, Solute effects on electromigration, *Phys. Rev.*, **B8**, 4534 (1973).

119. P. Benoist and G. Martin, Atomic model for grain boundary and surface diffusion, *Thin Solid Films*, **25**, 181 (1975).

120. R. Rosenberg, A. F. Mayadas, and D. Gupta, Grain boundary contributions to transport, *Surf. Sci.*, **31**, 566 (1972).

121. P. Guyot and J. P.'Simon, Theoretical aspects of the interaction between grain-boundaries and impurities, *J. de Phys.*, **36**, C4-141 (1975).

122. J. K. Howard and P. S. Ho, unpublished results.

123. M. Ohring and P. Singh, Role of alloy valence on electromigration in thin tin alloy films, *Thin Solid Films*, **31**, 253 (1976).

124. P. S. Ho and J. K. Howard, to be published.

125. I. A. Blech, Copper-electromigration in aluminum, *J. Appl. Phys.*, **48**, 473 (1977).

126. N. Van Doan, A new method for determination of the jump frequency ratios for diffusion in dilute f.c.c. alloys, *J. Phys. Chem. Solids*, **33**, 2161 (1972); see also reference 118.

127. F. M. d'Heurle, A. Gangulee, C. F. Aliotta, and V. A. Ranieri, Effects of Mg additions on the electromigration behavior of Al thin film conductors, *J. Electron. Mater.*, **4**, 497 (1975).

128. F. M. d'Heurle, A. Gangulee, C. F. Aliotta, and V. A. Ranieri. Electromigration of Ni in Al thin-film conductors, *J. Appl. Phys.*, **46**, 4845 (1975).

129. N. L. Peterson and S. J. Rothman, Impurity Diffusion in Aluminum, *Phys. Rev.*, **B1**, 3264 (1970).

130. W. C. Mullard, A. B. Gardner, R. F. Bass, and L. H. Slifkin. Self-Diffusion in Ag-Au Solid Solutions, *Phys. Rev.*, **129**, 617 (1963).

131. D. Gupta and K. Asai, Grain Boundary self-diffusion in evaporated Au films at low temperatures, *Thin Solid Films*, **22**, 121 (1974).

132. B. N. Agarwala, G. Digiacomo, and R. R. Joseph, Electromigration damage in aluminum-copper films, *Thin Solid Films*, **34**, 165 (1976).

133. F. M. d'Heurle, N. G. Ainslie, A. Gangulee, and M. C. Shine, Activation energy for electromigration failure in aluminum films containing copper, *J. Vac. Sci. Technol.*, **9**, 289 (1972).

134. A. J. Learn, Electromigration effects in aluminum alloy metallization, *J. Electron. Mater.*, **3**, 531 (1974).

135. B. N. Agarwala, L. Berenbaum, and P. Peressini, Electromigration induced failures in thin film Al-Cu conductors, *J. Electron. Mater.*, **3**, 137 (1974).

136. S. Mader and S. Herd, Formation of second phase particles in aluminum-copper alloy films, *Thin Solid Films*, **10**, 377 (1972).

137. G. A. Walker and C. C. Goldsmith, Precipitation and solid solution effects in aluminum-copper thin films their influence on electromigration, *J. Appl. Phys.*, **44**, 2452 (1973).

138. E. Hall, E. Philofsky, and A. Gonzales, Electromigration in Al–2 percent Cu thin films, *J. Electron. Mater.*, **2**, 333 (1972).

139. L. Braun, Electromigration testing—a current problem, *Microelectron. Reliab.*, **13**, 271 (1974).

140. M. Saito, H. Anayama, and S. Shikama. Reliability and failure analysis of semiconductor integrated logic circuits, *Rev. Electron. Comm. Lab.* (Japan), **21**, 339 (1973).

141. I. A. Blech, Electromigration and crevice formation in thin metallic films, *Thin Solid Films*, **13**, 117 (1972).

142. A. T. English, K. L. Tai, and P. H. Turner, Electromigration in conductor stripes under pulsed dc powering, *Appl. Phys. Lett.*, **21**, 397 (1972).

143. A. Bobbio, A. Ferro, and O. Saracco, Electromigration failure in Al thin films under constant and reversed dc powering, *IEEE Trans.*, **R-23**, 194 (1974).

144. J. D. Venables and R. G. Lye, "A Statistical Model for Electromigration Induced Failure in Thin Film Conductors," in *Proceedings of the 10th Annual Reliability Physics Symposium*, IEEE, New York (1972) p. 159.

145. R. A. Sigsbee, Electromigration and metallization lifetimes, *J. Appl. Phys.*, **44**, 2533 (1973).

146. R. A. Sigsbee, "Failure Model for Electromigration," in *Proceedings of the 11th Annual Reliability Physics Symposium*, IEEE, New York (1973), p. 301.

147. F. M. d'Heurle, Electromigration in thin films: the effects of solute atoms on grain boundary diffusion, *J. de Phys.*, **36**, C4-191 (1975).

148. P. S. Ho, Motion of inclusion induced by a direct current and a temperature gradient, *J. Appl. Phys.*, **41**, 64 (1970).

149. A. Bobbio and A. O. Saracco, On the spread of time-to-failure measurements in thin metallic films, *Thin Solid Films*, **17**, S13 (1973).

150. M. J. Attardo, R. Rutledge, and R. C. Jack, Statistical metallurgical model for electromigration failure in aluminum thin-film conductors, *J. Appl. Phys.*, **42**, 4343 (1971).

151. A. T. English, K. A. Tai, and P. A. Turner, Electromigration of Ti-Au thin film conductors at 180°C, *J. Appl. Phys.*, **45**, 3757 (1974).

152. G. A. Scogan, B. N. Agarwala, P. P. Peressini, and A. Brouillard, "Width Dependence of Electromigration Life in Al-Cu, Al-Cu-Si and Ag Conductors," in *Proceedings of the 13th*

Reliability Physics Symposium, IEEE, New York (1975), p.151.

153. M. Saito and S. Hirota, Effect of grain size on the life time of aluminum interconnections, *Rev. Electron. Comm. Lab.* (Japan), **22**, 678 (1974).

154. J. R. Black, Electromigration—a brief survey and some recent results, *IEEE Trans.*, **ED-16**, 338 (1969).

155. A. Gangulee and F. M. d'Heurle, Anomalous large grains in alloyed aluminum thin films. I. Secondary grain growth in aluminum-copper films, *Thin Solid Films*, **12**, 399 (1972); *idem*, II. Electromigration and diffusion in thin films with very large grains, *Thin Solid Films*, **16**, 227 (1973).

156. I. Ames, F. d'Heurle and R. Hortsmann, unpublished results.

157. A. J. Learn, Effect of structure and processing on electromigration-induced failure in anodized aluminum, *J. Appl. Phys.*, **44**, 1251 (1973).

158. M. Revitz and P. A. Totta, A Ta-Au-Ta thin film interconnection system for high current density integrated circuits, *Electrochem. Soc.*, **72**, 631 (1972).

159. J. K. Howard and P. S. Ho, unpublished results.

160. N. G. Ainslie, F. M. d'Heurle, and O. C. Wells, Coatings, mechanical constraints, and pressure effects on electromigration, *Appl. Phys. Lett.*, **20**, 172 (1972).

161. F. d'Heurle, A. Gangulee, and V. Ranieri, unpublished results.

162. T. Satake, Electromigration failure in Ni Cr thin-film stripes, *Appl. Phys. Lett.*, **23**, 496 (1973).

163. A. Gangulee, C. H. Bajorek, F. M. d'Heurle, and A. F. Mayadas, Long term stability of magnetoresistive bubble devices, *IEEE Trans.*, **MAG-10**, 848 (1974).

164. G. Sideris, Power from transitors, *Electronics*, **46**, 68 (1973).

165. A. K. Kakar, Electromigration studies on aluminum-copper stripes, *Solid State Technol.*, **16**, 47 (1973).

166. F. Fischer and J. Fellinger, "Failure Analysis of Thin-Film Conductors Stressed with High-Current Densities by Application of Matthiessen's Rule," in *Proceedings of the 15th Annual Reliability Physics Symposium*, IEEE, New York (1977), p. 250.

167. G. J. van Gurp, Electromigration in Al films containing Si, *Appl. Phys. Lett.*, **19**, 476 (1971).

168. F. M. d'Heurle and A. Gangulee, "Effects of Complex Alloy Additions on Electromigration in Aluminum Thin Films," in *Proceedings of the 11th Annual Reliability Physics Symposium*, IEEE, New York (1973), p. 165.

169. F. M. d'Heurle, The effect of copper additions on electromigration in aluminum thin films, *Met. Trans.*, **2**, 683 (1971).

170. F. M. d'Heurle, A. Gangulee, and V. Ranieri, unpublished results.

171. E. D. Hondros, "Grain Boundary Segregation: Assessment of Investigative Techniques," in *Grain Boundary Structure and Properties*, G. A. Chadwick and D. A. Smith, eds., Academic Press, London (1976), p. 165.

172. E. H. Bogardus, J. K. Howard, P. Peressini, and J. W. Philbrick, Ion implantation damage in thin metal films, *Appl. Phys. Lett.*, **18**, 77 (1971).

173. F. M. d'Heurle, R. Fiorio, A. Gangulee, V. Ranieri, and S. Zirinsky, "The Effect of Oxygen Ion Implantations on Electromigration in Al Thin Film Conductors," in *Proceedings of the 7th International Vacuum Congress*, Vienna (1977), p. 2123.

9

METAL–METAL
INTERDIFFUSION

J. E. E. Baglin

IBM Thomas J. Watson Research Center, Yorktown Heights, New York

J. M. Poate

Bell Laboratories, Murray Hill, New Jersey

9.1 INTRODUCTION

Many areas of thin film technology require a detailed understanding of the interdiffusion or interaction of metal films, and this is sufficient incentive for the study of such phenomena. We hope to show, however, that the phenomena are of intrinsic enough interest to merit study in their own right. There has been a great increase in the number of experimental studies of thin film interdiffusion since Weaver's review (1) of 1971. This increase is probably due to a combination of technological imperatives and the advent of the microscopic profiling techniques discussed in Chapter 6. Weaver has given a lucid account of the physical background of thin film diffusion, but it is noticeable that there were very few available examples of direct measurements of diffusion profiles. This could present a considerable handicap in elucidating or interpreting thin film phenomena.

What are the distinguishing features of interdiffusion in thin metal films? Most outstanding is the observation of mass transport on a large scale at low temperatures, which is due not only to the small distances, where observable diffusion times can be short, but also to the property that films can contain large numbers of defects. Defects, besides controlling the interdiffusion, may also give rise to levels of diffused or reacted materials that are not interpretable in terms of equilibrium phase diagrams.

While there is a wealth of experimental detail on the subject of thin film diffusion, there are no detailed theories appropriate for massive interdiffusion in thin films; in this review, we look for some of the simplifying physical parameters. We attempt to connect our knowledge of diffusion couples with that of the equilibrium binary phase diagrams. Our understanding of the interdiffusion of the binary systems that form solid solutions is probably the most complete, and we review these systems in detail first. We choose the most cogent examples of experiment and theory.

It appears logical to start the discussion with the ideal case of diffusion between single-crystal couples that form complete series of solid solutions; unfortunately, there are very few experimental examples in this category. This is followed by a discussion of polycrystalline couples that form solid solutions; there are many examples in this category which illustrate the competition between grain boundary and bulk diffusion. We shall briefly outline the theories of the diffusion mechanism. (Grain boundary diffusion, in the low-concentration limit, is discussed at length in Chapter 7.)

We then discuss the couples that form compound phases. There are many experiments in this intriguing area. The formation of compound phases can result in the formation of planar layered structures with rather well-defined interfaces, similar to the silicides of Chapter 10. Following a brief discussion of oxidation effects and diffusion barriers, we present a critical data summary of the thin metal film studies of the past ten years or so.

In this chapter we deal almost exclusively with the interdiffusion of vacuum-deposited thin films. The reason for this will be obvious; vacuum-deposited films are technologically important and can be prepared with a high degree of purity. Films deposited by processes such as electroplating can contain rather large concentrations of impurities and defects whose influence on the interdiffusion behavior may be difficult to unscramble. It should be possible, nevertheless, to generalize from the present results of clean vacuum-deposited films to the more complicated cases such as electroplated films.

The review is limited to diffusion between single-crystal or polycrystalline films. Diffusion between amorphous or highly disordered metal films is potentially a subject of much interest, but no experiments or theory in this area have been reported.

9.2   SOLID SOLUTIONS

9.2.1   Single-Crystal Films—Miscible

Single-crystal films are seldom found in practical situations. However, measurements of the interdiffusion of specially grown single-crystal films can provide unambiguous values for bulk diffusivities and activation energies. These values are of basic importance in interpreting the more complex behavior of practical polycrystalline films, as described later.

9.2.1.1   Concentration-Dependent Diffusivity $\bar{D}_{AB}$.   We begin by discussing the simplest, idealized case of binary diffusion: that in which heat treatment at temperature T leads to planar interdiffusion at the interface between two single-crystal films of metals A and B, which are fully miscible at that temperature. As interdiffusion proceeds, metal A will develop a profile of fractional atomic concentration $C_A(x)$ of the kind sketched on the left of Figure 9.1. Now, this chapter is concerned with interdiffusion where *substantial* mass transport of atoms of both species A and B may occur. Hence, we cannot enjoy the luxury of standard tracer-diffusion analysis in which the moving species A is known and its diffusivity D_A is directly measured in the pure host matrix B. Instead, we must try to derive D_A and D_B from the observable "mutual diffusivity" $\bar{D}_{AB}(x)$.

$\bar{D}_{AB}(x)$ describes the progress of A atoms at a particular depth x measured from the original interface, where the host matrix is a mixture of A and B atoms in the proportion $C_A(x):C_B(x)$. Here, exchange of two A atoms does not contribute to the observed diffusing flux of species A. However, in a random vacancy diffusion process, the probability of a fruitful replacement of a B atom with one of type A will be proportional to $C_B(x)$. Thus, the effective transport of species A from the interface, due to the combined migration

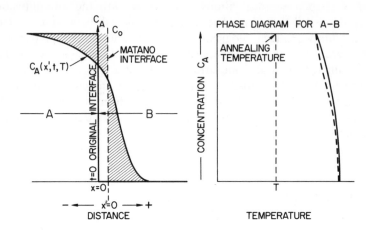

Figure 9.1. Single-crystal miscible species A and B interdiffuse at different rates. The resultant effective Matano interface is shown; it is the origin of the x' depth scale. Hatched areas are equal. The phase diagram indicates that there will be no discontinuities in $C_A(x', t, T)$ at temperature T (contrast Figs. 9.2, 9.13, and 9.14).

of both A and B atoms, is described by $\bar{D}_{AB}(x)$, where

$$\bar{D}_{AB}(x) = D_A C_B(x) + D_B C_A(x) \tag{1}$$

In order to deduce values of D_A and D_B from the observable concentration-dependent $\bar{D}_{AB}(x)$, subtle analysis is required. It would be nice to measure $\bar{D}_{AB}$ at several known concentrations (C_A, C_B) and then solve for D_A and D_B. However, most diffusion measurements are of a concentration profile covering a continuous range of compositions such as that shown in Figure 9.1. Ways have been devised for extracting $\bar{D}_{AB}(C_A, C_B)$ from profile data, and we proceed to describe one such method.

The time rate of change of concentration C of one constituent of a diffusion couple is given by Fick's second law:

$$\frac{\partial C(x, t)}{\partial t} = \frac{\partial}{\partial x}\left(D\,\frac{\partial C(x, t)}{\partial x}\right) \tag{2}$$

In the case of concentration-dependent $\bar{D}(x)$, this equation cannot be further reduced. Its exact analytic solution is not generally possible. However, it can be simplified if distances at any instant t are measured as x' relative to the Matano interface (2) at that time t instead of as x relative to the original interface. Figure 9.1 illustrates this situation. The Matano interface ($x' = 0$) is translated from the physical interface $(x=0)$ so as to satisfy the condition

$$\int_{C_A(x'=0)}^{C_0} x' \, dC_A = \int_0^{C_A(x'=0)} (-x') \, dC_A \tag{3}$$

that is, the hatched areas of the curve in Figure 9.1 are equal.

It can then be shown that $C_A(x', t)$ is a function only of the variable $\lambda = (x'/\sqrt{t})$. Equation 2 may then be written

$$-\frac{\lambda}{2} \frac{\partial C}{\partial \lambda} = \frac{\partial}{\partial \lambda}\left(\bar{D} \frac{\partial C}{\partial \lambda}\right)$$

Integrating with respect to λ gives

$$\bar{D}_{(C=C_1)} = -\frac{1}{2} \frac{d\lambda}{dC} \int_{C=0}^{C=C_1} \lambda \, dC$$

where

$$\int_{C=0}^{C=C_0} \lambda \, dC = 0$$

Choosing a particular time $t = t_1$, this becomes

$$\bar{D}_{(C=C_1)} = \frac{1}{2t_1}\left(\frac{dx'}{dC}\right)_{C=C_1} \int_{C=C_1}^{C=C_0} x' \, dC \tag{4}$$

where

$$\int_{C=0}^{C=C_0} x' \, dC = 0$$

Equation 4 provides the means whereby an experimental measurement of C versus depth in a diffused sample can be used to obtain values of a concentration-dependent $\bar{D}$. The slope and integral may be evaluated directly from an experimental C–x curve. Values of $\bar{D}$ obtained at different concentrations can then be inserted in Eq. 1 to give unique values for the individual diffusivities D_A and D_B at the temperature of the experiment. By repetition of the process at a series of temperatures, Arrhenius plots for D_A and D_B may be obtained, leading to experimental values for activation energies Q_A and Q_B.

In principle, then, it is possible to determine completely by experiment the interdiffusion parameters for a pair of fully miscible metals in contact in a single-crystal form.

Much less tractable are the cases discussed in later sections of this chapter, where grain boundary diffusion, formation of new phases, and changes in film morphology may completely obscure the bulk diffusion effects treated above. In most of those cases, however, interpretations of experimental data are likely to demand a knowledge of D_A, D_B, Q_A, and Q_B which describe the underlying bulk diffusion processes.

9.2.1.2 Factors Affecting D_A, D_B, Q_A, and Q_B. The quantities D_A, D_B, Q_A, and Q_B are generally well-defined fundamental quantities that completely specify the diffusion kinetics of a given single-crystal metal pair A and B in contact at temperature T. For this reason, we present their values explicitly wherever possible in the data compilation at the end of this chapter.

In some cases, D_A or Q_A can depend on the orientation, surface energy, chemical energy, or stress of the thin film during a particular observation. In order to illustrate how such a dependence might occur, we review some basic physical mechanisms which constitute simple diffusion.

Vacancy diffusion is most likely to occur in systems where high mutual solubility of the two metals leads to a substitutional solution after the metals diffuse under heat treatment. In this case, the ensemble of large numbers of random vacancy movements within the host lattice will cause an overall drift of atoms in such a way as to diminish initial concentration gradients (in the interface region). (Interstitial diffusion occurs in some systems (3), but it is relatively rare. It is described by expressions similar to those of the following discussion of vacancy diffusion.)

The bulk diffusivity of A in B, D_A, may be derived in terms of the factors governing the probability of atom jumps occurring to produce a net vacancy diffusion in a direction x normal to the interface. For example, in the special case of *self*-diffusion (A and B identical),

$$D = \alpha R \Lambda^2 \Gamma n_v \qquad (5)$$

where α is a "jump factor" indicating the geometric population of possible jump paths in the x-direction for the host lattice, R is a jump correlation factor (≈ 0.8 for self-diffusion in bcc and fcc crystals), Λ is the separation of atomic planes in the x-direction, and Γ is a jump frequency given by

$$\Gamma = v \exp\left(-E^*/kT\right) \qquad (6)$$

where v is the vibration frequency of atoms in the host lattice and E^* is the activation energy for the motion of the vacancy.

The equilibrium fraction of vacant sites, n_v, is given by:

$$n_v = \exp\left(S_v^f/k\right) \exp\left(-E_v^f/kT\right) \qquad (7)$$

where S_v^f is the entropy increase on forming a new vacancy and E_v^f is the energy needed to create a vacancy in the single crystal.

The substitution of Eqs. 6 and 7 in Eq. 5 serves to show the intrinsic origins of the Arrhenius equation:

$$D = D_0 \exp\left(-Q/kT\right) \qquad (8)$$

where

$$D_0 = \alpha R \Lambda^2 \gamma \exp (S_v^f/k)$$

and Q, the activation energy for the process, is

$$Q = E_v^f + E^*$$

Evidently, the quantities α, R, and Λ can be strongly dependent on the geometry of the host lattice. Hence, the value of D_0 could be dependent upon orientation of the single-crystal matrix relative to the diffusion direction. In the common fcc and bcc structures, D_0 for self-diffusion is not, in fact, orientation dependent. However, in hcp metals, D_0 is about twice as large for movements along the c-axis as it is for diffusion perpendicular to that axis (4).

Through the energy terms in Eqs. 5, 6, 7, and 8, the values of Q and D_0 can be altered by intrinsic stress in a film or by surface energy effects at the interface, or by large changes in chemical potential due to alloy formation. Stress effects are expected to be very small for bulk diffusion, in contrast to their potential significance in grain boundary diffusion which can sometimes influence the interaction of polycrystalline films (5).

9.2.2 Single-Crystal Films—Partially Miscible

Consider now the pair of metals L and M whose phase diagram appears at the right of Figure 9.2. At the temperature T of the experiment, there exists a maximum solubility C_{ML} of L in M and a maximum solubility $(1 - C_{LM})$ of M in L.

Interdiffusion of partially miscible single-crystal films exhibits the same

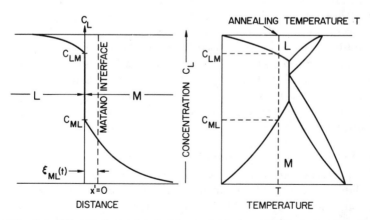

Figure 9.2. Interdiffusion of partially miscible metals (single-crystal films). At the temperature of anneal T, the limit of solubility of M in L is C_{ML} and the limit of solubility of L in M is $1 - C_{LM}$.

general behavior as that described for fully miscible metals. However, the interface concentrations are limited by C_{LM} and C_{ML}. Direct determination of the low-temperature region of a phase diagram like that of Figure 9.2, using diffusion profiles recorded at temperatures below the eutectic, would seem to be a sensitive, useful, and simple technique.

9.2.2.1 Movement of the Matano interface.

Kidson (6) has derived an expression for the displacement ξ_{ML} of the Matano interface in a system of this kind, after diffusion for a time t:

$$\xi_{ML} = 2\left(\frac{\overline{D}_{ML}K_{ML} - \overline{D}_{LM}K_{LM}}{C_{LM} - C_{ML}}\right) \cdot \sqrt{t} \tag{9}$$

where $K_{ML} = (dC/d\lambda)_{ML}$ evaluated at the M side of the physical interface and $\overline{D}_{ML}$ is the interdiffusion coefficient at the same point. Since C is a function only of λ, then $dC/d\lambda$ is a function of λ alone. But C has a constant value at the physical interface, and hence $dC/d\lambda \, (= K)$ must be a constant. It should be noted that although it is easy to observe experimentally that

$$\xi_{ML} = (\alpha_{ML})\sqrt{t} \tag{10}$$

where α_{ML} is constant at given temperature, the experimental determination of α_{ML} does not yield unique values for D_M and D_L. Furthermore, while α_{ML} is temperature dependent (via D_{LM} and D_{ML}), it does not follow a simple Arrhenius law unless one or other DK product is negligibly small. Thus, only if one of the DK terms dominates the expression for α_{ML} can a meaningful activation energy be obtained from this technique.

9.2.2.2 Evaluation of $\overline{D}(C)$.

The fact that here the C-versus-x' locus has a discontinuity at the physical interface does not invalidate (7) the expression for $\overline{D}(C)$ given in Eq. 4. Just as in the case of miscible films, a plot of concentration versus x', the distance from the Matano interface, obtained after a given diffusion time t, can be used to deduce values of $\overline{D}(C)$ versus C by numerical integration. The only difference here will be that $\overline{D}$-versus-C data cannot be obtained within the concentration range C_{ML} to C_{LM} (where zero contribution is made to the integral, since $dx'/dC = 0$).

In this system we expect to find that $C(x)$ is a function only of $x/\sqrt{t}$, since concentrations at the physical interface $(x = 0)$ are constant $(C_{ML}$ and $C_{LM})$.

9.2.3 Polycrystalline Films (Ag–Au, Pd–Au)

9.2.3.1 Experiment.

In contrast to the case of single-crystal films, there is available a wealth of experimental data on extensive diffusion in polycrystalline films. We shall first consider interdiffusion in polycrystalline couples that form complete series of solid solutions. One of the best-known examples of couples exhibiting complete solid solubility is Ag–Au, and in

Figure 9.3 we show an example of the dominant role that defects (e.g., grain boundaries) can play in low-temperature interdiffusion between such couples. One of the aims of this experiment of Kirsch et al. (8) was to elucidate the role of grain boundary diffusion, and this was achieved by comparing diffusion in single-crystal and polycrystalline couples.

Figure 9.3 shows backscattering spectra from self-supporting couples of Ag (1000 Å) and Au (800 Å). For couples such as Ag–Au where there is mass overlap in the backscattering spectra, it is possible to tailor the experiment so as to reveal separate diffusion profiles for the constituents (this procedure is described in Chapter 6). The ion beam is incident on the lighter (Ag) film; the profile of the higher-mass diffusing species then moves to higher energies, and vice versa for the lighter species. In the present case thicknesses were chosen to give complete overlap of Ag and Au peaks. The left side of the figure

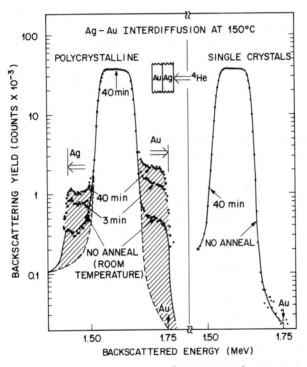

Figure 9.3.   Backscattering spectra from Ag (1000 Å) and Au (800 Å) couples for 1.9 MeV ^{4}He. Film thicknesses were chosen so that, for beam incident on Ag, complete overlap of the Ag and Au peaks occurs. Interdiffusion then results in the diffused Au moving to the higher-energy side and Ag to the lower-energy side. Position of surface Au is indicated by arrows. As-deposited spectra without annealing are shown by smooth curve. From reference 8.

shows the effect of annealing for the polycrystalline couple, and the right side shows the effect of annealing for the single-crystal couple which had the same film thicknesses as the polycrystalline couple. The preanneal spectra for the couples are given by the smooth lines.

It is evident that diffusion in the polycrystalline couples is radically different from that in single-crystal couples. For the polycrystalline couple, Au diffuses out through the Ag film to the surface (the arrow marked Au on the abscissa gives the energy for surface scattering from Au), and Ag similarly diffuses through the Au film. The hatched areas give the diffusion profiles, and it can be seen that they are practically flat throughout the thickness of the films. For example, after the anneal at 150°C for 40 min, interdiffusion has occurred at approximately the 10 at. % level.

No interdiffusion, however, is evident from the single-crystal couple after the anneal at 150°C for 40 min (the points for the spectrum after annealing lie exactly on the smooth line drawn through the preanneal spectrum). The fact that no interdiffusion is observed is consistent with extrapolated Ag–Au bulk or lattice diffusion coefficients (9), which predict interdiffusion to a depth of only 10^{-3} Å. The experimental depth resolution in the present measurement is ≈ 150 Å. Interdiffusion between the single-crystal couples is observed for anneals at 400 and 500°C. The profiles for Au diffusing into Ag and vice versa have the classic erfc shapes predicted for lattice diffusion and from which lattice diffusivities can be calculated, in good agreement with published bulk data (9).

It is obvious that diffusion in the polycrystalline couple is dominated by the short-circuiting effect of defects, and the following room temperature diffusion measurements show that grain boundaries play a major role. The polycrystalline couple exhibits interdiffusion as first measured at room temperature, one week after evaporation (this spectrum is marked "no anneal" in Fig. 9.3). The question whether this is a room-temperature diffusion process or one occurring during the film deposition, for example, is answered by the diffusion profiles of Figure 9.4 for Au diffusing into Ag. A sample was examined promptly after evaporation, and the diffusion was followed over the course of several months. It can be assumed that Au is diffusing along Ag grain boundaries at room temperature for the following reason. If indeed the saturation value of 2 at. % represents Au saturating Ag grain boundaries, then simple geometric arguments (assuming grain boundaries of width 5 Å) require columnar grains 500 Å in diameter. Transmission electron microscopy measurements did indeed indicate the grains to have mean diameters of about this value. This experiment is an example of how several measurements and techniques have to be used to draw conclusions regarding the diffusion mechanisms. It is just not experimentally possible at present to resolve material in the grain boundaries. For example, in the

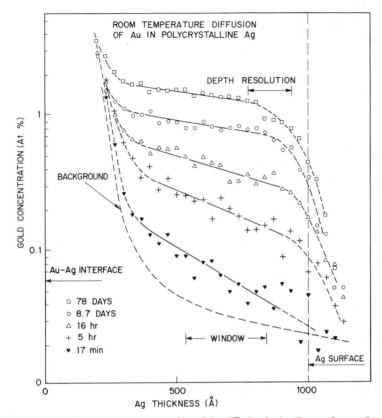

Figure 9.4.   Room temperature profiles of Au diffusing in Ag. From reference 8.

Ag–Au experiment, the ^{4}He beam had a diameter of 0.5 mm, and the back-scattering profiles thus represent lateral averaging over some 10^4 grains. The best beam diameters so far achieved ($\sim 1\ \mu m$) still fall far short of 5 Å.

9.2.3.2 Interpretation—Defect Enhanced Diffusion (Whipple Model). The previous Ag–Au results give a graphic demonstration of the high levels of interdiffusion that can occur at low temperatures in polycrystalline films as opposed to single-crystal films. It is obvious that low-temperature diffusivity between polycrystalline films cannot be described in terms of a single diffusion mechanism. How does one proceed from experimental measurements of polycrystalline diffusion profiles to an understanding of the diffusion mechanism? The starting point of an analysis usually follows the framework of the Whipple model (10) discussed at length in Chapter 7. Two diffusion coefficients are assumed at low temperatures: a rapid grain boundary diffusivity D_b and the usual, slower lattice diffusivity D_l. The shape and magnitude of diffusion profiles should then depend upon the relative magnitudes of D_b, D_l, and the density of grain boundaries. A polycrystalline couple which has received considerable attention is Pd–Au, and we shall use it as an illustration of the various courses taken in unscrambling the diffusion processes. Pd–Au forms a complete series of solid solutions, and interdiffusion in these polycrystalline couples is very similar to that in Ag–Au.

One of the earlier measurements of diffusion in the Pd–Au thin film couple is that of Lau and Sun (11). The degree of interdiffusion was determined by the change in intensity of the Pd(111) x-ray diffraction pattern. An activation energy of 0.85 eV was obtained and assumed to be indicative of a grain boundary process. Such x-ray techniques do not give a direct measure of the diffusion profiles. Direct measurements of the diffusion profiles have been obtained by Poate et al. (12) using Rutherford backscattering and Hall et al. (13) using Auger spectroscopy and Rutherford backscattering. An example from Hall et al. of the diffusion profile of Pd diffusing in Au is given in Chapter 6. One obvious feature of these diffusion profiles is the presence of essentially zero concentration gradients after each annealing cycle as in the case of Ag–Au in Figure 9.3. This leads to the plausible assumption that grain boundary diffusion is extremely fast compared with the lattice diffusion. In its simplest form, then, the problem should reduce to one of lattice diffusion into the interior of the grains with the grain boundaries constantly full of solute.

Another striking feature of the diffusion profiles is the large concentration levels (5 to 20 at. %) of the "plateaus" of Pd in Au. On geometric arguments only, a few at. % Pd could be lodged in the Au grain boundaries. It appears therefore that most of the impurity must have diffused into the grain interior. The lattice diffusivity of the Pd in Au at 175°C, for example, is 5×10^{-22} cm^2 s^{-1}. Thus, the vacancy diffusion lengths $\sqrt{Dt}$ are only 2 Å for 200 h at 175°C. This is clearly not enough to account for the average observed concentration of 20%. The model must therefore assume that transport into the interior of the grains is aided by a diffusion enhancement mechanism such as a high (nonequilibrium) concentration of vacancies or dislocations or the motion of grain boundaries, to be discussed later.

We shall therefore assume the general framework of the Whipple model but with the lattice diffusivity D_l replaced by a defect-enhanced lattice diffusivity D'_l. The Whipple model simplifies in the limit of $D_b \gg D_l$ to give the concentration as $C = C_1 + C_2$, where C_1 is due to diffusion at the interface:

$$C_1 = C_0 \operatorname{erfc}\left(\frac{y}{2(D'_l t)^{1/2}}\right) \tag{11}$$

and C_2 is due to sideways diffusion from the grain boundaries:

$$C_2 = C_0 \operatorname{erfc}\left(\frac{x-d}{2(D_l' t)^{1/2}}\right) \operatorname{erf}\left(\frac{y}{2(D_l' t)^{1/2}}\right) \tag{12}$$

where C_0 is the composition at $y=0$ (assumed to be independent of time) and d is the film thickness.

If we make the simplifying assumption that the grain sizes are uniform, $L \times L$, in the plane of the film, the average concentration in a plane is given by

$$\overline{C} = \frac{2}{L} \int_0^{L/2} C_1 \, dx + \frac{4}{L} \int_0^{L/2} C_2 \, dx \tag{13}$$

Provided that $L \gg 4(D_l' t)^{1/2}$, the integration gives

$$\frac{\overline{C}}{C_0} = \frac{8(D_l' t)^{1/2}}{L \pi^{1/2}} + \operatorname{erfc}\left(\frac{y}{2(D_l' t)^{1/2}}\right)\left\{1 - \frac{8(D_l' t)^{1/2}}{L \pi^{1/2}}\right\} \tag{14}$$

(The average concentration has to be evaluated because that is the parameter measured by the profiling techniques.) The first term describes the plateau whose rate of rise follows a $t^{1/2}$ dependence. This is shown in Figure 9.5, where the time dependence found by Lau and Sun has also been included. Initially, there is a $t^{1/2}$ dependence which of course must level off when the films become interdiffused to any great extent, that is, diffusion will stop when there are no concentration gradients.

Diffusion coefficients can be extracted from the plateau concentration

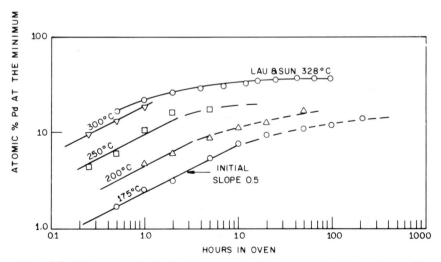

Figure 9.5.   Pd concentration in the plateau as a function of aging time. From reference 13.

(assuming $L = 1000$ Å), and they are shown in Figure 9.6 plotted against $1/T$. An activation energy of 0.69 eV is obtained for the process. A more exact solution to this problem is given by Poate et al. (12) taking into account the cumulative effect of diffusion into each grain from all its surfaces. The diffusivities obtained by the two methods are in good agreement.

The second term in Eq. 14 is appreciable only near the interface, and

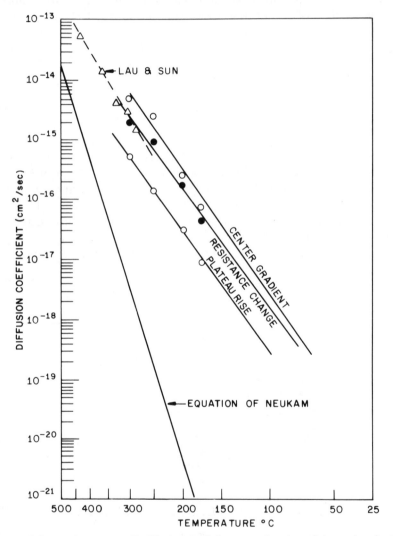

Figure 9.6. "Defect enhanced" diffusion coefficients as a function of time using the three independent methods discussed in the text. From reference 13.

from it D'_l can also be evaluated. Hall et al. (13) call this the "center gradient" method. These values are plotted on Figure 9.6 and give an activation energy of 0.83 eV. Hall and Morabito (14) have also devised a nomograph for extracting diffusion coefficients from erfc concentration gradients at interfaces.

One of the reasons for the interest in Pd–Au couples is to determine the change of resistivity caused by interdiffusion of Pd into Au thin film conductors. Hall et al. (13) measured such changes. If it is assumed that the change in resistivity of the Au film is directly proportional to the Pd concentration in the Au, and assuming diffusion models similar to the previous ones, then diffusivities can be deduced from the resistivity changes. These values are also plotted on Figure 9.6 and indicate an activation energy of 0.75 eV, which agrees rather well with those obtained from the "plateau rise" and "center gradient" methods. The diffusivities of Lau and Sun are also plotted on Figure 9.6.

In the previous analyses, it was assumed that grain boundary diffusion is infinitely fast. However, if we have an estimate of D'_l, it should be possible to evaluate $D_b\delta$ from the Whipple model. According to LeClaire (15), this is best evaluated from the equation

$$D_b\delta = 0.66 \left(\frac{\partial \ln \overline{C}}{\partial y^{6/5}}\right)^{-5/3} \left(\frac{4D'_t}{t}\right)^{1/2} \tag{15}$$

where δ is the grain boundary width and $(\partial \ln \overline{C})/(\partial y^{6/5})$ is the slope of the "plateau." Figure 9.7 shows a "Whipple" plot of $\ln \overline{C}$ versus $y^{6/5}$ for the interdiffusion profiles of Pd in Au at 250°C. Within 400 Å of the interface, erfc diffusion predominates—the apparent linearity of this part is due to the fact that $\ln [\text{erfc } C(y)]$ is almost linear with $y^{6/5}$ down to a concentration of 6%. The presence of grain boundary diffusion is shown by the breaks in the curves and the linear regions beyond 400 Å. The good linearity of the regions beyond 400 Å should not be taken as a validation of the $\frac{6}{5}$ power law since equally good agreement would be obtained by plotting $\ln C$ against $y^{1.0}$. Rather gross assumptions have also gone into the Whipple model, so it is probably naive to attribute too much physical significance to D_b values obtained from these $y^{6/5}$ plots. Using the "plateau rise" values of D'_l, a grain boundary activation energy of 0.87 eV is obtained, with the diffusivities being some four orders of magnitude greater than D'_l.

The Whipple model is based on the assumption of diffusion into semiinfinite media and was never intended to describe a finite thin film system. Moreover, only diffusion from a single grain boundary perpendicular to the interface is considered. Gilmer and Farrell (16, 17) and Holloway et al. (18) have recently calculated interdiffusion in polycrystalline films allowing for finite thickness effects. We shall consider the calculations of Gilmer and

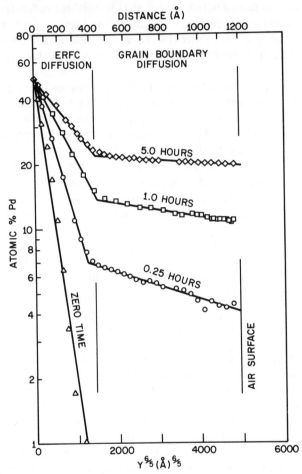

Figure 9.7.   Whipple plot of Pd profiles in Au at 250°C. From reference 13.

Farrell. Three limiting cases are considered in which the outer surface is (i) a barrier to diffusion, (ii) a plane with an infinite surface diffusion coefficient, and (iii) a zero concentration surface. Figure 9.8 shows their calculation of the average concentration $\overline{C}(y, t)$ plotted versus y/Y_0, where Y_0 is the film thickness and is equal to the spacing of the grain boundaries, $D_b/D_l = 10^4$ and $\delta/2Y_0 = 2.5 \times 10^{-3}$. Several values of the reduced time $T = D_l t/Y_0^2$ are used. The solid curves were calculated for the case of the outer surface being a perfect barrier to diffusion, and the dashed curves are the usual Whipple curves for a semi-infinite medium. Those results nicely demonstrate that near

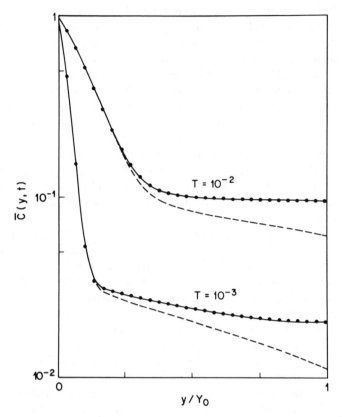

Figure 9.8. Calculations of Gilmer and Farrell (17) of the average concentration $\overline{C}(y, t)$ of diffused material as a function of the distance into the film y/Y_0 for several different values of the reduced time $T = D_l t/Y_0^2$.

the interface, both the finite and semi-infinite models give approximately equal average concentration. This is to be expected because, as mentioned previously, the bulk error function dominates. At larger y values, where grain boundary diffusion dominates, the two models deviate considerably because in the finite thickness model, the diffusant is being accumulated near the outer surface thereby giving rise to lower concentration gradients. Obviously, thin film diffusion data should be analyzed with such finite thickness models. We are faced, however, with the limitation in the Pd–Au case that the high levels of interdiffused material still cannot be explained by such a model. Moreover, the behavior of the surface, as a barrier or sink, has not been elucidated.

9.2.3.3 Mechanisms.

We considered the Pd–Au system at length because by careful measurements of the diffusion profiles and grain sizes it appears possible to arrive at a consistent picture of the diffusion process. The question remains as to the physical reality of the derived diffusivities. The values of the defect-enhanced diffusivities D'_l obtained by essentially three different analyses are in rather good agreement. The diffusivities obtained by Lau and Sun are also in quite good agreement with these D'_l values. More recently, Murakami et al. (19) have also used x-ray diffraction techniques, based on the methods of Houska (20), to determine the diffusivity of Pd–Au at $350°$C. Their values agree with those of Lau and Sun. It must be fortuitous to some extent that the x-ray diffusion coefficients and D'_l values agree, as the x-ray analyses assume only one diffusion mechanism. The agreement probably indicates, as we have already noted, that Pd transport is dominated by the diffusion into the grains.

What physical significance can be attached to D'_l values that are many orders of magnitude greater than extrapolations of the bulk vacancy diffusion coefficients (Fig. 9.6)? Both the resistivity and profiling measurements, combined with a knowledge of grain sizes, show that the diffused Pd is within the Au grains. Moreover, the sheet resistivity measurement indicates that the as-deposited films are highly defective. It can be conjectured therefore that diffusion into the grains is assisted by high concentrations of vacancies or defects. Simple estimates (12, 13) indicate, however, that unrealistically high densities of dislocations or vacancies are required to explain the interdiffused concentration. Balluffi and Blakely (21) have recently reviewed thin film interdiffusion and characterize the various situations that result from different classes of short circuits. It appears, however, from the thin film measurements that the only short circuits identified with confidence are high-angle grain boundaries. Unscrambling the defects that give rise to the D'_l values is obviously a very difficult problem.

The Ag–Au results of Kirsch et al. also throw some light on the defect-enhanced diffusion. The diffusion profiles of Ag diffusion in Au (and Au in Ag) are very similar to Pd–Au, being rather flat. The rate of rise of those plateaus (summing over the window shown in Fig. 9.4) is shown in Figure 9.9. For both the room temperature and annealing studies, there are initial very rapid rises followed by much slower interdiffusion. It appears as though the much slower interdiffusion rates can be explained by diffusion into the grain interior using the usual lattice diffusivity D_l. Similar behavior was previously observed by Cook and Hilliard (9) in their x-ray measurements of the change of satellite intensities following the anneal of multilayered Ag–Au films. They used the technique to determine bulk diffusivities D_l following the cessation of the fast initial interdiffusion.

One explanation of the fast initial interdiffusion is that grain growth occurs

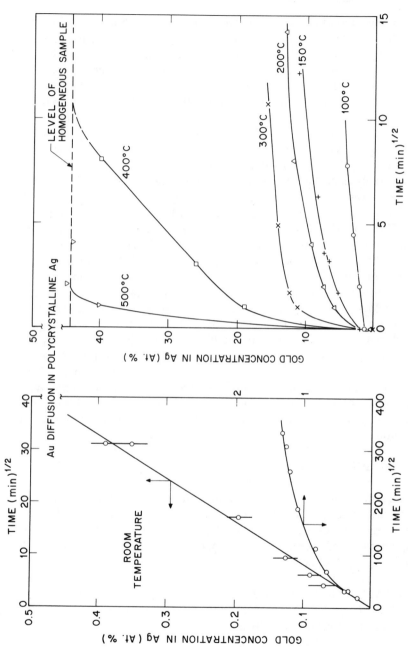

Figure 9.9. Time dependence of Au diffusing in polycrystalline Ag. Left side shows room temperature diffusion; right side, the annealing sequences. Au concentrations were obtained by averaging over windows as shown in Figure 9.4. From reference 8.

323

during annealing, thus dumping solute from the grain boundaries into the interior of the grains. The fact that in the Pd–Au example the center gradient solution agrees rather well with the resistivity and plateau rise solutions would tend to argue against the grain growth model. Nevertheless, there are two distinct time regimes in the Ag–Au interdiffusion which may be interpreted as (i) an initial very fast interdiffusion that must be explained in terms of hierarchies of defects and grain boundaries which saturate and thus bring solute deep into the interior of the grains, and (ii) after this process has occurred, a diffusion that takes place from the saturated defects into the interior of the grains by normal lattice diffusion.

In the analysis of the Pd–Au profiles, the Whipple model, with its rather gross approximations, was employed to determine grain boundary diffusivities. There are no other measurements of Pd grain boundary diffusion in Au. Probably the closest comparison we can get is the grain boundary self-diffusion in Au films as measured by Gupta (22). The diffusivities agree rather well at 150°C (Gupta gives $\delta D_b = 5 \times 10^{-21}$ cm^3 s^{-1} with $Q_b = 1$ eV; Hall et al. give $\delta D_b = 2 \times 10^{-21}$ cm^3 s^{-1} with $Q_b = 0.9$ eV). Determination of D_b can be important for those thin film properties that are *surface dependent*. Hall and Morabito (23) and Nelson and Holloway (24) have constructed methods for determining D_b values which depend on observation of the diffused material at the surface of the film. This method does not involve depth profiling but observation of the change of surface concentration of the diffusant. The assumption is made that the surface is a perfect sink, and the formalism of Crank (25) for the diffusion across a membrane is employed. The grain boundary diffusivities for Pd in Au as determined by this method are in reasonable agreement with those obtained by the Whipple model at a temperature of 150°C.

9.2.4   Interpenetration and Grain Growth

It is possible for the nucleation and development of new grains, or the growth of existing grains, to aid the penetration of atoms of one film into another. In fact, in the case of Cr and Cu having less than 0.1% mutual solubility and no compound phases, polycrystalline films have been observed (26) to intermix extensively after heat treatment at 550 to 750°C, apparently as a result only of the growth of new grains.

The diffusion profiles obtained by Baglin et al. for polycrystalline Cu (1050 Å) + Cr (1050 Å) films after a series of heat treatments are shown in Figure 9.10. These Rutherford backscattering results alone would give the impression of a normal interdiffusion process. However, scanning electron micrographs of a "diffused" film showed that 2000 Å segregated grains of Cu had developed in the upper layer, as shown in Figure 9.11. Single-crystal films identically heated did not display any significant interpenetration at all

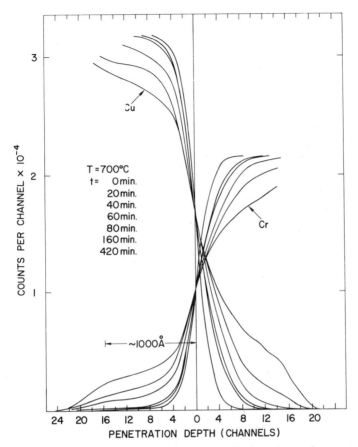

Figure 9.10. Intermixing of polycrystalline films of Cu (1050 Å) + Cr (1050 Å). Concentration profiles were obtained (36) from Rutherford backscattering profiles displayed here.

(see Fig. 9.12). It seems likely that the polycrystalline Cr–Cu films interpenetrate initially by grain boundary diffusion, which continues to supply the growth of new grains that subsequently nucleate at defects within the host film. However, the actual mechanism of migration has not yet been determined.

The Cu–Pb system (also not miscible) has been studied by Campisano et al. (27) using Rutherford backscattering analysis and TEM. They conclude that the *relative grain sizes* in the Cu and Pb films will determine whether Pb or Cu is the dominant moving species responsible for interpenetration of the f ms. This is the result of competition between processes of grain growth .n the host film (which reduces the available grain boundary surface

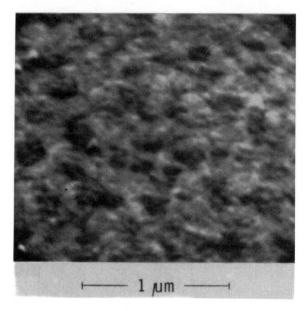

Figure 9.11. Scanning electron micrograph showing surface of Cu + Cr film after heat treatment for 2 h at 700° C. Grains of Cu (dark) are seen to be developing in the upper layer of the film.

for migration) and segregation of impurity within grain boundaries (which tends to inhibit further host grain growth).

9.2.4.1 Energetics.   Grain growth is one means whereby a film or couple seeks to reduce its total surface/interface potential energy, given the atomic mobilities (by heating) to achieve this. Other forms of energy may be stored in a film in the form of intrinsic stress, interface stress, or chemical bonding in the case of compounds. When any heat treatments are given to an experimental film, there will *always* be some tendency for the film to seek to minimize its stored energy—by a reorganization of grain sizes or interfaces, which has nothing to do with the diffusion mechanisms discussed in the first part of this chapter. Those were based on concentration gradients alone as the driving forces.

Clearly, it is important to check for the possible dominance of such processes as grain growth before attempting to interpret observed "interpenetration" data in terms of simple models restricted to planar interfaces, normal grain boundaries, and diffusion driven by concentration gradients alone. This check may be made by examining films with SEM or TEM. Departure from the simple $\sqrt{t}$ dependence of Eq. 10 is a clear warning of potential complications due to changing film morphology.

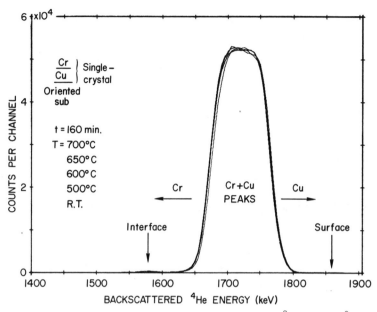

Figure 9.12. Backscattering spectrum from single-crystal Cu (1050Å)+Cr (1050Å) film (26). Heat treatment produced no substantial change in Cr and Cu distributions, in contrast to the effect on the polycrystalline sample of Figure 9.9.

9.2.4.2 Time Dependence. Farrell, Gilmer, and Suenaga (28) have analyzed film couples in which only grain boundary diffusion supplies atoms to the point of growth of a new phase. They establish the fact that, provided the array of host grain boundaries carrying the diffusant does not alter with time, the amount M of material diffused after a time t should be proportional to $\sqrt{t}$. However, any concurrent grain growth in the host film will produce a time-dependent reduction of diffusing flux, leading to a t^n dependence of M, where n is less than 0.5.

9.3 COMPOUND PHASES

9.3.1 Single-Crystal Films

We now consider the interaction of overlaid films of metals α and γ when heated to a temperature at which a compound phase β is stable. As in our discussion of solid solutions (Section 9.2), we first consider the limiting case of single-crystal films even though the physical growth of a continuous single-crystal layer of a compound phase is highly improbable in reality on account of lattice mismatch.

9.3.1.1 One Compound Layer. As in the case of miscible metals, the boundaries between zones of the phase diagram for the two metals will produce discontinuities in the concentration-versus-depth curves for species α or γ. This is illustrated in Figure 9.13, where the interdiffusion profile at the temperature T displays layers of stability containing the α, β, and γ phases, respectively. In this example, the simple compound phase β exists between concentration limits $C_{\beta\alpha}$ and $C_{\beta\gamma}$. The compound layer has formed by the mutual diffusion of α and γ atoms to interface sites where they lodge to extend the growth of the new intermediate layer.

It should be noted that portions of an unknown phase diagram could in principle be deduced from a series of diffusion profiles such as this, observed at a series of temperatures.

The width W_β of the β phase layer after a time t has been derived by Kidson (6):

$$W_\beta = \xi_{\beta\gamma} - \xi_{\alpha\beta}$$
$$= 2\left[\left(\frac{(DK)_{\gamma\beta}-(DK)_{\beta\gamma}}{C_{\beta\gamma}-C_{\gamma\beta}}\right)-\left(\frac{(DK)_{\beta\alpha}-(DK)_{\alpha\beta}}{C_{\alpha\beta}-C_{\beta\alpha}}\right)\right]\sqrt{t}$$
$$= B_\beta\sqrt{t} \tag{16}$$

(with the notation of Section 9.2.2). The simple $\sqrt{t}$ dependence should be observable as a test to show that growth of a particular new planar phase layer is being governed only by diffusion processes, as assumed here. B_β is a complicated function of four different diffusion coefficients and is temperature dependent. However, it does not follow an Arrhenius law in general;

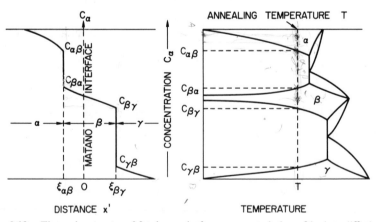

Figure 9.13. Three-phase system. Metals α and γ form compound phase β by interdiffusion at temperature T.

and hence an "activation energy" deduced for the compound formation from measurement of $W_\beta(t, T)$ can be meaningless unless by chance the component Q values are all identical or one DK term dominates.

9.3.1.2 Several Compound Layers.
Often, several intermediate compound phases can be produced at a given temperature. Kidson (6) has presented the general expression for the width of a layer of a phase j in a diffusing system made up of N phases separated by $N-1$ interfaces:

$$W_j = \xi_{j,j+1} - \xi_{j-1,j}$$
$$= 2\left\{\left[\frac{(DK)_{j+1,j} - (DK)_{j,j+1}}{C_{j,j+1} - C_{j+1,j}}\right] - \left[\frac{(DK)_{j,j-1} - (DK)_{j-1,j}}{C_{j-1,j} - C_{j,j-1}}\right]\right\}\sqrt{t}$$
$$= B_j\sqrt{t} \tag{17}$$

The *relative widths* of the layers of adjoining phases may depend in a complex way upon the miscibility gaps and the diffusivities of the migrating atoms in each phase. However, in general, the larger the diffusivity $\overline{D^j}$ within phase j, the faster layer j will grow (6). *All* phases allowed by the phase diagram at the temperature of the experiment should be present, in however small an amount. Often, however, some phases do not develop an observable thickness because of their low diffusivity.

Shatynski et al. (29) have derived an expression for the relative number of moles produced of *two* compound phases A_kB and A_mB growing at the junction of saturated terminal solid solutions A_jB and A_nB. Our notation is defined in Figure 9.14. Here, j, k, m, and n represent *average* atomic composi-

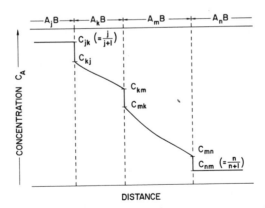

Figure 9.14. Formation of two compound phase layers at the junction of saturated terminal solid solutions A_jB and A_nB. A_kB and A_mB are average compositions for those layers. After Shatynski et al. (29).

tion for each layer. It is assumed that parabolic rate constants apply to the growth of each of the phases at an A–B interface. They derive, in particular, the framework for calculating the partitioning factor which is the *ratio of moles of A_kB formed to the moles of A_mB formed* after steady-state diffusion has been established. We quote the relevant expressions:

$$\frac{\text{no. of moles } A_kB}{\text{no. of moles } A_mB} = (e + \sqrt{e^2 + 4fg})/2g$$

where

$$e = \left(\frac{\Omega_k}{\Omega_m}\right)^2 \left[\frac{1}{k-m} h_A^m + \frac{m}{k-m} h_B^m\right] - \left[\frac{1}{k-m} h_A^k + \frac{k}{k-m} h_B^k\right]$$

$$f = \left(\frac{\Omega_k}{\Omega_m}\right)^2 \left[\left(\frac{1}{k-m} + \frac{1}{m-n}\right) h_A^m + \left(\frac{k}{k-m} + \frac{n}{m-n}\right) h_B^m\right]$$

$$g = \left(\frac{1}{j-k} - \frac{1}{k-m}\right) h_A^k + \left(\frac{j}{j-k} + \frac{m}{k-m}\right) h_B^k$$

and Ω_j is the molar volume of A, etc., and

$$h_A^k = -\Omega_k \int_{C_{kj}}^{C_{km}} D_A^k \, dC_A^k \tag{18}$$

In the special case where D_A^k is not appreciably concentration dependent, this could be simplified to

$$h_A^k \approx -\Omega_k D_A^k (C_{kj} - C_{km}) \tag{19}$$

Already, with two compound phases present in this example, no simple conclusion can be made about the relative growth of the two layers. However, it is evident that, given the phase boundary concentrations and individual diffusivities D_A and D_B, the relative growth of compound layers could be predicted. It may also be deduced from these expressions that the growth tends to favor a phase in which diffusivity is large and whose range of stable compositions is wide.

9.3.2　Polycrystalline Films

9.3.2.1 Grain Boundary Assisted Diffusion. If we consider the possibility that diffusing atoms could move along grain boundaries and then diffuse into the bulk of a host grain before joining a compound phase, the formation of a new compound by diffusion might be described with the models discussed previously. In this case the application would be similar to that for two metals forming a solid solution. The quantity of new phase formed in time t would then be proportional to $t^{1/n}$, where $2 < n < 4$ as a rule (30).

9.3.2.2 Chemical Activities. Up to this point, our discussion has been exclusively in terms of concentration gradient as a driving force for diffusion, with the diffusion rates always determining the rate of interaction. However, compound formation requires two steps: (i) migration of the relevant atoms to suitable interface locations, and (ii) capture and bonding of the new atoms into the lattice of the new compound phase.

In polycrystalline films especially, interface or grain boundary diffusion can sometimes deliver atoms to the interface of the growing phase faster than they can become bonded by it. In a multiphase system, this leads to a concentration profile consisting of a series of plateaus since, within each layer of a given phase, the concentration gradient is negligible. Diffusion is then governed mainly by chemical activities and the process is *reaction limited*. In that extreme case, the amount of new phase formed depends linearly on the length of time t for which the interfaces are exposed to reactive atoms. Hence, in this case the thickness of the layer of new phase will be directly *proportional to* t. This t dependence should enable an experimenter to distinguish between reaction-limited processes and grain boundary assisted processes, even though both can produce flat concentration profiles.

9.3.2.3 Grain Size; Film Morphology. As we have noted above, it is possible for a process of compound formation to be *diffusion limited* ($t^{1/2}$ dependence in the case of single-crystal films), or it may be *reaction limited* (t dependence) when grain boundary diffusion more than adequately provides the supply of reactive atoms to the new-phase interfaces. The latter effect clearly depends on the size and configuration of grains in the two films. Hence, the kinetics of compound formation may be sensitive to the morphology of the initial polycrystalline films and to *grain growth* effects. Grain growth can strongly influence the interface exposure for reactions to occur and limit the available paths for rapid diffusion. Experimental measurements ideally should always include a test of time dependence of the interaction, to aid in recognition of effects of temperature- and time-dependent changes of grain size in both source and growing layers.

9.3.3 Grain Growth

Farrell, Gilmer, and Suenaga (31) have treated the case of compound formation in a polycrystalline film couple. They point out that the formation of a compound phase at the growth interface significantly alters the conditions of diffusion from those of Whipple's model (where the large grain boundary diffusivity allows a large flux in the boundary from which there is a small leakage to form a solid solution in the grains on either side). For compound formation, a large flux must be delivered to the growth interface; however, in most cases a negligible loss to the bulk occurs. In that situation a steady state is established in which concentration within the grain boundary is almost constant, and the growth of the compound layer follows a $\sqrt{t}$ dependence, as it would for pure bulk diffusion.

9.3.3.1 Compound Formation During Grain Growth.

If the grain boundary volume of the compound layer is diminished with time due to the growth of compound-phase grains, the rate of development of the compound layer thickness $Y(t)$ will be slowed. An atomic mechanism for such a grain growth process has been detailed by den Broeder (32).

The model and notation of Farrell et al. (31) are shown in Figure 9.15. The growing compound layer has grains of width L and bulk diffusivity D_l separated by parallel equidistant grain boundaries of width $2a$ and diffusivity D_b, the boundaries all being perpendicular to the interface plane. It is assumed that the activated grain growth can be described by an equation of the form

$$\frac{\overline{L}^2}{\overline{L}} = \lambda(T)\, t^m \tag{20}$$

where $\overline{L}^2$ and $\overline{L}$ are mean values derived from a large number of grains in the sample and

$$\lambda(T) = \lambda_0 \exp(-Q_L/RT) \tag{21}$$

in which Q_L is an "effective" activation energy for grain growth.

The accompanying grain boundary diffusion is described by

$$D_b = D_{b0} \exp(-Q_b/RT)$$

The thickness Y_0 of the compound layer after time t is then given by

$$Y_0 \sim \left(\frac{D_b(T)}{\lambda(T)}\right)^{0.5} t^{(1-m)/2} \tag{22}$$

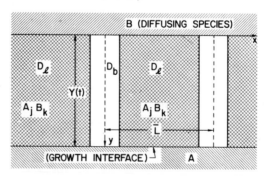

Figure 9.15.   Growth of polycrystalline compound layer A_jB_k. The notation is indicated for the formalism of Farrell, Gilmer, and Suenaga (31).

The apparent effective activation energy for the growth of the layer thickness Y_0 is therefore

$$Q_{eff} = (Q_b - Q_L)/2 \tag{23}$$

Grain growth in the compound film thus has the effect of reducing the time dependence of Y and decreasing the effective observed 'activation energy' significantly. Such cases have been reported in practice by Farrell et al. (31) in measurements on the formation of Nb_3Sn in polycrystalline films.

9.3.3.2 Change of Kinetics with Layer Thickness Y_0.

To obtain a linear t dependence of Y_0 in polycrystalline compound layers of *constant* grain size L, it must be assumed that grain boundary diffusivity D_b is always more than adequate to supply diffusant atoms to the reaction interface. In practice, when the compound thickness Y_0 is very small, the grain boundary diffusion path from the source layer becomes very short and this condition can be satisfied. Only for longer grain boundary paths will compound formation deplete the grain boundary concentration to some equilibrium value as assumed above, and the growth of Y_0 will become proportional to $\sqrt{t}$. The crossover between these two conditions can be deduced (31) from the following equation:

$$Y_0^2 + \frac{2D_{eff}}{k_1} Y_0 = \frac{2D_{eff}C_0}{C_1} t \tag{24}$$

where C_1 is the minimum local diffusant concentration in the grain boundary which is required for the initial growth of the compound phase, C_0 is the diffusant concentration in the grain boundary at the source layer ($y=0$), k_1 is (flux of diffusant across growth interface)/(grain boundary concentration of diffusant C_1), D_{eff} is the "effective diffusion coefficient of the grain boundary network" $= (a\bar{L}/\gamma \bar{L}^2)D_b$ where a is the width of a grain boundary, and γ is a constant factor depending on grain geometry. Clearly, interface kinetics will become rate limiting when the second term of Eq. 24 is much larger than the first, that is, $k_1 \ll D_{eff}/Y_0$. In that case Y_0 is linear with time:

$$Y_0 = k_1(C_0/C_1)t \tag{25}$$

At the other extreme, the first term will dominate if $k_1 \gg D_{eff}/Y_0$. Then,

$$Y_0 = \sqrt{\frac{2D_{eff}C_0}{C_1}} \sqrt{t} \tag{26}$$

Hence, for fast interface kinetics or large Y_0, we obtain a $\sqrt{t}$ dependence of Y_0.

9.3.4 Experimental Examples

Detailed kinetic information about compound formation can now be derived, in principle, by proper application of the foregoing analytic techniques to experimental data which have been obtained using well-characterized samples in situations tailored to test specific theoretical interpretations. However, little work has so far been published which goes much beyond evaluation of $\bar{D}$ and $\bar{Q}$, coupled with checks on $\sqrt{t}$ dependence. The following experimental examples serve to illustrate the simplicities and complexities of some representative compound-forming systems.

9.3.4.1 Al–Au. Due to its importance in semiconductor device conducting metallurgy, the Al–Au system has been the subject of considerable attention. At temperatures below $250°C$, the phase diagram (33) would imply the stability of the compound phases Au_4Al, Au_5Al_2, Au_2Al, $AuAl$, and $AuAl_2$. All except $AuAl$ have been found in measurements on interdiffusing Al–Au couples. However, not all phases form at one time, and in fact the entire reaction path and the final phase structure depend systematically on the reaction temperature and the proportions of Al and Au in the initial binary film.

Campisano et al. (34) observed interactions in polycrystalline films using backscattering profiles to see phase boundary movements and using x-ray diffraction to identify the phase present. The schematic diagram of Figure 9.16 shows the sequence of alloy formation which they deduced. The first interaction during heat treatments at $\sim 100°C$ produced a layer of Au_2Al via a brief intermediate stage of Au_5Al_2 formation. This process dominates until all the Al or all the Au has been consumed to form Au_2Al. At this point, if there remains unconsumed Al, then further heating at $125°C$ will cause the conversion of Au_2Al to Au_5Al_2. Further annealing at $175°C$ then brings about the formation of Au_4Al, as seen by backscattering and verified by x-ray diffraction. If, after production of the Au_2Al layer, an excess of Au remains unconsumed, then heating at $230°C$ causes conversion to $AuAl_2$, as shown in Figure 9.17. This proceeds to completion, given sufficient Al.

In all cases the thickness of a developing compound layer was shown to be proportional to $\sqrt{t}$, which implies that all the interactions were diffusion limited. Plots showing the time dependence of the formation of Au_2Al and $AuAl_2$ are shown in Figure 9.18. Data from runs made at a series of temperatures are shown on the Arrhenius plot (Fig. 9.19), from which can be derived the interdiffusion coefficients and activation energies for each of the compound formation processes. Campisano et al. obtain for

Au_2Al: $D_0 = 1.4 \text{ cm}^2 \text{ s}^{-1}$, $Q = 1.03 \text{ eV}$

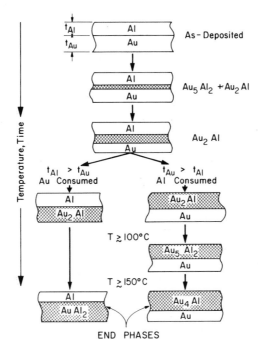

Figure 9.16. Schematic diagrams showing the phase formation in Au–Al thin film couples. The indicated temperatures are those typically required for the formation of phases in films several thousand Å thick after an anneal time of the order of 1 h. Initially, Au_5Al_2 and Au_2Al are formed, with a transition to Au_2Al at higher temperatures. When the Au layer is consumed, Au_5Al_2 is formed at temperatures $\geqslant 100°C$; and at higher temperatures ($\geqslant 150°C$) Au_4Al is formed. From reference 34.

$AuAl_2$: $D_0 = 2.0 \text{ cm}^2 \text{ s}^{-1}$, $Q = 1.2$ eV

Poate, Kirsch and Eibschutz (35) have compared the diffusion in polycrystalline films with that in a system of a single-crystal Au film (1200 Å) overlaid by polycrystalline Al (4000 Å). The rate of diffusion for the single-crystal case is at least an order of magnitude smaller than that in polycrystalline films, indicating the dominant dependence on grain boundaries as transport paths in this system. Both Au_2Al and subsequently $AuAl_2$ grew with $\sqrt{t}$ time dependence, indicating that in all samples below 200°C, diffusion rates limit the reaction of Al with Au to form Au_2Al and that of Au with Al to form $AuAl_2$. It is interesting to note that even the rate of the latter reaction was much smaller in the film which had started with single-crystal Au and had been converted to Au_2Al. Evidently, the Au_2Al grows rela-

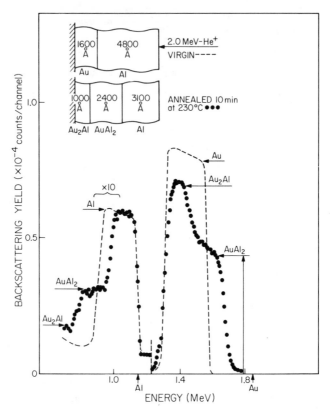

Figure 9.17. Backscattering energy spectrum showing the formation of the AuAl$_2$ phase at the Al–Au$_2$Al interface at 230°C. Arrows indicate the calculated spectrum heights for both Au$_2$Al and AuAl$_2$. From reference 34.

tively free of the defects such as grain boundaries, which assist diffusion in polycrystalline films.

The observations of Poate et al. (35) imply the dominance of a grain boundary assisted mechanism for diffusion in Au–Al polycrystalline samples. Campisano et al. (34) have unraveled a complex phase formation sequence. In both experiments, a detailed data analysis would require that the layer thickness expressions of Kidson (6) or Shatynski et al. (29) should be incorporated into the grain boundary assisted diffusion formalism of Gilmer and Farrell (16, 17), allowing also for intermediate grain growth. This would be a task of monumental complexity. Even in such a complex model, it would remain valid to make the simple test for a $\sqrt{t}$ dependence of compound phase

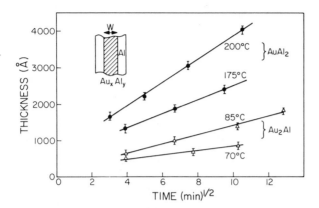

Figure 9.18. Compound thickness vs. square root of annealing time for different annealing temperatures for Au_2Al (triangles) and $AuAl_2$ (solid circles). From reference 34.

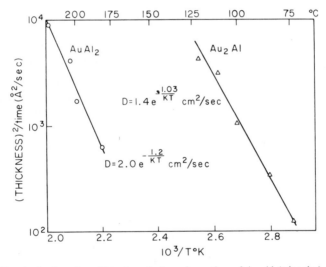

Figure 9.19. Arrhenius plot of kinetics of phase formation of Au_2Al (triangles) and $AuAl_2$ (circles). Values of the activation energies are given in eV. From reference 34.

formation as an indication that the process is diffusion limited. Poate et al. and Campisano et al. found this to be the case at all stages of the reaction.

9.3.4.2 Al–Ni. The previous Au–Al data present a beautiful example of the growth of layered compounds similar, for example, to silicide growth discussed in Chapter 10. We now present an example of a couple (Al–Ni) whose interaction is dominated by the irregular growth of grains of the compound phase.

When a polycrystalline Al–Ni film couple is heated to 275°C, the compound phase Al_3Ni begins to form in segregated grains near the interface. With continued heating, these grains grow, forming islands which ultimately join to form a rough, discontinuous layer. Observations on a pair of 1500 Å films by Baglin and d'Heurle (36) using Rutherford backscattering are shown in Figure 9.20. It is evident, especially in the Al profile, that separated compound grains develop to $\gtrsim 1000$ Å in diameter within the Al film, producing an increasingly sloped concentration gradient during 12 h of annealing at 275°C. The conclusion is supported by SEM photographs of the reacted film (Fig. 9.21), in which chemical etching of the unreacted Al can be seen to expose (white areas) crystallites of Al_3Ni. During this initial part of the interaction, the amount of Al_3Ni formed is approximately proportional to $\sqrt{t}$. This indicates a diffusion-limited process, probably dominated by the grain boundary diffusion of Ni atoms to sites within the Al grain boundaries where the Al-rich compound is nucleated. After 26 h at 275°C, the Al_3Ni evidently forms a continuous layer, at which point the reaction becomes much slower. Apparently, the diffusion in boundaries between Al_3Ni grains is slow. Furthermore, the SEM photographs imply that the continuous Al_3Ni film will have very large grains, and it may thus represent a diffusion barrier similar to a deposited single-crystal film.

A system similar to Al–Ni is Al–Sb. After heat treatment of polycrystalline Al–Sb systems at 600°C, Lyttle et al. (37) observed the growth of a very irregular layer of crystallites of AlSb at the interface. Their appearance is not unlike that of the Al_3Ni discussed above.

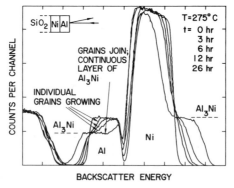

Figure 9.20.   Backscattering spectra showing successive stages of the interaction of polycrystalline films of Ni (1500 Å) + Al (1500 Å) at 275°C. From reference 36.

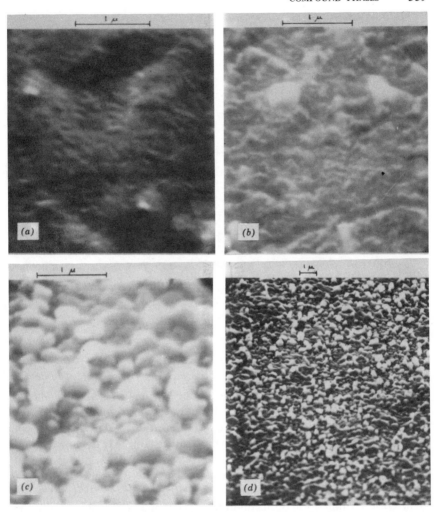

Figure 9.21. Growth of Al$_3$Ni grains after heat treatment of Ni (1500 Å) + Al (1500 Å) films at 275°C for 21 h. For frames (*b*), (*c*), and (*d*), part or all of the unreacted Al was etched away chemically to expose the new compound layer: (*a*) Surface (not etched); (*b*) Al partially etched; (*c*) Al fully etched; (*d*) Al fully etched. From reference 36.

9.3.4.3 Al–Hf. A striking example of the peculiar way in which the formation of a compound phase from thin films can develop has been recorded by Lever et al. (38). The initial system consisted of 900 Å Al on 1000 Å Hf on 6000 Å Al on SiO$_2$. All the films were polycrystalline, with grain sizes supposedly of the order of the film thickness. Figure 9.22 shows typical backscattering spectra following a sequence of 400°C heat treatments. They

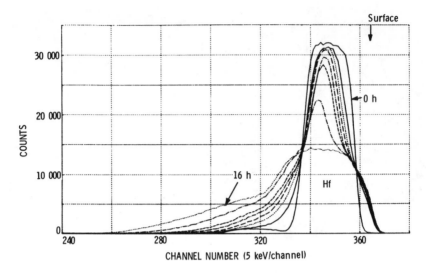

Figure 9.22. Backscattering of ^{4}He from Al–Hf–Al samples annealed at $400°$C for 0.0, 0.5, 0.75, 1, 1.5, 2, 4, and 16 h. Only the Hf portion is shown. The peak in the zero time curve between channels 300 and 330 is due to zirconium impurity. From reference 38.

indicate the presence of Hf in both the upper and the lower Al layers. Combining these data with analysis by Auger electron spectroscopy and TEM, the authors arrived at the following model for the main processes, shown schematically in Figure 9.23. After heat treatment, it was found that a layer of HfAl$_3$ had formed near the surface, consuming all the 900 Å Al film except for some residual surface Al which seemed to be stable. Meanwhile, Hf had diffused into the Al grain boundaries, where spikes of HfAl$_3$ nucleated and grew. Upon further heating, Al was seen to penetrate through the Hf layer, where it lodged to continue the formation of HfAl$_3$ at the Hf–HfAl$_3$ interface.

The clear asymmetry in diffusion in this system may be attributable to the difference in initial grain size between the 6000 Å Al layer and the thinner surface layer of Al.

9.3.4.4 Summary Comments. Most thin metal film systems used for measurement of diffusion kinetics or for industrial applications are polycrystalline. The intrinsic bulk diffusion parameters for a given metal couple, although being the quantities of fundamental identity with the metals in question, may seldom be the parameters which govern the interaction of practical metal films. We have seen how defect assisted diffusion and the kinetics of growth of grains of both host films and new compound phases can determine the entire course of the interaction of a polycrystalline film

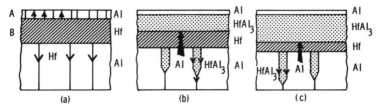

Figure 9.23. Schematic representation of the Al–Hf–Al reaction as observed by Lever et al. (38). Small arrows signify grain boundary diffusion in (a) and interphase diffusion in (b) and (c) as the reaction progresses.

couple. These processes depend directly on the morphology—grain sizes and film thickness—of the original metal films. They probably are also dependent on stresses and texture in those films.

9.4 OXIDATION AND CHEMICAL EFFECTS

Many thin film experiments are not carried out in good vacuum. Indeed, films used in devices may undergo heat cycling in anything but benign environments. There appear to be two areas where chemical effects are important: during evaporation when the films are highly reactive and during annealing. Belser (39) showed how the presence of an oxide at the interface between many Al-based couples could severely inhibit interdiffusion. The oxide was introduced, for example, by opening the evaporation chamber for a few minutes between evaporations. With rapid sequential evaporation, however, no effects due to interfacial oxides could be observed. In most experiments it is tacitly assumed that if interface oxides might cause problems, the less reactive metal should be evaporated first. For example Campisano et al. (34) in their Al–Au studies evaporate Au first to minimize Al_2O_3 formation at the interface.

The effects that can result from annealing in an active ambient were demonstrated for Ti–Pd and Ti–Au couples by Poate et al. (12). The results of air and vacuum anneals are compared in the backscattering spectra of Figures 9.24 and 9.25. Although the Ti–Au interdiffusion is faster than that for Ti–Pd for vacuum annealing, the general effects are similar. If the couples are annealed in air, however, their behavior is markedly different. For Ti–Au, the Ti out-diffuses along grain boundaries to the free surface of the Au where it is oxidized. The surface oxidation of Ti causes continued enhanced diffusion by creating a chemical potential sink. In this way a thick Ti oxide (~ 400 Å) can be grown for the annealing conditions shown in the figure. Very little interdiffusion, however, is observed in the Ti–Pd couple annealed in air. It is suggested that several competing processes can occur simultaneously during air annealing: (i) interdiffusion of the Ti and Pd, (ii)

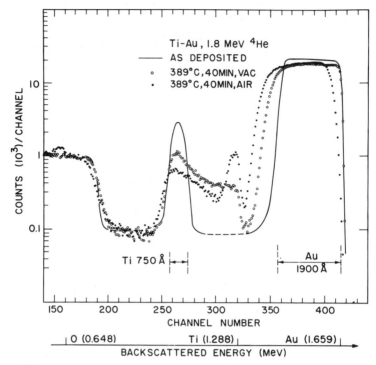

Figure 9.24. Backscattering spectra from Ti–Au couples. The energy scale on the abscissa indicates the backscattered energies from surface Au, Ti, and O atoms. From reference 12.

diffusion of oxygen through Pd, and (iii) oxidation of Ti near the Ti–Pd interface. As soon as the oxidation reaches a critical stage, most likely when a continuous oxide layer is formed along grain boundaries near the interface, further diffusion of Pd and Ti is blocked. Therefore, oxygen enhances diffusion in the Ti–Au couple but blocks diffusion for Ti–Pd.

Recently, Chang and Quintana (40) have observed interesting behavior for the diffusion of Au through Pt films. Annealing in inert N_2 and He atmospheres produced greater interdiffusion than comparable vacuum anneals. The mechanism for this behavior is not known.

It is difficult to generalize on these various chemical effects. However, in the data summary we indicate some experiments where such effects were reported.

9.5 DIFFUSION BARRIERS

In many technological applications it is desirable to minimize interdiffusion or reaction between metal couples. This can be achieved by interposition of a

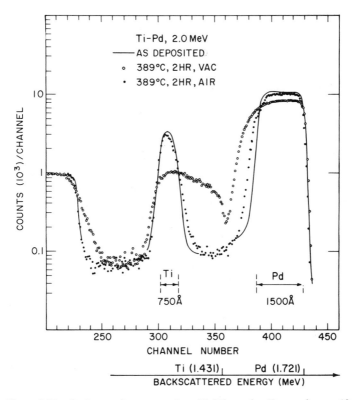

Figure 9.25. Backscattering spectra from Ti–Pd couples. From reference 12.

thick continuous oxide but here we will consider only metallic barriers. Few systematic studies have been carried out, but from previous discussions some general observations can be made. It is obvious that in the low-temperature regime a single crystal layer would act as an effective barrier. Such barriers are extremely difficult to construct, and other approaches must be tried. For example, as discussed in Chapter 2, intermetallic layers can act as effective diffusion barriers. (In Chapter 10 it is shown that silicide layers can act as diffusion barriers; in Chapter 11 it is similarly shown that a PtAs$_2$ layer can act as a barrier to Ga out-diffusion.) It is also to be expected, from simple lattice diffusion arguments, that the highest melting point or refractory metals should act as barriers. For example, Au diffusivity is low in thin films of W (41), Rh (42), and Pt (43). Nickel is also used as a thin film diffusion barrier in Cu–Au couples (44–46).

The mechanisms that cause some metal films to act as barriers can be quite complex. Figure 9.26 shows a comparison between the Ti–Pd–Au and Ti–Rh–Au systems (12, 42). There is much less interdiffusion between Rh and Au than between Pd and Au. Rh, the highest melting point component, is acting as the best barrier. This is indeed to be expected from measurements on Rh–Au and Pd–Au couples (42, 47).

In both Ti–Rh–Au and Ti–Pd–Au there is no interdiffusion between the Ti and Pd or Rh. This is intriguing. Although little diffusion is expected for the Ti–Rh couple (42), diffusion should occur in an isolated Ti–Pd couple at these temperatures (12). In the ternary system it is believed that the Au–Pd interdiffusion is so rapid that the diffusion paths for Ti are blocked. Measurements of grain boundary diffusion in bulk polycrystalline samples have shown that grain boundary diffusivities may be greatly enhanced or reduced by solute effects; this field has been reviewed by Gleiter and Chalmers (48).

It is to be expected that solute effects will play an important role at low temperatures in thin film interdiffusion owing to the dominance of defect diffusion. In Chapter 8, several examples are given of the way in which solute effects can impede self-diffusion in thin films.

9.6   DATA SUMMARY FOR METAL–METAL FILM INTERACTIONS

In this section we attempt to summarize low-temperature diffusion data which have been reported for well-characterized thin film samples. We are not attempting to include self-diffusion data or data obtained with bulk or thick samples, or measurements made at high temperatures. This collection refers specifically to effects seen in thin films and does not pretend to be a comprehensive source of diffusion data in general.

Ag–Al.   Weaver and Brown (49), using optical reflectivity and TEM techniques, observed formation of the Ag_2Al phase alone at temperatures up to 240°C for air anneals; deduced $Q = 1.15$ eV, $D_0 = 9.0$ cm^2 s^{-1} (D (136°C) $= 9 \times 10^{-14}$ cm^2 s^{-1}) for growth of phase. Picraux (50) observed parabolic growth of Ag_2Al phase with reasonably sharp interfaces, using backscattering, and found D values similar to those of Weaver and Brown; also observed effects of substantial grain boundary diffusion. Westmoreland and Weisenberger (51), using backscattering, observed diffusion barrier effects of interfacial oxides; showed 500 Å Cr to be effective diffusion barrier between Ag and Al for $T < 300°$C.

Ag–Au.   This couple forms complete series of solid solutions. Cook and Hilliard (9), using x-ray modulation techniques of DuMond and Youtz (52), demonstrated that bulk diffusivities could be measured of order 10^{-20} cm^2 s^{-1}; determined $Q_l = 1.7$ eV with D_l (200°C) $= 10^{-20}$ cm^2 s^{-1}. Tu (53) demonstrated by x-ray techniques the role of grain boundaries for mass transport in Pb–Ag–Au samples at 200°C. Wood et al. (54), using surface

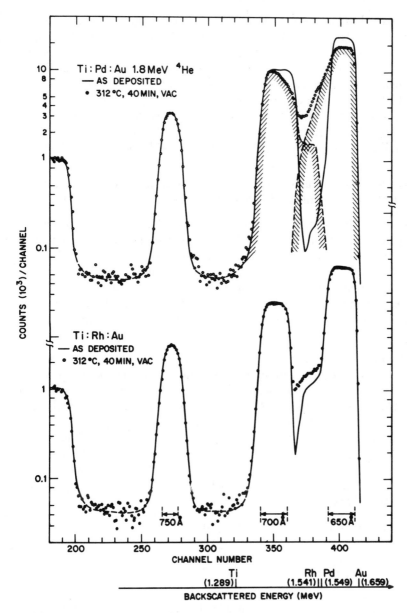

Figure 9.26. Backscattering spectra of Ti–Pd–Au and Ti–Rh–Au films before and after annealing. From references 12 and 42.

analysis by ESCA and ion milling, deduced D_b of Au through Ag as 4.3×10^{-17} cm^2 s^{-1} at 141°C. Czanderna and Summermatter (55), using ISS and ion milling, deduced D_b of Ag through Au at room temperature of 2×10^{-18} cm^2 s^{-1}. Meinel et al. (56), using Auger surface analysis of Ag diffusion through 66 Å epitaxial Au, measured $D\,(192°C) = 10^{-20}$ cm^2 s^{-1} with $Q = 1.12$ eV. Kirsch et al. (8) (see Section 9.2.3) compared diffusion in single-crystal and polycrystalline couples using backscattering and TEM; determined room temperature $D_b \approx 10^{-14}$ cm^2 s^{-1} of Au in Ag; at 300°C observed fast initial lattice diffusion $D_l' \approx 10^{-15}$ cm^2 s^{-1} followed by $D_l \approx 10^{-18}$ cm^2 s^{-1}.

Ag–Au–Pd.   Murakami et al. (57), using x-ray diffraction techniques, obtained four elements of diffusion matrix between 250 and 320°C.

Ag–Cu.   This system has very limited miscibility. DiGiacomo et al. (58), using an electron microprobe, measured the diffusion from Cu dots on Ag films at temperatures of 200, 250, and 300°C; determined $D_0 = 3.8 \times 10^{-8}$ cm^2 s^{-1} and $Q = 0.5$ eV which could be due to grain boundary or surface diffusion. Murakami and DeFontaine (59), using x-ray diffraction techniques, studied interdiffusion between 18,000 Å Ag and 94,000 Å Cu films at temperatures of 600°C; deduced $\bar{D} = 1.45 \times 10^{-12}$ cm^2 s^{-1} for the Ag-rich phase.

Al–Au.   In polycrystalline films, heat treatment at 80°C leads to formation of the phase Au$_2$Al until one of the metal films has been exhausted. If there remains excess Al, subsequent heating at above 150°C allows formation of AuAl$_2$ in place of the Au$_2$Al. If there is excess Au in contact with Au$_2$Al, heating above 100°C produces conversion of the compound to Au$_5$Al$_2$. Further heating above 150°C produces Au$_4$Al. The layer growth is parabolic in time for each phase. The phase AlAu is not seen. This picture has been presented by Campisano et al. (34) (see Section 9.3.4) following work in films ranging from 1000 to 7000 Å thick. They used backscattering and x-ray diffraction. They derived for

Au$_2$Al: $\bar{D}_0 = 1.4$ cm^2 s^{-1}, $\bar{Q} = 1.03$ eV

AuAl$_2$: $\bar{D}_0 = 2.0$ cm^2 s^{-1}, $\bar{Q} = 1.2$ eV

These values are in close agreement with those of Weaver and Brown (60) and Takei and Francombe (61). Poate et al. (35) found similar behavior in Au (4000 Å) + Al (1200 Å) self-supporting polycrystalline films, observing alloy formation using backscattering. In similar single-crystal films, Au$_2$Al followed by AuAl$_2$ formed, as above, but the growth rates were about five times smaller. Grain boundary assisted diffusion was inferred. Weaver and Parkinson (62) used typically 2000 Å layers of Au and Al (polycrystalline) observing changes in optical reflectivity, adhesion, and resistivity upon annealing. At 50°C, a partly disordered Au_2Al layer was formed, which became ordered upon further heating. For the formation process, they report $\bar{D}_0 \approx 200$ cm^2 s^{-1}, $\bar{Q} = 1.02$ eV. At 130°C, they observed formation of $AuAl_2$, with $\bar{D}_0 = 0.011$ cm^2 s^{-1}, $\bar{Q} = 0.94$ eV. After annealing at higher temperatures (321 to 399°C), Onishi et al. (63) identified Au_2Al_5, using EPMA, x-ray diffraction, and microhardness tests. They infer for

Au in Au_5Al_2: $D_0 \approx 0.04$ cm^2 s^{-1}, $Q_{Au} \approx 0.97$ eV

Al in Au_5Al_2: $D_0 \approx 5 \times 10^{-3}$ cm^2 s^{-1}, $Q_{Al} \approx 0.94$ eV

although results of their different techniques are not consistent with each other. Kolesnikov et al. (64), working at 400 to 500°C, found an effective activation energy for total alloy formation of 0.45 eV and measured diffusivities for various ratios of Au:Al thickness.

   Al–Co.   Howard et al. (65) report finding Co_4Al_{13} after heat treating Al (5000 to 7000 Å) + Co (1200 to 1400 Å) samples at 350 to 400°C. Quantity of alloy $\sim \sqrt{t}$. Techniques: backscattering, TEM, x-ray diffraction, sheet resistivity.

   Al–Cr.   Howard et al. (65) observed formation of $CrAl_7$ after heat treating Al (5000 Å) + Cr (1200 Å) couples at temperatures of 300 to 450°C. Films were polycrystalline, grain sizes initially 5000 Å Al, 200 Å Cr. They used backscattering, x-ray, and sheet resistance measurements, supported by TEM study of grain sizes; they observed a $\sqrt{t}$ dependence of the amount of $CrAl_7$ formed, interdiffusion coefficient $\bar{D}_0 = 9.7 \times 10^5$ cm^2 s^{-1}, $\bar{Q} = 1.9 \pm 0.1$ eV.

   Al–Cu.   Wildman et al. (66), using Auger depth profiling techniques and Whipple analysis, determined δD_b (175°C) for Cu in Al to be 4.3×10^{-20} cm^3 s^{-1}. D_0 was determined from initial part of Auger profile and found to be in good agreement with extrapolated tracer data (67).

   Al–Hf.   Lever et al. (38) and Howard et al. (65) (see Section 9.3.4) have observed formation of $HfAl_3$ in Al (900 Å) + Hf (100 Å) and Al (5000 to 7000 Å) + Hf (1200 to 1400 Å) after heat treatment at 350 to 400°C. When the Al film had large grains (>5000 Å), needle-like precipitates of $HfAl_3$ were formed along underlying Al grain boundaries, and their development was nonparabolic in time. When the Al grain size was initially small (<2000 Å), a planar $HfAl_3$ layer developed. Techniques were AES, backscattering, and TEM.

Al–Ni. Baglin and d'Heurle (36) (Section 9.3.4) reported the formation of Al_3Ni in rough granular form at the interface between 1500Å Ni and 1500 Å Al polycrystalline films at temperatures of 200 to 300°C. Backscattering and x-ray diffraction were used. Activation energy for the process was 1.7 eV. At 400°C, some Al_3Ni_2 was observed.

Al–Pd. Howard et al. (65) report finding $PdAl_3$ after heat treating Al (5000 to 7000 Å) + Pd (1200 to 1400 Å) couples at 250 to 350°C. Quantity of alloy ~ $\sqrt{t}$. Techniques: backscattering, TEM, x-ray, sheet resistivity.

Al–Pt. Murarka et al. (68) have observed the formation of Pt_2Al_3, $PtAl_2$, $PtAl_3$, and $PtAl_4$ in samples consisting of Al (1500 Å) + Pt (2000 Å) after heat treatments in the temperature range of 200 to 500°C. Possibly, Pt_5Al_3 and Pt_3Al_2 were also observed. Phases were identified using x-ray diffraction, supported by measurements of sheet resistivity and by SEM and optical micrographs. Intermetallics comparatively richer in Pt formed first, and further heating or higher temperatures caused Al-rich phases to grow at the expense of the initial phases. Pt_2Al_3 formed initially (observable after 2 h at 250°C), its activation energy being 1.3 to 1.5 eV. The rate of interaction to form Pt_2Al_3 was found to be strongly sensitive to the annealing ambient. Reaction was several times slower under vacuum or in air than in Ar, He, or forming gas ambients. It was suggested that an oxide layer may form during vacuum or air anneals, which may then inhibit further diffusion. The Al-rich phases all have molecular volumes substantially larger than those of Al or Pt. Murarka et al. suggest that this is the reason for a buildup of compressive stress during growth of the Pt_2Al_3 layer, followed by detachment and shattering of the compound film. Howard et al. (65) found $PtAl_4$ after heat treating Al (5000 to 7000 Å) + Pt (1200 to 1400 Å) couples at 250 to 350°C. Quantity of alloy ~ $\sqrt{t}$. Techniques: backscattering, TEM, x-ray, sheet resistivity. Comer (69) used electron diffraction to observe 500 Å films of Al and Pt after 400°C annealing. Several intermetallics were identified.

Al–Ta. Howard et al. (65) found $TaAl_3$ after heat treating Al (5000 to 7000 Å) + Ta (1200 to 1400 Å) couples at 450 to 575°C. Quantity of alloy not parabolic in t. Techniques: backscattering, TEM, x-ray, sheet resistivity.

Al–Ti. Only the compound phase Al_3Ti was observed to form after 350 to 475°C heat treatments of Al (10000 to 17000 Å) + Ti (600 to 1500 Å). Bower (70) studied its growth using backscattering, x-ray diffraction, resistivity measurements, and SEM. Observed diffusion coefficient for the process: $D_0 = 0.15$ cm^2 s^{-1}, $\overline{Q} = 1.85$ eV. Compound growth ~ $\sqrt{t}$. Howard et al. (65) also observed formation of Al_3Ti at 325 to 475°C, the phase growth having parabolic time dependence. They used backscattering, TEM, x-ray, and sheet resistivity measurements. They report finding a phase $Al_{23}Ti_9$ also at these temperatures. It is said to revert to Al_3Ti at ~440°C.

Al–Zr.   Howard et al. (65) found $ZrAl_3$ after heating Al (5000 to 7000 Å) + Zr (1200 to 1400 Å) couples at 350 to 450°C. Quantity of alloy not parabolic in t. Techniques: backscattering, TEM, x-ray, sheet resistivity.

Au–Cr.   All studies appear to show importance of grain boundary diffusion; surface oxidation of Cr can be driving force for diffusion. The following references are presented in chronological order: Kenrick (71) observed Cr grain boundary diffusion for $T < 300°C$ using TEM replica techniques. Rairden et al. (72) measured $D_0 = 1.1 \times 10^{-5}$ cm^2 s^{-1} and $\overline{Q} = 1.1$ eV from resistivity changes and air anneals. Thomas and Haas (73) determined diffusivities from work function changes of 100 Å Au films on Cr. They found D_0 dependent on film structure ($D_0 = 3.5 \times 10^{-6}$ to 3.2×10^{-7} cm^2 s^{-1}) with common $\overline{Q} = 1.2$ eV. Hirvonen et al. (38), using backscattering, observed buildup of surface Cr oxide for Ar anneals up to 450°C. Nelson and Holloway (24), using ISS and Auger depth profiling, compared vacuum and oxidizing anneals and showed Cr_2O_3 at surface to be driving force for grain boundary diffusion with grain boundaries being mass transport path. Determined $D_0 = 3.8 \times 10^{-3}$ cm^2 s^{-1} and $Q_b = 1.1$ eV; useful review of Cr–Au data. Alternative analysis (18) gave the lumped grain boundary diffusion coefficients $\delta D_0 = 8.6 \times 10^{-4}$ cm^3 s^{-1} and $\overline{Q} = 1.1$ eV. Barcz et al. (75), using backscattering, compared interdiffusion in Cr–Au and NiCr–Au couples for air anneals; claimed observation of Cr_3Au phase, but there is no evidence of this phase from equilibrium phase diagrams (76).

Au–Cu.   Equilibrium phase diagram shows complete series of solid solutions with presence of ordered compounds of Cu_3Au, CuAu, and $CuAu_3$. Many thin film studies, but there is some disarray as to the diffusion mechanisms and role of the ordered compounds. References are presented in chronological order: DuMond and Youtz (52), in one of the first thin film interdiffusion experiments, determined room temperature diffusivity of 5×10^{-20} cm^2 s^{-1}, using x-ray modulation techniques. Alessandrini and Kuptsis (77), using electron microprobe techniques, deduced $\overline{Q} = 0.8 \pm 0.2$ eV between 200 and 300°C. Pinnel and Bennett (45), using electron microprobe techniques, determined interdiffusion in micron-thick Au films plated on Cu for air annealing. Boltzmann-Matano analysis in the range of 150 to to 750°C gave $D_0 = 1.5 \times 10^{-5}$ cm^2 s^{-1} and $\overline{Q} = 1.03$ eV. Deduced that at temperatures lower than 150°C, diffusion occurs predominantly along defects with $Q \approx 0.5$ eV.

Tu and Berry (78) used x-ray diffraction analysis of samples annealed in silicone oil baths. In the range of 160 to 220°C, Cu_3Au reflections were observed followed by $CuAu_3$. Metallographic observations indicated layered structure $Cu|Cu_3Au|CuAu_3|Au$ with sharp interfaces; $\overline{Q} = 1.56 \pm 0.17$ eV deduced from temperature dependence of linear thickening rate. Borders (79), using backscattering, suggested that interdiffusion in the range of 200 to 500°C takes place by rapid saturation of grain boundaries. No evidence was found for layers of $CuAu_3$ or Cu_3Au; $\overline{Q} = 1.35$ to 1.5 eV with room temperature diffusivity of Cu through Au $\approx 10^{-17}$ cm^2 s^{-1}. Campisano et al. (80, 81), using backscattering, present conclusions similar to those of Borders. In the range of 200 to 300°C, deduce $\overline{Q} = 1.05$ eV and $D_0 = 3.2 \times 10^{-4}$ cm^2 s^{-1}.

Tompkins and Pinnel (46), by Auger depth profiling and contact resistance, measured interdiffusion in Co-hardened 2.5 μm electroplated Au films; found diffusion coefficients several orders of magnitude higher than those of previous measurements in the range of 75 to 150°C. This was thought to be due to sinking action of chlorine at surface. Hall et al. (82) measured interdiffusion in evaporated and electroplated couples using Auger depth profiling and resistivity changes. For air annealing, over the range of 50 to 250°C, the diffusivities for the different films agreed well and could be represented by $\overline{D}_0 = 2.9 \times 10^{-3}$ cm^2 s^{-1} and $\overline{Q} = 1.21$ eV. Auger profiles from a Cu ribbon which had been Au-plated and stored for 18 years at room temperature gave $\overline{D} \leqslant 2 \times 10^{20}$ cm^2 s^{-1}. A useful summary of the review by Butrymowicz et al. (83) is given.

Au–Fe. Ziegler et al. (84), using backscattering and x-ray diffraction techniques, showed that interdiffusion at 350°C for Fe in Au approaches equilibrium solubility levels (~ 10 at. %). However, Au concentration in Fe at 350°C is ~ 1 at. %, which is approximately an order of magnitude higher than equilibrium solubility. Annealing couple in vacuum furnace with air leak, $\sim 10^{-5}$ torr, resulted in movement of all Fe in film to surface as oxide.

Au–Ga. Simic and Marinkovic (85), using x-ray diffraction techniques, observed interdiffusion of Au and Ga films at room temperature. If more than 60% Ga is present the compound, $AuGa_2$ is formed immediately and remains stable even after a year. For lower Ga concentrations, $AuGa_2$ is the first phase formed, which then reacts with free Au to form Au_2Ga or Au_7Ga_2.

Au–Mo. Solid solubility of Mo in Au is ~ 0.5 to 1.0 at. % in the range of of 200 to 800°C and solubility of Au in Mo is exceedingly small (33). Backscattering measurements of Harris et al. (86) reveal ~ 2 to 3 at. % Au in Mo as-deposited films. Grain boundary diffusion assumed to occur during deposition. Annealing Ti–Mo–Au at 600°C for 20 min produced substantial interdiffusion. The Mo film, however, appeared little changed in thickness although 30% of the Au was located below the Mo and 50% of the Ti above the Mo. It was suggested that the Mo grain boundaries were acting as membranes for the Ti–Au interaction.

Au–Ni. Couple displays miscibility gap. All diffusion measurements show dominance of grain boundary diffusion. Backscattering and x-ray diffraction show average interdiffused concentration approaching equilibrium values at 350°C. No evidence in any of measurements for intermediate phases. Richards and McCann (87), using x-ray techniques, deduced that microcrystalline alloy aggregates are formed in grain boundary of each layer of 20,000 Å electroplated films for anneals between 400 and 750°C. Aggregates become richer in diffused metal until composition at the miscibility boundary is reached. Supersaturation and subsequent precipitation then occurs. Rairden et al. (72) observed oxidation of Ni on surface of Au for air anneals and found that the resistance increase for Ni in Au (at 450°C) is much less than for Cr in Au.

Nenadovic et al. (88), using TEM and electron microprobe techniques, observed surface diffusion at 450°C with $\bar{D} \approx 1 \times 10^{-10}$ cm^2 s^{-1}. No evidence of intermediate phases. Judy and Koliwad (89), using optical reflectivity techniques, observed for Ni diffusion in Au (340 to 440°C) change in activation energy from 0.6 to 1.9 eV with increasing temperature. Ziegler et al. (84) determined, from backscattering, interdiffused concentrations in 1000 Å films with grain sizes ≈ 1000 Å. Au appears to penetrate Ni films at 0.1 at. % level during evaporation, remaining at this level for 150 and 250°C anneals after 100 min. Complete penetration of the Ni film occurs at 350°C with concentration levels of 0.5 at. % (equilibrium solubility ≈ 0.9 at. %). Ni diffusion appears faster in Au with concentration at 250°C of 0.8 at. %; at 350°C, concentration reaches equilibrium solubility ≈ 8 at. %.

Gangulee (90), using x-ray diffraction techniques, measured interdiffused levels at 350°C very similar to those obtained by Ziegler et al. (84). Barcz et al. (75), using backscattering in Au (700 Å)|Ni (100 Å)|Au (700 Å), observed nearly complete migration of Ni film to outside surface at 350°C; oxidation assumed to be driving force.

Au–Pd. This couple forms a complete series of solid solutions. All experiments demonstrate importance of grain boundary diffusion. Measured concentrations of interdiffused material for temperature <400°C are too high to be explained by model of grain boundary diffusion followed by simple lattice diffusion into interior of grains. Backscattering (12, 47), Auger (13) and x-ray (11, 19) measurements of interdiffusion are discussed in Section 9.2.3. Boyko et al. (91–93) have measured surface diffusion in very thin Pd–Au couples with an activation energy of 0.5 eV.

Au–Pt. Sinha et al. (43) studied Pt diffusion through Au using electrical resistivity, x-ray diffraction, and TEM techniques and found $D_0 = 6.6 \times 10^{-8}$ cm^2 s^{-1} and $Q = 0.95$ eV in the temperature range of 325 to 600°C. Nowak and Dyer (94), however, by optical reflectivity techniques obtained $D_0 \approx 10^{-3}$ cm^2 s^{-1} and $Q = 1.65$ eV for air anneals in the temperature range of 450 to 550°C. Chang and Quintana (95) measured diffusion of Au through Pt using Auger spectroscopy and ion milling and found $Q_b = 0.96$ eV for annealing in

N_2 ambient in the temperature range of 250 to 350°C. They found (40) that Au diffusion is 15 and 30 times faster in 1 atm. N_2 and 1 atm. He, respectively, than under vacuum at 350°C.

Au–Rh. Solid solubility of elements is extremely limited (76), $\leqslant 1$ at. % at 1000°C. Backscattering and TEM measurements (42, 47) show Au saturating Rh grain boundaries, at room temperature, with $D_b \approx 3.5 \times 10^{-17}$ cm^2 s^{-1}. Very little interdiffusion observed even for annealing at 490°C. Amount of interdiffusion probably due to grain growth.

Au–Ta. Christou and Day (96) and Tisone and Lau (97) observed Ta–Au interdiffusion using x-ray diffraction, electron microscopy, and electrical resistance measurements. Tisone and Lau investigated both bcc Ta–Au and β Ta–Au interdiffusion. Christou and Day concluded that the TaAu phase is formed at temperatures above 450°C ($\overline{Q} = 1.54$ eV); grain boundary diffusion of Au into Ta occurs at temperatures below 350°C ($Q_b = 0.41$ eV). Tisone and Lau conclude that in the lower temperature region Au diffuses into Ta and that at higher temperatures Au is transformed into the high-resistivity TaAu phase with $Q_b = 1.52$ eV, for both bcc Ta–Au and β Ta–Au. For bcc Ta–Au, no compound formation was observed for temperatures below 511°C; for β Ta–Au, the TaAu phase was observed for temperatures above 347°C.

Au–Ti. Tisone and Drobek (98), using TEM and glancing-angle electron diffraction techniques, showed that for couples with approximately equal thickness of Au and Ti, the Au_2Ti intermetallic phase formed first and when Au was depleted, the AuTi phase formed. As diffusion proceeded, the AuTi disappeared and $AuTi_3$ began to form. For thicker couples Au_4Ti phase was observed in initial stages of diffusion. Backscattering studies of Poate et al. (12) (see Section 9.4) gave no evidence for layered growth of phases; x-ray studies did show evidence for $TiAu_4$ and $TiAu_2$. Thick surface Ti oxides observed for air anneals.

Be–Cu. Myers et al. (99) implanted thin ($\leqslant 300$ Å) layers of Cu into single-crystal Be; for other test samples they deposited by electron beam 3000 Å Cu on Be single-crystal layers. Heat treatment at 700 to 1000°C produced intermixing of the films, which was observed by backscattering. Diffusion progressed as $\sqrt{t}$ in both cases. Values of D_0 and Q were found to depend on whether the Be crystals were aligned with the a- or c-axis normal to the film. For movement of Cu in Be, they found for

a-axis: $D_{a0} = 0.416$ cm^2 s^{-1}, $Q = 2.00 \pm 0.10$ eV

c-axis: $D_{c0} = 0.381$ cm^2 s^{-1}, $Q = 2.05 \pm 0.10$ eV

Cr–Cu. Polycrystalline films of Cu (1050 Å) + Cr (1050 Å) were observed to intermix extensively upon heat treatment in the range of 550 to 750°C.

Baglin et al. (26) (see Section 9.2.4) used backscattering and SEM to study the process. At these temperatures Cr and Cu are almost completely immiscible, and extensive grain nucleation and growth within grain boundaries are thought to be responsible for the mixing. Effective activation energies deduced for the process are

$$Q \text{ (Cu into Cr)} = 2.4 \pm 0.4 \text{ eV}$$
$$Q \text{ (Cr into Cu)} = 2.0 \pm 0.4 \text{ eV}$$

Similar treatment of single-crystal films of Cr and Cu produced no such interpenetration; however, a surface/interface layer developed on each film, presumably owing to diffusion in the boundaries between the large grains.

Cr–Pt. Several intermetallic alloy phases can be produced in Cr + Pt at temperatures below 900°C. Danyluk et al. (100) report a study with AES of polycrystalline Pt (2000 Å) + Cr (1400 Å) at temperatures of 700 to 850°C. No identification is made of the intermetallic phase formed. However, its growth is described by $D_0 = 1.02 \times 10^{-2}$ cm^2 s^{-1}, $\overline{Q} = 1.69$ eV. d'Heurle and Baglin (101), using x-ray diffraction and backscattering with Cr (900 to 4000 Å) + Pt (2000 Å) polycrystalline films, have identified the cubic phases Cr$_3$Pt and CrPt$_3$ after heat treatments above 420°C. At 950°C, it became possible also to identify, with x-ray diffraction, the noncubic phase CrPt.

Cu–Pb. Cu and Pb are immiscible at temperatures below 327°C. Campisano et al. (27) have used backscattering and TEM to study the interaction of polycrystalline Cu and Pb films at 220 to 250°C. With various film thicknesses, they observe grain growth in the Cu film and migration of Pb through the Cu layer. Wilson (102) reports similar effects with $\overline{Q} = 0.52 \pm 0.07$ eV.

Cu–Sn. Tu (103) has reported observing interdiffusion of films of Cu and Sn and formation of two compound phases. At temperatures above 60°C, Cu$_3$Sn was formed; at temperatures within the range between −2 and 100°C, Cu$_6$Sn$_5$ was formed, apparently by interstitial diffusion of Cu into Sn. The growth of Sn whiskers in films containing Cu is interpreted as the relaxation of internal stress built up in the Sn due to the interdiffusion process at room temperature. Tu identified new phases with a Seeman-Bohlin x-ray spectrometer and with SEM.

In–Pb. Lahiri (104) studied polycrystalline films of In (200 to 3000 Å) + Pb (5000 Å), using x-ray diffraction to identify phases and orientation. The films displayed (111) fiber texture. Homogeneous solid solutions were observed to form after 24 h at 20°C; deduced that $\overline{D}$ (20°C) = 7×10^{15} cm^2 s^{-1}. SEM showed no significant change in (5000 Å) Pb grain size even after alloy formation was completed. Rapid homogenization was attributed to an interface-controlled mechanism or to diffusion through dislocations.

Pd–Ti. Tisone and Drobek (98) used polycrystalline films of Pd (600 to 10000 Å) + Ti (600 to 10000 Å) heat treated in the temperature range of 200 to 500°C. Recrystallization of Pd grains occurs rapidly upon heating, while the compound phase TiPd (B19) is formed at the grain boundaries of the Pd. Subsequently, the phase Ti_3Pd (A15) develops in a ~ 100 Å layer at the film interface and inhibits further growth of the TiPd phase. Observations were made by TEM and by glancing-angle electron diffraction with successive layered chemical etching of the film. Poate et al. (12), using backscattering, measured substantial interdiffusion for vacuum anneals; at 312°C Ti was distributed uniformly through Pd at 10 at. % level. No evidence for layered compounds. Air annealing produced no interdiffusion due to oxide growth at interface.

Pt–Ti. DeBonte et al. (42), using backscattering, report finding very little interdiffusion at $T \leqslant 400°C$. At 490°C, they observed the growth of a layer of rather uniform composition (Ti + 10 at. % Pt). Tisone and Drobek (98), using electron diffraction and TEM, report the identification of the compounds TiPt (B19) and Ti_3Pt (A15) formed in thin film diffusion between 100 and 400°C. The Ti_3Pt is said to form a stable thin layer at the Ti–TiPt interface, which acts as a good diffusion barrier.

Rh–Ti. DeBonte et al. (42), using backscattering and TEM, found no discernible interdiffusion at $T \leqslant 400°C$.

REFERENCES

1. C. Weaver, in *Physics of Thin Films*, Vol. 6, M. H. Francombe and R. W. Hoffman, Eds., Academic Press, New York (1971).
2. L. Boltzmann, *Ann. Phys., Lpz.*, **53**, 959 (1894).
3. W. K. Warburton and D. Turnbull, *Thin Solid Films*, **25**, 71 (1975).
4. L. A. Girifalco, *Atomic Migration in Crystals*, Blaisdell, New York (1964), p. 69.
5. A. Gangulee, *Phil. Mag.*, **22**, 865 (1970).
6. G. V. Kidson, *J. Nucl. Mater.*, **3**, 21 (1961).
7. W. Jost, *Diffusion in Solids, Liquids and Gases*, Academic Press, New York (1952), pp. 76 and 77.
8. R. G. Kirsch, J. M. Poate, and M. Eibschutz, *Appl. Phys. Lett.*, **29**, 772 (1976).
9. H. E. Cook and J. E. Hilliard, *J. Appl. Phys.*, **40**, 2191 (1969).
10. R. T. P. Whipple, *Phil. Mag.*, **45**, 1225 (1954).
11. S. S. Lau and R. C. Sun, *Thin Solid Films*, **10**, 273 (1972).
12. J. M. Poate, P. A. Turner, W. J. DeBonte, and J. Yahalom, *J. Appl. Phys.*, **46**, 4275 (1975).
13. P. M. Hall, J. M. Morabito, and J. M. Poate, *Thin Solid Films*, **33**, 107 (1976).
14. P. M. Hall and J. M. Morabito, *Surf. Sci.*, **54**, 79 (1976).
15. A. D. Le Claire, *Brit. J. Appl. Phys.*, **14**, 351 (1963).
16. G. H. Gilmer and H. H. Farrell, *J. Appl. Phys.*, **47**, 3792 (1976).
17. G. H. Gilmer and H. H. Farrell, *J. Appl. Phys.*, **47**, 4373 (1976).

18. P. H. Holloway, D. E. Amos, and G. C. Nelson, *J. Appl. Phys.*, **47**, 3769 (1976).
19. M. Murakami, D. deFontaine, and J. Fodor, *J. Appl. Phys.*, **47**, 2850 (1976).
20. C. R. Houska, *Thin Solid Films*, **25**, 451 (1975).
21. R. W. Balluffi and J. M. Blakely, *Thin Solid Films*, **25**, 363 (1975).
22. D. Gupta, *Thin Solid Films*, **22**, 121 (1974).
23. P. M. Hall and J. M. Morabito, *Surf. Sci.*, **59**, 624 (1976).
24. G. C. Nelson and P. H. Holloway, ASTM, Special Technical Publication (STP) No. 596, Philadelphia (1976), p. 68.
25. J. Crank, *The Mathematics of Diffusion*, Oxford University Press, London (1970), p. 48.
26. J. E. E. Baglin, V. Brusic, E. Alessandrini, and J. Ziegler, in *Application of Ion Beams to Metals*, S. T. Picraux, E. P. EerNisse, and F. L. Vook, Eds., Plenum Press, New York (1974), p. 169.
27. S. U. Campisano, E. Costanzo, G. Foti, and E. Rimini, in *Ion Beam Surface Layer Analysis*, Vol. 1, O. Meyer, G. Linker, and F. Käppeler, Eds., Plenum Press, New York (1976), p. 397.
28. H. H. Farrell, G. H. Gilmer, and M. Suenaga, *J. Appl. Phys.*, **45**, 4025 (1974).
29. S. R. Shatynski, J. P. Hirth, and R. A. Rapp, *Acta Metall.*, **24**, 1071 (1976).
30. J. D. Baird, *J. Nucl. Energy*, **A11**, 81 (1960).
31. H. H. Farrell, G. H. Gilmer, and M. Suenaga, *Thin Solid Films*, **25**, 253 (1975).
32. F. J. A. den Broeder, *Acta Metall.*, **20**, 319 (1972).
33. M. Hansen and K. Anderko, *Constitution of Binary Alloys*, 2nd ed., McGraw-Hill, New York (1958).
34. S. U. Campisano, G. Foti, E. Rimini, S. S. Lau, and J. W. Mayer, *Phil. Mag.*, **31**, 903 (1975).
35. J. M. Poate, R. G. Kirsch, and M. Eibschutz, to be published (1977).
36. J. E. E. Baglin and F. M. d'Heurle, in *Ion Beam Surface Layer Analysis*, Vol. 1, O. Meyer, G. Linker, and F. Käppeler, Eds., Plenum Press, New York (1976), p. 385.
37. J. Lyttle, P. Dembrowski, and L. Castleman, *Met. Trans.*, **2**, 303 (1971).
38. R. F. Lever, J. K. Howard, W. K. Chu, and P. J. Smith, *J. Vac. Sci. Technol.*, **1**, 158 (1977).
39. R. B. Belser, *J. Appl. Phys.*, **31**, 562 (1960).
40. C. C. Chang and G. Quintana, *Appl. Phys. Lett.*, **29**, 453 (1976).
41. R. L. Ruth, J. M. Poate, and A. K. Sinha, unpublished data.
42. W. J. DeBonte, J. M. Poate, C. M. Melliar-Smith, and R. A. Levesque, *J. Appl. Phys.*, **46**, 4284 (1975).
43. A. K. Sinha, T. E. Smith, and T. T. Sheng, *Thin Solid Films*, **22**, 1 (1974).
44. M. Antler, *Plating*, **57**, 615 (1970).
45. M. R. Pinnel and J. E. Bennett, *Met. Trans.*, **3**, 1989 (1972).
46. H. G. Tompkins and M. R. Pinnel, *J. Appl. Phys.*, **47**, 3804 (1976).
47. W. J. DeBonte and J. M. Poate, *Thin Solid Films*, **25**, 441 (1975).
48. H. Gleiter and B. Chalmers, in *Progress in Materials Science*, "High Angle Grain Boundaries," Vol. 16, B. Chalmers, Ed., Pergamon Press, New York, (1972), Chapter 4.
49. C. Weaver and L. C. Brown, *Phil. Mag.*, **17**, 881 (1968).
50. S. T. Picraux, Proc. VI International Vac. Congress, *Jap. J. Appl. Phys. Suppl.*, **2**, 657 (1974).
51. J. E. Westmoreland and W. H. Weisenberger, *Thin Solid Films*, **19**, 349 (1973).
52. J. DuMond and J. P. Youtz, *J. Appl. Phys.*, **11**, 357 (1940).
53. K. N. Tu, *J. Appl. Phys.*, **43**, 1303 (1972).
54. J. K. Wood, J. L. Alvarez, and R. Y. Maughan, *Thin Solid Films*, **29**, 359 (1975).
55. A. W. Czanderna and R. S. Summermatter, *J. Vac. Sci. Technol.*, **13**, 386 (1976).
56. K. M. Meinel, M. Klaus, and H. Bethge, *Thin Solid Films*, **34**, 157 (1976).
57. M. Murakami, D. deFontaine, J. M. Sanchez, and J. Fodor, *Acta Metall.*, **22**, 709 (1974).

58. G. DiGiacomo, P. Peressini, and R. Rutledge, *J. Appl. Phys.*, **45**, 1626 (1974).
59. M. Murakami and D. deFontaine, *J. Appl. Phys.*, **47**, 2857 (1976).
60. C. Weaver and L. C. Brown, *Phil. Mag.*, **8**, 1379 (1963).
61. W. J. Takei and M. H. Francombe, *Solid State Electron.* **11**, 205 (1968).
62. C. Weaver and D. T. Parkinson, *Phil. Mag.*, **22**, 377 (1971).
63. M. Onishi and K. Fukumoto, *J. Jap. Inst. Met.*, **38**, 148 (1974).
64. D. P. Kolesnikov, A. F. Andrushko, and Ye. I. Sukhinina, *Fiz. Met. Metall.*, **34**, 529 (1972).
65. J. K. Howard, R. F. Lever, P. J. Smith, and P. S. Ho, *J. Vac. Sci. Technol.*, **13**, 68 (1976).
66. H. S. Wildman, J. K. Howard, and P. S. Ho, *J. Vac. Sci. Technol.*, **12**, 75 (1975).
67. N. L. Peterson and S. J. Rothman, *Phys. Rev.*, **B1**, 3264 (1970).
68. S. P. Murarka, I. A. Blech, and H. J. Levinstein, *J. Appl. Phys.*, **47**, 5175 (1976).
69. J. J. Comer, *Acta Crystallogr.*, **17**, 444 (1964).
70. R. W. Bower, *Appl. Phys. Lett.*, **23**, 99 (1973).
71. P. S. Kenrick, *Nature*, **217**, 1249 (1968).
72. J. R. Rairden, C. A. Neugebauer, and R. A. Sigsbee, *Metall. Trans.*, **2**, 719 (1971).
73. R. E. Thomas and G. A. Haas, *J. Appl. Phys.*, **43**, 4900 (1972).
74. J. K. Hirvonen, W. H. Weisenberger, J. E. Westmoreland, and R. A. Meussner, *Appl. Phys. Lett.*, **21**, 37 (1972).
75. A. Barcz, A. Turos, and L. Wielunski, in *Ion Beam Surface Layer Analysis*, O. Meyer, G. Linker, and F. Käppeler, Eds., Plenum Press, New York (1976), p. 407.
76. F. A. Shunk, *Constitution of Binary Alloys*, 2nd Supplement, McGraw Hill, New York, (1969).
77. E. I. Alessandrini and J. D. Kuptsis, *J. Vac. Sci. Technol.*, **6**, 647 (1969).
78. K. N. Tu and B. S. Berry, *J. Appl. Phys.*, **43**, 3283 (1972).
79. J. A. Borders, *Thin Solid Films*, **19**, 359 (1973).
80. S. U. Campisano, G. Foti, F. Grasso, and E. Rimini, *Thin Solid Films*, **19**, 339 (1973).
81. S. U. Campisano, G. Foti, F. Grasso, and E. Rimini, *Thin Solid Films*, **25**, 431 (1975).
82. P. M. Hall, J. M. Morabito, and N. T. Panousis, *Thin Solid Films*, **41**, 341 (1977).
83. D. B. Butrymowicz, J. R. Manning, and M. R. Read, *J. Phys. Chem. Ref. Data*, **3**, 527 (1974).
84. J. F. Ziegler, J. E. E. Baglin, and A. Gangulee, *Appl. Phys. Lett.*, **24**, 36 (1974).
85. V. Simic and Z. Marinkovic, *Thin Solid Films*, **34**, 179 (1976).
86. J. M. Harris, E. Lugujjo, S. U. Campisano, M. A. Nicolet, and R. Shima, *J. Vac. Sci. Technol.*, **12**, 524 (1975).
87. J. Richard and W. McCann, *J. Vac. Sci. Technol.*, **6**, 644 (1969).
88. T. Nenadovic, Z. Fotiric, B. Djuric, O. Nesic, T. Dimitrijevic, and R. Sofrenovic, *Thin Solid Films*, **12**, 411 (1972).
89. M. M. Judy and K. M. Koliwad, *J. Electron. Mater.*, **2**, 331 (1973).
90. A. Gangulee, *Jap. J. Appl. Phys.*, **Suppl. 2**, Part 1, 621 (1974).
91. B. T. Boyko, L. S. Palatnik, and M. V. Lekedeva, *Fiz. Met. Metall.*, **25**, 845 (1968).
92. B. T. Boyko and M. V. Lebedeva, *Fiz. Met. Metall.*, **29**, 603 (1970).
93. L. S. Palatnik, B. T. Boyko, and M. V. Lebedeva, *Dokl. Akad. Nauk SSSR*, **213**, 141 (1973).
94. W. B. Nowak and R. N. Dyer, *J. Vac. Sci. Technol.*, **9**, 279 (1972).
95. C. C. Chang and G. Quintana, *Thin Solid Films*, **31**, 265 (1976).
96. A. Christou and H. Day, *J. Appl. Phys.*, **44**, 3386 (1973).
97. T. C. Tisone and S. S. Lau, *J. Appl. Phys.*, **45**, 1673 (1974).
98. T. C. Tisone and J. Drobek, *J. Vac. Sci. Technol.*, **9**, 271 (1972).
99. S. M. Myers, S. T. Picraux, and T. S. Prevender, *Phys. Rev.*, **B9**, 3953 (1974).
100. S. Danyluk, G. E. McGuire, K. M. Koliwad, and M. G. Yang, *Thin Solid Films*, **25**, 483 (1975).

101. F. M. d'Heurle and J. E. E. Baglin, *Proceedings of the 7th International Vacuum Congress and 3rd International Conference on Solid Surfaces*, Vienna (1977).
102. J. M. Wilson, *Phil. Mag.*, **27**, 1467 (1973).
103. K. N. Tu, *Acta Met.*, **21**, 347 (1973).
104. S. K. Lahiri, *Thin Solid Films*, **25**, 279 (1975).

10

SILICIDE FORMATION

K. N. Tu

IBM Thomas J. Watson Research Center, Yorktown Heights, New York

J. W. Mayer

California Institute of Technology, Pasadena, California

10.1   INTRODUCTION

The recent impetus for studying silicide formation is due to the requirements placed on the performance of integrated circuits. It now appears that one of the limiting factors in achieving high production yield and high device realiability lies in the choice of the metallization scheme. The problem arises from the requirements for a controlled contact at the semiconductor–metal interface and for a low-resistance metal that does not lead to a voltage drop across the contact. One must realize that there are practical reasons for the choice of a particular metallization scheme. A concept that is feasible under research laboratory conditions may not be adaptable to production requirements where high yield is required. The metallization must also be compatible with the processing temperature and packaging structure. Since every production line has a somewhat different process control, it is difficult to give specific details on all the constraints imposed on the metallization. Instead, we give some general guidelines for the selection of metals used to make contacts.

Figure 10.1 shows schematically an FET with a heavily doped source and drain and a lightly doped channel connecting the source and drain. The contacts to the source and drain and to the channel may have a different structure, but ideally the same metallization is used for both. In early devices, the junction regions made by diffusion were located several microns below the surface. There were not many constraints on metallization other than adhesion, barrier height, low electrical resistance, and high corrosion resistance. In these aspects aluminum was ideal. Because of the relatively large device dimensions, the dissolution of Si into Al and the consequent penetration of Al into Si around the periphery of the contact caused relatively

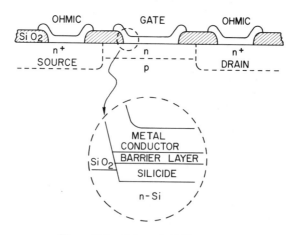

Figure 10.1.   Schematic FET structure.

few problems in shorting the junction. However, as device dimensions decreased and junction depths became shallow, it was obvious that the Al penetration problem could no longer be tolerated.

The design philosophy next introduced was to use a silicide-forming metal (1–4) to produce a uniform layer with controlled barrier height and good adhesion to Si and then a second metal, typically Al, Au, or W, to provide the low-resistance path. The thickness of the silicide layer is typically less than 1000 Å, and only a few hundred Å of Si is consumed in forming the silicide. The purpose of the silicide layer is not only to provide a good Schottky barrier but also to prevent the reaction between Al and Si. But unfortunately, Al also reacts with the silicide layer (5, 6), and hence in some cases a diffusion barrier layer must be interposed between the silicide and the Al. This structure is shown in the insert in Figure 10.1. In order to give a feeling of how such a three-layer structure is fabricated, we show some typical processing steps in Figure 10.2. Again we emphasize that each production line may use a slightly different temperature and sequence, and the lithography, etching, and pattern definition steps also differ from one production line to another. We show only the general principles in the following.

The metallization process steps used in forming the gate contact to the channel region of an FET are shown in Figure 10.2. Before deposition, Figure 10.2a, there is a native oxide layer on the Si surface in the window opened by etching the thick oxide layer. It is this native oxide layer which has caused many of the problems associated with contact metallization. One of the requirements for any contact metal is that the metal must be able to penetrate this oxide layer whose thickness may range between 15 and 50 Å. It has been found that some metals, such as Pt and Pd, can readily penetrate this thin oxide layer to form silicide layers.

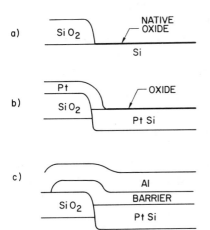

Figure 10.2. Schematic metallization steps in forming a silicide contact on Si: (a) oxide window before Pt deposition; (b) sintering to complete PtSi formation; (c) contact structure after barrier and Al deposition.

We shall illustrate the metallization scheme by describing the formation of platinum silicide, which is probably the most generally used contact material in integrated circuits. During the deposition of Pt, the substrate is sometimes maintained at an elevated temperature around 300°C and subsequently sintered at an elevated temperature of 550°C to complete the silicide formation. After sintering, the structure, shown in Figure 10.2b, is composed of a thin silicide layer (about 1000 Å thick for a 500 Å Pt layer) with a thin oxide on top. This thin oxide layer has formed during either deposition or sintering owing to the presence of oxygen in the ambient (7). Since silicon is a necessary ingredient in the formation of this oxide layer and since Pt does not react with thick SiO_2 layers, the layer is formed only above the silicide region but not on the Pt layer over the thick oxide. The excess Pt over the thick SiO_2 layer is removed by aqua regia etching. The thin oxide layer prevents dissolution of the silicide during the aqua regia etching step.

In the next general step, the thin oxide layer is removed with a buffered HF solution, and the barrier and the Al layer are then deposited. In general, another heat treatment step around 450°C is required at this point to penetrate any oxide layer between the metal films in order to produce a low-resistance contact. It is because of this second heat treatment step that the barrier layer is introduced to prevent the Al from penetrating the Pt silicide layer. The barrier layer is often a refractory metal such as Cr, Ti, or V which can form compounds with both Al and Si.

The general requirement for any contact is that there must be a uniform and limited reaction between the two layers in contact. A reaction is required to make sure that there is intimate contact and adhesion. We use the word "limited" to imply that there is no long-range atomic migration that could lead to film penetration or junction shorts. The requirement for uniformity arises from the fact that deep penetration can start at localized points due to interface instability. If such localized penetration does occur, severe pitting will result since all the reactions produced must originate from a localized region. Figure 10.3 shows two SEM pictures which reveal pitting around the periphery of a contact area. These pictures were obtained from an Al-metallized Si wafer which was sintered at 450°C for 1 h in N_2 ambient and then etched to remove the metal electrode.

In the following sections, we neglect many of the problems encountered in the lithography used in pattern formation in order to emphasize the major features of silicide formation: mechanisms of phase formation, kinetics of reaction in terms of time and temperature, reactions with oxide layers, and influence of oxygen in the ambient.

Although we have emphasized the technological aspects of silicide formation, the reactions between transition metals and silicon are of interest in

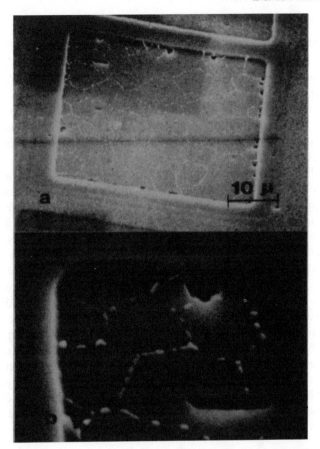

Figure 10.3. SEM micrographs of spike formation around the periphery of a contact area. From K. Takahata et al., Extended Abs. No. 305, the 105th Electrochem. Soc. Meeting, Las Vegas, October 1976.

their own right. In contrast to the case of bimetallic thin film reactions, in silicide formation one deals with a fine-grained metallic thin film in contact with a single-crystal, covalently bonded semiconductor. One might anticipate that the crystal orientation and the microstructure of silicide might play a role. Since it is known that silicide formation can occur at a temperature as low as $100°C$ (8, 9), it is also interesting to investigate the physical mechanism responsible for breaking the covalent bonds in Si. At this low reaction temperature, the growth phase may be determined by the reaction kinetics rather than thermodynamic driving forces.

10.2 ANALYSIS OF SILICIDE FORMATION

10.2.1 Sample Preparation

In this section we adopt a different viewpoint from that for device production and consider the most basic aspects of silicide formation. Rather than follow the constraints imposed by production requirements with narrow openings in the oxide layer and the associated stress and coverage problems around the periphery of the openings, we deal with large-area deposition. Then, too, we assume that enough effort has been made to remove interfacial oxide layers and to achieve oxygen-free thin film deposition. In subsequent heat treatments, either a vacuum better than 10^{-6} torr or a purified flowing atmosphere of He or other inert gas is used. Although all these precautions may not lead to an impurity-free silicide and may not reproduce the exact situation in the production line, they do make it possible to achieve reproducible experimental results. We return to the discussion of practical aspects of silicide formation at the end of this section.

Typically, the thin films are deposited in a electron beam deposition system under dry vacuum conditions. Although elevated temperature substrate conditions are sometimes employed, often the substrates are maintained at ambient temperature during deposition. The film thickness generally ranges from 500 to 2000 Å. The substrates are in most cases Si wafers with a polished surface of device quality such as those used in integrated circuit device fabrication. Immediately before being placed in the deposition chamber, the wafers are immersed in a buffered HF solution and rinsed by deionized H_2O and blown dry. In certain situations, the wafers are sputter etched before metal deposition when it is suspected that the native oxygen on the wafer may interfere with the silicide formation (10).

10.2.2 Experimental Analysis Techniques

The analysis techniques used to determine the structure of deposited films and phase changes resulting from heat treatment are described in Chapters 5 and 6. It is obvious from those chapters that it is advantageous to apply more than one analysis technique to characterize any given film structure. Rather than describing all the different techniques again, we only mention some of the more common techniques used today in the majority of silicide studies. We do not describe these techniques in detail but try to cover some of the strong points and limitations of these techniques from the standpoint of information that can be gained in respect to silicide formation

Identification of phases is provided by glancing-angle x-ray diffraction. The diffraction patterns can be obtained either in a camera or a diffractometer (11, 12). The advantages of a camera are simplicity and speed. As shown in Figure 10.4, which is an x-ray diffraction pattern of a NiSi film (12), a Read

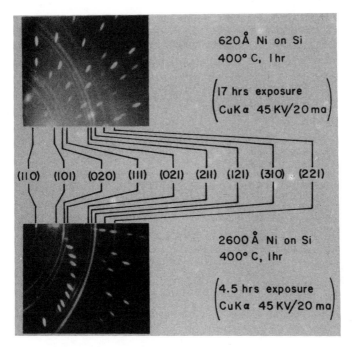

620Å Ni on Si
400° C, I hr

$\left(\begin{array}{c}\text{17 hrs exposure} \\ \text{CuK}\alpha \;\; 45\,\text{KV/20 ma}\end{array}\right)$

(110) (101) (020) (111) (021) (211) (121) (310) (221)

2600Å Ni on Si
400° C, I hr

$\left(\begin{array}{c}\text{4.5 hrs exposure} \\ \text{CuK}\alpha \;\; 45\,\text{KV/20 ma}\end{array}\right)$

Figure 10.4. X-Ray diffraction pattern of a NiSi film obtained with a Read camera. From reference 12.

camera has sufficient sensitivity to identify phases a few hundred Å thick present in films a few thousand Å thick. Quantitative information about peak intensity and line broadening as well as phase identification can be provided by a diffractometer. Figure 10.5 shows a glancing-angle x-ray diffraction spectrum from a Si wafer on which Ni was deposited and then heat treated at 250°C for 24 h. The Ni_2Si phase can be identified by more than 12 of its reflections. Also present in the spectrum are reflections of the unreacted Ni, but no reflections of other phases can be detected.

The sensitivity of both glancing-angle x-ray techniques is approximately 200 Å. Consequently, in utilization of such diffraction techniques we are studying the growth phase of a silicide that has a thickness dimension of over 200 Å. This growth phase is not necessarily the first phase nucleated at the metal–silicon interface during the initial heat treatment. To study the very early stage of growth, transmission electron microscopy (TEM) could be used. At present, such a study has not been carried out, and the TEM technique has been used primarily in microstructure measurement and identification of defects and epitaxial orientation relations (13–15). Figure 10.6 shows transmission electron micrographs and diffraction patterns of

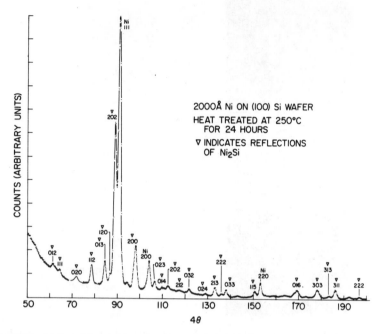

Figure 10.5. X-Ray diffraction spectrum of a heat-treated Ni film on Si obtained with Seemann-Bohlin x-ray diffractometer. From reference 16.

thin Ni_2Si films grown on (100) and (111) Si. In this case, attempts were made to correlate differences in microstructure with different growth rates on (100) and (111) Si.

To obtain growth kinetics and the chemical composition of the phases, it is customary to use Rutherford ion backscattering (16) or Auger electron spectroscopy (AES) combined with sputtering (17). The advantage of Rutherford backscattering is that it provides a fast measurement of the thickness and composition of the growth phase. From spectra such as shown in Figure 10.7, one can measure the energy width of the phase to determine its thickness and the ratio of the spectrum heights in the Ni and Si signals to give the composition ratio. Details of the analysis are given in Chapter 6 and references therein. A limitation of backscattering is that it is difficult to make a positive identification of composition of phases less than 200 Å thick when they are located at the metal–Si interface several thousand Å below the top surface of the metallic film. In this respect again, we investigate the growth phase of silicide formation rather than the nucleation of the silicide.

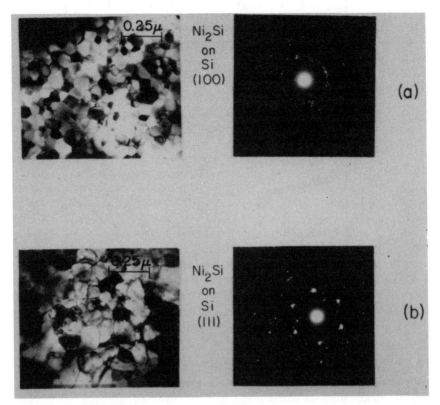

Figure 10.6. Transmission electron micrographs and diffraction patterns of Ni$_2$Si formed by 1600 Å of Ni (*a*) on (100) Si and annealed at 300°C for 5 h and (*b*) on (111) Si and annealed at 325°C for 580 min. The average grain size of Ni$_2$Si is about 600 Å in (*a*) and 1300 to 1400 Å in (*b*). The diffraction pattern in the latter case indicates that the Ni$_2$Si is textured. From reference 15.

The backscattering technique is not sensitive to the presence of low-mass impurities in the silicide. This information as well as the thickness and composition of silicide can be obtained from AES measurements. The Auger depth profiles of a sample of Ni on Si before and after the reaction at 250°C for 1 h to form Ni$_2$Si are shown in Figure 10.8. In these profiles there are problems associated with establishing the absolute depth scale and ensuring that the sputtering process does not change the composition of the eroded surface (see Chapter 6). On the other hand, it is possible to look at the shifts in the peak position in the Auger spectrum to obtain information on the chemical bonds of the species. To date, the fullest advantage of this approach has not been taken.

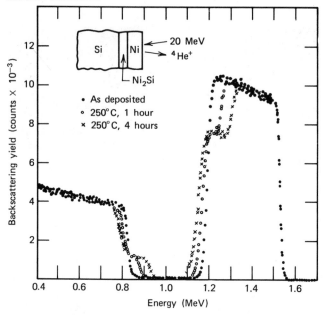

Figure 10.7. Rutherford backscattering spectra of a 2000 Å Ni film on Si before and after annealing at 250°C for 1 and 4 h. From reference 16.

10.2.3 Examples of Silicide Formation

Three examples of silicide formation are given in the following. The formation of nickel silicides is discussed first, followed by hafnium silicide and vanadium silicide. The growths of the silicides in these three examples are quite different. The first phase formation in the nickel case is a nickel-rich silicide, Ni_2Si. In the case of hafnium, it is the monosilicide, HfSi, that forms first. However, vanadium is found to react with Si to form disilicide, VSi_2, without being preceded by the formation of a monosilicide or a metal-rich silicide.

10.2.3.1 Nickel on Silicon.   The formation of nickel silicide between a film of Ni and a Si wafer exhibits a sequential growth of three phases: Ni_2Si, NiSi, and $NiSi_2$ (18). Figure 10.9 shows examples in which Ni_2Si and NiSi are formed. At higher temperatures, $NiSi_2$ is formed; this phase has been shown to grow epitaxially on Si. The growth of the phase Ni_2Si is initiated at the interface between Ni and Si at temperatures from 200 to 350°C, as shown in Figure 10.7. The growth kinetics of Ni_2Si follow a parabolic relation between the thickness of Ni_2Si and the annealing time, as shown in Figure 10.10. One also observes that the growth rate of Ni_2Si at 275°C is a factor of 2 greater on (100) oriented Si substrates than on (111) substrates. TEM

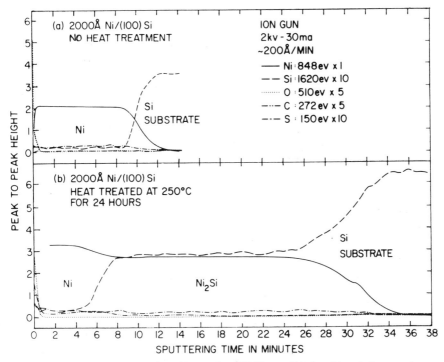

Figure 10.8. Auger depth profiles of a sample of Ni film on Si before (a) and after reaction at 250°C for 24 h (b) to form the silicide Ni₂Si. From reference 16.

studies have indicated, as shown in Figure 10.6, that there is a difference in the microstructure of silicide formed on (100) and (111) Si (15). The activation energy of the formation of Ni_2Si has a value of 1.5 ± 0.2 eV over the temperature range of 200 to 325°C (Fig. 10.11). This figure shows that the activation energies do not differ remarkably between silicides grown on (100) and (111) Si.

Similar studies of kinetics have not been carried out for the growth of the two higher-temperature phases, NiSi and $NiSi_2$. The transformation of Ni_2Si into NiSi is initiated at the interface between Si and Ni_2Si. The transformation is found to accompany a stress change from compression in Ni_2Si to tension in NiSi (19). The stress level is about 10^{10} dynes cm^{-2}. The formation of NiSi is very fast at temperatures above 350°C, and the phase is stable up to 750°C. At temperatures higher than 750°C, epitaxial growth of $NiSi_2$ has been found on (111), (110), and (100) Si. The epitaxial orientation relationship has been determined by channeling studies using MeV He ions and by reflection electron diffraction (18).

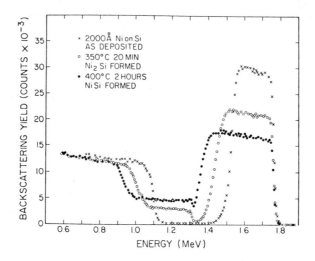

Figure 10.9. Backscattering spectra of a sample of Ni on Si showing the sequential formation of Ni_2Si and NiSi. From reference 18.

10.2.3.2 Hafnium on Silicon.

The reaction of Hf with Si represents a case where the metal-rich silicides shown in the equilibrium phase diagrams are not the growth phases observed (10). The reaction occurs at a much higher temperature than that for the growth of metal-rich silicides, such as Ni_2Si, and the first phase formed is the monosilicide HfSi. The monosilicide is stable from 525 to 700°C, and at temperatures from 750 to 990°C it transforms to the disilicide $HfSi_2$.

The growth kinetics for the formation of HfSi are shown in Figure 10.12 for a 1500 Å Hf film on (100) and (111) Si. The growth follows a parabolic relation, and there is no pronounced influence of the substrate orientation on the formation kinetics. The activation energy for HfSi growth was determined to be 2.5 eV, as shown in Figure 13. This activation energy is about 1 eV larger than that found for Ni_2Si.

The formation of $HfSi_2$ is different from that of HfSi, which forms as a uniform layer between the hafnium and silicon. The disilicide $HfSi_2$ does not form a uniform layer but rather grows in randomly localized regions throughout the HfSi (10). Because of this nonuniform growth, it was not possible to determine the growth kinetics.

10.2.3.3 Vanadium on Silicon.

The reaction of V with Si produces the disilicide as the only growth phase (20, 21). The initial growth of VSi_2 follows a linear growth rate as shown in Figure 10.14. For longer times there is a deviation from the linear growth rate. Because it was found that the

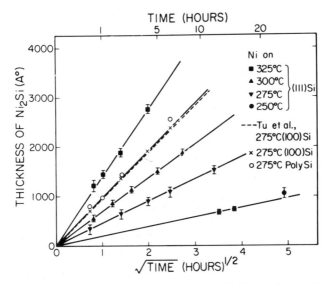

Figure 10.10.   Data of parabolic growth of Ni_2Si. From reference 15.

presence of oxygen can slow down the growth rate of VSi_2, it was suggested that the deviation shown in Figure 10.14 was due to contamination of oxygen introduced into the film during deposition and subsequent anneals. A general consequence of a linear growth behavior is that the linear region will gradually change to a parabolic region when the growth becomes limited by diffusion through the growth layer. However, in the case of VSi_2 growth, experiments with films containing different impurities showed that the departure from linear growth was due to the presence of impurities.

The formation of vanadium silicide represents the situation where the end phase, the disilicide, is also the dominant growth phase. In the other examples cited in the above, the formation of disilicide was preceded by the formation of a monosilicide or a metal-rich silicide. Of the examples shown, the growth of VSi_2 was the only case in which a linear rather than a parabolic growth rate was found.

10.2.4   Summary of Silicide Formation

In the phase diagrams of metal–Si systems, in general, there are more than three silicides present. One general observation is that not all the equilibrium phases are present as dominant growth phases during silicide formation in thin film systems. We do not imply that some of the equilibrium phases can not nucleate but only that they do not grow to macroscopic dimensions greater than the few hundred Å which can be detected by the techniques

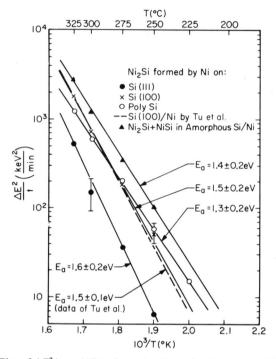

Figure 10.11. Plot of $\Delta E^2/t$ vs. $1/T$ to determine the activation energy of growth of Ni_2Si. From reference 15.

mentioned in Section 10.2.2. The examples shown in the preceding sections were chosen as representative of the three broad classes of silicide formation observed in structures composed of metal thin films on single-crystal silicon: the metal-rich silicide, typically M_2Si, the monosilicide MSi, and the disilicide MSi_2.

Table 10.1 shows the general pattern of silicide formation. The silicides listed in the right-hand columns are chosen as specific examples of the phase formation for the elements shown in the left column (with the exception of Zr, whose behavior has not been examined). The phase formation proceeds from left to right in each row, with the more metal-rich silicide forming first. In all cases the final phase (most right-handed entry in each row in the table) represents the end phase between Si and the metal shown in the equilibrium phase diagram. For Mg, only one phase is found in the phase diagram. For the refractory metals, many phases exist in their phase diagram (22), yet only the disilicide is found.

10.2.4.1 Metal-Rich Silicides. A survey of silicide formation in metal-rich silicides is given in Table 10.2. One notes that the growth kinetics follow a

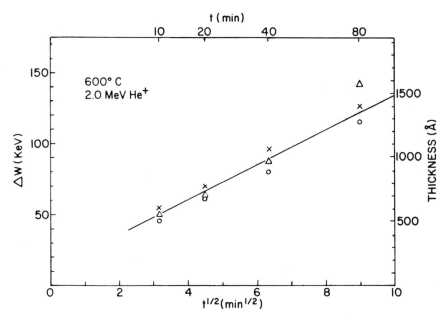

Figure 10.12. Width ΔW of the plateaus in the Hf peak as a function of (time)$^{1/2}$ for three 1500 Å Hf films deposited on $\langle 100 \rangle$ ($\times$) and $\langle 111 \rangle$ (O) Si at $\sim 200°C$ and on $\langle 100 \rangle$ Si at $\sim 400°C$ ($\triangle$). From reference 10.

parabolic law with an activation energy around 1.5 eV. The formation temperature for all five silicides starts around 200°C, with the notable exception of Co_2Si, which starts forming at 350°C. At the present time, we do not know the reason for this high formation temperature and note that silicide formation of Co_2Si has been checked independently in three different laboratories (Caltech, IBM, and Phillips). On the other hand, all the surface preparation and deposition systems were basically the same, and it is possible that an alternative preparation technique might lead to a lower formation temperature.

10.2.4.2 Monosilicides. The formation of monosilicides shown in Table 10.3 is also characterized by a parabolic growth rate but with an activation energy of 1.6 to 2.5 eV. The latter number appears higher than the general trend. The formation temperature for monosilicides is higher than that for the metal-rich silicide and generally occurs above 350°C. A notable exception is PdSi, which forms above 700°C. This must imply that the precusory phase Pd_2Si is extremely stable. We comment that in at least three cases, Ni, Co, and Pt, it is possible to have the metal-rich silicide formed as a layer between the metal and the monosilicide. In other cases, such as Hf and Rh, the

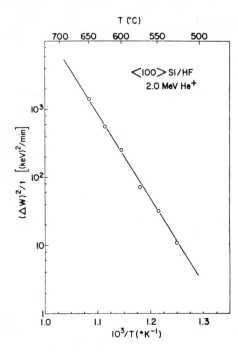

Figure 10.13. Plot of $(\Delta W)^2/t$ vs. $1/T$ to determine the activation energy of growth of HfSi. From reference 10.

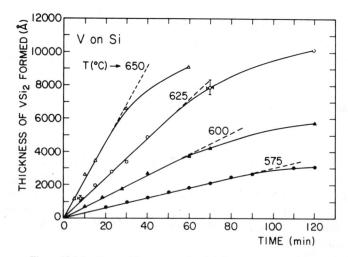

Figure 10.14. Data of linear growth of VSi_2. From reference 21.

TABLE 10.1. SILICIDE FORMATION BY CONTACT REACTION

Elements	Silicides		
Mg	Mg$_2$Si (only phase)		
Pt, Pd	Pt$_2$Si	PtSi (end phase)	
Ni, Co	Ni$_2$Si	NiSi	NiSi$_2$ (end phase)
Ti, (Zr), Hf		HfSi	HfSi$_2$
Fe, Rh, Mn			
V, Nb, Ta			VSi$_2$
Cr, Mo, W			

monosilicide is the first growth phase observed in the silicide. In the case of Ti, it is the disilicide that is generally observed as the growth phase. However, x-ray diffraction has recently indicated that TiSi is formed prior to the growth of the disilicide (23).

10.2.4.3 Disilicides. Table 10.4 summarizes the formation of the disilicide phase observed in the growth of silicide layers. The formation temperature is around 600°C, except for CrSi$_2$, which forms around 450°C. The activation energy of formation is high, ranging from 1.7 to 3.2 eV. The disilicides of three metals, Ni, Co, and Fe, are found to grow epitaxially on Si substrates. In these cases, the disilicide has a close lattice match with the Si.

The disilicide of the refractory metals tends to have a linear growth curve in the early stage of reaction when the disilicide is the first growth phase. It is difficult in some cases to obtain a clean measurement of the growth kinetics because of the influence of oxygen contamination as found for VSi$_2$ and CrSi$_2$. For refractory metals, the presence of a thin interfacial oxide layer between the metal and Si can impede silicide formation, as has been directly shown in the case of Cr on Si.

10.2.5   Practical Aspects

In this section, we would like to return to a few practical aspects of silicide formation that are encountered either in the production environment or in the choice of a metallization scheme. We do not cover all aspects since each device has its own particular requirements. Further, we do not cover some areas such as corrosion, adhesion, and line definition (the latter is important in lithographic and lift-off steps). Instead, we treat two topics about which

TABLE 10.2. FORMATION OF METAL-RICH SILICIDE M_2Si

Silicide	Formation temperature, °C	Activation energy of growth, eV	Growth rate	Crystal structure	Melting point, °C	Density,[a] g cm^{-3}	References
Ni_2Si	200 ~ 350	1.5	$t^{1/2}$	orthorhombic $PbCl_2$	1318	7.23	15, 16, 19
Pd_2Si	100 ~ 700	1.3–1.5	$t^{1/2}$	hexagonal Fe_2P	1330		4, 13, 14, 17 29, 30, 31
Pt_2Si	200 ~ 500	1.1–1.6	$t^{1/2}$	tetragonal $CuAl_2$	1100		32, 33, 34
Mg_2Si	⩾200			cubic CaF_2	1102		35
Co_2Si	350 ~ 500	1.5	$t^{1/2}$	orthorhombic $PbCl_2$	1332		36, 37

[a]Density of Si is 2.33 g cm^{-3}.

TABLE 10.3. FORMATION OF MONOSILICIDE MSi

Silicide	Formation temperature, °C	Activation energy of growth, eV	Growth rate	Crystal structure	Melting point, °C	Density,[a] g cm^{-3}	References
PtSi (end phase)	≥300	1.6	$t^{1/2}$	orthorhombic MnP	1229		1, 32, 33, 34, 38
PdSi (end phase)	≥700			orthorhombic MnP	1100		31
NiSi	350–750			orthorhombic MnP	992		18, 19, 39
CoSi	425–500	1.9	$t^{1/2}$	cubic B20	1460	6.5	36, 37
FeSi	450–550	1.7	$t^{1/2}$	cubic B20	1410		40
RhSi	377			cubic B20			41, 42, 1
HfSi	550–700	2.5	$t^{1/2}$	orthorhombic FeB			10
TiSi	500			orthorhombic FeB	1760	4.21	23
MnSi	400–500			Cubic B20	1275		43

[a]Density of Si is 2.33 g cm^{-3}.

377

TABLE 10.4. FORMATION OF DISILICIDE MSi_2

Silicide	Formation temperature, °C	Activation energy of growth, eV	Growth rate	Crystal structure	Melting point, °C	Density,[a] g cm^{-3}	References
$TiSi_2$	600			orthorhombic	1540	3.9	23, 44
$ZrSi_2$	700			orthorhombic	1700		1,43
$HfSi_2$	750			orthorhombic	1950		10
VSi_2	600	2.9, 1.8	t and $t^{1/2}$	hexagonal	1750		20, 21
$NbSi_2$	650			hexagonal	1930		27, 45
$TaSi_2$	650			hexagonal	2200		46, 46a
$CrSi_2$	450	1.7	t	hexagonal	1550	−4.91	43, 44
$MoSi_2$	525	3.2	t	tetragonal	2050		44, 47, 47a
WSi_2	650	3	t and $t^{1/2}$	tetragonal	2165		48, 47a
$NiSi_2$	750			cubic (CaF_2)	993		18
$CoSi_2$	550			cubic (CaF_2)	1326		36, 37
$FeSi_2$	550			tetragonal	1212	4.54	40
$MnSi_2$	800			tetragonal	1150		43

[a]Density of Si is 2.33 g cm^{-3}.

there have been some published work related to the basics of silicide formation as discussed in the preceding sections.

10.2.5.1 Oxygen Ambient. Oxygen and water vapor are universal contaminants in deposition systems and thermal processing ambients. We have earlier pointed out that the presence of oxygen in a metal film can retard the growth of the silicide. A more general phenomenon is the presence of some oxygen in the heat treatment processing atmosphere. Since silicon oxidizes readily, one may anticipate that oxide layers are formed at the top surface of silicide layer when oxygen or water vapor is present.

It has been clearly established that oxide layers are formed near the top surface of PtSi (24). As we have pointed out in Section 10.1, the presence of this oxide layer is often used to protect the PtSi during etching processes.

If excessive amounts of oxygen are present in the heat treatment ambient, it is possible to form an oxide layer well below the Pt surface (7). The structure of the silicide layer is shown in Figure 10.15 for silicide layers either formed in an oxygen ambient or after annealing in oxygen. The presence of such oxide layers below the Pt surface can lead to undesirable side effects such as poor adhesion and poor electrical contact. This outer Pt and oxide layer must be etched away before the next metal layer is deposited. The formation of oxide layers is a rather general phenomenon during silicide formation, and one

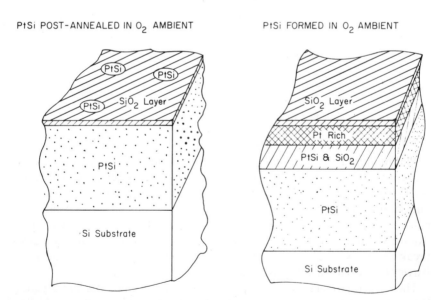

Figure 10.15.   Schematic diagrams of oxide layer formation during the growth of PtSi in an oxygen ambient or postannealed in oxygen. From reference 7.

must anticipate to include oxide layer removal steps in the development of the metallization steps.

10.2.5.2 Reaction of Oxide Layer.   Many of the transition elements can form silicides when in contact with SiO_2 as well as Si. We can broadly classify the reaction between metals and SiO_2 as shown in Table 10.5. Gold shows no tendency to form oxide, and Au–silicides are unstable. However, Au in contact with thin oxide layers can cause dissolution of the oxide when heated above the Au–Si eutectic point. The mechanism is that Au penetrates the thin oxide and reacts with the underlying Si layer to form a eutectic in liquid form which wets the SiO_2 and leads to dissolution (25, 26). Aluminium, on the other hand, can form an oxide layer, and when Al is deposited on SiO_2, strong adhesion is obtained. We suspect that this is due to the reaction between Al and SiO_2 to form strong Al–O bonds at the interface.

TABLE 10.5. METAL–SiO_2 REACTION

Metallic element M	Metal oxide	Metal silicide
Au[a]	no	no
Al, Sn, Pb	yes	no
Pt	no	yes
Fe, Co, Ni Cr, Mo, W V, Nb, Ta Ti, Zr, Hf	yes	yes

[a]Forms a metastable silicide (66).

Platinum and Pd themselves do not form oxide layers easily, and consequently their adhesion to thick layers of SiO_2 is rather poor. However, when these metals are in contact with a thin native oxide on Si, silicide formation can still occur at low temperatures. In these cases we believe the reaction is similar to that with Au in that the Pt or Pd can penetrate the thin oxide layer and form a silicide at the Si–oxide interface and at the same time break up the oxide. However, if the oxide is too thick, it will form a diffusion barrier to prevent silicide formation.

The refractory metals such as Cr, Ti, and V form both oxides and silicides. When deposited on thick oxide layers, they develop strong adhesive bonds. In fact, Cr and Ti are often used as a glue layer between the oxide layer and less adhesive metals. On the other hand, since these metals do form oxide layers, the metal oxide becomes a diffusion barrier for further reaction between the metal and SiO_2. It has been shown that silicide layers can be formed on thick SiO_2 layers but generally at temperatures 100 to 200°C above that where the metal reacts with bare Si, namely, 700 to 900°C rather than 600°C. In addition (20, 21, 27), the silicide formed is generally metal rich, for example, V_3Si is formed rather than VSi_2. Table 10.6 summarizes the reaction products of Ti, V, and Nb on Si and SiO_2 substrates. Marker experiments (see next section) on the reaction between V and SiO_2 has shown that V is moving species in V_3Si formation (28) rather than Si, as was found in VSi_2. We believe that this represents a rather general condition where the metal species must diffuse through the growing metal-rich silicide so that the metal atoms can reach the reaction front at the silicide–SiO_2 interface.

TABLE 10.6. COMPARISON OF REACTION PRODUCTS OF Ti, V, AND Nb ON Si AND SiO_2 SUBSTRATES HEAT TREATED UNDER VACUUM[a]

Elements		Si	SiO_2	
			Intermediate layer	Surface layer
Ti	Backscattering	Ti:Si=0.5	Ti:Si=1.6	Ti:O~1
	X-Ray diffraction	$TiSi_2$	Ti_5Si_3	unidentified
	Reaction temperature, °C	>500	>700	
V	Backscattering	V:Si=0.5	V:Si=3	V:O~1
	X-Ray diffraction	VSi_2	V_3Si	$V_5Si_3 + V_2O_5$
	Reaction temperature, °C	>500	>700	
Nb	Backscattering	Nb:Si~1.7	Nb:Si~1.7	Nb:0~1
	X-ray diffraction	$NbSi_2$	Nb_3Si (fcc)	unidentified
	Reaction temperature, °C	>650	>900	

[a]From reference 27.

10.3 MARKER MOTION

10.3.1 Ideal Marker

In our discussion so far we have not identified whether Si or metal atoms move across the silicide layer during the growth of the layer. The fact that growth occurs implies that one or both of the species are transported to either the metal–silicide or Si–silicide interface. However, it is impossible to determine the identity of the diffusing species without some form of marker experiments.

The initial experiments with a diffusion marker were carried out in bulk samples by Kirkendall and Smigelskas (49). In these bulk diffusion couples, the faster of the diffusing species is identified by the direction of displacement of embedded wire markers mils in diameter. With thin films a few thousand Å thick, the concept of embedded wire marker must be scaled down in dimensions. One requires markers that are a few hundred Å in dimension that do not influence the kinetics of silicide growth. Ideally, the marker should also be inert, that is, should not react with Si or the metal at the silicide growth temperature and should remain immobile as the diffusing species streams by.

An additional constraint is that the marker should be located within the silicide layer to avoid possible influence due to the presence of interfaces. As shown in Figure 10.16, if the metal atoms are the diffusing species, the marker will be displaced toward the top surface of the film. Conversely, if Si atoms are predominantly the moving species, the marker will be displaced deeper into the sample. If both species are moving, it is possible to have relatively small displacement of the marker during the silicide growth. Under this condition, as shown by Darken's analysis (50), the relative displacement of the marker can be used to determine the difference in the fluxes of the two diffusing species.

10.3.2   Interface Drag

If the marker is located near the interface, the marker can be dragged along by the interface irrespective of the direction of the net flux of the diffusing

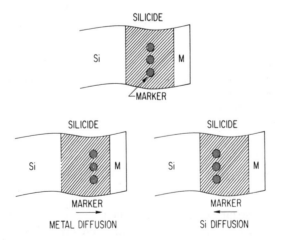

Figure 10.16.   Schematic diagram of marker motion in silicide formation.

species. Figure 10.17 shows an idealized case with a spherical marker located at the interface between the Si and the silicide layer (50a). Since the interface moves as a result of reactions occurring at the interface, the question whether or not the marker can be buried within the silicide depends upon the relative magnitude of interfacial energies σ per unit area in the vicinity of the marker. If the following energy equation is satisfied, the marker will be dragged by the advancing interface:

$$2\pi r^2 \sigma_1 + \pi r^2 \sigma_2 \leqslant 2\pi r^2 \sigma_3 \tag{1}$$

where r is the radius of the spherical marker and σ_1, σ_2, and σ_3 are the specific interfacial energies between Si and marker, Si and silicide, and silicide and marker, respectively. In other words, even if the metal atoms are the diffusing species, the marker can be dragged deeper into the sample if too much energy is required to break away from the interface. An associated phenomenon is found in oxidation of silicon where n-type dopants are dragged by the advancing oxide front deeper into the sample (51, 51a). This snow plot effect is caused by the small solubility of the dopant within the oxide layer. However, at the relatively low temperatures of silicide formation it is not appropriate to consider the solubility of the marker, whereas in high-temperature oxidation the segregation effect can play a significant role.

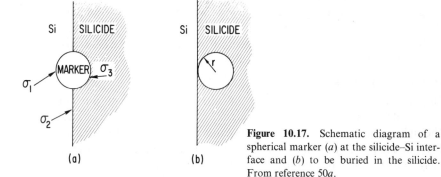

Figure 10.17. Schematic diagram of a spherical marker (a) at the silicide–Si interface and (b) to be buried in the silicide. From reference 50a.

Let us consider a marker located at the Si–silicide interface that moves with the interface. We shall show that it is impossible to determine which species is moving across the silicide layer during growth. If Si is the dominant diffusing species, growth of the silicide occurs at the metal–silicide interface and some motion of Si occurs at the boundary of the marker, as shown in Figure 10.18a. This is the situation predicted even if no marker drag occurs. If the metal atoms are the diffusing species and interface drag occurs, silicide growth

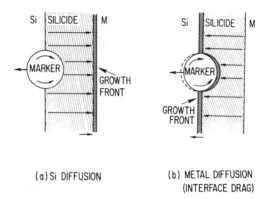

Figure 10.18.   Schematic diagram showing the dragging of a diffusion marker by the silicide–Si interface: (*a*) Si is the dominant diffusing species; (*b*) metal atom is the dominant diffusing species. From reference 50*a*.

takes place at the silicon–silicide interface and also at the marker–silicide interface. To accomplish this, some Si transport must occur again around the boundary of the marker, as shown in Figure 10.18*b*. The net result, however, is that the marker is displaced deeper into the sample as would have occurred if Si atoms were the diffusing species across the silicide layer.

Similar arguments can be applied to the metal–silicide interface. If marker drag occurs, it is still not possible to determine which species is moving across the silicide. These arguments imply that if the marker position coincides with the interface position during silicide growth, one must be cautious about interpreting the identity of the diffusing species.

10.3.3   Actual Experiments

In practice, there are experimental difficulties associated with introducing a detectable marker into a thin film structure. One of the difficulties is to achieve the right dimension of the marker, and another is to avoid forming an oxide layer at the position of the marker. As an example of the latter case, it is not easy to form a silicide, to introduce a marker, and then to redeposit a metal for subsequent reaction. This approach is generally tried first. However, it has been observed that the reaction stops at the ubiquitous oxide layer located at the interface between the silicide and the newly deposited metal. We can anticipate that sputter cleaning before metal deposition could be used to overcome the problem associated with the presence of an oxide. To date, two concepts have been applied successfully to inert marker studies: implantation of inert gas atoms (52) and deposition of a discontinuous W layer (53). A third approach, the use of radioactive Si markers, is discussed in Section 10.3.3.3.

10.3.3.1 Implanted Marker.   In this work inert gas atoms such as Ar or Xe are implanted into the Si surface. The Si sample must then be annealed to temperatures above 600°C to reorder the implanted layer and remove the major amount of implantation-induced defect structures. During this heat treatment, it has been shown for the case of Xe that Xe bubbles of 50 to 100 Å in diameter are formed (54). After these implantation and heat treatment steps, the Si surface must be etched to remove any oxide or hydrocarbon layers. Following the surface cleaning step, a metal layer is deposited on the sample. Figure 10.19 shows in the upper portion the backscattering spectra from a Xe marker-implanted sample before and after the formation of Ni_2Si. After silicide formation, the Xe marker is buried within the silicide and displaced toward the surface. The lower portion of Figure 10.19 shows the amounts of the displacement of the Xe marker versus the silicide thicknesses. These results show that interface drag did not occur and that Ni is the dominant diffusing species.

If Si is the moving species, when the advancing silicide front reaches the implanted marker, the marker will move deeper into the sample at the position of the interface. To determine the identity of the moving species, it is then necessary to implant markers into the deposited metal layer. Then, during silicide growth, as the silicide–metal interface advances to the marker, if the marker becomes buried, its position will be shifted deeper into the sample, indicating that Si is the dominant diffusing species.

With high-energy implantation, 0.5 to 10 MeV, it is possible to partially form the silicide and then implant through the remaining metal into the silicide. To date, there have been no reported results of this method of marker implantation within the silicide layer.

One of the disadvantages in using backscattering techniques to detect the implanted marker is that the atomic mass of the marker atoms must be selected so that the signal from the marker occurs in a region in the energy spectrum that is free from the signal from the Si or the metal atoms. It is also necessary to select the right thickness of the deposited metal to avoid signal interference. Preliminary experiments have shown that these constraints can be lifted by the use of Auger depth profiling measurements.

10.3.3.2 Deposited Marker.   Another approach is to deposit small and isolated islands of an inert metal on the Si surface followed by a cleaning of the Si surface and deposition of the silicide-forming metal. The islands of the metal marker must have sufficient adherence to the Si so that they remain intact during the surface cleaning step. These islands must also be small in size and should not react with either the Si or the silicide-forming metal at the growth temperature.

Van Gurp et al. (53) deposited a thin Sn film which forms islands upon

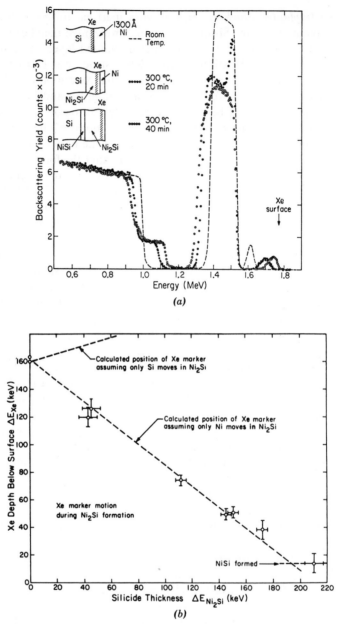

Figure 10.19. (*a*) Marker motion during the formation of Ni₂Si. From reference 16. (*b*) Amount of Xe marker displacement against the silicide thickness. From reference 16.

deposition, then deposited a 30 Å W film, and finally dissolved the Sn islands. This left patches of W in the regions between the original Sn islands. After this, Co was deposited on the sample surface, and the formation of Co_2Si was carried out at temperatures from 400 to 500°C. The growth kinetics in this sample were the same for those samples without a W layer, except that the growth started after an incubation period of 1 to 10 min. During formation of the silicide layer, the position of W marker shifted toward the surface at the position of the Co_2Si–metal interface. This experiment shows that it is possible to deposit a metal film which does not interfere with the silicide formation. However, further experiments are required to show that interface drag effects do not control the marker motion.

10.3.3.3 Radioactive Marker.

A tracer technique using radioactive ^{31}Si has been applied to study the nature of mass transport during the solid phase epitaxial growth of Si on $\langle 100 \rangle$ Si with amorphous Si as a source layer and Pd_2Si compound as an intermediate layer (55). The tracer was obtained by neutron activation in a nuclear reactor and has a half-life of 2.62 h by emitting beta radiation. By determining the position change of the tracer before and after the epitaxial growth, it is found that the supply of Si to the growth comes indirectly from the dissociation of the Pd_2Si rather than directly from the amorphous Si. For readers who are interested in this technique, a full description is presented in Chapter 12.

Whether the tracer Si can be used as a diffusion marker in silicide formation is an interesting question, and some preliminary results of using tracer Si as a diffusion marker in the formation of Pd_2Si, Ni_2Si, and Co_2Si have been obtained (56). These results showed that the tracer Si has taken part in the reaction of silicide formation, so it is not an inert marker. Nevertheless, it is found that at low temperatures if the interdiffusion of Si takes place by grain boundary diffusion, the tracer Si within silicide grains apparently remains inert. Preliminary experiments (56) indicate that it is possible to differentiate mass transport in the lattice from that along grain boundaries by determining the position change of the tracer atoms. The use of radio tracers is in the developmental stage; however, it appears that the information gained from tracer studies along with that from inert markers offers considerable insight into the transport mechanism.

10.3.3.4 Intermediate Silicide Layer.

There have been several cases where an intermediate silicide layer is formed such as NiSi between Ni_2Si and Si or Pd_2Si between Si and $CrSi_2$ (57). It is tempting to describe the formation of the outer silicide layer, for example, $CrSi_2$ on top of Pd_2Si, as being due to the transport of Si through the intermediate silicide, Pd_2Si in this case. This conclusion cannot be supported because it is possible that the intermediate silicide can dissociate giving up Si atoms to form $CrSi_2$ and releasing Pd atoms to diffuse through the remaining Pd_2Si to form a new layer of Pd_2Si at the $Si–Pd_2Si$ interface. Such a dissociation mechanism is discussed in Chapter 12.

10.3.4 Diffusing Species

Table 10.7 lists the silicides in which the diffusing species have been identified by means of marker experiments. From the table, one notices that metal atoms diffuse in the cases where metal-rich silicides M_2Si are formed. In fact, for the three silicides Ni_2Si, Mg_2Si, and Co_2Si, the metal is the dominant diffusing species. Conversely, in the case where the disilicide is formed, silicon is the diffusing species. We believe that the same pattern will be found in other disilicides if marker experiments are carried out in such cases.

For the case of monosilicides, Si is again the dominant diffusing species. For the three cases listed, the monosilicide was the first phase formed. Marker experiments have not yet been performed in monosilicides that form after the metal-rich silicide.

There is a disadvantage to the silicide growth when one species dominates in the diffusion. Vacancy accumulation should occur at the Si–silicide interface if Si is the diffusing species. This might lead to poor silicide adhesion and impaired electrical contact. Such an effect has been found with W on SiGe alloys (60). However, we are not aware of other studies of adhesion carried out on disilicides.

TABLE 10.7. MARKER EXPERIMENTS IN SILICIDE FORMATION

Element	Silicide	Dominant diffusing species	References
Ni	Ni_2Si	Ni	16, 52
Mg	Mg_2Si	Mg	35
Co	Co_2Si	Co	53
Pd	Pd_2Si	Pd, Si	35
Pt	Pt_2Si	Pt, Si	9, 34
Fe	FeSi	Si	40
Hf	HfSi	Si	58
Rh	RhSi	Si	59
Ti	$TiSi_2$	Si	35
V	VSi_2	Si	35

10.3.5 Marker Analysis

In the last section, the application of marker studies has been used mainly to identify the dominant diffusing species during silicide formation. It is well known from Darken's analysis (50) of marker motion in bulk diffusion couples (49, 61, 62) that the intrinsic diffusion coefficient of each species can be determined from marker measurements. Therefore, more information than identification of the dominant diffusion species can be extracted from a marker study. For details of Darken's analysis, readers are referred to the original paper (50), to Chapter 5 in Shewmon's book (63), or to Manning's treatment (64). In extending Darken's analysis to marker studies in silicide formation, we realize that there are two outstanding features of silicide formation that require attention (50a). First, we are dealing with a diffusion couple that forms compounds rather than solid solution. At the interfaces where discontinuity of structure and composition occurs, the lattice is being destroyed and created at the same time, so the continuity equation or Fick's second law cannot be applied there. Instead, we must use a growth equation. Second, the interdiffusion coefficient $\tilde{D}$ in Darken's analysis is measured typically by the use of a Boltzmann-Matano solution, so the equation

$$\tilde{D} = D_1 \frac{C_2}{C} + D_2 \frac{C_1}{C} \qquad (2)$$

can be combined with the marker motion equation,

$$v = \frac{1}{C} (D_1 - D_2) \frac{dC_1}{dx} \qquad (3)$$

to unravel D_1 and D_2 from $\tilde{D}$. Here, D_1 and D_2 are the intrinsic diffusion coefficient of component 1 and 2 in the alloy of composition $C = C_1 + C_2$, respectively, where C_1 and C_2 are the concentration of component 1 and 2 in alloy C. The quantity v is the drift velocity measured by a marker in a coordinate system with the origin located at one end of the couple where no interdiffusion takes place. However, in a diffusion-controlled growth of a silicide, we do not measure $\tilde{D}$ by the use of a Boltzmann-Matano solution, but rather we measure typically a kinetic parameter A from the rate of growth of the layer based on the relation

$$x^2 = At \qquad (4)$$

where x and t are layer thickness and time of growth, respectively. In general, A is similar to a diffusion coefficient; nevertheless, it does not equal $\tilde{D}$. Thus, we must find out how A is related to the intrinsic diffusion coefficients D_1 and D_2. We should point out that Kidson (65) has shown an analysis of layer growth and obtained an analytic solution of A as a function of $\tilde{D}$. Although it is a rather complicated solution, it relates A to D_1 and D_2 indirectly through

$\tilde{D}$. In the following, we take Kidson's approach of layer growth and extend it to include marker motion.

In Figure 10.20 a schematic phase diagram and concentration profile of the metallic element during the formation of a silicide layer between the metal and Si is shown. In the figure and equations that follow, α, β, and δ represent the Si, silicide, and metal phase, respectively. Also the superscript 1 stands for metal and 2, for silicon. At the interface between α and β, the growth equation takes the following form:

$$(C^1_{\beta\alpha} - C^1_{\alpha\beta})\left(-\frac{d\xi_{\alpha\beta}}{dt}\right) = J^1_{\beta\alpha} - J^1_{\alpha\beta} \tag{5}$$

where

$$J^1_{\beta\alpha} = -D^1_\beta\left(\frac{dC^1_{\beta\alpha}}{dx}\right)_{\alpha\beta} + v_{\alpha\beta}C^1_{\beta\alpha} \tag{6}$$

and

$$J^1_{\alpha\beta} = -D^1_\alpha\left(\frac{dC^1_{\alpha\beta}}{dx}\right)_{\alpha\beta} + v_{\alpha\beta}C^1_{\alpha\beta} \tag{7}$$

are the flux terms. The term $d\xi_{\alpha\beta}/dt$ is the growth velocity of the $\alpha\beta$ interface and $v_{\alpha\beta}$ is the drift velocity of the silicide layer as a whole measured at the $\alpha\beta$ interface. The quantities D^1_α and D^1_β are intrinsic diffusion coefficients of the metal in α and β, respectively. Substituting Eqs. 6 and 7 into Eq. 5,

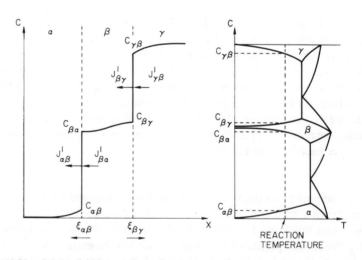

Figure 10.20.   Schematic binary phase diagram and the concentration profile of metallic elements during silicide formation. From reference 50a.

we have

$$-\frac{d\xi_{\alpha\beta}}{dt}=\frac{-D_\beta^1\left(\dfrac{dC_{\beta\alpha}^1}{dx}\right)_{\alpha\beta}+D_\alpha^1\left(\dfrac{dC_{\alpha\beta}^1}{dx}\right)_{\alpha\beta}}{C_{\beta\alpha}^1-C_{\alpha\beta}^1}+v_{\alpha\beta} \tag{8}$$

Similarly, at the $\beta\gamma$ interface, we have

$$\frac{d\xi_{\beta\gamma}}{dt}=\frac{-D_\gamma^1\left(\dfrac{dC_{\gamma\beta}^1}{dx}\right)_{\beta\gamma}+D_\beta^1\left(\dfrac{dC_{\beta\gamma}^1}{dx}\right)_{\beta\gamma}}{C_{\gamma\beta}^1-C_{\beta\gamma}^1}+v_{\beta\gamma} \tag{9}$$

Since the concentration variation across the silicide layer is not large, the drift velocities $v_{\alpha\beta}$ and $v_{\beta\gamma}$ are practically equal. Then, if we follow Kidson's analysis and take

$$K_{ij}^1=\frac{dC_{ij}^1}{d\lambda}$$

to be constant at the ij interface, where $\lambda=x/\sqrt{t}$, we obtain, by combining Eqs. 8 and 9 and integrating, the following equation:

$$(\xi_{\beta\gamma}-\xi_{\alpha\beta})=2\sqrt{t}\left[\frac{-(D_\gamma K)_{\gamma\beta}^1+(D_\beta K)_{\beta\gamma}^1}{C_{\gamma\beta}^1-C_{\beta\gamma}^1}+\frac{-(D_\beta K)_{\beta\alpha}^1+(D_\alpha K)_{\alpha\beta}^1}{C_{\beta\alpha}^1-C_{\alpha\beta}^1}\right]+x_m \tag{10}$$

where $\xi_{\beta\gamma}-\xi_{\alpha\beta}=W_\beta$ is the layer thickness of the β phase and x_m is the marker displacement ($2vt=x_m$).

Equation 10 can be simplified in certain special cases. In the case where one species is found to be the dominant diffusion species, such as Ni during the growth of Ni_2Si, the growth takes place mainly at one of the interfaces, the $\alpha\beta$ interface. Then, we can assume $J_{\alpha\beta}^1=J_{\beta\gamma}^1$ and neglect the first term on the right-hand side of Eq. 10. Furthermore, if a plot of $\ln(W_\beta-x_m)$ versus $1/T$ shows that the relation is linear, it means either the activation energies are about the same or one of the terms is dominant. Now, if we assume that $D_\beta^1K_{\beta\alpha}^1$ is dominant over $D_\alpha^1K_{\alpha\beta}^1$, we obtain a simplified form of Eq. 10:

$$W_\beta-x_m=2t^{1/2}\frac{-D_\beta^1K_{\beta\alpha}^1}{C_{\beta\alpha}^1-C_{\alpha\beta}^1} \tag{11}$$

When the same assumptions are applied to Kidson's equation of layer growth (65), we again obtain a simplified form for the layer growth where one species dominates the diffusion:

$$W_\beta=2t^{1/2}\frac{-\tilde{D}_\beta K_{\beta\alpha}^1}{C_{\beta\alpha}^1-C_{\alpha\beta}^1} \tag{12}$$

where $\tilde{D}_\beta$ has the following form (see Eq. 2):

$$\tilde{D}_\beta = D_\beta^1 \frac{C_\beta^2}{C_\beta} + D_\beta^2 \frac{C_\beta^1}{C_\beta}$$

and

$$C_\beta = C_\beta^1 + C_\beta^2$$

By subtracting Eq. 11 from Eq. 12, we have

$$v = \frac{C_\beta^1}{C_{\beta\alpha}^1 - C_{\alpha\beta}^1} \frac{1}{C_\beta} (D_\beta^1 - D_\beta^2) \left(\frac{dC_{\beta\alpha}^1}{dx}\right)_{\alpha\beta} \tag{13}$$

We note that $C_\beta^1 \cong C_{\beta\alpha}^1 - C_{\alpha\beta}^1$, so Eq. 13, in fact, has the same form as Eq. 3. What is being shown here is that Eqs. 11 and 12 can be reduced back to Eqs. 2 and 3.

Combining Eqs. 11 and 12, we can determine D_β^1 and D_β^2 provided that the other parameters can be measured. Among them, the measurement of $K_{\beta\alpha}^1$ may require some clarification. Since it represents the driving force of the diffusion, it is directly related to the formation energy ΔH of silicide in the case of silicide formation. The formation energy of silicide will be discussed in the next section.

Using the simple relation that $\mu = RT \ln C$, where μ is chemical potential in the usual sense, we obtain

$$K_{ij} = \frac{\sqrt{t}}{RT} C_{ij} \left(\frac{\Delta\mu}{\Delta x}\right)_{ij} \tag{14}$$

For a first-order approximation, we can take $\Delta\mu$ to be ΔH, the formation energy of silicide, as given in Table 10.9. Then, in the case of growth of Ni_2Si ($\Delta H = 11.2$ kcal mole^{-1}) as reported in Figure 2a in reference 52, we found (50a) the diffusivities of Ni and Si in Ni_2Si at 325°C to be

$$\begin{aligned} D_{Ni} &= 1.7 \times 10^{-14} \text{ cm}^2 \text{ s}^{-1} \\ D_{Si} &= 0.6 \times 10^{-14} \text{ cm}^2 \text{ s}^{-1} \end{aligned} \tag{15}$$

which show that in Ni_2Si the Ni atoms diffuse faster than the Si atoms by a factor of 3. If marker motion data at other temperatures become available, we can then determine the diffusion activation energy and preexponential factor for the Ni and Si atoms separately. On the other hand, using the activation energy of 1.5 eV given in Table 10.2 for interdiffusion in Ni_2Si, we obtain a diffusivity at 325°C of the same order as those given in Eq. 15 provided that the preexponential factor is about 0.1 to 0.01. These values seem reasonable in view of the crude treatment given above.

10.4   MECHANISM OF SILICIDE FORMATION

10.4.1   Introduction

From the viewpoint of phase transformation, silicide formation as presented in this chapter concerns the reaction between two solid phases in direct contact to form ordered intermetallic compounds at temperatures well below the formation of any liquid phase. The reaction is unique in that it occurs between two different kinds of solids: the substrate is a covalently bonded single crystal and the thin film is metallic and fine-grained. From the summary given in Section 10.2.4, one finds the formation of three classes of silicides: metal-rich silicides, monosilicides, and disilicides. Typically, with some scattering of data, they start to form around 200, 400, and 600°C, respectively. The mechanism of silicide formation must take into account the large temperature differences.

Since the metallic films are fine-grained, we expect that a fast mass transport of metallic atoms can occur at their grain boundaries (see Chapter 7). Even for those high melting point refractory metals, this also seems plausible because the reaction temperature of 600°C is not so high that bulk diffusion necessarily dominates over grain boundary diffusion.

The observation that all the first phase disilicides of refractory metals start to form around 600°C is striking in view of the large variation in the melting points of the metals. This suggests that the supply of refractory metal atoms is not rate limiting. Otherwise, a much wider range of formation temperatures should have been found. On the other hand, the formation of metal-rich silicides at 200°C, which is about 0.3 of the absolute melting point of Si, raises the question about the mechanism whereby Si atoms can break away from their lattice at such a low temperature. It seems then that the supply of Si is crucial to silicide formation for both the disilicides and metal-rich silicides. The mechanism of Si supply at high temperatures might very well be different from that at low temperatures. At temperatures higher than 600°C, the phonon energies in a Si atom are probably high enough to enable Si atoms to break away from Si surfaces. At temperatures as low as 200°C, the phonon energies are not sufficient to dissociate the covalent bonds in Si, so other mechanisms to free a Si atom from its lattice must be invoked.

While the kinetic data show the large variation in silicide formation behavior and allow us to classify the silicides accordingly, the mechanisms of formation cannot be understood without also knowing the thermodynamic data of silicides, since the driving force behind the formation comes from a free-energy change. The free-energy change is related to the formation energy of silicides, and usually there exist more than one silicide in a metal–Si binary system and their formation energies are different. For example, the equilibrium binary phase diagram of Ni–Si is shown in Figure 10.21. The

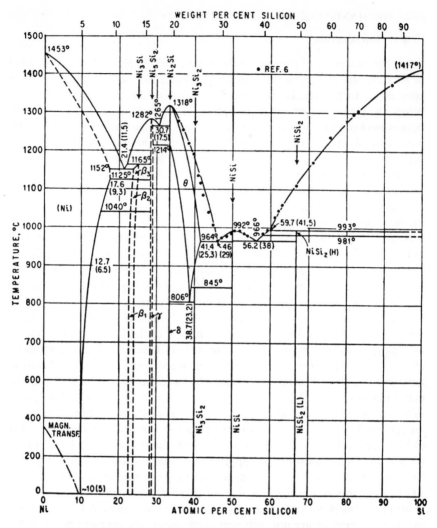

Figure 10.21. Binary phase diagram of Ni–Si. From reference 22.

394

diagram shows that there are six Ni–Si intermetallic compounds that can exist below the lowest eutectic point in the system. The compound which is next to Si is $NiSi_2$ and is the stable phase (the end phase) to be expected when the reaction is complete. Since Ni is shown to have a solubility of Si up to several percent, the formation of the end phase could be preceded by the solution of Si in Ni or by the formation of the other phases. However, substitutional solution can only occur when lattice diffusion (via vacancies) takes place. Hence, at low reaction temperatures the formation of the other phases is more likely. Here, a difficult question arises as to whether we can predict the first phase nucleation and growth. In the following, we first discuss the thermodynamic data, then the kinetic mechanism of silicide formation, and finally the problems of predicting first phase nucleation.

10.4.2 Thermodynamic Data of Metal–Si Systems

10.4.2.1 The End Phase. The end phase of silicides which is stable with Si for various metallic elements was first given systematically by Lepselter and Andrews using the periodic table (1). Figure 10.22 is a reproduction of their table with a minor change: the Au is now shown to form a metastable metal-rich silicide with Si (66). A trend that can be seen in Figure 10.22 is that from left to right the end phase shows an increase in its metal concentration—disilicide, monosilicide, metal-rich silicide, and finally metal with a small amount of Si as solute. An obvious correlation to the trend is the lowering of melting point of these end phases; but other than that, any correlation with the mechanism of silicide formation is not at all clear at present.

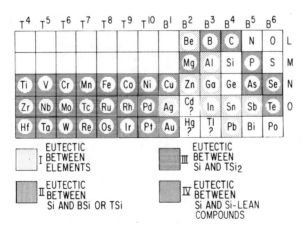

Figure 10.22. Partial periodic table showing the metallic elements and their end phase that is stable with Si. From reference 1.

10.4.2.2 Solubility and Diffusivity.   The solubility of Si in metallic elements can be simply presented by using the Darken-Gurry plot (67), which is reproduced in Figure 10.23. The metallic elements are classified into four groups according to their solubility of Si, as given in the upper right corner. Elements such as Ni, Co, Fe, and V that possess a large solubility of Si are scattered close to Si in the plot, and those that are immiscible with Si are found to locate far away from Si. Solubility data are important in estimating whether there is dissolution before silicide formation. In reacting Ni and V with Si, no noticeable solution of Si in these metals was found preceding the silicide formation. But Si was found to dissolve into Fe film up to 25 at. % at 400°C before the formation of FeSi takes place (40).

The plot cannot be used to deal with the solubility of metallic elements in Si. One reason is that the plot was constructed by using the atomic radius of Si of 12 coordinations. It is known that Si has an extremely small solubility of metallic elements, with no exceptions. Yet, even these low solubility limits are hard to reach at low temperatures if the solute has to take substitutional sites in Si by a vacancy mechanism of diffusion. This is due to the high formation energy of the vacancy in Si. On the other hand, there are metals that are known to be fast diffusants in Si, with an activation energy of about 1 eV, and that might dissolve into Si at low temperatures (68). Table 10.8 lists the available data of self-diffusion and some solute diffusion in Si. Most of these data were obtained by tracer techniques using Si samples at an equilibrium state with no excess point defects. The table shows that the noble and near-

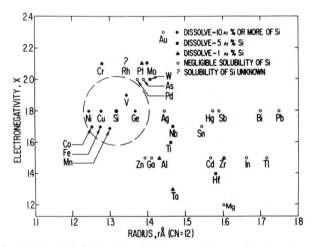

Figure 10.23.   Darken and Gurry plot of electronegativity vs. atomic radius to show the solubility of Si in various metals. From reference 67.

TABLE 10.8. SELF-DIFFUSION AND SOLUTE DIFFUSION COEFFICIENT IN Si

Element	D cm^2 s^{-1}	References
Cu	$4.7 \times 10^{-3} \exp(-0.43/kT)$ (300 to 700°C)	70
Ag	$2 \times 10^{-3} \exp(-1.6/kT)$ (1100 to 1350°C)	71
Au	$2.4 \times 10^{-4} \exp(-0.39/kT)$ (700 to 1300°C)	72
Al	$8 \exp(-3.47/kT)$ (1085 to 1375°C)	73
Si	$1.81 \times 10^4 \exp(-4.86/kT)$ (900 to 1300°C)	74, 72
As	$60 \exp(-4.2/kT)$ (850 to 1150°C)	75
Cr	$<10^{-8}$ (1200°C)	76
Mn	$>2 \times 10^{-7}$ (1200°C)	77
Fe	$>5 \times 10^{-6}$ (100 to 1115°C)	76
Ni	1.57×10^{-7} (800°C)	78
Pt	similar to Au in Si	79

noble metals can diffuse in Si much faster than Si itself. The fast diffusion can occur at a temperature as low as 200°C, and it is most likely a dissociative interstitial diffusion. No diffusion data of refractory metals in Si are available. One measurement of Ti diffusion in Ge showed that it is slower than Ge self-diffusion (69). Because of their high number of valence electrons and their large atomic size, refractory metals are expected to dissolve and diffuse substitutionally in Si (Cr might be an exception). Since substitutional diffusion will require an activation energy comparable to that of self-diffusion, refractory metals therefore are not expected to dissolve into Si at temperatures such as 200°C.

10.4.2.3 Formation Energy of Silicide.   Table 10.9 lists the free energy of formation of silicides expressed in units of kcal g-atom^{-1} (80–82). The middle column in the table shows the silicides of Ti and Zr which are characterized by a larger formation energy of monosilicide than that of disilicide. The right-hand column lists silicides of the rest of the refractory metals which are characterized by having no monosilicide (except CrSi) and also by the large formation energy of the disilicide. For example, VSi$_2$ is found to have the largest formation energy per g-atom among all the silicides in Table 10.9. Compared to other disilicides, the formation energy seems surprisingly high. From the summary given in Section 10.2.4, the temperature of disilicide formation for these metals is about 600°C. At this temperature, if we assume that the kinetics are fast, namely, not a limiting factor so that energy change will dictate the selection of silicide formation, the disilicide is favored. Indeed, the disilicide is the only observed growth phase for these metals. For silicides of Ti, Zr, and Hf, the formation of TiSi and HfSi occurs at temperatures above

TABLE 10.9. FREE ENERGY OF FORMATION ΔH OF SILICIDES[a]

Silicide	ΔH, kcal g-atom^{-1}	Silicide	ΔH kcal g-atom^{-1}	Silicide	ΔH, kcal g-atom^{-1}
Mg_2Si	6.2	Ti_5Si_3	17.3	V_3Si	6.5
		TiSi	15.5	V_5Si_3	11.8
FeSi	8.8	$TiSi_2$	10.7	$V Si_2$	24.3
$FeSi_2$	6.2				
		Zr_2Si	16.7	Nb_5Si_3	10.9
Co_2Si	9.2	Zr_5Si_3	18.3	$NbSi_2$	10.7
CoSi	12	ZrSi	18.5, 17.7		
$CoSi_2$	8.2	$ZrSi_2$	12.9, 11.9	Ta_5Si_3	9.5
				$TaSi_2$	8.7, 9.3
Ni_2Si	11.2, 10.5	HfSi			
NiSi	10.3	$HfSi_2$		Cr_3Si	7.5
				Cr_5Si_3	8
Pd_2Si	6.9			CrSi	7.5
PdSi	6.9			$CrSi_2$	7.7
				Mo_3Si	5.6
Pt_2Si	6.9			Mo_5Si_3	8.5
PtSi	7.9			$MoSi_2$	8.7, 10.5
RhSi	8.1			W_5Si_3	5
				WSi_2	7.3

[a]Data from references 80, 81, and 82.

500°C. The formation of these monosilicides is not only favored by their high energy of formation but also by their simpler crystal structure than that of disilicides. The unit cell of monosilicide of Ti, Zr, and Hf (FeB type) has eight atoms, while the unit cell of their disilicide (C49 type) has 12 atoms. Although the energy of formation of these two units cells are about the same, the monosilicide is kinetically favored because of the lesser number of atoms involved in the nucleation and growth. In a few cases, the silicide M_5Si_3 (where M stands for the refractory metals) is shown in Table 10.9 to have a higher formation energy than that of monosilicide or disilicide. But the formation of this silicide can only occur when the supply of Si is limited; otherwise, it is obvious that the formation of five molecules of monosilicide or disilicide is energetically more favorable than the formation of one molecule of M_5Si_3.

The left-hand column in Table 10.9 gives the formation energy of silicides of other transition metals and Mg. Here, no obvious trends can be given based on the energies. However, the metal-rich silicides are characterized by their low temperature of formation. At a low temperature, whether atoms can jump or not often plays a more significant role in controlling the formation than the available energy change. The mechanism of formation of these metal-rich silicides is discussed later.

The results of this section show that the empirical data on silicide formation are consistent with tabulated free energies. The correlations are not strong enough to predict silicide formation in specific cases, but general trends can be formulated.

10.4.2.4 Interfacial Energy. During single phase silicide formation, the three interfaces of our concern are the metal–silicon, metal–silicide, and silicide–silicon interfaces. The nucleation of a silicide can be better understood if we know the magnitude of these interfacial energies, yet unfortunately there are no such data in the literature. Also, the growth rate of a silicide may depend on the structure of its interfaces. Since the Si substrate is a single crystal, the silicide that grows epitaxially on the Si could develop a coherent or semicoherent interface with the Si, depending on the amount of mismatch. For example, Pd_2Si is known to grow epitaxially on the (111) surface of Si with a very small mismatch, so their interface can be regarded as coherent. A coherent interface is low in free energy; this may be one of the reasons for the rather easy formation of Pd_2Si during the deposition and its rather high stability with respect to the transformation to PdSi. On the other hand, the epitaxial growth raises a question concerning how the interface can maintain its coherency while moving into the (111) Si.

10.4.3 Kinetic Mechanism of Silicide Formation

The crucial step in the kinetics of silicide formation is how to maintain the supply of Si by breaking the bonds in the substrate. For the formation of disilicide of refractory metals, the bond breaking probably takes place at weak spots on the Si surface such as at kinks and ledges. Thus, a high reaction temperature is required to supply to a phonon the needed energy to free the surface atoms. The supply of Si atoms is limited by the rate the bonds are broken. Hence, the thickening rate of disilicide at the early stage of growth is linear, and marker studies show that Si atoms are the dominating diffusing species. For the formation of metal-rich silicide of near-noble metals, the bond breaking cannot rely on phonon energy alone because the formation temperature is too low; instead, an interstitial mechanism has been proposed in which the interstitial metal atom assists in bond breaking (9).

The interstitial mechanism is based on the correlation that metals that can diffuse interstitially in Si are the metals that can react with Si at low temperatures. The consequences of a metal atom jumping into an interstitial site in Si are that there is an increase in the number of nearest neighbors of the surrounding host atoms and at the same time there is a vacancy left behind. The increase in the number of nearest neighbors of a Si atom weakens its bonds due to charge transfer. The weakened bonds can be regarded as being transformed from a covalent type to a metallic type. In a band picture, charge transfer from a saturated covalent bond means the formation of a hole in the valence band. Because a hole is not localized at a bond as is mandated by the uncertainty principle, the bond breaking of one particular Si atom may require the combined effect of several interstitials near the interface. However, since the interface has a larger free energy than the Si lattice, it is possible that under the driving force of reaction the interface can take a higher concentration of interstitials. The presence of the interstitials then provides the release of the Si atoms at temperatures around $200°C$.

The interstitial mechanism depends on a continuous supply of metal atoms to the silicide–silicon interface to keep the reaction going. Thus, it is no surprise to find from marker studies that metal atoms are the diffusing species in the formation of all metal-rich silicides. Without free metal atoms to form interstitials, the growth will stop. The growth rate then depends on how fast the metal atoms can reach the interface. It is a diffusion-limited growth, and so it obeys the parabolic rule. Also, because a metal-rich silicide tends to favor the diffusion of the metal atoms, the growth of a metal-rich silicide is selected (9).

A similar interstitial mechanism has been used implicitly by Buckley and Moss (13) to explain the initial epitaxial growth of Pd_2Si on (111) surface of Si. Figure 10.24 is a reproduction of Figure 5a in reference 13. It shows the basal plane of Pd_2Si which has a hexagonal unit cell of dimensions $a = 13.055$ Å and $c = 27.490$ Å and containing 288 atoms. The dark circles, arranged in a hexagonal net, represent Si atoms in the silicide. The net is identical to the (111) plane of Si if the dotted, but missing, Si atoms at position A are included. For the formation of the first layer of Pd_2Si on (111) Si, we quote the following sentence, "The Pd atoms drop into the three positions around A and force the central silicon atom out of the way moving it up between them so that it nests on top of them to form the second plane of the silicide" (13). It is obvious that the three positions around A are the interstitial sites in Si. What is also implicit in reference 13 is that the subsequent growth requires the same

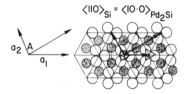

Figure 10.24. Atomic arrangement in the basal plane of Pd_2Si. From reference 13.

action of interstitial Pd atoms in order to remove Si from its lattice, therefore the diffusion of Pd through Pd_2Si is important. Furthermore, as we discussed before, the diffusion of Si alone will lead to vacancy condensation forming voids at the Pd_2Si–Si interface and will cause a poor adhesion and impair the contact. Yet, Pd_2Si is known to show good adhesion and contact on (111) Si. This indicates that the formation of Pd_2Si cannot take place by the diffusion of Si alone, and indeed marker studies show that both Pd and Si are diffusing species.

We have presented two very different pictures of growth of silicides, one for the disilicide of refractory metals and another for the metal-rich silicide of near-noble metals. It should be pointed out that some silicides seem to fall on the borderlines of these two mechanisms. Notably, $CrSi_2$ can form at 450°C (a low temperature for disilicide formation) and Co_2Si forms above 350°C (a high temperature for metal-rich silicide formation). It seems that until we understand how Si bonds are broken in these cases, the fine detail of the formation of these silicides cannot be given.

For the monosilicides, we recall that they can be the first growth phase or their formation can be preceded by the metal-rich phase as shown in Table 10.1. In the latter cases, the formation will be influenced by the dissociation of the preceding phase. When a monosilicide becomes the first growth phase, the mechanism of supply of Si is not clear for all the metals that fall into this class, partly because the kinetic data are not complete. We expect that the interstitial mechanism works for metals such as Fe and Rh. Although the formation temperature of FeSi at 450°C is higher than expected, it is possible that the difference in the diffusion of metal atoms through a metal-rich silicide and through a monosilicide has caused the higher formation temperature of the latter. But for Ti, Zr, and Hf, while the mechanism may depend mainly on phonon energy, we suspect that there may be a small contribution to bond breaking that comes from a direct, rather than on interstitial, interaction between Si and these metal atoms. The interaction is quite strong, as can be seen from the very high formation energy of their silicides as given in Table 10.9. Yet the details of such a direct interaction are not clear at all.

10.4.4 Prediction of First Phase Nucleation

Walser and Bené (83) have published a rule about the first phase nucleation in silicon–transition metal planar interfaces. The concept behind the rule is that an amorphous phase is assumed to form at the interface between the Si and metal during the deposition of the metal which resembles a fast quenching of metallic atoms onto the Si surface. Phillips in Chapter 3 refers to this phase as a glassy membrane. Upon annealing at the silicide formation temperature, the silicide which has a concentration near that of the amorphous phase will nucleate first. The concentration of the amorphous phases is to be found near a deep eutectic point in the phase diagram, based on an earlier suggestion of Cohen and Turnbull (84). Then, the rule of Walser and Bené is to take the highest melting point silicide neighboring the deepest eutectic point (the most stable congruently melting silicide) in the binary phase diagram as the first nucleated phase.

There are several aspects of the rule and its predicted phases which require further clarification. First, the concept is based on the formation of a stable amorphous phase with a rather narrow concentration range at the interface between Si and metal, yet it is known that the evaporated binary amorphous alloys can exist over a very wide concentration range (85). Next, for the most well-established amorphous alloy of $Pd_{81}Si_{19}$, it is known that the alloy upon heat treatment transforms into the crystalline Pd_3Si rather than the Pd_2Si as predicted (86). The rule when applied to the Pt–Si system is somewhat ambiguous; it is difficult to choose between Pt_3Si and Pt_2Si. Both are on the same side to the deepest eutectic point in the diagram. While Pt_2Si has a higher melting point, Pt_3Si is closer to the eutectic concentration. Furthermore, the prediction does not apply to the cases of Ti–Si and Mn–Si where monosilicide is the observed first growth phase rather than the predicted $TiSi_2$ and Mn_5Si_3. Finally, of course, one more difficulty is that we have been concerned with growth phase in silicide formation, as it has been shown by Kidson's analysis (65) that a nucleated phase could be wiped out in the growth stage due to unfavorable kinetic parameters of the growth. Yet the prediction is for nucleation. Therefore, whether the first growth phase can be taken to be the first nucleated phase remains to be proven.

10.5 CONCLUSIONS

In conclusion, this chapter was intended to give an overview of silicide formation. Although we obtain a very systematic pattern of silicide formation, it is clear that further work must be carried out to understand both the physical mechanism involved in silicide growth and the influence of other parameters such as stress and impurities. A glaring omission is the absence of studies of silicide formation where impurity-free metal films are deposited under ultrahigh vacuum on clean Si surfaces. We know (21, 87) that impurities influence formation kinetics. However, we do not know if studies of clean systems will force major revisions of the patterns we have presented.

The present overview strongly suggests that more work is required to clarify the microstructure of silicides in terms of defects and also to clarify the nature of the first phase that nucleates. Many of the models that we have suggested need verification. At the least, a systematic attempt should be made to observe the "glassy-membrane" at the metal-semiconductor interface and the interstitial mechanism should be tested on the metal-Ge system. There have been only a few systematic studies (88, 31) of first-phase formation of germanides. A more detailed analysis of marker motion should be pursued over a wider range of silicides, and tracer studies should be included to give insight into the transport mechanisms. We also note that there have been only a few studies of silicides as diffusion barriers. As implied by Figures 10.1 and 10.2, reaction and interdiffusion barriers play a crucial role in thin film technology. Here, too, there is a major need for further work to understand the nature of the barrier.

Our view, then, of silicide formation is that only the general features have been sketched. With the urgency of technology to create controlled thin film structures, we believe that further investigations will lead to new insights in silicide formation.

ACKNOWLEDGMENT

The authors thank their many colleagues at IBM and Caltech for their suggestions and contributions. Many of the concepts grew out of joint projects between the two groups. The Caltech work was supported in part by AFCRL (D. E. Davies).

REFERENCES

1. M. P. Lepselter and J. M. Andrews, in *Ohmic Contact to Semiconductors*, B. Schwartz, Ed., The Electrochemical Society, New York (1969), p. 159.
2. T. Kawamura, D. Shinoda, and H. Muta, *Appl. Phys. Lett.*, **11**, 101 (1967).
3. N. G. Anantha and K. G. Ashar, *IBM J. Res. Dev.*, **15**, 442 (1971).
4. C. J. Kircher, *Solid State Electron.*, **14**, 507 (1971).
5. H. H. Hosack, *Appl. Phys. Lett.*, **21**, 256 (1972).
6. A. K. Sinha and T. E. Smith, *J. Appl. Phys.*, **44**, 3465 (1973).
7. R. J. Blattner, C. A. Evans, Jr, S. S. Lau, J. W. Mayer, and B. M. Ullrich, *J. Electrochem. Soc.*, **122**, 1732 (1975).
8. J. W. Mayer and K. N. Tu, *J. Vac. Sci. Technol.*, **11**, 86 (1974).
9. K. N. Tu, *Appl. Phys. Lett.*, **27**, 221 (1975).
10. J. F. Ziegler, J. W. Mayer, C. J. Kircher, and K. N. Tu, *J. Appl. Phys.*, **44**, 3851 (1973).
11. R. Feder and B. S. Berry, *J. Appl. Crystallogr.*, **3**, 372 (1970).
12. S. S. Lau, W. K. Chu, J. W. Mayer, and K. N. Tu, *Thin Solid Films*, **23**, 205 (1974).
13. W. D. Buckley and S. C. Moss, *Solid State Electron.*, **15**, 1331 (1972).
14. S. S. Lau and D. Sigurd, *J. Electrochem. Soc.*, **121**, 1538 (1974).
15. J. O. Olowolafe, M-A. Nicolet, and J. W. Mayer, *Thin Solid Films*, **38**, 143 (1976).
16. K. N. Tu, W. K. Chu, and J. W. Mayer, *Thin Solid Films*, **25**, 403 (1975).
17. D. J. Fertig and G. Y. Robinson, *Solid State Electron.*, **19**, 407 (1976).
18. K. N. Tu, E. I. Alessandrini, W. K. Chu, H. Kraütle, and J. W. Mayer, *Jap. J. Appl. Phys.*, Suppl. **2**, Pt. 1, 669 (1974).
19. V. Koos and H. G. Neumann, *Phys. Stat. Solidi*, **A29**, K115 (1975).
20. K. N. Tu, J. F. Ziegler, and C. J. Kircher, *Appl. Phys. Lett.*, **23**, 493 (1973).
21. H. Kraütle, M-A. Nicolet, and J. W. Mayer, *J. Appl. Phys.*, **45**, 3304 (1974).
22. M. Hansen, *Constitution of Binary Alloys*, McGraw-Hill, New York (1959); R. P. Elliott, *Constitution of Binary Alloys, First Supplement*, McGraw-Hill, New York (1965); F. A. Shunk, *Constitution of Binary Alloys, Second Supplement*, McGraw-Hill, New York (1969).

23. H. Kato and Y. Nakamura, *Thin Solid Films*, **34**, 135 (1976).
24. M. J. Rand and J. F. Roberts, *Appl. Phys. Lett.*, **24**, 49 (1974).
25. E. I. Alessandrini, D. R. Campbell, and K. N. Tu, *J. Appl. Phys.*, **45**, 4888 (1974).
26. K. N. Tu and S. Libertini, *J. Appl. Phys.*, **48**, 420 (1977).
27. H. Kraütle, W. K. Chu, M.-A. Nicolet, J. W. Mayer, and K. N. Tu, *Proceedings of the International Conference on Applications of Ion Beams to Metals, Albuquerque, New Mexico*, Plenum Press, New York (1974), p. 193.
28. W. K. Chu, private communication.
29. R. W. Bower, R. E. Scott, and D. Sigurd, *Solid State Electron.*, **16**, 1461 (1973).
30. D. Sigurd, W. van der Weg, R. Bower, and J. W. Mayer, *Thin Solid Films*, **19**, 319 (1974).
31. G. A. Hutchins and A. Shepala, *Thin Solid Films*, **18**, 343 (1973).
32. A. Hiraki, M.-A. Nicolet, and J. W. Mayer, *Appl. Phys. Lett.*, **18**, 178 (1971).
33. H. Muta and D. Shinoda, *J. Appl. Phys.*, **43**, 2913 (1972).
34. J. M. Poate and T. C. Tisone, *Appl. Phys. Lett.*, **24**, 391 (1974).
35. W. K. Chu, S. S. Lau, J. W. Mayer, H. Müller, and K. N. Tu, *Thin Solid Films*, **25**, 393 (1975)
36. G. J. Van Gurp and C. Langereis, *J. Appl. Phys.*, **46**, 4301 (1975).
37. S. S. Lau, private communication.
38. A. K. Sinha, R. B. Marcus, T. T. Sheng, and S. E. Haszka, *J. Appl. Phys.*, **43**, 3637 (1972).
39. J. M. Andrews and F. B. Koch, *Solid State Electron.*, **14**, 901 (1971).
40. S. S. Lau, J. S.-Y. Feng, J. O. Olowolafe, and M.-A. Nicolet, *Thin Solid Films*, **25**, 415 (1975).
41. D. J. Coe, E. H. Rhoderick, P. H. Gerzon, and A. W. Tinsley, in *Metal–Semiconductor Contacts*, Institute of Physics Conf. Ser. No. 22, London and Bristol (1974), p. 74.
42. J. F. Ziegler, private communication.
43. K. E. Sundström, S. Petersson, and P. A. Tove, *Phys. Solid State*, **A20**, 653 (1973).
44. R. W. Bower and J. W. Mayer, *Appl. Phys. Lett.*, **20**, 359 (1972).
45. C. J. Kircher and J. F. Ziegler, private communication.
46. A. G. Revesz, private communication.
46a. A. Christou and H. M. Day, *J. Electron. Mater.*, **5**, 1 (1976).
47. B. I. Fomin, A. E. Gershinskii, and E. I. Cherepov, *Talanta*, in press.
47a. B. Oertel and R. Sperling, *Thin Solid Films*, **37**, 185 (1976).
48. L. D. Locker and C. D. Capio, *J. Appl. Phys.*, **44**, 4366 (1973).
49. E. O. Kirkendall, *Trans. AIME*, **147**, 104 (1942); A. D. Smigelskas and E. O. Kirkendall, *Trans. AIME*, **171**, 130 (1947).
50. L. S. Darken, *Trans. AIME*, **175**, 184 (1948).
50a. K. N. Tu, *J. Appl. Phys.*, **48**, 3379 (1977).
51. F. Brown and W. D. Mackintosh, *J. Electrochem. Soc.*, **120**, 1096 (1973).
51a. H. Müller, J. Gyulai, W. K. Chu, J. W. Mayer, and T. W. Sigmon, *J. Electrochem. Soc.*, **122**, 1234 (1975).
52. W. K. Chu, H. Kraütle, J. W. Mayer, H. Müller, M.-A. Nicolet, and K. N. Tu, *Appl. Phys. Lett.*, **25**, 454 (1974).
53. G. J. van Gurp, D. Sigurd, and W. F. van der Weg, *Appl. Phys. Lett.*, **29**, 159 (1976).
54. S. Mader and K. N. Tu. *J. Vac. Sci. Technol.*, **12**, 501 (1975).
55. R. Pretorius, Z. L. Liau, S. S. Lau, and M.-A. Nicolet, *Appl. Phys. Lett.*, **29**, 598 (1976).
56. R. Pretorius and S. S. Lau, private communication.
57. J. O. Olowolafe, M.-A. Nicolet, and J. W. Mayer, *J. Appl. Phys.*, **47**, 5182 (1976).
58. C. J. Kircher, J. W. Mayer, K. N. Tu, and J. F. Ziegler, *Appl. Phys. Lett.*, **22**, 81 (1973).
59. W. K. Chu, private communication.
60. J. A. Borders and J. N. Sweet, *J. Appl. Phys.*, **43**, 3803 (1972).
61. L. C. Correa da Silva and R. Mehl, *Trans. AIME*, **191**, 155 (1951).
62. D. R. Campbell, K. N. Tu, and R. E. Robinson, *Acta Met.*, **24**, 609 (1976).

63. P. G. Shewmon, *Diffusion in Solids*, McGraw-Hill, New York (1963).
64. J. R. Manning, *Diffusion Kinetics for Atoms in Crystals*, Van Nostrand, Princeton, N.J. (1968).
65. G. V. Kidson, *J. Nucl. Mater.*, **3**, 21 (1961).
66. A. K. Green and E. Bauer, *J. Appl. Phys.*, **47**, 1284 (1976).
67. L. S. Darken and R. W. Gurry, *Physical Chemistry of Metals*, McGraw-Hill, New York (1953).
68. S. M. Hu, in *Atomic Diffusion in Semiconductors*, D. Shaw, Ed., Plenum Press, London and New York (1973).
69. V. I. Tagirov and A. A. Kuliev, *Sov. Phys. Solid State*, **4**, 196 (1962).
70. R. N. Hall and J. H. Racette, *J. Appl. Phys.*, **35**, 379 (1964).
71. B. I. Boltaks and S. Y. Hsüch, *Sov. Phys. Solid State*, **2**, 2383 (1961).
72. W. R. Wilcox and T. J. LaChapelle, *J. Appl. Phys.*, **35**, 240 (1964).
73. D. Navon and V. Chernyshov, *J. Appl. Phys.*, **28**, 823 (1957).
74. R. F. Peart, *Phys. Stat. Solidi*, **15**, K119 (1966).
75. B. J. Masters and J. M. Fairfield, *J. Appl. Phys.*, **40**, 2390 (1969).
76. C. B. Collins and R. O. Carlson, *Phys. Rev.*, **108**, 1409 (1957).
77. R. O. Carlson, *Phys. Rev.*, **104**, 937 (1956).
78. H. P. Bonzel, *Phys. Stat. Solidi*, **20**, 493 (1967).
79. R. F. Bailey and T. G. Mills, "Semiconducting Silicon," in *Proceedings of 1st International Symposium* (Electrochem. Soc., New York (1969) p. 481.
80. C. T. Lynch, *Handbook of Materials*, CRC Press, Cleveland (1974).
81. C. J. Smithells, *Metals Reference Book* 4th ed., Vol. I, Plenum Press, New York (1967).
82. A. W. Searcy and L. N. Finnie, *J. Amer. Ceram. Soc.*, **45**, 270 (1962).
83. R. M. Walser and R. W. Bene, *Appl. Phys. Lett.*, **28**, 624 (1976).
84. M. H. Cohen and D. Turnbull, *Nature*, **189**, 131 (1961).
85. S. Mader, *Thin Solid Films*, **35**, 195 (1976).
86. D. Gupta, K. N. Tu, and K. W. Asai, *Phys. Rev. Lett.*, **35**, 796 (1975).
87. J. B. Bindell, J. W. Colby, D. R. Wonsidler, J. M. Poate, D. K. Conley, and T. C. Tisone, *Thin Solid Films*, **37**, 441 (1976).
88. M. Wittmer, M.-A. Nicolet, and J. W. Mayer, *Thin Solid Films*, in press.

11

METAL–COMPOUND SEMICONDUCTOR
REACTIONS

A. K. Sinha and J. M. Poate

Bell Laboratories, Murray Hill, New Jersey

11.1 INTRODUCTION

All devices based on compound semiconductors need metal contacts either to form an active Schottky barrier or an ohmic contact. In the past, various heat treatments have been utilized to produce ohmic contacts or Schottky diodes at the metal–compound semiconductor interface (1). However, only recently have systematic attempts been made to correlate the electrical effects with the chemistry of the interface. In this chapter we review such work and also show how the recently acquired information has led to the synthesis of effective contact metallizations. Nearly all of this work has been concerned with contacts on GaAs and GaP.

Figure 11.1 shows the relative positions in the periodic table of the components of important III–V compound semiconductors ($A_{III}B_V$), of typical substitutional dopants, and of various metals commonly used for contact formation. From this figure, it may be seen why the divalent elements Be and Zn are p-type dopants whereas the hexavalent S and Te are n-type dopants and Si and Ge (tetravalent) are amphoteric. From the known information

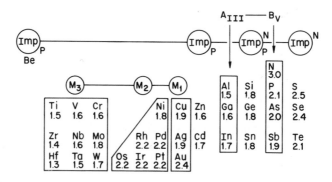

Figure 11.1. Portion of the periodic table showing components of important $A_{III}B_V$ compound semiconductors, common n- and p-type dopants (Imp) and various groups of metals (M_1, M_2, M_3) used to form contacts. Numbers indicate Pauling electronegatives.

about the phase equilibria between a given metal and the components A_{III} and B_V, one can anticipate a wide range of possible thermally induced reactions, including eutectics and intermetallic compounds and extended solid solutions (2–4). What is observed in practice, however, depends upon a combination of thermodynamic as well as kinetic considerations. Often, true equilibrium is never established in a thin film system; metastable phases are quite commonly observed (5).

It is possible to calculate, from first principles using, for example, the pseudopotential approach, the potential energy of a certain arrangement of atoms characterizing a given alloy phase. However, great difficulties arise in assessing the relative stabilities (energies) of various phases. The differences in the energies of various metallic phases as estimated from thermodynamic measurements (7) are small (tens of meV), compared to the calculated bonding energies (a few eV), and to the various corrections which must be applied to the latter. Due to the extensive experimental studies on metal–

metal interactions and to the rules of alloying behavior formulated by Hume-Rothery and co-workers (8), it is possible to establish certain empirical concepts of alloy phase stability. These are stated in terms of the size factor, electron concentration, and electronegativity difference effects.

Atomic size is commonly expressed in terms of atomic radius, which is taken to be half the distance of closest approach in the crystal structure of the pure element. Large size differences ($> 14\%$) preclude extensive solid solubility. In many cases atoms cannot be assumed to be hard incompressible spheres, and allowance must be made for the complex interdependence between the size and nature of atomic coordination around an atom (9).

The average electron concentration (electron:atom ratio) was first used to characterize the well-known Hume-Rothery electron compounds in alloys of Cu, Ag, and Au (8). Assuming a rigid band model, the stability of certain intermetallics may be related to an electron concentration dependent interaction between the Fermi surface and the Brillouin zone. However, even when the density of states at the Fermi level is known from, say, low-temperature heat capacity measurements, the problem remains of evaluating the true electron concentration since most transition metals have variable valencies.

Electronegativity, a dimensionless number, represents the power of an atom to attract electrons. An increase in electronegativity difference $\Delta\chi$ increases the tendency toward intermetallic compound formation. A large electronegativity difference also imparts a greater ionicity to an A–B type of bond due to electron transfer. Pauling electronegativities (10) were defined in terms of heats of formation. According to Pauling, the difference between the heat of formation of a given $A^+ : B^-$ bond and the average heat of formation $H_{A^+:B^-}$ of covalent A:A and B:B bonds represents the excess (ionic) bonding energy due to electronegativity difference $\Delta\chi$, and it is proportional to the square of $\Delta\chi$. Thus,

$$\Delta\chi \propto \left[H_{A^+:B^-} - \tfrac{1}{2}(H_{A:A} + H_{B:B}) \right]^{1/2}$$

Pauling, as well as Phillips (11) whose definition is based on the characteristic band gaps, have derived a most useful set of self-consistent electronegativity values for various elements as shown in Figure 11.1. By convention, most intermetallics as well as compound semiconductors are represented in the form of an AB combination where B atoms are more electronegative than A atoms. For many compound semiconductors the bonding orbitals are centered on more electronegative B atoms, and the antibonding ones are centered on the electropositive A atoms (cations) (11). Experimentally, a transition has been observed from predominantly covalent to ionic bonding at $\Delta\chi > 0.6$ (12).

There seem to be no well-established criteria for reaction between a metal and a compound semiconductor. In the case of the $M–A_{III}B_V$ systems, we have adopted an empirical approach based upon electronegativities in rationalizing the various types of reactions. The metals of interest in this review can be divided into three groups as shown in Figure 11.1. Group M_1 consists of noble metals, which are very electronegative. For these metals, the dominant form of interaction with a compound semiconductor $A_{III}B_V$ involves the outdiffusion of the electropositive component A_{III} into the metal. Group M_2 consists of near-noble metals near the end of the transition series. These metals are also quite electronegative; in addition, they display a strong tendency to attain the stable d^{10} electronic structure through electron transfer (8, 9). These metals form stable compounds with the metalloid B_V and also with the metals A_{III}. The metal–metal (M_2–A_{III}) interaction may result in ordered close-packed structures where the coordination is 12 for both A and M atoms and there are no A–A nearest neighbors (13). The third group, M_3, consists of early transition metals and Al, which have relatively small electronegativities. The interfaces of these metals with $A_{III}B_V$ are relatively inert.

11.2   OUTDIFFUSION EFFECTS

11.2.1   Metallurgy

The outdiffusion effects are best illustrated with reference to the Au–GaAs interface. Experimental evidence of thermally induced migration of Ga from GaAs into and through Au films has been obtained by several groups (14–19) using Rutherford backscattering, Auger spectroscopy, or secondary ion mass spectroscopy with sputter sectioning for depth profiling. Figure 11.2 shows backscattering spectra (15, 16) for an 800 Å Au film on n-GaAs annealed at 250°C and 350°C in air. At 250°C, the reaction results in (a) indiffusion of Au into GaAs to a depth of ~ 1500 Å as shown by the low-energy tail of the Au spectrum of Figure 11.2 and (b) outdiffusion of Ga to the surface of Au as shown by the peak at a backscattered energy of 1.59 MeV expected for surface Ga. That this peak is indeed due to Ga atoms on the surface of Au was confirmed by the observation that its position did not change upon increasing the effective thickness of Au by tilting the sample with respect to the incident beam. The shaded area under the Ga peak corresponds to 10^{16} atoms cm^{-2}, or about three monolayers. This Ga is almost certainly in the form of Ga_2O_3; its presence in devices has been known to cause difficulties during subsequent thermocompression bonding operation.

Similar but greatly enhanced effects are seen for samples annealed at 350°C and higher temperatures which lead to a very disordered atomic arrangement at the interface. Figure 11.3 shows the energy spectra obtained by Gyulai et al. (4) for a 400 Å Au film annealed at 400 and 500°C in N_2 for various times. At 400°C, the exponential tails on the Au distribution are clearly evident. The spectrum for the sample annealed at 500°C for 12 min

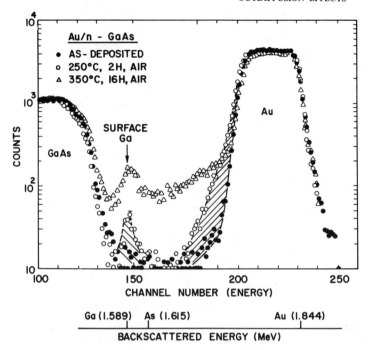

Figure 11.2.   Rutherford backscattering spectra of 2 MeV ^{4}He$^+$ ions for Ga outdiffusion through 800 Å of Au on GaAs. From references 15 and 16.

represents the case where all the Au has penetrated into the GaAs. It was concluded that the deeply penetrating Au tail (~ 300 Å in Fig. 11.3) was relatively independent of process time or temperature after the surface Au layer was consumed.

Because of a low-temperature eutectic in the Au–Ga–As system (4), localized melting may occur for samples annealed at 500°C. It is possible therefore that the 500°C spectra may result from localized erosion and pitting rather than uniform diffusion (see Chapter 6). Evidence of localized pitting has been obtained, using SEM techniques, by both Gyulai et al. (14) and Todd et al. (17) in the range of 450 to 600°C.

Rutherford backscattering is not the ideal technique for distinguishing between Ga and As, and Auger spectroscopy in conjunction with Ar ion sputtering has been employed to study Ga diffusion through Au. A depth profile from the work of Robinson (18) is shown in Figure 11.4 for an anneal at 372°C. Ga and O have accumulated at the surface with much smaller amounts of As. No Ga or As is present in the bulk of the film. Similar results were previously obtained by Todd et al. (17) for anneals at 450°C.

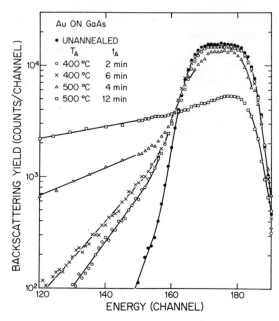

Figure 11.3. Rutherford backscattering spectra of 2 MeV ^{4}He$^+$ ions for 400 Å Au film on GaAs subjected to various alloying treatments. Only the Au profile is shown. From reference 14.

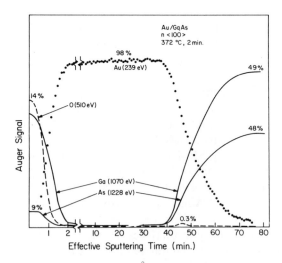

Figure 11.4. Auger depth profile for ~1000 Å Au on GaAs sample after heat treatment at 372°C for 2 min. From reference 18.

Presence of free Ga on the surface of Au is also responsible for the observation that evaporated Au films on GaAs and GaP can melt at temperatures ($\sim 350°C$) considerably below the Au–GaAs/GaP eutectic temperature (20). These lower melting temperatures obviously correspond to that of the Au–Ga eutectic ($341°C$) (2).

The tendency of Ga to outdiffuse through Au and form Ga_2O_3 on the Au surface correlates with an increase in electronegativity in the sequence $Ga \rightarrow As \rightarrow Au \rightarrow O$. This suggests that the electropositive Ga atoms are seeking the most electronegative element. The surface oxide initially acts as a free energy sink for incoming Ga atoms; however, as more Ga accumulates on the Au surface, it is possible that some free Ga may remain underneath the oxidized Ga_2O_3 layer.

Although the relative kinetics of Ga outdiffusion and Au indiffusion have not been established, it seems reasonable to assume that the primary effect is thermal dissociation of GaAs at the interface with Au. This liberates Ga whose migration leaves a relatively large number of vacancies. These Ga vacancies may be expected to be electrically active. If Au were to predominantly occupy vacant Ga sites, then depending on whether its valence state is greater or less than $3+$, it may also act as a donor or an acceptor type of impurity (21). It should be noted, however, that channeling measurements by Gyulai et al. (14) gave no evidence that the diffused Au was residing predominantly on substitutional or tetrahedral interstitial sites.

In summary, the GaAs at the interface is affected in the following ways: (i) the material near the interface has an off-stoichiometric composition with $Ga/As < 1$; (ii) the effective surface doping level may be different from that in the bulk due to Au indiffusion; (iii) the conductor–semiconductor interface is quite diffuse over many interatomic distances. These physical changes correlate with rather profound variations in the electrical behavior of reacted Au–GaAs contacts.

There appear to be only two studies of the interaction of Au films with GaP. Backscattering and channeling studies by Ruth et al. (22) showed no interaction between 1000 Å Au film and GaP after anneals under vacuum at $350°C$ for 2 h. Considerable interdiffusion was observed under the same annealing conditions for 1000 Å Au film on GaAs (Figs. 11.2 and 11.3). It is interesting that the as-deposited Au film on GaP displayed some epitaxy which was improved on annealing. Considerable Ga outdiffusion through thick Au-based contacts (Au–Si and Au–Be) on GaP has been observed for anneals at $600°C$ (23).

11.2.2 Electrical Effects

On n-GaAs, Au initially forms a nearly ideal Schottky barrier as shown by the I–V and C–V data of Figure 11.5. The barrier height is ~ 0.9 V and the

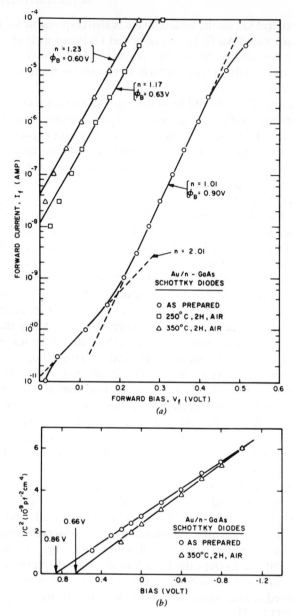

Figure 11.5. Effect of heat treatment on the electrical behavior of Au–*n*-GaAs Schottky diodes: (*a*) forward *I–V* characteristics; (*b*) *C–V* characteristics. From reference 16.

idealiy factor n is ~1.0. The sample used for these measurements consisted of S-doped GaAs with a donor concentration of ~10^{16} atoms cm^{-3}. Heat treatment appreciably degrades the apparent barrier height ϕ_B from ~0.9 V to ~0.6 V, whereas the parameter n increases from ~1.0 to ~1.2. Such increased values of n have been interpreted as being caused by enhanced thermionic field emission which is favored by a small ϕ_B and/or physical irregularities at the interface as a result of interdiffusion.

For more heavily doped n^{+}-GaAs ($N_D \sim 2 \times 10^{18}$ cm^{-3}), even the initial depletion layer width is small enough to be comparable with the extent of the diffuse conductor–semiconductor interface. This situation would favor field emission on a micro three-dimensional scale and consequently a tendency toward ohmic behavior. Such a tendency is illustrated in Figure 11.6, where forward and reverse I–V data are shown on a log–log scale (21). An ohmic contact will be represented by a straight line with slope of 45°. The arrows

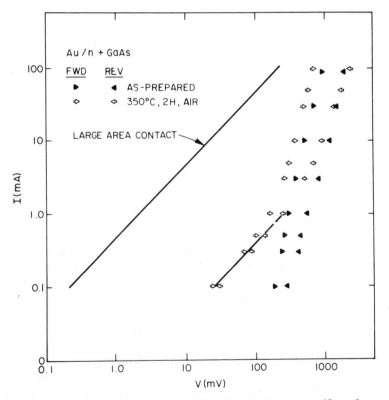

Figure 11.6. Current voltage characteristics of Au–n^{+}-GaAs ($N_D = 1.5 \times 10^{18}$ cm^{-3}) contacts having areas of 0.5 cm^2 (solid line) and 5×10^{-4} cm^2 (data symbols). From reference 21.

pointing to right represent forward currents, and those to the left represent reverse currents. For as-deposited contacts, the forward current is always greater than the reverse current, suggesting rectifying behavior. However, upon heat treatment there is a clear tendency toward development of an ohmic contact at low fields. On an empirical basis, achieving good ohmic contacts to n-GaAs requires adding Ge (n^+ dopant) to the Au. Gold and Ge are either coevaporated or deposited in separate films on the GaAs substrate in their eutectic composition.

Heat treatment of Au contacts to p^+-GaAs is more effective in achieving ohmic contacts than of Au on n^+-GaAs. That is probably due to a combination of the diffuse interface effect and a further reduction in the depletion layer width due to a smaller barrier height of Au on p-GaAs ($\phi_B \lesssim 0.5$ eV).

11.3   COMPOUND FORMATION

11.3.1   Chemistry

Compound formation will be illustrated with respect to Pt–GaAs contacts. One interesting consequence of a dominant metal–metalloid (Pt–As) reaction is that the reaction products are arranged in the form of fairly well-defined layers. This was revealed through a combined Rutherford backscattering analysis and x-ray diffraction study of heat-treated Pt–GaAs samples (15, 16, 22) as well as depth profiling using ion-induced x-rays (24).

The layered structure can be demonstrated (22) quite directly from Rutherford backscattering spectra using rather thin Pt films so that the scattering peaks from the reaction products do not overlap. Figure 11.7 shows the spectra from a 950 Å Pt film on GaAs before and after reaction at 425°C. The high-energy portion of the reacted spectrum—channels 360 to 430— arises almost entirely from backscattering from Pt atoms in the compounds $PtAs_2$ and PtGa. Assuming the presence of only these two compounds, their relative amounts can then be determined (see Chapter 6). The hatched areas shows the profiles for scattering from Pt in PtGa and $PtAs_2$; the profiles for scattering from Ga and As were obtained from the Pt profiles. The peak at channel 340 is just due to the overlap of the Ga and As scattering profiles. The outer 900 Å consists almost entirely of PtGa, while the underlying 800 Å consists of 75 at. % $PtAs_2$ and 25 at. % PtGa. The very small amount of $PtAs_2$ indicated in the surface layer of PtGa is at the level of experimental error. The ion-induced x-ray studies (24), however, also showed a very small concentration of As in the outermost PtGa layer.

The Rutherford backscattering spectra also show that the original GaAs interface moves as a result of this reaction. The amount of GaAs consumed is nearly twice that of the reacted Pt, in accord with the following reaction:

$$3Pt + 2GaAs \rightarrow PtAs_2 + 2PtGa$$

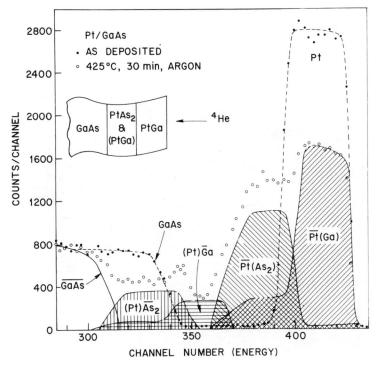

Figure 11.7. Rutherford backscattering spectra of ~950 Å Pt film on GaAs reacted to form PtAs₂ + PtGa. Shaded regions represent scattering profiles for the individual constituents, for example, $\overline{Pt}$ (Ga) represents the scattering profile for Pt in PtGa. From reference 22.

where the densities of the intermetallics are assumed to be those estimated from x-ray lattice parameter measurements (25).

Direct confirmation of a layered structure in Pt–GaAs samples has also been provided by transmission electron microscope (TEM) examination of a cross section of the interface. Figure 11.8 shows a dark-field TEM micrograph of a 3000 Å Pt film on GaAs which was annealed at 300°C for 216 h in air. Crystal structure analysis, using selected area diffraction techniques, showed that the outermost layer was unreacted Pt; this was followed by a layer of fine-grained Pt–Ga compounds and then by a layer of predominantly PtAs₂ near the GaAs interface.

Chemical depth profiles of reacted Pt–GaAs samples have also been obtained by Auger spectroscopy (27, 28) and secondary ion mass spectroscopy (19) using sputter sectioning. Unfortunately, quantitative analysis of the data is quite complicated due to the multicomponent nature of the system and associated differential sputtering effects (see Chapter 6). A major advant-

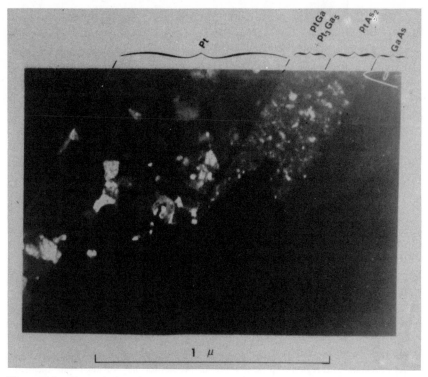

Figure 11.8. Transmission electron micrograph (dark-field) showing a cross section of the layered structure in a reacted sample of 3000 Å Pt on GaAs. From reference 26.

age of the Auger spectroscopy and SIMS techniques is their capability to reveal significant variations in the concentration of light atoms such as oxygen through the reacted layer. For example, Auger analysis of Pt–GaAs samples annealed in air (27) indicated that the reaction products are covered by a layer of predominantly Ga oxides.

In considering the reasons for appearance of a layered arrangement of PtGa–PtAs$_2$–GaAs in reacted Pt–GaAs samples, it should be realized that two distinct effects are involved: (*a*) rapid migration of Ga into Pt and (*b*) formation of a very stable compound PtAs$_2$ at the GaAs interface. The latter effect releases Ga from GaAs, whereas the driving force for Ga outdiffusion seems to arise from the large electronegativity difference between Ga and Pt. Although Ga and Pt have apparently similar atomic radii for 12-fold coordination of 1.411 and 1.587 Å, respectively (25), the favorable size factor for solid solution formation is offset by the relatively large electronegativity difference between Ga and Pt ($\chi_p = 1.6$ and 2.2, respectively).

The effect of the electrochemical factor is evident in the earlier stages of Pt–GaAs reaction when x-ray diffraction shows that Pt_3Ga (Cu_3Au-type) forms. This phase has an ordered close-packed cubic structure (coordination number 12) containing no Ga–Ga nearest neighbors. The high coordination is indicative of metallic-type bonding. As the reaction progresses and more Ga becomes available, PtGa becomes the more dominant Pt–Ga compound. PtGa is also cubic (FeSi type) but the (equivalent) local atomic arrangement around each Ga or Pt atoms consists of seven closer neighbors of opposite kind and six of like kind at a slightly farther distance.

By contrast, the $PtAs_2$ structure (FeS_2 type) displays elements of directional, probably covalent, bonding, especially in the case of As atoms. Each As atom is surrounded by three Pt and one As atom in a tetrahedral arrangement, whereas each Pt atom is octahedrally surrounded by six As atoms (29). Such a tendency toward covalent bonding may be a reason for higher stability of the $PtAs_2$ phase. Moreover, the presence of $PtAs_2$ at the GaAs interface leads to a stable Schottky barrier, which is of considerable importance in the case of several GaAs devices.

Another interesting example of layered reaction products has been observed in the case of Pt–Ti–n-GaAs (30). Figure 11.9 shows the Rutherford backscattering spectra for a sample containing ~ 1000 Å Pt on ~ 500 Å Ti on n-GaAs. At 350°C, there is no reaction at the Ti–GaAs or the Pt–Ti interface. But on annealing at 500°C, Ti tends to form TiAs, which leads to a reduction in the peak height of Ti. The TiAs layer is apparently a poor barrier for a subsequent Pt–GaAs interaction (presumably by Ga and Pt diffusion), which leads to the formation of $PtAs_2$ and Pt_3Ga. This has been inferred from x-ray analysis data as well as from the reduced height of the original Pt peak and its split into two peaks. Interestingly, there is an appreciably separation between the two Pt peaks suggesting that the two layers of Pt_3Ga and $PtAs_2$ are not contiguous. The sequence of reaction products thus appears to be Pt_3Ga–TiAs–$PtAs_2$–GaAs.

Other systems where compound formation has been shown to be dominant include Pd–GaAs, Pt–GaP (22), and Rh–GaAs (31). In the first two systems, annealing in the range of 250 to 350°C produces layers of $PdAs_2$ and PtP_2, respectively, at the semiconductor interface in conjunction with various layered Pd–Ga and Pt–Ga compounds. Annealing Rh–GaAs at 450°C reveals a RhAs layer at the GaAs interface with an outermost layer of RhGa.

Recent work by Robinson (18, 32) on the Ni–Ge–Au–GaAs system also indicates a dominant metal–metalloid interaction, especially at lower temperatures. Samples were prepared by sequentially evaporating a Au–12 wt. % Ge alloy followed by ~ 200 Å of Ni. Owing to fractionation effects during deposition, the alloy film (~ 500 Å) consisted of a layer of Au adjacent to GaAs followed by a layer of Ge. At temperatures below 360°C, there was a

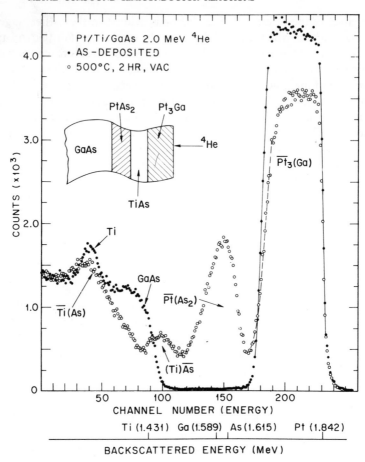

Figure 11.9. Rutherford backscattering spectra of 2 MeV ⁴He⁺ ions from Pt (1000 Å)–Ti (500 Å)–GaAs. Bars over individual elements indicate scattering from that particular element in the compound. From reference 30.

rapid outmigration of Ga, whereas Ni migrated to the interface, probably forming a NiAs intermetallic. This had the effect of increasing the barrier height from that characteristic of Au, Ge–n-GaAs ($\sim$0.68 V) to that associated with a Ni–n-GaAs interface ($\sim$0.83 V). At successively higher temperatures (up to 600°C), a liquid eutectic phase forms between Au–Ge and between Au–GaAs. Ni–As apparently improves the wettability of the liquid and prevents it from balling up. On cooling, this liquid precipitates a layer of n^+ Ge-doped GaAs near the interface which promotes ohmic conduction.

11.3.2 Kinetics

The kinetics of reaction at the Pt–GaAs interface have been investigated by Kumar (33). He utilized x-ray diffraction to establish the endpoint of reaction, at which time PtAs$_2$ and PtGa were the only products present in the metallic layer. It was found that initially the reaction rates are rapid enough to consume all of the Pt film and take the reaction to completion provided the Pt film is thin enough (i.e., less than ~ 2000 Å). For thicker Pt films, the initial rapid reaction phase is followed by sharply reduced kinetics, that is, the reaction has a tendency to become "self-limiting." This aspect is illustrated in Figure 11.10, where the time for completion of reaction at 300°C is shown to increase rapidly with increasing film thickness. The apparent self-limiting nature of the reaction is due to the formation of a continuous layer of PtAs$_2$ at the interface which then acts as a barrier to further rapid outdiffusion of Ga from GaAs.

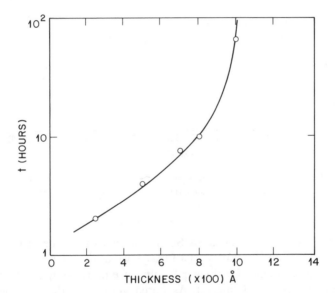

Figure 11.10.   Time for complete reaction at 350°C for Pt films of various thicknesses on GaAs. From reference 33.

The temperature dependence of Pt–GaAs reaction kinetics is shown in Fig. 11.11 in the form of an Arrhenius plot of time for completion of the reaction for 500 Å-thick Pt on GaAs. The activation energy for this reaction is rather high at about 2.3 eV, about twice that commonly observed for grain

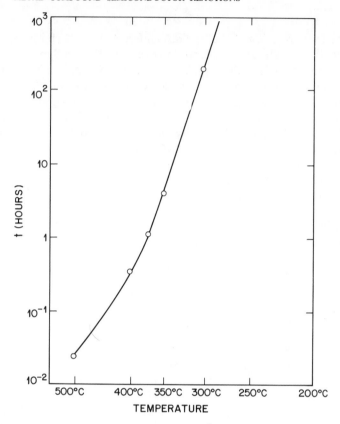

Figure 11.11.   Time for complete reaction for a 500 Å-thick Pt film on GaAs annealed at various temperatures (Arrhenius plot). From reference 33.

boundary diffusion of metal atoms. Such a high activation energy is apparently asociated with the formation of a $PtAs_2$ layer at the interface.

The reaction kinetics for thicker Pt films are more complicated but they are of greater practical interest. As noted above, for Pt film thicknesses greater than 2000 Å, the reaction will not readily go to completion, even above 400°C. In the initial stages, some Ga migrates out of GaAs and distributes itself nearly uniformly throughout the Pt film. This has the effect of decreasing the lattice parameter and increasing the resistivity of the overlying Pt film (33, 34). The average concentration of Ga in Pt solid solution for a given temperature anneal is an inverse function of thickness. Next, a layer of $PtAs_2$ forms at the interface, and for thinner layers Ga continues to migrate through $PtAs_2$ into the overlying Pt. However, this Ga now tends to be confined as a layer of Pt_3Ga, and subsequently of $PtGa$, over the $PtAs_2$ layer. The outermost layer is largely unreacted $Pt(Ga)$ solid solution. The kinetics are now relatively slow because they are limited by the rate of Ga diffusion through the intermetallic $PtAs_2$ layer. However, the GaAs continues to be consumed at a slow but finite rate. Another way to follow the reaction is through electrical measurements of the position of the conductor–semiconductor interface.

Coleman et al. (35) studied the kinetics of GaAs consumption in thick Pt–GaAs couples where the n-GaAs had a nonuniform doping profile, namely, lo–hi–lo (5×10^{15}–10^{17}–5×10^{15} atoms cm^{-3}) used for high-efficiency IMPATT diodes. In such samples the highly doped region acts as a marker whose position can be monitored through differential capacitance voltage measurements using an automatic profilometer. It was shown that the amount of reacted GaAs increases with the square root of time over the temperature range of 300 to 400°C, signifying that the process was diffusion limited. The thermal activation energy for this process was found to be 1.6 eV, which is smaller than that (~ 2.3 eV) reported by Kumar (33) for the case (500 Å Pt) where kinetics were limited by the formation and growth of the $PtAs_2$ layer.

11.3.3 Electrical Effects

The electrical effects of metallurgical interactions in Pt–n-GaAs Schottky diodes have been the subject of several investigations (15, 16, 36). For both partially and fully reacted samples, the barrier height ϕ_B tends to stabilize at a value which is ~ 0.05 V above the barrier height initially measured for as-deposited Pt. Such a stabilization is consistent with the conclusion that the new ϕ_B is associated with the $PtAs_2$–n-GaAs interface, and it is not much affected by the kind of Pt–Ga compounds in the outer layers. Presence of $PtAs_2$ at the GaAs interface also seems to correlate with a significant recombination component in the forward current at low bias. Moreover, Murarka (36) has shown that annealing Pt–n-GaAs diodes in air induces nonideal behavior and a decrease in the barrier height; this was ascribed to the presence of donor-like oxygen atoms near the interface.

With p-GaAs, Pt forms good ohmic contacts, especially after sintering as shown in Fig. 11.12 (21). This has been attributed to the combination of a relatively small barrier height ($\phi_B \lesssim 0.5$ V) and creation of an oxide-free diffuse interface as a result of the sintering operation.

In the case of Pt–Ti–GaAs, the postulated sequence of reaction products (Pt_3Ga–$TiAs$–$PtAs_2$–GaAs) appears to be consistent with the electrical behavior of Pt–Ti–n-GaAs Schottky diodes (30). In the as-prepared state and after a 350°C anneal, these diodes behave the same as Ti on GaAs, with a ϕ_B of ~ 0.81 V and an ideality factor n of ~ 1.05. But after a 500°C

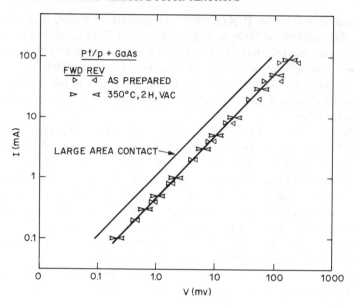

Figure 11.12. Current–voltage characteristics of Pt–p^+-GaAs ($N_A = 2 \times 10^{18}$ cm^{-3}) contacts having areas of 0.5 cm^2 (solid line) and 5×10^{-4} cm^2 (data symbols). From reference 21.

anneal, the electrical measurements show a substantially increased ϕ_B and n, nearly equal to those expected from a dominant presence of PtAs$_2$ at the n-GaAs interface.

11.4 RELATIVELY INERT INTERFACES

If the main driving force for low-temperature dissociation of GaAs is the interaction between Ga atoms and an overlying electronegative metal as observed in the two previous cases, then metals with low electronegativities should be relatively inert. Such a case of a relatively inert metal interface with a compound semiconductor is best illustrated with reference to the W–GaAs contacts. Figure 11.13 shows the Rutherford backscattering spectra of ~850 Å W on GaAs in the as-deposited state and after annealing at 500°C for 2 h in vacuo (15). Clearly, there is no evidence of interdiffusion within the available depth resolution of ~150 Å. The sensitivity of the technique is such that a concentration of the order of 10^{13} atoms cm^{-2} would be detectable within this resolution. This conclusion is supported by electrical measurements on W–n-GaAs Schottky diodes which show a stable ϕ_B, within about 0.04 V, even after extended thermal aging at 350 or 500°C. In contrast with the case of Pt–Ti–n-GaAs diodes, the electrical characteristics of Au–W–n-GaAs and Pt–W–n-GaAs Schottky diodes (see Chapter 13)

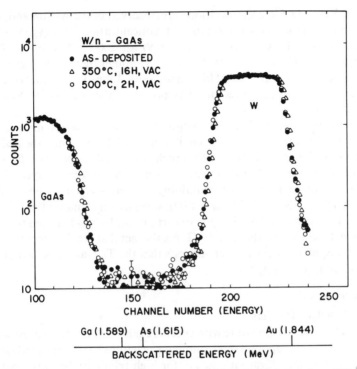

Figure 11.13.   Rutherford backscattering spectra of 2 MeV ^{4}He$^+$ ions from 850 Å W on *n*-GaAs. From reference 15.

remain very stable with thermal aging at up to 500°C, signifying that W acts as a good diffusion barrier between GaAs and Au or Pt (37).

In the introduction we suggested that Al, with its relatively small electronegativity of 1.5, should be classified with the M_3 group of metals that have relatively inert interfaces. Indeed, Kim et al. (19) have compared the reactivities of Al, Au, Au–Ni, and Pt on GaAs using SIMS and found that only the Al–GaAs system is metallurgically stable at temperatures up to 500°C. Johnson et al. (38) have also examined the thermal aging of Al films on GaAs. After annealing at 400 or 500°C for 1 h in vacuo, the 1250 Å Al films are covered with erosion centers that expose the GaAs substrate. The Schottky barrier height increases from 0.73 eV in as-deposited contacts to 0.82 eV after the 500°C anneal. Transmission electron microscopy studies reveal the presence of Al precipitates in the near surface region of the GaAs. There is also indication of AlAs crystallites at the interface. However, depth profiles, obtained by Auger spectroscopy and ion milling, show little interdiffusion or reaction. Very low Ga concentrations are observed in the Al film and at the surface in conjunction with significant concentration of oxygen. Depth profiles obtained by Christou and Day (39) for 2000 Å Al films on GaAs annealed at 450°C show similar behavior. Although there is no large inter-diffusion or reaction between Al films and GaAs, Johnson et al. (38) conclude that the thin interfacial layer of AlAs produces the increase in Schottky barrier height.

Recently, Wada et al. (40), using Auger spectroscopy and ion milling, have investigated the interaction between Ti films and GaAs. They showed that the Ti–GaAs interface is thermally much more stable than the Pt–GaAs interface. However, at 480°C, Ti reacts with GaAs to form a layer of TiAs $+Ti_5As_3$ followed by a layer containing $Ti_2Ga_3 + Ti_5Ga_4$; the resulting barrier height ϕ_B is ~0.8 V (n is ~1.09), substantially smaller than that for $PtAs_2$–n-GaAs ($\phi_B \geqslant 0.9$ V). In spite of certain similarities between the inter-actions in Ti–GaAs and those in Pt–GaAs, the fact that much higher tempera-tures are required for the former suggests that the Ti–GaAs interface should be classified as relatively inert.

11.5   APPLICATIONS

11.5.1   Schottky Barriers

In this section we will illustrate with specific examples how an understanding of various metal–compound semiconductor reactions has contributed toward development of metallization schemes for high-reliability Schottky barriers and ohmic contacts.

Until recently, the n-GaAs IMPATT diodes (Fig. 11.14) employed Pt Schottky barriers overlaid with plated Au which was then bonded to a diamond heat sink (41, 42). A major mode of Schottky barrier degradation was found to be migration of Au toward the GaAs interface during device operation (43). This problem was minimized by making the Pt layer suffic-iently thick (>6000 Å). It was subsequently realized that high-efficiency GaAs IMPATT diodes can be best made using tailored doping profiles in

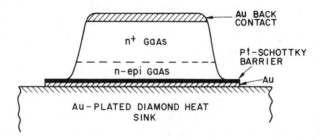

Figure 11.14.   Schematic cross section of a GaAs IMPATT diode. From reference 41.

GaAs (e.g., hi–lo or lo–hi–lo) (44). For such structures, the use of thick Pt film imposes certain basic limitations. Even after the Schottky barrier is initially stabilized by forming $PtAs_2$ at the interface, at the high operating temperatures ($\sim 200°$C) Ga continues to migrate out of GaAs through $PtAs_2$ at a very slow but finite rate. The $PtAs_2$ layer gradually builds up in thickness, causing a gradual consumption of the GaAs epitaxial layer. The rate of GaAs consumption at $\sim 200°$C has been estimated as being ~ 200 Å $(10^4$ h$)^{-1}$ (45). Even such a small change cannot be tolerated for high-efficiency, high-reliability diodes (expected lifetime $> 10^5$ h) since their performance and operating characteristics depend critically on the exact doping profile (i.e., thickness) of the epi-GaAs. On the other hand, one cannot employ the simple expedient of replacing Pt with relatively inert metals such as W or Ti since the ϕ_B values of the latter are too small for effective device operation at 200°C.

An effective solution to the above problem consists of fully reacting a thin Pt film on GaAs to produce a layered structure of $PtGa$–$PtAs_2$–GaAs; this reacted structure is then used in conjunction with a diffusion barrier, for example, W, followed by a Au bonding layer (37). In actual practice, a W–Pt–Au type of metallization is preferred, where the second Pt layer protects the W from oxidation during further processing and also provides a better surface for Au plating (45). Since the Pt is fully reacted, the GaAs interface stays stationary during device operation. The W layer prevents undesirable outdiffusion of Ga and indiffusion of Au. The situation is similar to that encountered in silicon integrated circuit technology where sintered PtSi contacts have been extensively utilized as a thermodynamically stable contact to Si (46) (see also Chapters 2 and 10).

The effectiveness of the above metallization scheme has been demonstrated by the reliability data obtained by Mahoney on IMPATT diodes (45). He monitored the junction movement by an automatic doping profilometer and Schottky barrier degradation by reverse I–V measurements. As shown in Figure 11.15, the use of sputtered W over sintered Pt contacts essentially eliminated aging due to GaAs consumption and also slowed down the tendency of reverse I–V characteristics to soften as a result of Au penetration.

An additional advantage of sputtered W films is that their stresses can be made compressive by suitable choice of deposition parameters (47). This property was utilized by Rozgonyi et al. (48) to compensate tensile stresses in electroplated Au films that caused excessive wafer curvatures, cracking, and yield loss during chip separation. The result of stress compensation was improved yield, very low reverse-bias leakage currents, and good rf performance.

Berenz et al. (49) have recently carried out a study, by SIMS depth profiling and x-ray diffraction, of the efficacy of various "barrier" metals between the

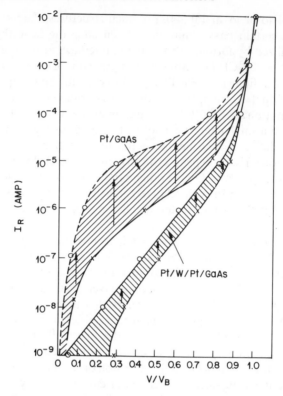

Figure 11.15. Effect of thermal aging on reverse current–voltage characteristics for hi–lo IMPATT diodes metallized with Pt (upper) and Pt–W–Pt (lower). From reference 45.

Pt contact and Au film. Tungsten and Hf are demonstrated to be clearly effective in preventing Ga and As outdiffusion. Titanium and Mo are not as effective in preventing Ga outdiffusion.

11.5.2 Ohmic Contacts

Another important area of recent improvement has been that of ohmic contacts to n-GaAs. Traditionally, formation of ohmic contacts requires an alloying step in which a thin film alloy is melted on the surface of n-GaAs. Alloy contacts which have been successfully employed are Ag–In–Ge (50–52) and Au–Ge (53–55). Recrystallization from the melt results in an n^+ region near the semiconductor–conductor interface, which facilitates formation of an ohmic contact. However, melting is a generally undesirable process step since it can lead to balling up problems and poor dimensional control in, for example, GaAs FET structures (56) (Figure 11.16).

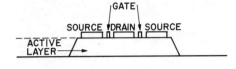

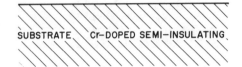

BUFFER LAYER, LOW-DOPED, 1–12 μm THICK

Figure 11.16. Schematic cross section of a GaAs field effect transistor. From reference 50.

One approach to this problem is to evaporate a top layer of Ni to prevent balling of the Au–Ge alloy (57, 58). Section 11.3.1, on the chemistry of compound formation, discussed the work of Robinson (18, 32) where Ni was observed to diffuse through the Au–Ge to form NiAs at the interface. These results are in apparent disagreement with the work of Ohata and Ogawa (59) who observed only Au–Ga grains at the interface. Wittmer et al. (60) have thrown some light on this complex metallization by investigating interdiffusion in Au–Ge–Ni thin films on inert substrates (SiO_2). They find that the Ni acts as a sink for Ge with Ge diffusing out of the Au layer to the Ni layer where it forms stable compounds. This work shows the importance of the Ni and Ge relative thicknesses in determining the diffusion processes. It should be noted, however, that the five-element GaAs–Au–Ge–Ni system has all the potential for behaving very differently from the three-element Au–Ge–Ni system.

A new ohmic contact metallization scheme has been developed which requires sintering (i.e., solid-state interdiffusion) rather than melting (21). The as-deposited metallization is Pd–Ge–n-GaAs which, upon sintering, reacts to form Pd_2Ge, $PdAs_2$ and $PdGa$. Ohmic behavior (Fig. 11.17) apparently results from a combination of the doping action of Ge donor atoms and fast in-diffusion kinetics of Pd in GaAs. The metallurgical junction so formed is comparable in depth to the depletion layer width; this favors ohmic behavior through micro three-dimensional field emission (61).

Unfortunately, it has not yet been possible to devise sintered ohmic contacts to n-GaP. Formation of "alloyed" contacts to GaP LEDs (62) (Fig. 11.18) involves melting a Au–Si alloy on n-GaP and a Au–Be alloy on p-GaP. With such contacts, Ga migration through Au is a potential reliability problem which can cause electrical degradation of the contact and also poor bondability of the Au due to formation of a Ga_2O_3 layer on the Au surface.

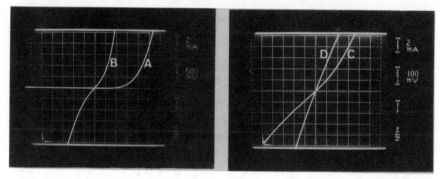

Figure 11.17. Current–voltage characteristics of progressively sintered Pd (500 Å)–Ge (500 Å) contacts to n-GaAs (10^{16} cm^{-3}): (A) as deposited; (B) sintered at 500°C for $\frac{1}{2}$ h; (C) sintered at 500°C for 1 h; (D) sintered at 500°C for 2 h. From reference 21.

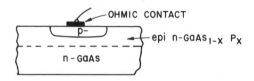

Figure 11.18. Schematic cross section of a monolithic GaAs$_{1-x}$P$_x$ light-emitting diode display. From reference 63.

This problem was effectively solved by addition of a Ga diffusion barrier of W or Ti between the alloyed layer and the final Au bonding layer (63).

REFERENCES

1. B. Schwartz, Ed., *Ohmic Contacts to Semiconductors*, Electrochemical Society, New York (1969).
2. M. Hansen, *Constitution of Binary Alloys*, McGraw-Hill, New York (1958).
3. R. P. Elliott, *Constitution of Binary Alloys, First Supplement*, McGraw-Hill, New York (1965).
4. M. B. Panish, *J. Electrochem. Soc.*, **114**, 516 (1967).
5. A. K. Sinha, B. C. Giessen, and D. E. Polk, "Crystalline and noncrystalline Solids" in *Treatise on Solid State Chemistry*, Vol. 3, N. B. Hannay, Ed., Plenum, New York (1976), p. 1.
6. V. Heine and D. Weaire, in *Solid State Physics*, Vol. 24, H. Ehrenreich, F. Seitz, and D. Turnbull, Eds., Academic Press, New York (1970), p. 249.
7. L. Kaufman, in *Phase Stability in Metals and Alloys*, P. S. Rudman, J. Stringer and R. I. Jaffee, Eds., McGraw-Hill, New York (1967), p. 125.
8. W. Hume-Rothery and G. V. Raynor, *Structure of Metals and Alloys*, Institute of Metals, London (1962).
9. A. K. Sinha, in *Progress in Materials Science*, B. Chalmers, J. W. Christian, and T. B. Massalski, Eds., Pergamon Press, Oxford (1972), p. 79.
10. L. Pauling, *Nature of the Chemical Bond*, Cornell University Press, Ithaca (1960).
11. J. C. Phillips, *Bonds and Bands in Semiconductors*, Academic Press, New York (1973).

12. S. Kurtin, T. C. McGill, and C. A. Mead, *Phys. Rev. Lett.*, **22**, 1433 (1969).
13. A. K. Sinha, *Trans. AIME*, **245**, 237 (1969); *idem*, **245**, 911 (1969).
14. J. Gyulai, J. W. Mayer, V. Rodriquez, A. Y. C. Yu, and H. J. Gopen, *J. Appl. Phys.*, **42**, 3578 (1971).
15. A. K. Sinha and J. M. Poate, *Appl. Phys. Lett.*, **23**, 666 (1973).
16. A. K. Sinha and J. M. Poate, Proc. 6th International Vacuum Congress, *Jap. J. Appl. Phys.*, Suppl. 2, Pt 1, 841 (1974).
17. C. J. Todd, G. W. B. Ashwell, J. D. Speight, and R. Heckingbottom, *Inst. Phys. Conf. Ser.* (London), **22**, 171 (1974).
18. G. Y. Robinson, *J. Vac. Sci. Technol.*, **13**, 884 (1976).
19. H. B. Kim, G. G. Sweeney, and T. M. S. Heng, *Inst. Phys. Conf. Ser.* (London), **24**, 307 (1975).
20. H. Nakatsuka, A. J. Domenico, and G. L. Pearson, *Solid State Electron.*, **14**, 849 (1971).
21. A. K. Sinha, T. E. Smith, and H. J. Levinstein, *IEEE Trans. Electron. Devices*, **ED22**, 218 (1975).
22. R. L. Ruth, J. M. Poate, and A. K. Sinha, unpublished data (1975).
23. W. Brantley, B. Schwartz, V. G. Keramidas, G. W. Kammlott, and A. K. Sinha, *J. Electrochem. Soc.*, **122**, 434 (1975).
24. L. C. Feldman and P. J. Silverman, in *Ion Beam Surface Layer Analysis*, O. Meyer, G. Linker, and F. Käppeler, Eds., Plenum Press, New York (1976), p. 735.
25. W. B. Pearson, *Handbook of Lattice Spacings and Structures of Metals*, Vol. 2, Pergamon Press, Oxford (1967).
26. P. M. Petroff and T. T. Sheng, private communication.
27. C. C. Chang, S. P. Murarka, V. Kumar, and G. Quintana, *J. Appl. Phys.*, **46**, 4237 (1975).
28. M. C. Finn, H. Y. P. Hong, W. T. Lindley, R. A. Murphy, E. B. Owens, and A. J. Strauss, paper presented at AIME Electronic Materials Conf., Las Vegas (1973), unpublished.
29. W. B. Pearson, *The Crystal Chemistry and Physics of Metals and Alloys*, Wiley-interscience, New York (1972).
30. A. K. Sinha, T. E. Smith, M. H. Read, and J. M. Poate, *Solid State Electron.*, **19**, 489 (1976).
31. C. J. Todd, J. D. Speight, G. W. B. Ashwell, and R. Heckingbottom, ECS Extended Abstracts, No. 119, Spring Meeting, Toronto, (1975), p. 274.
32. G. Y. Robinson, *Solid State Electron.*, **18**, 331 (1975).
33. V. Kumar, *J. Phys. Chem. Solids*, **36**, 535 (1975).
34. S. P. Murarka, *Solid State Electron.*, **17**, 985 (1974).
35. D. J. Coleman, W. R. Wisseman, and D. W. Shaw, *Appl. Phys. Lett.*, **24**, 355 (1974).
36. S. P. Murarka, *Solid State Electron.*, **17**, 869 (1974).
37. A. K. Sinha, *Appl. Phys. Lett.*, **26**, 171 (1975).
38. N. M. Johnson, T. J. Magee, and J. Peng, *J. Vac. Sci. Technol.*, **13**, 838 (1976).
39. A. Christou and H. M. Day, *J. Appl. Phys.*, **47**, 4217 (1976).
40. O. Wada, S. Yanagisawa, H. Takanashi, *Appl. Phys. Lett.*, **29**, 263 (1976).
41. J. C. Irvin, D. J. Coleman, W. A. Johnson, I. Tatsuguchi, D. R. Kecker, and C. N. Dunn, *Proc. IEEE*, **59**, 1212 (1971).
42. W. R. Wisseman, D. W. Shaw, R. L. Adams, and T. E. Hasty, *IEEE Trans. Electron. Devices*, **ED21**, 317 (1974).
43. J. C. Irvin, in *Proceedings of the 4th Biennial Cornell Electronic Engineering Conference*, Cornell University Press, Ithaca (1973), p. 287.
44. S. Su and S. M. Sze, *IEEE Trans. Electron. Devices*, **ED20**, 541 (1973).
45. G. E. Mahoney, *Appl. Phys. Lett.*, **27**, 613, (1975).
46. M. P. Lepselter, *Bell Syst. Tech. J.*, **40**, 233 (1966).
47. R. S. Wagner, A. K. Sinha, T. T. Sheng, H. J. Levinstein, and F. B. Alexander, *J. Vac. Sci. Technol.*, **11**, 582 (1974).

48. G. A. Rozgonyi, J. V. DiLorenzo, and E. Heinlein, *J. Electrochem. Soc.*, **121**, 426 (1974).
49. J. J. Berenz, G. J. Scilla, V. L. Wrick, L. F. Eastman, and G. H. Morrison, *J. Vac. Sci. Technol.*, **13**, 1152 (1976).
50. R. H. Cox and H. Strack, *Solid State Electron.*, **10**, 1213 (1967).
51. R. H. Cox and T. E. Hasty, in *Ohmic Contacts to Semiconductors*, B. Schwartz, Ed., Electrochemical Society, New York (1969), p. 88.
52. T. Sebestyen, H. Hartnagel, and L. H. Herron, *Electron. Lett.*, **10**, 372 (1974).
53. N. Braslau, J. B. Gunn, and J. L. Staples, *Solid State Electron.*, **10**, 381 (1967).
54. J. S. Harris, Y. Nannichi, G. L. Pearson, and G. F. Day, *J. Appl. Phys.*, **40**, 4575 (1969).
55. S. Knight and C. Paola, in *Ohmic Contacts to Semiconductors*, B. Schwartz, Ed., Electrochemical Society, New York (1969), p. 102.
56. B. S. Hewitt, H. M. Cox, H. Fukui, J. V. Di Lorenzo, W. O. W. Schlosser, and D. E. Iglesias in "GaAs and Related Compounds," Institute of Physics Conf. Series 33-A (1977).
57. W. D. Edwards, W. A. Hartman, and A. B. Torrens, *Solid State Electron.*, **15**, 387 (1972).
58. K. Heime, U. König, E. Kohn, and A. Wortmann, *Solid State Electron.* **17**, 835 (1976).
59. K. Ohata and M. Ogawa, in *12th Annual Proceedings of the IEEE Reliability Physics Symposium*, Las Vegas (May 1974).
60. M. Wittmer, R. Pretorius, J. W. Mayer, and M.-A. Nicolet, to be published.
61. A. K. Sinha, *J. Electrochem. Soc.*, **120**, 1967 (1973).
62. A. A. Bergh and P. J. Dean, *Light Emitting Diodes*, Clarendon Press, Oxford, (1976).
63. W. A. Brantley, B. Schwartz, V. G. Keramidas, A. K. Sinha, and G. W. Kammlott, *J. Electrochem. Soc.*, **122**, 1152 (1975).

12

SOLID PHASE EPITAXY

S. S. Lau

California Institute of Technology, Pasadena, California

W. F. van der Weg

Philips Research Laboratories, Oosterringdijk 18,
Amsterdam, The Netherlands

12.1   INTRODUCTION

12.1.1   Basic Concept

The basic goal of solid phase epitaxy (SPE) is to achieve epitaxial growth on single-crystal substrates by means of solid-state reactions. Generally speaking, SPE can be divided into two catagories: (i) SPE processes that require transport media such as a metal or a compound layer, (ii) SPE processes that require no transport medium, such as reordering or recrystallization of amorphous semiconductor layers obtained by ion implantation. (The general phenomenon of epitaxy has been reviewed in references 1 and 1a.)

12.1.2   SPE with Transport Medium

For SPE processes of the first type, it is basically a precipitation process of crystals out of solid solution. Precipitation has been a phenomenon long studied in physical metallurgy. These metallurgical principles are though to apply to the vicinity of the semiconductor–metal interfaces.

The configuration used in this type of SPE process usually consists of a semiconductor substrate onto which a thin metal layer (≈ 1000 to $5000\,\text{Å}$) is deposited, for example, thin Al layers ($\approx 1000\,\text{Å}$) deposited on Ge single-crystal substrates (2, 3). As the thin film composite structure is brought up in temperature but well below any eutectic melting temperatures, dissolution of semiconductor into the metal will take place until the solubility limit is reached. The composite structure is then cooled slowly to promote precipitation and growth of the dissolved atoms onto the substrate. Epitaxial layers thus formed will incorporate some metal atoms into the regrown layer. By proper choice of the metal layer, the regrown semiconductor layer will be doped with the proper dopant, as in the case of Al–Ge (xtal) where the epitaxial layer is now p-type doped with Al atoms.

An extended configuration for this type of SPE process is to deposit again a layer of amorphous semiconductor onto the metal layer so that the composite now has a structure of semiconductor (amorphous)–metal–semiconductor (crystal), as in the case of Ge (a)–Al–Ge (xtal). During the annealing processing steps of this structure, the amorphous semiconductor, because of its higher free energy, has a tendency to dissolve and diffuse across the metal layer and crystallize epitaxially onto the substrate (4, 5). The advantage of using this configuration is that the thickness of the epitaxial layer is essentially limited only by the thickness of the deposited amorphous semiconductor layer.

Semiconductors such as Ge and Si usually have thin native oxide layers on the surfaces. The presence of this oxide layer prevents uniform interactions between the semiconductor and the metal layer. As a result, the growth layer may have an island structure rather than a uniform and continuous epitaxial layer across the substrate surface. To promote uniform interactions between the semiconductor and the metal layer at the interfaces, metals that react easily with semiconductors without being impeded by the native oxide at low temperatures are often used as transport media, such as Pd in the Si (a)–Pd–Si (xtal) system. In this case a compound is first formed between the metal and the semiconductor in the initial low-temperature step of the annealing procedure to achieve "clean" surface conditions. The compound layer is then used as a transport medium upon further annealing. For this type of SPE sample there is evidence that more complex transport mechanisms are involved.

One of the major advantages of SPE processes is the relatively low growth temperatures ($\approx 500°C$). This low-temperature advantage makes SPE attractive and potentially useful in semiconductor and solar cell technologies. Problems generally associated with high-temperature processes such as autodoping and dopant profile modification can be avoided by this low-temperature process. Also, solid phase epitaxy is fully compatible with photolithographic techniques. The metal (or compound) layers, which are located on the surface of the epitaxial layer after transport, usually form good ohmic contacts with the epitaxial layers. In fact, one convenient and sometimes only method to form ideal metal–semiconductor ohmic contacts is to use the concept of SPE as in cases of high-purity Ge (3) and GaAs (6). The use of SPE also gives an extra dimension of control of the geometry and areas for deposition on the samples. This can be achieved by mechanical masking or ion beam deposition techniques. This type of convenient control is difficult to obtain in liquid or vapor phase epitaxy.

12.1.3 SPE Without Metal Transport Media

In the case of SPE without a transport medium, the amorphous semiconductor layer is obtained by either deposition techniques or by ion implantation into the single-crystal substrates to a sufficiently high dose, followed by a heat treatment to recrystallize the amorphous layer epitaxially on the substrate. For a complete epitaxial regrowth of the amorphous layers, it is important that the interface between the amorphous layer and the crystalline substrate is free of any oxide or "dirty" layers. This condition is readily obtained in the ion-implanted samples since the interface between amorphous and crystalline materials has never been exposed. It is the SPE process in ion-implanted layers that we discuss in more detail in this chapter.

The motivation to study the regrowth behavior is due to the current trend in ion implantation for device applications, where high energy, high current ($> 10\,\mu A\ cm^{-2}$), and high dose (10^{15} to $10^{16}\ cm^{-2}$) are required. This trend is caused by interest in heavily doped regions for bipolar structures and buried collector layers and for greater sample throughput in production runs. The dose is sufficiently high so that amorphous layers are formed. The epitaxial regrowth behavior of the implanted layers plays important roles in the electrical properties of these layers used in device applications. It is therefore understandable that there is substantial interest in this process.

In the following sections we attempt to review briefly the results available in the literature and also report on some of the more recent advances in the field of solid phase epitaxy.

12.2   SPE WITH TRANSPORT MEDIUM

12.2.1   Metal–Semiconductor Reactions; General Considerations

In general, we can divide the reactions of semiconductors with a metal layer into two classes. In first class are interactions in which compounds (e.g., silicides or germanides) are formed such as Pt–Si (see Chapter 10). In the second class we group those semiconductor–metal combinations which form eutectic systems and do not form intermetallic compounds. Examples of the second class are the Si–Au and the Ge–Al systems.

A salient feature of interactions between semiconductors and thin metal layers is that solid phase reactions occur at very low temperatures, that is, well below eutectic temperatures, or, in the compound-forming case, well below reported temperatures of compound formation in bulk diffusion couples.

12.2.2   The Al–Ge System

12.2.2.1 Introduction.   The phase diagram between Al and Ge is relatively simple, where a eutectic composition of 2.8 at. % of Ge is formed at 424°C. The investigation of solid phase epitaxy in this system started in about 1970; the motivation was to form thin blocking and injection contacts on very high-purity Ge radiation detectors (2). For proper operation of the Ge detectors, it is necessary to form these contacts at temperatures low enough to prevent a deterioration of the starting high-purity material. p-Type contacts on Ge were formed by evaporating a 500 to 1000 Å-thick Al film onto a high-purity Ge single-crystal substrate (net dopant concentrations between 2×10^{10} and 5×10^{11} cm^{-3}), followed by heat treatment at 380°C for 1 h in a N_2 atmosphere and a slow cooling period. n-Type contacts were made by a sequential deposition of 200 to 400 Å Sb, 200 to 400 Å Ge, and 500 to 1500 Å of Sb, followed by the same heat treatment.

The I–V characteristics of the nuclear particle detector thus obtained in combination with measurements of reverse recovery times showed that the contacts formed in this way were indeed of the blocking and injecting types. Backscattering analysis (see Chapter 6) showed that intermixing of the various components had taken place during the heat treatment. This initial success stimulated a more detailed study of the Ge–Al system.

12.2.2.2 The Regrowth Process. The solid phase growth of Ge from evaporated Al layers was demonstrated in the following way (7). An Al layer about 6000 Å thick was evaporated into a clean Ge substrate. The sample was then heated at temperatures between 120 and 400°C for times between 15 min and 2 h and then quenched to room temperature. A first result is that the measured Ge concentration at the two temperatures (0.18 at. % at 250°C and 0.78 at. % at 350°C) agrees well with the previously reported equilibrium solid solubility of Ge in Al. A second result is that the solution and growth process is a reversible one, since the concentration of Ge dissolved into the Al depends only on the final temperature treatment, irrespective of previous temperature cycles. It is important to realize at this point that the maximum thickness that can be obtained for a regrown layer in this way is dictated by the solubility of Ge into Al at the processing temperature and by the Al film thickness. For example, only 60 Å of Ge would be dissolved from a single-crystal substrate into a 5000 Å Al film at 300°C.

12.2.2.3 Electrical Properties of the Regrown Layer. The 0.2 at. % solubility of Al in the regrown Ge layer causes a p^+ character of this layer, and this effect has been put to advantage as a simple way of fabricating diodes and transistors. In one case (8), the solid phase reaction of Al on n-type Ge at 310°C was used to form the emitter of a transistor.

In order to elucidate further the origin of the p-type doping of the regrown Ge layers, Hall effect measurements were performed on samples which had been heat treated at 400° for 30 min and at 250°C for 60 min. Typical values obtained for sheet resistivity ρ_s and Hall mobility μ_H were 22 Ω^{-1} and 16 cm^2 V^{-1} s^{-1}, respectively. This Hall mobility agrees well with reported measurements on heavily doped Ge crystals. Furthermore, the concentration of holes in the layer did not depend on process time at 400°C. This latter observation rules out the possibility that the p-type character was caused by diffusion of Al into the Ge substrate.

12.2.2.4 Growth by Dissolution and Precipitation. The morphology of reacted Ge–Al samples was studied by several methods combining backscattering and channeling techniques with transmission electron microscopy, scanning electron microscopy, and electron microprobe analysis. In the experiments described above (Sections 11.2.2.2 and 12.2.2.3) the Ge substrate always was a single crystal. In these cases, channeling measurements showed that the regrown layers were epitaxial on the substrate. For example, in a structure consisting of single-crystal Ge, having a ⟨111⟩ oriented surface with an evaporated layer of 1.1 μm Al, good epitaxial regrowth was observed after annealing to 400°C (30 min) and subsequent slow cooling (15°C min^{-1}). Channeling measurements with 400 keV He$^+$ ions, using the ⟨111⟩ axis of the substrate, indicated that the grown layer had the same axis of orientation and the same threefold symmetry as the substrate. The value of the minimum yield, being 5%, gave a further indication of the good single-crystal quality of the regrown layer. In conclusion, therefore, the reported measurements lead to a simple model for the regrowth process, which is also illustrated in Figure 12.1. Basically, upon heating, substrate atoms dissolve into the metal layer until the solubility limit is reached. Upon cooling, this solid solution becomes supersaturated, which precipitates an epitaxial regrowth of dissolved atoms back onto the substrate.

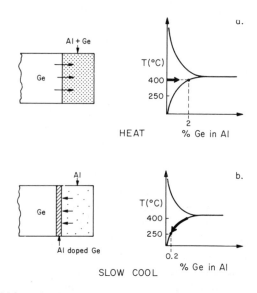

Figure 12.1.   Model for solid phase epitaxial growth: (a) Ge rapidly dissolves and diffuses into the Al film until the solubility limit, 2 at. % at 400°C, is reached. (b) When the sample is cooled slowly to 250°C and held there, the solubility of Ge in the Al film drops to 0.2 at. %, with the majority of the excess Ge growing onto the surface of the Ge substrate. The Ge layer incorporates Al while it is growing. From reference 3.

The situation is quite different, however, when we consider the interactions between an amorphous semiconductor layer and an evaporated metal film. This is borne out by a series of experiments of Ottaviani et al. (4, 5, 9, 10) in which amorphous Ge–Al samples were prepared by vacuum deposition of these two elements on Si, Ge, or C substrates. Heat treatment of such a sample at temperatures as low as 100°C caused the Ge to move into the metal film. One can summarize the behavior of amorphous Ge in contact with an Al film in the following way. A low-temperature annealing cycle carries the amorphous semiconductor into its lower-energy state, that is, the crystalline state by way of dissolution of Ge into the metal film, followed by the precipitation of semiconductor crystallites out of a solid solution. The driving force for this dissolution and precipitation reaction is thought to be provided by the higher free energy of the amorphous material, while the metal layer only acts as a solvent medium. For example, the heat of transformation from amorphous Ge thin film to crystalline Ge is found to be between 2.75 and 3.8 kcal mole^{-1} (10a, 10b). Therefore, a driving force of this order of magnitude is expected for these reactions. Although fairly large crystallites can be formed in this way, they appear not to be oriented with respect to the crystal substrates.

12.2.2.5 The Ge–Al–Ge System.

So far we have seen that one can obtain epitaxial regrowth of Ge by first dissolving it into an Al film and also that one can form large crystallites of Ge in Al when one uses amorphous Ge as a starting material. The first process is limited by the solid solubility of Ge into Al, which results in very thin regrown layers, while the second process does not produce epitaxial growth. It is therefore natural to combine these two methods to achieve SPE growth in larger thicknesses. For this reason the system consisting of single-crystal Ge with sequentially evaporated layers of Al and amorphous Ge was studied (11). Sample preparation consisted of evaporating Al films with thicknesses between 0.3 and 1.5 μm and without breaking vacuum, and depositing Ge films with thicknesses between 0.3 and 1 μm on Ge wafers preetched, rinsed, and dipped in HF. These structures were heated to 300°C and quickly cooled down. Visual comparison of such a sample before and after anneal readily shows that areas previously covered with Ge have an Al color after annealing. Backscattering spectra obtained from such a sample are shown in Figure 12.2. Apparently, Ge atoms originally situated in the outermost amorphous layer have been transported through the Al layer and have regrown on the crystalline Ge substrate. Channeling measurements with 210 keV protons indicate that the regrown layer is epitaxial with the substrate. This layer is well ordered, as evidence by a minimum yield of 5%. From the backscattering data of Figure 12.2 it can be calculated that 81% of the initial amorphous Ge layer was migrated to the substrate while the remaining 19% is distributed in the Al film. One also infers from this figure that the thickness of the regrown layer is about 5000 Å. It is of utmost importance to take great care in surface preparation for obtaining reproducible results. Reproducible regrowth could be obtained only when the substrate Ge crystal was first sputtered with 1.5 keV Ar before the evaporating steps (in the same vacuum system) (12). If the surface is not properly cleaned, a nonuniform layer or a polycrystalline layer may result as discussed in Section 2.2.4.

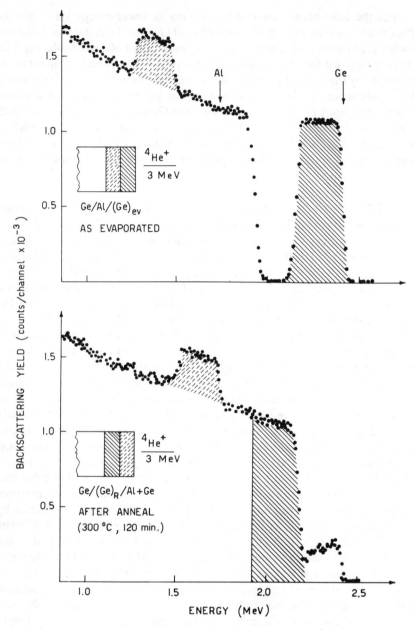

Figure 12.2. Backscattering spectra for 3 MeV ^{4}He$^+$ incident on a 4970 Å Ge layer evaporated on a 5160 Å Al layer deposited on crystalline substrate. Upper portion of the figure shows the as-deposited case. Lower portion shows the spectrum after anneal at 300°C for 2 h. Dashed area represents the regrowth layer, having a thickness of 4800 Å. Arrows indicate the energy corresponding to scattering from atoms at the surface. From reference 11.

It is also shown (12) that the epitaxially grown layer is electrically doped. Hall effect measurements on Van der Pauw structures showed that the Al, which is present in the regrown Ge layer, causes p^+-type doping with a carrier concentration of about 10^{21} cm^{-3}. The measured Hall mobility values range from 20 to 25 cm^2 V^{-1} s^{-1} in the temperature range of 90 to 220°K, which agrees with earlier measurements on heavily doped Ge.

In conclusion, it has been shown that crystal Ge–Al–amorphous Ge structures can be fruitfully employed to achieve SPE growth of doped layers. It remains, however, to study in a systematic way important aspects such as the influence of various impurities on the SPE process and the growth kinetics. Some of these problems have been investigated in more detail in the crystal Si–Pd–amorphous Si system (see Section 2.7).

12.2.3   The Al–Si System

12.2.3.1 Introduction.   This system, consisting of a Si crystal with evaporated Al layer, behaves in many respects like the Ge–Al system. One notable difference, however, is the role of the Si–Al interface as a reaction-limiting factor. It appears that it is experimentally very difficult to achieve uniform SPE growth because of the almost unavoidable presence of oxides on the original Si surface even with sputter cleaning. (See Chapter 1 and reference 13 for a general discussion of the Al–Si system.)

12.2.3.2 Interaction of Single-Crystal Si with Evaporated Al.   Much of the knowledge about reactions occurring in the crystal Si–Al system is derived from the work of McCaldin et al. We summarize only some of the results here. An initial observation (14) was that the solubility of Si into an evaporated Al layer (3 μm thick) agreed with the reported values for Si in bulk Al. The diffusivity of Si into the Al layer, however, was considerably larger (more than one order of magnitude) than the diffusivity of Si into wrought Al. The authors (14) ascribe this enhancement in the thin film to a high concentration of imperfections such as dislocations and grain boundaries in the Al film. Regrowth of Si was obtained by evaporating 10 to 20 μm Al onto ⟨111⟩ oriented Si substrates, followed by an annealing treatment of 5 min at 540°C, slow cooling to 150°C, and quenching the samples to room temperature (15). A typical example of Si mesa growth obtained in this way, using a Si crystal with oxide cuts, is shown in Figure 12.3. This SEM picture shows a characteristic island structure of the growth of Si from Al films. It appears that near the boundaries of oxide cuts a preferred growth is obtained. The Si islands have lateral dimensions of several μm and heights up to 2000 Å. It is in fact possible to obtain a single connected Si mesa by using very small (a few μm) oxide cuts. The favored growth of Si near the oxide boundaries was explained in terms of the stress in the Al film because of the volume decrease occurring

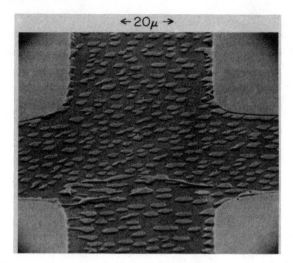

Figure 12.3. Silicon mesas grown onto the exposed substrate Si (cross region) of the prototype integrated circuit, as exhibited by SEM after removal of the Al metallization. Tilt angle, 65°. The metallization had been provided with ∼1.2 at. % Si by coevaporation. Oxide from the photolithographic preparation is still present in the four corners. From reference 15.

when Si dissolves into the Al layer. Application of external force during annealing of a crystalline Si–evaporated Al sample showed that dissolution of Si was enhanced in regions of high pressure, while preferential growth occurred in areas with less stress (15). The typical island growth of Si from an Al layer was also confirmed in another investigation (16).

One way to achieve a relatively clean Si–Al interface is to enhance the dissolution of Si into Al (17). In this way one hopes to provide a clean interface, which can act as a starting point for laterally homogeneous epitaxial growth of Si. Indeed, recently planar growth of Si was obtained using this principle (see Section 12.2.3.4).

12.2.3.3 Interaction of Polycrystalline Si with Evaporated Al.

In order to understand the growth mechanism of Si crystallites in contact with metal layers, it is worthwhile not only to consider the interaction of single-crystal Si with metals but also the behavior of polycrystalline or amorphous Si contacting a metal layer. We shall presently concentrate on the interactions between polycrystalline Si and evaporated Al. This combination was studied in some detail by Nakamura et al. (18–20) by SEM, backscattering, Auger electron spectrometry, and x-ray diffraction. For these experiments, Si wafers were oxidized to a thickness of about 1000 Å and covered by a 0.3 to 0.5 μm-thick layer of polycrystalline Si formed by chemical vapor deposition

(CVD) at 640°C. This structure was covered by an evaporated Al layer 1.2 μm thick. In some of the experiments the Al layer thickness was reduced by etching.

A case in which the initial poly-Si thickness (0.5 μm) was smaller than the Al layer thickness is shown in Figure 12.4. SEM pictures obtained at various annealing temperatures show that in this case large precipitates develop from the poly-Si layer. These precipitates were identified as Si crystallites by x-ray diffraction with a Read camera. The crystallites have different sizes, and their height equals the original Al thickness. It also appears from the SEM photographs that the crystallites are fairly narrow at their base where they contact the SiO_2 substrate.

A different situation develops when the poly-Si layer is thicker than the Al layer. In this case it appears that upon annealing, a continuous layer of Si develops. Depth profiles of Al and Si as obtained with Auger electron spectrometry in combination with sputtering or alternatively by He backscattering show that the outer layer of a sample consists of Si. This layer resides over a mixed layer consisting of Al and Si. Backscattering also showed that the thickness of the outer Si layer equals the initial Al layer thickness.

The result of these studies is schematically indicated in Figure 12.5. The conclusion is that during annealing the poly-Si dissolves and forms precipitates in the Al layer. These crystallites will grow until the outer surface is reached. If there is enough supply of Si, these precipitates will continue to grow laterally (Figs. 12.5a to 12.5c) and otherwise will remain as pillar-like structures in the Al (Figs. 12.5d to 12.5f). Later work (19) showed that this grain growth also occurs for poly-Si layers in contact with Ag and Au. The interesting conclusion of this work is that not only amorphous Si in contact with a metal layer tends to crystallize but that the decrease in free energy (or grain boundary energy) provides enough driving force to induce grain growth in the poly-Si case at low temperatures.

12.2.3.4 The Si–Al–Si System.   By analogy to the Ge–Al–Ge system described in Section 2.2.5, and in view of the results described in the previous section, it seems relatively straightforward to try to achieve SPE growth in the system consisting of Si (a)–evaporated Al–Si (xtal). However, these experiments were only partially successful. It was not possible to obtain large-area uniform growth of Si in contact with an Al film. This was thought to be mainly due to the experimental difficulty of achieving sufficient clean Si surface before Al evaporation. This can also be inferred from Figure 12.3, where growth does occur, but in isolated mesas.

Nevertheless, it has been shown (21) that the Si (a)–Al–Si (xtal) concept can be successfully applied to produce SPE in small areas. These experiments started by producing dissolution pits by heating structures consisting of

Figure 12.4. Scanning electron micrographs of samples with 1 μm Al on 0.5 μm poly-Si: (*a*) before annealing, and after annealing for 30 min at (*b*) 400°C and (*c*) 440°C. The Al was chemically removed before examination in all cases. The incident angle of the electron beam with respect to the normal of the sample is 60°. From reference 18.

444

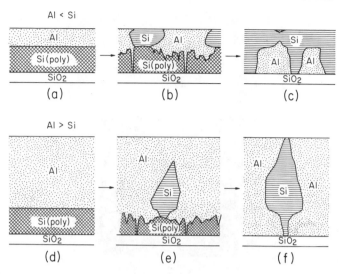

Figure 12.5.   Schematic diagrams showing the changes that occur during annealing: (*a*), (*b*), (*c*) for samples with original Al layer thinner than the poly-Si layer; (*d*), (*e*), (*f*) for samples with original Al layer thicker than the poly Si layer. From reference 18.

n-type ⟨111⟩ Si substrates covered by an oxide with photolithographically produced cuts covered by an evaporated Al layer. At annealing temperatures of about 500°C (up to 20 min) dissolution of Si into Al proceeds in such large quantities that the Al–Si interface recedes into the substrate, under the oxide window. The significance of this step is to move the Si–Al interface away from impurities at the original surface. Subsequently, a Si layer (thickness between 900 and 5000 Å) was evaporated on this structure without breaking vacuum, followed by annealing at 475 to 525°C for 10 to 20 min. These processes, together with SEM photographs of the surface, are shown in Figure 12.6. The evaporated Si had reacted with the Al, with the result that the original dissolution pit was filled with grown Si, while the outer layer consisted of Al with Si precipitates. The pit is so smoothly filled with Si, that it only becomes visible by reverse biasing of the sample (the grown layer is *p*-type because of doping with Al). An analysis of the geometry of the dissolution pits at various stages in the reaction with the evaporated Si layer leads to some interesting conclusions about the growth mechanism.

First, during the formation of the pit the dissolution reaction itself is very fast and causes faceting of the dissolution pit. The pit is bounded by the inclined {111} planes. In the second stage, however, where the pit is filled in by the evaporated Si, it appears that transport of Si to the growth site dominates the reaction. In the central portions of the pit the growth of Si

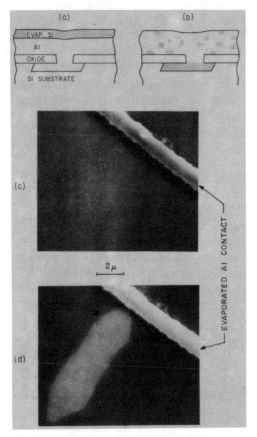

Figure 12.6. Typical p-type growth region in an n-Si substrate, as displayed by voltage contrast in the SEM. Schematics illustrate the processing : (a) dissolution pit is established and amorphous Si layer has been laid down ; (b) after heat treatment, evaporated Si has been redistributed so as to fill the dissolution pit exactly. After removal of oxide and Al and Si inclusions, the specimen appears featureless in the SEM at (c), where a metal contact line has been added but is grounded in the n-Si substrate. (d) When a negative bias of 6 V is applied to the metal line, however, the p-Si growth becomes evident ; p-Si has a relatively low electrical barrier against the metal and follows it into reverse bias against the n-Si substrate. From reference 21.

occurs most rapidly; the growth velocity was found to vary inversely with the distance from the oxide window. The authors interpret this to indicate that Si growth occurs much faster than Si dissolution.

Furthermore, it is likely that vacancies are created while the evaporated Si is dissolving into the Al, because of the considerable volume decrease of Si atoms during dissolution. These vacancies could migrate to the substrate and facilitate growth there because they are used for the volume increase of Si coming out of the solid solution.

Recently, Majni and Ottaviani (21a) have achieved large-area uniform growth of Si layers by solid phase epitaxy using the Si (a)–Al–Si (xtal) configuration. In this case, $\langle 100 \rangle$ n-type Si was used as substrate. After the removal of thermally grown oxide (~ 5000 Å) from the substrate surface, Al (500 to 7000 Å) and Si (4000 to 7000 Å) were electron-beam evaporated onto the substrates under a vacuum of about 2 to 6×10^{-7} torr (ion pump system). The evaporation rates were about 30 to 60 Å s^{-1}. The Si epitaxial layer (grown at 450 to 530°C) has p-type conductivity with a carrier concentration of 2×10^{18} cm^{-3} and a Hall mobility of 65 cm^2 V^{-1} s^{-1} at room temperature. It is believed that these results are obtained because of better surface preparation techniques, and the Si (a)–Al–Si (xtal) system is a suitable system to achieve SPE growth.

A final remark which might serve to illustrate the practical importance of regrowth in the Si–Al system pertains to the electrical properties of regrown layers. It has been noted (22) that the barrier height of Schottky contacts of Al to n-Si varies considerably with annealing treatments. SEM and microprobe analysis showed that between the Al contact and the substrate a thin p-type layer had been formed during the heat treatment. This was interpreted as being a regrown layer doped with Al. The conclusion of these authors (22) was that with the Al–Si metallization no true Schottky contact to n-Si can be made. The grown p-type Si layer always forms a quasi-p–n junction and, depending on the processing temperature, causes a shift in the observed work function. Recent work by Card (23) also confirmed this conclusion, and pointed to fact that oxide layers present at the Si–Al interface contain positive charges which influence the barrier height.

12.2.4 The Si–Ag System

Attempts to obtain SPE growth in the Si (a)–Ag–Si (xtal) system were mostly unsuccessful. Heat treatment of such a system at 700°C produced a mixed layer of Ag with crystalline Si precipitates. Channeling measurements of a sample consisting of a $\langle 110 \rangle$ Si substrate covered with respectively 2000 Å Ag and 1350 Å amorphous Si and annealed at 700°C for 1 h showed that approximately half of the Si crystallites were epitaxial with the substrate. From SEM photographs it was found that the Si islands occupied approximately 60% of the surface area. As in the Si–Al case, the oxide layer at the original Si surface presumably prevents uniform epitaxial growth. This fact, together with the very low solid solubility of Si in Ag (which rules out the possibility of using extensive dissolution as a way to clean the interface), causes the Si–Ag system to be an unlikely candidate for SPE purposes.

12.2.5   The Si–Au System

This system has been extensively studied by Hiraki et al. (24–27). One of the first striking observations was the rapid migration of Si atoms through a Au layer at a temperature as low as 150°C (Au–Si eutectic temperature is 375°C). This was readily apparent from the growth of a silicon oxide layer on top of a Au layer which was evaporated onto a Si crystal (21) during heat treatment (~ 250°C) in air. The migration of the Si into and through the Au layer cannot be described by dissolution and diffusion, as in the Ge–Al or Si–Al cases. This can be seen from the fact that the solid solubility of Si into Au is extremely low. Furthermore, interstitial diffusion of Si in Au also seems an unlikely proposition (28). Recently, however, the following mechanism for the Si migration was proposed (27). In this model it was speculated that the presence of Au on a clean Si surface relaxes the interfacial energy and leads to a diffuse structure of the interface. The electronic (or chemical) nature of Si in this diffuse interface is thought to be metallic. Silicon atoms can, then, be ejected from the interface and migrate through the Au layer.

As with the other metal–semiconductor systems, it is interesting to consider the Si (a)–Au–Si (xtal) as a vehicle for epitaxial growth on the substrate. One would expect the same problems as in the Ag case described in Section 2.4, that is, surface oxide and low solubility of the semiconductor in the metal film. Some of these objections have been recently overcome by a variation on the experimental SPE technique described so far (29). In these studies the metal film was deposited at a substrate temperature of 125°C, and furthermore the amorphous Si film was deposited on this structure at a temperature of 600°C. The first step has the advantage of already partially "cleaning" the substrate interface before and during the deposition of the metal. The hot evaporation of Si causes an immediate reaction with the metal causing substrate Si atoms and evaporated Si atoms to meet at a "fresh" nucleation interface. SEM micrographs and Auger spectroscopy showed that the surface layer of samples thus prepared consisted of agglomerations of Au in a Si host. The Au could be removed from the surface region by either etching or sputtering. Figure 12.7 shows a depth profile of various elements in the reacted sample obtained by Auger electron spectroscopy and sputtering. One notes that the Au signal is confined to the surface and that beyond this surface region the Si signal quickly becomes constant. The nature of this grown Si layer was investigated by etching off the layer containing Au and contaminations. The exposed surface then was studied by reflection electron diffraction. The diffraction pattern is characteristic of a single-crystal epitaxial film. The growth process is not clear at present, since during the second evaporation step the temperature is above the Si–Au eutectic temperature.

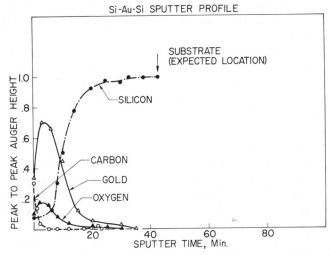

Figure 12.7. Auger sputter profile of the Si–Au–Si structure after reaction. From reference 29.

12.2.6 Summary of SPE in Simple Eutectic Systems

Epitaxial growth by solid solution in eutectic systems is achieved by a sequence of reactions. Dissolution takes place during an annealing treatment, while upon cooling the metal layer becomes supersaturated and recrystallization on the substrate takes place. This is observed in the Ge–Al and Si–Al systems.

Crystal growth can also take place at constant (elevated) temperatures by utilizing a second evaporated layer consisting of amorphous semiconductor material. In this case, the intermediate metal layer acts both as solvent and transport medium.

Driving forces for epitaxial growth often originate from the reduction in free energy associated with the amorphous→crystalline transition. Once the amorphous semiconductor has crystallized inside the metal layer, further growth on the substrate surface is believed to be due to processes analogous to Oswald ripening (30). In this case, smaller crystallites of semiconductor in the metal layer have a higher chemical potential than the larger particles, that is, the equilibrium concentration of the semiconductor near a small crystallite is higher than that near a large crystallite. If we assume that the crystallites are spherical and obey the laws of ideal solution, the difference in chemical potential (or equilibrium concentration) is given by

$$\Delta\mu = 2V_m\sigma \left(\frac{1}{r_1} - \frac{1}{r_2}\right) \simeq \Delta C$$

when μ is the chemical potential, C is the equilibrium concentration, V_m is the molar volume, σ is the interfacial energy, r_1 is the radius of crystallite 1, and r_2 can be taken as the radius of the substrate ($1/r_2 = 0$). This equation indicates that a relative concentration gradient exists between the crystallites and the substrates, and as a result the substrate grows at the expense of the crystallites. However, it is not clear at present if the growth process is diffusion or reaction limited, which may lead to different time dependence of growth of the epitaxial layer.

12.2.7 The Pd–Si System

12.2.7.1 Introduction. As mentioned previously, the Si substrate surface often contains a thin, native oxide layer even after careful chemical cleaning steps. This thin oxide layer often prevents uniform interactions between the substrate and the layers deposited on top of the substrates. For example, Sankur and McCaldin (31) found that dissolution of crystalline and amorphous Si substrates into thin films of evaporated Au was nonuniform along the plane of the surface and dependent on crystalline orientation of the substrate.

One possible cause for this localized dissolution behavior is due to the different native oxide thicknesses (or coherency) on various substrates. Si can migrate into the Au film only through disrupted points in the native oxide layers which may have different geometry and/or density for oxide formed on substrates of different orientations. When a thin Pd layer (≈ 200 Å) was deposited between Si substrate and the Au film, the dissolution of Si was observed to increase and the interactions were much more uniform across the surface. It was speculated that the effect of this intermediate Pd layer was to disrupt the surface oxide (or dirty) layer during the Pd_2Si formation, thus paving the way for Si to come in contact with Au at a larger number of locations. In fact, the reactivity between Pd and Si is so high that an interposing thin oxide layer formed by boiling the Si substrate in H_2O or in HNO_3 or H_2O_2 cannot stop the reaction between Si and Pd to form a layer of Pd_2Si between the Pd layer and the Si substrate.

In case of SPE, one can utilize this high reactivity between Pd and Si to prepare a "clean substrate" surface for a more uniform nucleation and growth process for SPE, provided that Pd_2Si allows the transport of Si atoms from the amorphous Si–Pd_2Si interface to the Pd_2Si–Si substrate interface. This idea of using Pd_2Si reaction to "clean" the interface was first put to use in SPE by Canali et al. in 1974 (11). They observed that the transport of Si was greatly facilitated when a thin Pd layer was used with the Si (a)–Ag–Si (xtal) structure. Consistent regrowth was obtained through a 5 μm oxide window with a Si (a)–Pd–Si (xtal) structure using processing steps of first forming the Pd_2Si layer at 280°C and then increasing the temperature to 600°C to facilitate Si transport and epitaxial growth. Encouraged by these observations, there have been a number of systematic investigations of SPE with the Si (a)–Pd–Si (xtal) structure (32–34). This system is perhaps the most extensively investigated system for SPE to date. A more detailed description of the processing steps and results is given here.

12.2.7.2 General Results.   Silicon wafers $\langle 100 \rangle$ oriented with resistivity between 1 and 10 Ω cm^{-1} were used. The cleaning process was to first grow a thin oxide layer by placing the samples in a $H_2O_2 + NH_4OH + H_2O$ solution (RCA process) and then immersing the samples in a dilute HF solution to remove the oxide. This solution was then quenched with water, and the samples were immediately loaded into the evaporation chamber equipped with ion pumps. Palladium layers 300 to 1500 Å thick were e-gun evaporated onto the wafers at a rate of 2 to 5 Å s^{-1}. Amorphous Si layers 3000 to 10,000 Å thick were then evaporated onto the Pd layer at a rate of ~ 10 to 30 Å s^{-1} at pressures of about 5 to 10^{-7} torr. It was found later that faster evaporation rates of Si (≈ 100 Å s^{-1}) gave more reproducible results. The process cycle was to first heat the samples to 280°C for 20 min to form Pd$_2$Si and then to increase the temperature to between 500 and 600°C at a rate of 0.2 to 5°C min^{-1}.

The formation of Pd$_2$Si and the transport of Si through the silicide layer were studied by MeV ^{4}He$^+$ backscattering techniques (discussed in Chapter 6 of this volume), scanning electron microscopy (SEM), and glancing-angle x-ray diffraction techniques (Read camera, see Chapter 5).

The general features of solid phase growth of Si are shown in a series of SEM micrographs (Fig. 12.8) of cleaved surfaces of a sample after various processing conditions. In the initial condition (Fig. 12.8a), 9000 Å amorphous

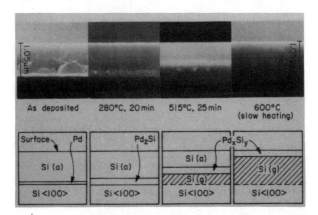

Figure 12.8.   SEM micrographs of cleaved surfaces of a sample after various processing conditions: (a) virgin; (b) 280°C, 20 min; (c) 515°C, 25 min; (d) 600°C. From reference 32.

Si was deposited on 1500 Å Pd. After heat treatment at 280°C, 2000 Å Pd_2Si formed at the interface by consuming both amorphous and substrate Si. In the second heating stage, in which the sample was heated to 515°C, 2300 Å Si was transported through the silicide layer. In the micrograph (Fig. 12.8c) it can be seen that there are inclusions in the grown Si layers. Subsequent investigations by MeV $^4He^+$ backscattering suggest that these inclusions contain Pd. In the final stage (Fig. 12.8d), all the remaining amorphous Si has been transported through the silicide layer which now appears in the top surface.

12.2.7.3 Silicide Formation.   The silicide formation and the transport kinetics were analyzed in more detail by backscattering techniques. Figure 12.9 gives the backscattering spectra for the sample shown in Figure 12.8 in the as-deposited (virgin) condition and after heat treatment.

After annealing the sample at 280°C, the Pd signal decreases in amplitude and increases in width. The ratio of the height of the Pd signal (at 1.4 MeV) to the height of the Si signal (at 0.8 MeV) indicates the formation of a silicide layer between the Si substrate and amorphous Si with Pd-to-Si composition ratio of 2 to 1. Glancing-angle x-ray diffraction techniques indicates that this layer is the Pd_2Si compound. The spread of the Pd signal to both lower and

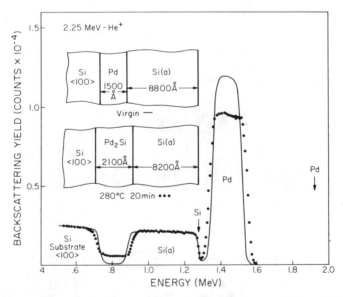

Figure 12.9. The 2.25 MeV $^4He^+$ backscattering spectra for the sample shown in Figure 12.8 in the as-deposited condition and after the formation of Pd_2Si at 280°C. The thicknesses given in the inset are based on bulk density and stopping cross section values. From reference 32.

higher energies shows that both the crystal substrate and the amorphous layer were partially consumed in the silicide formation. This behavior of Pd forming silicide with both crystalline and amorphous Si in nearly equal amounts is thought to have significance in achieving uniform epitaxial layers. Since silicide formation seems to "dissolve" interfacial oxides and other undesirable impurities, a "clean" substrate surface for subsequent epitaxial growth is provided as a result of Pd_2Si formation.

12.2.7.4 Transport Kinetics and the Transient Region. The transport kinetics of the Si can be investigated readily by backscattering techniques, and the concept is shown schematically in Figure 12.10. When a Si (a)–Pd_2Si–Si (xtal) structure is heated in the temperature range of 450 to 600°C, amorphous Si (dotted region in Fig. 12.10) starts to transport across the silicide layer and to grow onto the Si substrate (grown Si layer is denoted by a shaded region in Fig. 12.10). In the backscattering spectrum, the transport of Si can be deduced from the shifts in edges of the Si signal from A to A′

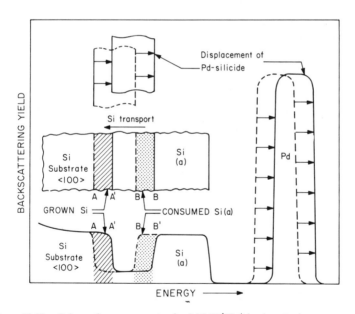

Figure 12.10. Schematic energy spectra for 2 MeV $^4He^+$ backscattering measurements from a thin film composite of crystal Si–Pd_2Si–amorphous Si. Dashed lines in the spectra denote the as-deposited condition. Solid lines denote the conditions after heat treatment in the temperature range of 500 to 600°C. Some of the amorphous Si (dotted region) dissolves into the Pd_2Si layer and migrates to the Si substrate. The crystal Si–silicide interface acts as a nucleating surface for the growth of Si (shaded area). From reference 32.

and B to B′. In the process of consuming amorphous Si and transporting Si through the silicide layer, there is a displacement of the Pd silicide layer toward the sample surface. The displacement can be deduced from the shift in the Pd signal toward higher energies as shown in the schematic back-scattering spectrum in Figure 12.10. The backscattered yield of the Pd signal for a partially transported sample is often lower than that predicted for Pd_2Si, suggesting a phase change or an accumulation of Si in the Pd_2Si layer (hence labeled as Pd_xSi_y in Fig. 10.8). Glancing-angle x-ray diffraction experiments indicated that the excess amount of Si in the silicide layer is in the form of crystalline precipitates and that the silicide layer remains essentially to be Pd_2Si. The fiber texture of the Pd_2Si layer also disappeared after the Si transport (35).

The time dependence of the amount of Si transported through a $\langle 100 \rangle$ Si substrate at 475°C is shown in Figure 12.11. There are two distinct regions in the curve. In the transient region the rate is fast; then, after a certain transported amount, the transport changes abruptly into a slower rate (steady-state region) and usually maintains this rate until the end of the whole process. In both of the transient and steady-state regions, rates can be approximated by straight lines, and the ratio of these two slopes is about 8. Similar ratios of 5 to 10 are also found for other samples made at different times and annealed at various temperatures.

The dependence of transport kinetics on the Pd silicide thickness is also shown in Figure 12.11. Two thicknesses of Pd_2Si layers were used (1800 and 500 Å). It was found that the corresponding transport rates in the transient region as well as in the steady-state region for samples with two different Pd_2Si thicknesses are the same (35, 37). The sample with the thinner Pd_2Si layer, however, reaches the steady-state region faster than that with the

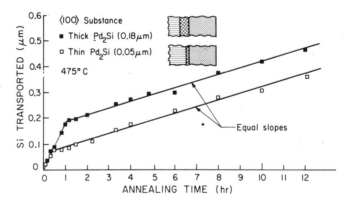

Figure 12.11.   Equal slopes of transport curves for two different Pd_2Si thicknesses.

thicker Pd_2Si layer. Since the thickness ratio of Pd_2Si for the two samples is more than a factor of 3, it is therefore possible to infer from Figure 12.11 that the overall concentration gradient of Si atoms across the Pd_2Si layers plays little or no role in determining the transport rates, that is, the rate-determining step in the transport kinetics is not likely to be diffusion controlled.

The correlation between the thickness of the transient region and the thickness of the original Pd_2Si layer is shown in Figure 12.12. It is interesting to note that for each sample the thickness of its initial transient growth is equal to the thickness of its Pd_2Si layer. Since the transport kinetics have been investigated only for the Si (a)–Pd–Si (xtal), it is not clear at present if other SPE systems, such as the Ge (a)–Al–Ge (xtal) system, which do not involve any compound formation have a similar correlation.

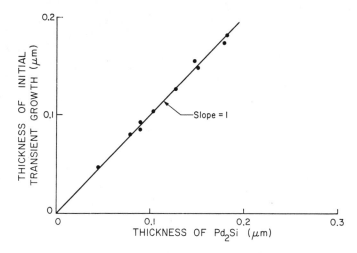

Figure 12.12.   Linear relation between the thickness of Pd_2Si layer and the thickness of the corresponding initial transient growth. From reference 33.

Backscattering and SEM analysis of the growth process show that the Si grown layers in the transient region are not uniform and have island structures. As the sample is annealed to the beginning of the steady-state region, the islands become interconnected and the grown layers become uniform in thickness. Figure 12.13 shows the time sequence of samples in the transient region and in the beginning of the steady-state region. The Pd silicide and the remaining top amorphous Si had been removed by HF from the samples prior to SEM examinations. These micrographs show that the growth of Si

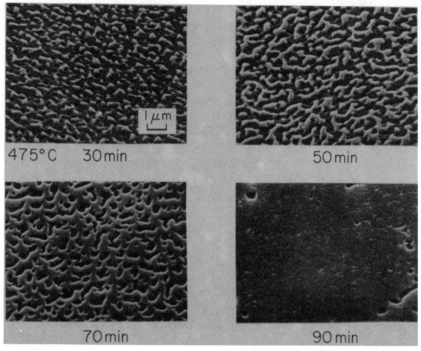

Figure 12.13. SEM photos showing Si islands grown on the substrate surface in the initial transient stage and a uniform Si grown layer that is formed just after the completion of the initial transient stage. From reference 33.

starts with islands of dimensions of about 0.2 μm in height and in width and that subsequently these islands grow laterally and join together. Toward the end of the transient region a uniform Si grown layer is gradually formed. These SEM observations are consistent with backscattering results. The growth process is also described schematically in Figure 12.14 according to backscattering and SEM analysis.

It should be noted that while the growth characteristics described above are general and applicable to all samples, the transport rate may be different from sample to sample. In fact, only samples that are prepared at the same time will have the same rates. Samples made at different times may have different transport rates, yet the general features remain unchanged. This interesting behavior is discussed further in the following sections.

12.2.7.5 Impurity Effects on Transport Kinetics.   Auger electron spectrometry analysis of the SPE samples generally shows only a barely detectable amount of oxygen in the evaporated Si layers (detection limit of oxygen for

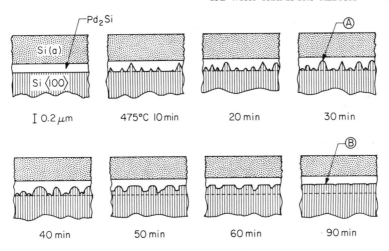

Figure 12.14. Schematic picture describing the growth of islands and the formation of a uniform layer for the sample of Figure 12.13. From reference 33.

AES is ~ 0.01 to 0.1%). The only other detected impurity is carbon. The carbon concentration varies from sample to sample and is usually less than about 1 at. % (38). It is likely that the carbon in the amorphous Si layers comes from the vitreous carbon lining for the Si evaporation charge, since residual gas analysis shows that little or no carbon is present in the vacuum before evaporation and an easily detectable amount is present after evaporation (35, 37).

There seems to be an apparent correlation between the transport rate of Si atoms and the carbon concentration in the amorphous Si layer (38), as shown in Figure 12.15. The transport rate decreases exponentially, about three orders of magnitude, with the carbon concentration in the amorphous Si layer. These nine experimental points in Figure 12.15 represent results obtained from samples prepared in a period of 12 months. Within this time period the evaporation rates and the residual vacuum may vary considerably; yet the transport rate correlates reasonably well with one single parameter, namely, carbon. This observation could very well account for the varying transport rates of samples prepared at different times.

12.2.7.6. Activation Energy.   The rate of crystal growth u can be expressed (38a) by

$$u = u_0 \exp\left(-Q/kT\right)\left[1 - \exp\left(-\Delta G_v/RT\right)\right]$$

where u_0 is a pre-exponential factor, Q is the activation energy for one atom to leave the amorphous phase, cross the interface, and incorporate to the

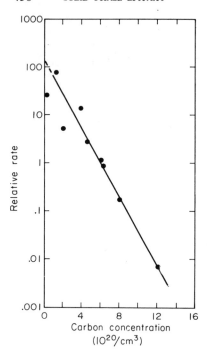

Figure 12.15. Relative transport rate of SPE in the steady-state region as a function of carbon concentration in the amorphous Si layers.

crystalline phase; and ΔG_v is the change of chemical free energy per mole. For crystallization of amorphous solids, the temperature of anneal is far below the melting point. In this case $\Delta G_v \gg RT$ and the rate equation can be expressed (38b) as

$$u \cong u_0 \exp\left(-Q/kT\right)$$

The transport rate of amorphous Si in the Si (a)–Pd–Si (xtal) system is observed to obey this Arrhenius-type behavior.

An activation energy of ∼4 eV is obtained in both the transient and steady-state regions (37). This is because the ratio between the transient transport rate and the steady-state transport rate remains relatively constant throughout the temperature range of investigation. It should be noted that the transport rates of samples prepared at different times may vary, while the activation energy for Si transport remain unchanged for all samples. Therefore, it is likely that the transport mechanism remains unchanged for samples containing varying amounts of carbon in the amorphous Si layer. Apparently, the number of carbon atoms incorporated in the Si layer only affects the pre-exponential term in the Arrhenius equation.

This activation energy of 4 eV is relatively high compared to an activation energy of 2.3 eV obtained by SPE process for ion-implanted samples in a similar reaction temperature range (the SPE process for ion-implanted samples is discussed in Section 3.2). However, there is evidence that the transport process is a combination of thermally activated processes in series. For example, the process of nucleation of islands and the lateral growth of each individual island may each have characteristic activation energies, giving rise to a total activation energy of 4 eV (39, 40).

12.2.7.7 Final Stage.   With further heat treatment all the amorphous Si is consumed and the silicide layer is displaced to the surface. Figure 12.16 shows the backscattering spectra for a sample in which a 8200 Å layer of Si

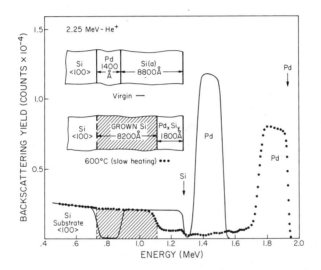

Figure 12.16.   Backscattering spectra for a sample heated to 600°C at heating rate of 2°C min^{-1}. A Si layer 8200 Å thick is grown on the substrate. The small signal of Pd indicates an average concentration of 2 at. % of Pd incorporated in the grown layer. From reference 32.

is grown and the Pd silicide is now located on the sample surface. In the backscattering spectra at energies between 1.3 and 1.6 MeV, the spectrum height indicates an average concentration of 2 at. % of Pd incorporated in the grown Si layer. This concentration of incorporated Pd varies as the processing cycle changes. Fast heating of the sample to the growth temperature generally led to a relatively large amount of Pd being included in the grown layer. For example, the amount of Pd is two to five times higher in a sample heated at 50°C min^{-1} than in a sample heated at 2°C min^{-1} (growth temperature 600°C). It is generally found that the amount of Pd included in the grown layer is about 1 at. % or less with a heating rate of $\approx 0.2°C$ min^{-1}. The amount of Pd inclusion can also be reduced by lowering the growth temperature (37). A growth temperature of 450°C usually leads to less Pd inclusions than a growth temperature of 600°C.

12.2.7.8 Epitaxy. Glancing-angle diffraction patterns taken on grown layers with the silicide layer removed show single-crystal characteristics. For more quantitative information on the epitaxial nature of the grown layer, channeling measurements were made. Figure 12.17 shows the random and aligned spectra of a sample at a heating rate of 0.2°C min^{-1} containing Pd inclusions less than 1 at. %. The channeling was performed after the removal of the Pd silicide layer. The aligned yield indicates that the grown layer is single-crystal in nature rather than oriented polycrystalline.

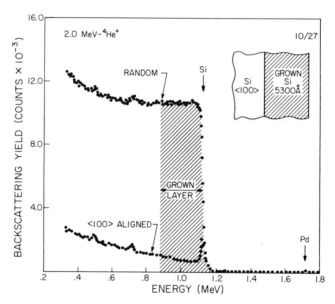

Figure 12.17. Random and aligned spectra for a slowly heated sample after the removal of Pd silicide. The aligned yield is about 7% of the random yield, which is greater than the calculated minimum yield of 3% and the measured yield of 4% on an etch-polished Si sample. The increased yield is partially caused by the thin oxide film on the surfaces formed during etching with aqua regia. From reference 32.

In channeling measurements on samples with Pd inclusions exceeding 1 at. %, the aligned yield increases to 10 to 20% of the random value. More recently, investigations (37, 37a) showed that the minimum yield in the transient growth region is often lower than that in the steady-state growth region.

This result is further confirmed by the presence of high twin density in the grown layer after the transient growth region for some samples, as shown in Figure 12.18.

12.2.7.9 Electrical Doping of the SPE Layer.

An attempt has been made to electrically dope the SPE layer by Sb atoms (34). The sample initially consists of an upper layer of amorphous Si ($\sim 1\ \mu$m thick), a very thin intermediate layer of Sb (nominally 5 Å), and a thin lower layer of Pd (~ 500 Å), all electron-gun deposited on top of a single crystal (1 to 10 Ω cm, p-type, $\langle 100 \rangle$ orientation). The structure of this type of sample is essentially the same as those discussed in previous sections, only with an additional thin layer of Sb interposing between the amorphous Si layer and the Pd layer. After a heating cycle to induce solid phase epitaxial growth (similar to those mentioned above), electrically active atoms are found to be incorporated into the SPE layer.

Elemental distribution profiles in depth were examined by Auger electron spectrometry (AES). Figure 12.19 shows depth profiles for Si, Pd, and Sb observed from a sample after Pd_2Si formation (top) and after the final growth stage (bottom). As Figure 12.19 shows, the Sb is essentially localized at the interface between the amorphous Si and Pd_2Si after heat treatment at 280° C. After the higher-temperature anneal, the amorphous Si has been transported through the Pd_2Si layer to the substrate and the Sb atoms are incorporated into the grown layer during the transport process. The width of the Si distribution is estimated to be about 2000 Å in the grown layer. The amount of Pd trapped in the grown layer (Pd inclusions) was determined by electron microprobe and AES to be about 1 at. %.

The electrical conductivity and the I–V characteristics of the SPE layer against the p-type substrate are shown in Figure 12.20. The schematic on top of the figure shows a cross section of a mesa structure for Hall-effect measurements employing the van der Pauw configuration. Rectifying behavior was observed between two Pd silicide contacts and the indium back contact on the p-type substrate (mode A), while ohmic behavior and low resistance were observed between the two Pd silicide contacts on top of the SPE layer (mode B). These I–V characteristics show that the SPE layer has good electrical conductivity and has rectifying behavior against the p-type substrate. Hall voltage polarity shows that the SPE layer has n-type conductivity. Hall effect measurements at room temperature give a sheet resistivity of approximately 100 $\Omega\ \square^{-1}$, an electron Hall mobility of approximately 50 cm^2 V^{-1} s^{-1}, and a surface concentration Ns of electrons of approximately 1×10^{15} cm^{-2}. The Hall mobility observed in the SPE layer is in general agreement with results reported for heavily doped n-type materials.

The lateral uniformity of the p–n junction between the SPE layer and the

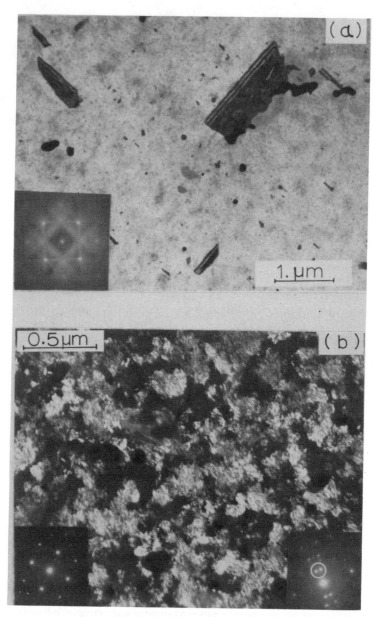

Figure 12.18. (a) TEM micrograph of a Si layer grown at the initial stage at the condition that two types of twins are diffracted. Insert is a selected area diffraction (SAD) pattern for the ⟨001⟩ direction. (b) TEM micrograph (dark-field) of a layer grown at later stages on ⟨100⟩ Si substrate. Insert on the left of the micrograph is a SAD of the matrix. Insert on the right is a SAD at the condition at which the dark-field micrograph is taken, i.e. the diffraction spots within the indicated circle. From reference 37a.

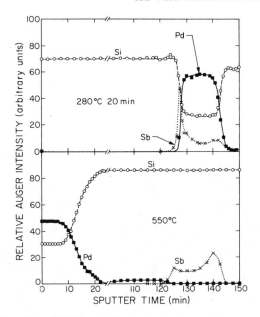

Figure 12.19. Depth profiles of Si, Pd, and Sb obtained by Auger electron spectroscopy for a sample consisting originally of a top layer of amorphous Si (~ 1 μm), a very thin intermediate layer of Sb (~ 5 Å), and a bottom layer of Pd (~ 500 Å) on a single-crystal substrate in the $\langle 100 \rangle$ orientation: *top*, after Pd_2Si formation by annealing at 280°C for 20 min; *bottom*, after solid phase epitaxial growth of the amorphous Si upon annealing to 550°C. The vertical sensitivity is different for each element and each figure. From reference 34.

underlying *p*-type substrate was examined with a scanning electron microscope operated in the electron beam-induced current (EBIC) mode (Fig. 12.21). The presence of a laterally continuous junction is indicated by the uniform collection of induced current in the space charge region between the *n*-type grown layer and the *p*-type substrate across the area shown. The junction appears continuous across dimensions limited only by the size of the samples.

12.2.7.10 Orientation Dependence. The results discussed so far are for solid phase epitaxial growth on $\langle 100 \rangle$ substrates. The growth characteristics on near $\langle 110 \rangle$ oriented Si ribbon substrates ($\sim 10.5°$ off from $\langle 110 \rangle$ orientation) are the same as those on $\langle 100 \rangle$ substrates. The transport rate in the transient region for $\langle 110 \rangle$ orientation is somewhat less than that for $\langle 100 \rangle$ orientation, the transport rates in steady-state region for both orientations being essentially the same. The thickness of the transient growth region for a $\langle 110 \rangle$ substrate again corresponds to the thickness of the initial Pd_2Si layer

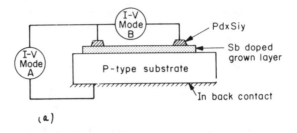

(a)

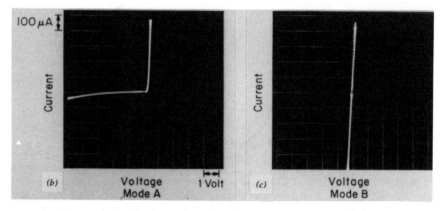

Figure 12.20. (a) Sample geometry, and the current–voltage characteristics showing (b) a rectifying junction between the SPE layer and the substrate (mode A) and (c) the ohmic behavior of the SPE layer itself (mode B). From reference 34.

thickness, as observed for the ⟨100⟩ substrates. The single-crystal nature of the grown layer was established by glancing-angle x-ray techniques. The fact that the SPE process can be used with silicon ribbon substrates may have significance in low-cost solar cell technology.

The SPE process on ⟨111⟩ substrates has very different characteristics from those of growth onto ⟨100⟩ or ⟨110⟩ Si substrates. This is because Pd_2Si layers formed on ⟨111⟩ substrates have single-crystal characteristics and are epitaxial with respect to the ⟨111⟩ Si substrate, while Pd_2Si layers formed with amorphous Si have random polycrystalline structures (36). The growth rate of Pd_2Si on single-crystal Si substrates has been reported to be essentially the same as that on amorphous Si (36). Therefore, about half of the Pd_2Si layer near the Pd_2Si–Si (xtal) interface in the Si (a)–Pd_2Si–Si (xtal) structure for SPE is epitaxial with respect to the ⟨111⟩ substrate; the other half of the Pd_2Si layer, near the Si (a)–Pd_2Si interface, is random polycrystalline. Samples with this structure have especially interesting possibilities. This is because after the Si atoms transport across the first half of the polycrystalline

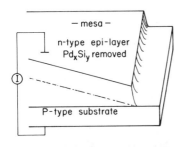

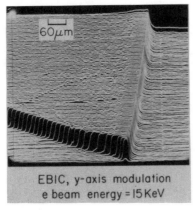

Figure 12.21. (*Top*) Schematic of experimental set-up for (*bottom*) scanning electron micrograph obtained in the electron beam-induced current mode of a mesa-etched SPE layer. The Pd silicide layer was chemically removed before mesa etching. The SEM scan is taken with the electron beam-induced current generating the *y*-deflection. From reference 34.

Pd$_2$Si layer, they now "face" an epitaxial Pd$_2$Si layer which is perhaps just as good a substrate for the epitaxial growth of the Si layer. In fact, a Si layer does nucleate and grow onto the second half of the Pd$_2$Si layer, as shown in Figure 12.22. The Pd signal is split into two locations after the final growth stage, centering around 1.7 MeV and 1.3 MeV. In marked contrast to the behavior observed with $\langle 100 \rangle$ and $\langle 110 \rangle$ substrates, the epitaxial Pd$_2$Si layer remains in contact with the underlying Si substrate and does not serve as a transport medium. The Pd signal at higher energies (~ 1.7 MeV) comes from the polycrystalline Pd$_2$Si which is now displaced to the surface of the sample. The Si signal between ~ 0.7 MeV and ~ 1.1 MeV comes from the grown Si layer. This structure of Si (a)–Pd–Si $\langle 111 \rangle$ enables the growth of a Si layer on a good electrical conductor (i.e., Pd$_2$Si), and this opens up the possibility of forming heterostructures by SPE processes.

The crystal quality of the Si layers grown on epitaxial Pd$_2$Si has not been as good as that of Si layers grown on $\langle 100 \rangle$ Si substrates so far. The grown layer often contain mosaic structures. The reason for the poorer crystal quality for the grown layers is, perhaps, related to the mosaic nature of the epitaxial Pd$_2$Si layer.

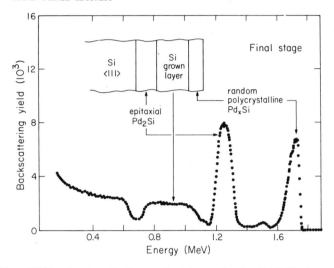

Figure 12.22. Spectrum for heteroepitaxial growth using $\langle 111 \rangle$ Si substrate.

12.2.8   The Ni–Si System

One other system taking advantage of the formation of a silicide that has met relative success in solid phase epitaxial growth is the Si (a)–Ni–Si (xtal) system. Thin layers of Ni deposited on Si substrates first form a Ni_2Si phase at temperatures as low as $\sim 200°C$ (see Chapter 10 in this volume). In the temperature range of ~ 450 to $800°C$, the NiSi phase is the stable phase for thin film system.

The idea of using Ni (or Ni silicide) as a transport medium for SPE process is similar to that of using Pd silicide. Nickel forms a compound with Si at low temperatures (Ni_2Si) and has a stable phase (NiSi) over the temperature range where the SPE process is expected to occur. The formation of the low-temperature phase (Ni_2Si) is used to "clean" the Ni silicide–Si (xtal) interface, and the NiSi phase is used for transporting amorphous Si.

The transport kinetics and epitaxy of the Ni–Si system have been investigated (40), although far less extensively as those of the Pd–Si system. The general growth behavior is similar to that of the Pd–Si system, while the degree of epitaxy of the grown layer has not been as perfect. A more extensive investigation of the Ni–Si system, however, may show that as far as epitaxy is concerned there is no fundamental limitation in using this system. Table 12.1 gives a comparison of SPE with Pd and Ni; the numbers given are typical values.

TABLE 12.1. COMPARISON OF SPE WITH Pd AND Ni

	Pd	Ni
Substrate	$\langle 100 \rangle$	$\langle 100 \rangle$
Metal film thickness, Å	~1000	~1000
Si (a) film thickness, μm	~1	~1
Formation of Transport Medium		
Typical temperature, °C	280	350
Typical time, min	30	30
Composition by backscattering	Pd/Si=2/1	Ni/Si=1/1
Structure by x-ray	Pd_2Si	NiSi
Transport of Si		
Typical temperature, °C	~500	~590
Transient transport rate, Å s^{-1}	~10	~2
Form	islands	no islands (?)
Steady-state rate, Å s^{-1}	~1	~1
Structure of grown layer	epitaxial	epitaxial (poor)
Apparent activation energy, eV	~4	~4

12.2.9 Discussion of SPE in Compound-Forming Systems

Recently, the transport mechanism in the Si (a)–Pd–Si (xtal) system was investigated by a tracer technique using ^{31}Si formed by neutron activation in a nuclear reactor (42). The results showed that the Pd silicide layer plays an active role in the transport of amorphous Si and that the amorphous Si does not seem to migrate across the silicide layer interstially or through the grain boundaries of Pd silicide. A likely transport mechanism involves a dissociation step of the Pd silicide at the crystalline Si–Pd silicide interface freeing Si atoms to grow epitaxially on the substrate while new Pd silicide is formed at the Pd silicide–amorphous Si interface. If this bond breaking (or dissociation) mechanism governs the transport, then the bond energy of the silicide layer plays a significant role in determining the success of solid phase epitaxial growth. To date, only the Si (a)–Pd–Si (xtal) system has been investigated using tracer technique; it remains to be seen if the dissociation mechanism is applicable to the Si(a)–Ni–Si (xtal) or other systems.

Although the tracer experiment gives strong indication of the active role played by the Pd silicide layer, the understanding of the transport mechanism is far from satisfactory. For example, it is not clear what the correlation is between the dissociation mechanism and the activation energy of 4 eV mentioned above. However, based on bond energy consideration, a few elements with relatively low silicide heats of formation (i.e., Pt (37), Co (41), etc.) have been tried successfully to induce solid phase epitaxial growth. More recent experiments (41a) have shown a strong correlation between silicide heats of formation and the rate of material transport through the silicide. Investigations on other systems with tracer technique are needed to confirm the generality of the dissociation model. There remain a great many areas in this field to be explored, for example, transmission electron-microscopic investigation of the epitaxial layers, electrical behavior of the metal silicide inclusions in the grown layer, techniques to electrically dope the grown layer uniformly, and the origin of the transient region, just to name a few. Further experimental investigations and theoretical considerations are necessary to gain a more complete picture of solid phase epitaxy in these systems.

12.3 SPE WITHOUT A TRANSPORT MEDIUM

12.3.1 Introduction

The concept of using a silicide layer as a transport medium for solid phase epitaxy is to prepare a "clean" substrate surface for the nucleation and growth processes of the grown layer, as discussed previously. If the substrate surface can be kept clean and free of oxide or other undesirable "dirty" layers, amorphous layers (or layers with sufficiently small grains) of the same material as the substrate should grow epitaxially after deposition and proper heating cycles. It is also expected that amorphous layers of different materials, but with reasonable lattice matching with the substrate in its crystalline form, should also recrystallize as epitaxial layers. Epitaxial growth investigations of this type have been intensive (43–47). It involves generally loading single-crystal substrates in ultrahigh vacuum systems and heating the substrates at $\approx 1200°C$ or above under vacuum to achieve clean surface conditions. After the surface cleaning process, layers are then deposited with or without substrate heating. Deposition methods of vacuum evaporation (46), pyrolysis of silane (47), ion sputtering (44), and sublimation (41) have been used. For a complete list of epitaxial systems of this type, see Chapter 9, Volume B, in reference 1.

There is, however, another method of producing an oxide-free interface without involving ultrahigh vacuum systems and heating the substrates to high temperatures. Thin amorphous layers can be formed by implanting to a sufficiently high dose in semiconductor crystals (such as Si, Ge, and GaAs) (48). Ion implantation can provide an amorphous layer that is in intimate contact with the underlying substrate thus eliminating any contamination at the interface. It is the solid phase epitaxial growth of the amorphous layers obtained by ion implantation that is discussed in this section. These amorphous layers play important roles in the electrical properties of the devices (49). A survey of recent advances in this field is attempted here.

12.3.2   SPE in Si

12.3.2.1   General Results.

It is generally accepted that amorphous elemental semiconductor layers on crystalline substrates regrow epitaxially at temperatures about half of the melting point, namely, $\sim 500°$C for Si and $\sim 300°$C for Ge (50). Recent studies have shown that the epitaxial regrowth of Si-implanted amorphous layers on $\langle 100 \rangle$ and $\langle 110 \rangle$ oriented Si have a well-defined activation energy and a linear growth rate (51, 52). In these investigations the amorphous layers were formed on Si single crystals of resistivity 2 to 10 Ω cm having $\langle 100 \rangle$, $\langle 110 \rangle$, and $\langle 111 \rangle$ orientation. A sequence of Si implants between 50 and 250 keV was used to ensure a relatively flat implant profile in the samples and to minimize impurity effects. The total dose was 8×10^{15} Si atoms cm^{-2}. The substrate was held at liquid nitrogen during implantation to minimize beam annealing effects. The thickness of the resulting layers was ~ 4600 Å. The regrowth anneals were performed in the temperature range of 400 to 625°C under vacuum. The regrowth process was investigated by channeling effect measurements with 2 MeV ^{4}He$^+$ ions. Figure 12.23 shows $\langle 100 \rangle$ aligned backscattering spectra on $\langle 100 \rangle$ Si samples for a sequence of isothermal anneal cycles at 550°C. In the aligned spectra there are two regions of differing backscattered yield. The high yield is due to backscattering from the amorphous layer. The low yield region corresponds to scattering from the underlying single-crystal region.

The regrowth of the amorphous layer is represented by the movement of the rear edge of the high-yield portion of the spectra toward the front edge. From the spectra in Figure 12.23 it is seen that the thickness of the amorphous layer decreases with increasing annealing time. Measurements along other crystal directions confirm that the amorphous layer regrows epitaxially on the single-crystal Si substrates.

For the $\langle 110 \rangle$ substrate orientation, the regrowth behavior is very similar to that on $\langle 100 \rangle$ Si. The regrowth rate, however, is reduced by about a factor of 2.

For the $\langle 111 \rangle$ case, one can distinguish two regions of different growth behavior, as shown in the $\langle 111 \rangle$ aligned backscattering spectra in Figure 12.23 and the thickness of the regrown layer-versus-time plot in Figure 12.24. In the first region, marked A in the figure, the interface is uniform. The growth rate in this region is a factor of 25 slower than that for the $\langle 100 \rangle$ orientation. In region B, the growth rate is fast but the interface is very nonuniform. This nonuniformity is also shown in the SEM micrograph in Figure 25a. Here, the amorphous layer of a partially annealed sample was removed by immersing the sample in HF. Similar etching procedures on $\langle 100 \rangle$ orientation did not reveal the presence of such irregularities.

The presence of high concentration defects in the $\langle 111 \rangle$ sample was confirmed by transmission electron microscopy measurements. Figure 12.25b

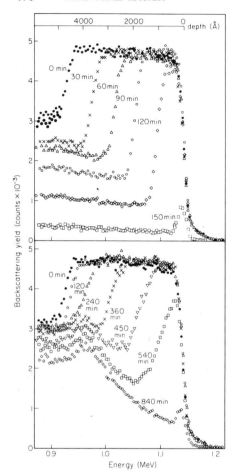

Figure 12.23. Aligned spectra for 2 MeV ⁴He ions incident on silicon samples implanted at LN₂ temperature, preannealed at 400°C for 60 min and annealed at 550°C. Upper portion shows ⟨100⟩ spectra on a sample with the surface normal 0.3° off the ⟨100⟩ axis. Lower portion shows ⟨111⟩ spectra on a sample with the surface normal 0.3° off the ⟨111⟩ axis. The depth scale is calculated assuming the bulk density of Si. From reference 52.

shows a bright-field micrograph of a sample annealed at 600°C for 100 min. This figure indicates the presence of heavy faults (stacking and/or twins) and dislocations. These faults were not observed on ⟨100⟩ samples, although a low density of dislocation was present.

12.3.2.2 Orientation Dependence of Growth Rate. The effect of orientation on the regrowth rate of amorphous Si on Si (xtal) has also been investigated by varying the orientation of the substrate from ⟨100⟩ to ⟨110⟩ to increments of 5° (53). The results are shown in Figure 12.26. The regrowth rate of the amorphous layer at 550°C decreases from about 100 Å min⁻¹ to 5 Å min⁻¹ as the orientation varies from ⟨100⟩ to ⟨111⟩. The regrowth rate,

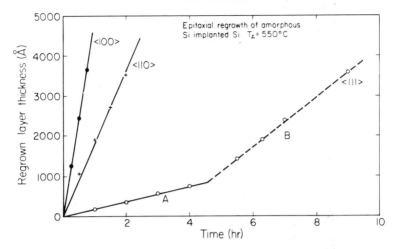

Figure 12.24. Regrowth layer thickness vs. time for Si samples annealed at 550°C. From reference 52.

then, increases again to about 60 Å min^{-1} as the $\langle 110 \rangle$ orientation is approached. The faults in the regrown layer start to appear within about 16° of the $\langle 111 \rangle$ orientation approaching from the $\langle 100 \rangle$ direction. These faults show up in the channeling measurements, as shown in Figure 12.23.

In Figure 12.27 the regrowth rates on various Si substrate orientations and on Ge substrates (discussed in Section 12.3.3) are plotted as a function of reciprocal temperature. The slower rate (region A in Fig. 12.24) of uniform growth for $\langle 111 \rangle$ orientation is used here. The activation energy associated with the regrowth process is about 2.3 eV (51–53) as compared to an apparent activation energy of 4 eV for the SPE process in the Si (a)–Pd–Si (xtal) system. This suggests that the growth processes in these two types of SPE process are probably different.

12.3.2.3 Impurity Effects. Since thin, native oxide layers on substrate surfaces prevent epitaxial regrowth of deposited amorphous layers, it is likely that a small amount of impurities in amorphous layers obtained by ion implantation would lead to different regrowth characteristics. The impurity effect of the regrowth kinetics has been studied (56, 57). Amorphous layers on $\langle 100 \rangle$ substrates were obtained by implanting Si into the substrates at liquid nitrogen temperature, similar to the procedure mentioned previously. Different impurity atoms were then implanted into the amorphous Si layers also at LN_2 temperature. The energies for the impurity implants were chosen such that distributions of the impurity atoms are well within the

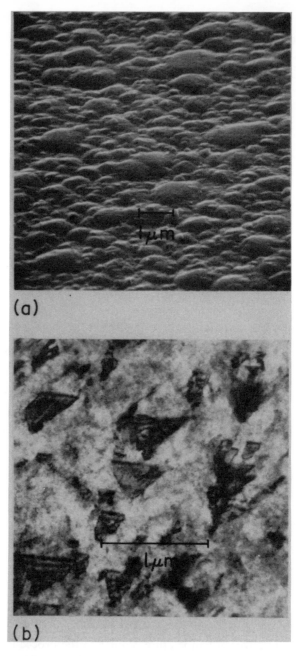

Figure 12.25. (*a*) SEM view of a ⟨111⟩ oriented sample partially annealed at 600°C with the amorphous layer removed by HF. (*b*) TEM micrograph (35 K ×) of a ⟨111⟩ sample annealed at 600°C for 100 min. From reference 52.

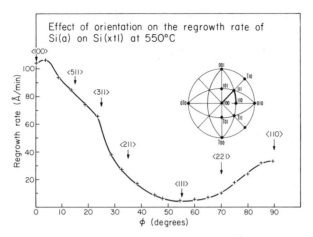

Figure 12.26. Regrowth rate of Si samples as a function of substrate orientation. From reference 53.

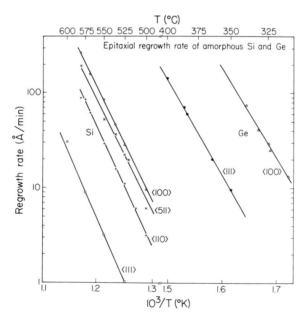

Figure 12.27. Regrowth rates of various Si and Ge orientations vs. reciprocal temperatures. The activation energy for Si samples is 2.3 eV; the activation energy for Ge samples is 2.0 eV. From references 51 and 52.

473

amorphous Si layer. Some impurities were implanted with a range of energies and doses such that the impurity profile is relatively constant over a large thickness; others were single energy implants. The regrowth rates for different impurity implants were determined by annealing the samples at 550°C. The regrowth rates were then correlated with the impurity concentration. Representative values of the regrowth rates are given in Table 12.2.

TABLE 12.2.   REGROWTH RATES

Impurity	Concentration, (atoms cm^{-3}) × 10^{-20}	Regrowth rate, Å min^{-1}
^{11}B	2.5	1028[a]
^{12}C	1.8	55
^{14}N	2.5	7.7
^{16}O	2.4	9.0
^{20}Ne	2.8	3.5
^{28}Si	—	85
^{31}P	2.5	527[a]
^{40}Ar	1.5	2.4
^{74}Ge	2.0	87
^{75}As	2.0	480[a]

[a]Extrapolated value from 500°C.

It is clear from Table 12.2 that not all impurities retard the regrowth rate. Boron, phosphorus, and arsenic appear to accelerate the regrowth rate, carbon and oxygen cause moderate rate reduction, while argon and neon almost stop the regrowth process. This observation is rather surprising; because of the chemical inertness of the noble gases one would not expect any chemical effects to slow down the regrowth process. The figures given in Table 12.2 are typical values. The regrowth rates as a function of impurity concentration in the amorphous layers are discussed here in two examples, namely, phosphorus and oxygen.

With phosphorus impurity, the regrowth rate is faster than that for layers without phosphorus. Figure 12.28 shows the concentration profile of phosphorus in the amorphous layer as well as the regrowth thickness for different temperatures and times (57). It is clear from the figure that at low concentrations ($\sim 0.5 \times 10^{20}$ cm^{-3}) the regrowth rate for the phosphorus-implanted sample is only slightly faster than that of the sample with only Si implants. The regrowth rate, then, accelerates as the phosphorus concentration increases from 1 to 2×10^{20} cm^{-3}. The regrowth rate accelerates to a

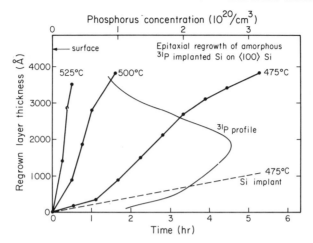

Figure 12.28. Nonlinear growth rate of Si samples amorphatized by Si ion implantation and then implanted with phosphorus. Dashed line corresponds to the regrowth of an amorphatized sample without the ^{31}P implantation and annealed at 475°C (growth rate = 3.2 Å min^{-1}). Thin line shows the phosphorus profile. The zero depth corresponds to the original amorphous–crystalline interface. From reference 57.

relatively constant value as the phosphorus profile goes through a maximum. After this fast regrowth rate region, the rate levels off as the phosphorus concentration decreases.

With oxygen impurity in the amorphous layer (56), the regrowth rate is retarded. Figure 12.29 shows a regrowth thickness-versus-oxygen concentration plot. The regrowth rate is seen to be retarded by the presence of oxygen. The concentration levels of the implanted oxygen in the amorphous layer exceed the solubility limit of oxygen in crystalline Si ($\sim 2 \times 10^{17}$ to 2×10^{18} cm^{-3} at ~ 950 to $1400°$C), and it is possible that oxides or suboxides were formed in the amorphous layers.

The regrowth rate of oxygen-implanted samples can be plotted as a function of oxygen concentration as shown in Figure 12.30. The regrowth rate decreases exponentially with oxygen concentration, similar to the effect of carbon on the transport rate of Si in the Si (a)–Pd–Si (xtal) system (see Fig. 12.15).

Another observation regarding the impurity effect is the fact that although impurities might either accelerate (P in implanted samples) or retard (C in Si (a)–Pd–Si (xtal) samples) the growth rate, the apparent activation energies for these SPE processes, however, remain relatively unchanged compared with those for samples without impurities.

For boron- and arsenic-implanted amorphous Si layers on $\langle 111 \rangle$ sub-

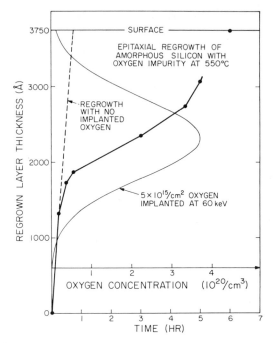

Figure 12.29. Thickness of regrown layer of Si samples amorphatized by Si ion implantation and then implanted with oxygen plotted vs. time, in hours, for 550°C anneals. Dashed line shows the epitaxial regrowth of a sample without the oxygen implantation. Thin line shows the oxygen profile. From reference 57.

strates, thermal history of the samples seems to influence the regrowth behavior (54, 55). A two-step annealing procedure of first heating the samples at 550°C and then at 950°C substantially reduces the residual damage in the regrown layer. The reason for this is not clear at present.

12.3.3 SPE in Ge

12.3.3.1 Silicon Implants into Ge Substrates. Channeling effects of MeV ^{4}He${}^+$ were also used to investigate the SPE processes in Ge substrate implanted with Si ions (50). Amorphous layers on $\langle 100 \rangle$ and $\langle 111 \rangle$ substrates were obtained by LN$_2$ temperature implants of Si with multiple energies of 50 to 250 keV, in steps of 50 keV. The amorphous layers start to regrow epitaxially at about half the meltingpoint temperature ($\sim 300°$C). The regrowth rates for both $\langle 100 \rangle$ and $\langle 111 \rangle$ substrates are linear with time. These observations are analogous to those found for $\langle 100 \rangle$ Si substrates. The regrowth behavior on $\langle 111 \rangle$ substrates is very similar to that on $\langle 100 \rangle$ sub-

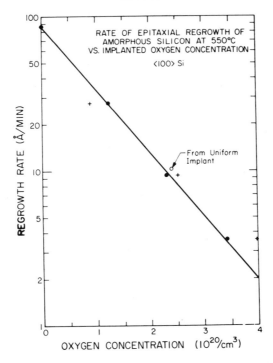

Figure 12.30.   Regrowth rate of $\langle 100 \rangle$ Si samples amorphatized by Si ion implantation and then implanted with oxygen as a function oxygen concentration in the amorphous layer. From reference 57.

strates. The regrowth rate, however, is about a factor of 10 slower in the $\langle 111 \rangle$ orientation. In contrast to $\langle 111 \rangle$ Si samples, there are no observable slow and fast regrowth regions for $\langle 111 \rangle$ Ge samples.

An activation energy of 2.0 eV was obtained for the regrowth process in Ge, as compared to an activation energy of 2.3 eV for Si samples (Fig. 12.27).

12.3.3.2 Germanium Implants into Ge Substrates.   Csepregi et al. have also investigated the SPE processes in Ge substrates implanted with Ge ions (58). Multiple implant energies and LN_2 implant temperature were used, similar to the case of Si implants into Ge. It was found that the regrowth behavior of Ge-implanted amorphous Ge layers is rather similar to that of Si-implanted Ge amorphous layers. The regrowth layers of $\langle 100 \rangle$, $\langle 110 \rangle$, and $\langle 111 \rangle$ samples all show little or no residual damage from channeling measurements, as compared to the presence of high residual damage in $\langle 111 \rangle$ Si substrates implanted with Si (see Section 12.3.2).

The growth kinetics of these three orientations, however, exhibit nonlinear growth rates at later stages. Samples with thinner amorphous layers show deviation from linear growth at shorter anneal times. When the initial linear portion of the growth curves was used in an Arrhenius plot, an activation energy of 2.0 eV was obtained for all three orientations, similar to the case of Ge samples implanted with Si (see Section 12.3.3.1). The initial growth rate of the $\langle 100 \rangle$ orientation at a given temperature is a factor of ~ 2 greater than that of the $\langle 110 \rangle$ and a factor of ~ 20 greater than that of the $\langle 111 \rangle$ orientation.

12.4 CONCLUSIONS

In this chapter the SPE processes in ion-implanted amorphous layers as well as those involving transport layers are discussed. The single most important parameter in determining the success of achieving epitaxial growth is perhaps the interface. A number of different techniques to obtain "clean" interface conditions have been investigated, resulting in varying degrees of success.

In the case of SPE processes involving transport media, many of the roles played by the transport layers, by the different methods of amorphous layer deposition, and by the impurities have not been thoroughly investigated and are not well understood, although some of the metallurgical principles governing these reactions seem relatively straightforward.

For regrowth of ion-implanted amorphous layers of Si and Ge, the "oxide-free" surface condition is readily met. An understanding of the many interesting regrowth characteristics is, however, still in its infancy. The situation with regrowth of compound semiconductors such as GaAs and GaP (59–64) is much more complex than that of Si and Ge. No attempt to review the subject is made here until a much better understanding is gained.

ACKNOWLEDGMENTS

One of the authors (S.S.L.) is most grateful to Prof. Dr. J. Kistemaker and Dr. F. W. Saris for their cooperation and encouragement, and to the FOM Institute of Atomic and Molecular Physics, Amsterdam, The Netherlands, for a fellowship during which this chapter was written. The partial financial support of the Office of Naval Research (L. Cooper and D. Ferry) is gratefully acknowledged. Thanks are also due to our colleagues for stimulating discussions and encouragement, especially to Drs. L. Csepregi, E. F. Kennedy, J. W. Mayer, M.-A. Nicolet, T. W. Sigmon, M. Wittmer, R. Pretorius, Z. L. Liau, J. Mallory, and R. Gorris, without whose cooperation this chapter could not have been written, and to Drs. G. Van Gurp, K. N. Tu, J. O. McCaldin, and T. Vreeland, Jr., for very valuable comments.

REFERENCES

1. J. W. Matthews, Ed., *Epitaxial Growth*, Vols. A and B, Academic Press, New York (1975).

1a. I. H. Khan, *Handbook of Thin Film Technology*, L. I. Maissel and R. Glang Eds., McGraw-Hill, New York (1970), Chapter 10.

2. G. Ottaviani, V. Marrello, J. W. Mayer, M.-A. Nicolet, and J. M. Caywood, *Appl. Phys. Lett.*, **20**, 323 (1972).

3. V. Marrello, J. M. Caywood, J. W. Mayer, and M.-A. Nicolet, *Phys. Stat. Solidi*, **A13**, 531 (1972).

4. G. Ottaviani, D. Sigurd, V. Marrello, J. W. Mayer, and J. O. McCaldin, *J. Appl. Phys.*, **45**, 1730 (1974).

5. D. Sigurd, G. Ottaviani, H. J. Arnal, and J. W. Mayer, *J. Appl. Phys.*, **45**, 1740 (1974).

6. T. Sebestyen, H. Hartnagel, and L. H. Herron, *Electron. Lett.*, **10**, 373 (1974).

7. J. M. Caywood, A. M. Fern, J. O. McCaldin, and G. Ottaviani, *Appl. Phys. Lett.*, **20**, 326 (1972).

8. A. M. Fern and J. O. McCaldin, *Proc. IEEE*, **60**, 1018 (1972).

9. G. Ottaviani, D. Sigurd, V. Marrello, J. O. McCaldin, and J. W. Mayer, *Science*, **180**, 948 (1973).

10. D. Sigurd, G. Ottaviani, V. Marrello, J. W. Mayer, and J. O. McCaldin, *J. Non-Crystall. Solids*, **12**, 135 (1973).

10a. H. S. Chen and D. Turnbull, *J. Appl. Phys.*, **40**, 4214 (1969).

10b. M. L. Rudee, *Thin Solid Films*, **12**, 207 (1972).

11. C. Canali, J. W. Mayer, G. Ottaviani, D. Sigurd, and W. F. Van der Weg, *Appl. Phys. Lett.*, **25**, 3 (1974).

12. G. Ottaviani, C. Canali, and G. Majni, *J. Appl. Phys.*, **47**, 627 (1976).

13. R. W. Bower, *Appl. Phys. Lett.*, **23**, 99 (1973).

14. J. O. McCaldin and H. Sankur, *Appl. Phys. Lett.*, **19**, 524 (1971).

15. H. Sankur, J. O. McCaldin, and J. Devaney, *Appl. Phys. Lett.*, **22**, 64 (1973).

16. J. Basterfield, J. M. Shannon, and A. Gill, *Solid State Electron.*, **18**, 290 (1975).

17. J. S. Best and J. O. McCaldin, *J. Appl. Phys.*, **46**, 4071 (1975).

18. K. Nakamura, M.-A. Nicolet, J. W. Mayer, R. J. Blattner, and C. A. Evans, Jr., *J. Appl. Phys.*, **46**, 4678 (1975).

19. K. Nakamura, J. O. Olowolafe, S. S. Lau, M.-A. Nicolet, and J. W. Mayer, *J. Appl. Phys.*, **47**, 1278 (1976).

20. K. Nakamura, S. S. Lau, M.-A. Nicolet, and J. W. Mayer, *Appl. Phys. Lett.*, **28**, 277 (1976).

21. R. L. Boatright and J. O. McCaldin, *J. Appl. Phys.*, **47**, 2260 (1976).

21a. G. Majni and G. Ottaviani, *Appl. Phys. Lett.*, **31**, 125 (1977).

22. T. M. Reith and J. D. Schick, *Appl. Phys. Lett.*, **25**, 524 (1974).

23. H. C. Card, *IEEE Trans. Electron. Devices* **ED-23**, 538 (1976).

24. A. Hiraki, M.-A. Nicolet, and H. W. Mayer, *Appl. Phys. Lett.*, **18**, 178 (1971).

25. A. Hiraki, E. Lugujjo, M.-A. Nicolet, and J. W. Mayer, *Phys. Stat. Solidi*, **A7**, 401 (1971).

26. A. Hiraki, E. Lugujjo, and J. W. Mayer, *J. Appl. Phys.*, **43**, 3643 (1972).

27. A. Hiraki and M. Iwami, *Jap. J. Appl. Phys.*, **Suppl. 2**, Part 2, 749 (1974).

28. J. O. McCaldin, *J. Vac. Sci. Technol.*, **11**, 990 (1974).

29. J. E. Davey, A. Christou, and H. M. Day, *Appl. Phys. Lett.*, **28**, 365 (1976).

30. D. Turnbull, *Solid State Phys.*, **3**, (1956).

31. H. Sankur and J. O. McCaldin, *J. Electrochem. Soc.*, **122**, 565 (1975).

32. C. Canali, S. U. Campisano, S. S. Lau, Z. L. Liau, and J. W. Mayer, *J. Appl. Phys.*, **46**, 2831 (1975).

33. Z. L. Liau, S. U. Campisano, C. Canali, S. S. Lau, and J. W. Mayer, *J. Electrochem. Soc.*, **122**, 1696 (1975).

34. S. S. Lau, C. Canali, Z. L. Liau, K. Nakamura, M.-A. Nicolet, J. W. Mayer, R. Blattner, and C. A. Evans, Jr., *Appl. Phys. Lett.*, **28**, 148 (1976).

35. S. S. Lau, unpublished results.

36. R. W. Bower, R. E. Scott, and D. Sigurd, *Solid State Electron.*, **16**, 1461 (1973).

37. Z. L. Liau, private communication.

37a. W. Tseng, Z. L. Liau, S. S. Lau, M.-A. Nicolet, and J. W. Mayer, to be published.

38. R. Blattner, C. A. Evans, Jr., Z. L. Liau, and S. S. Lau, to be published.

38a. M. E. Fine, *Introduction to Phase Transformation in Condensed Systems*, Macmillan, New York (1964).

38b. W. Koster, to be published in *Advan. Coll. Interfac. Sci.*

39. Z. L. Liau, S. S. Lau, M.-A. Nicolet, and J. W. Mayer, in *Proceedings of ERDA-NSF Photovoltaic Conversion Conference*, Lake Buena Vista, Florida (1976).

40. Z. L. Liau, S. S. Lau, M.-A. Nicolet, and J. W. Mayer, in *Proceedings of the 1st ERDA Semiannual Solar Photovoltaic Conversion Program Conference*, UCLA, Los Angeles, California, July 22–25 (1975).

41. M. Wittmer and E. F. Kennedy, private communication.

41a. S. S. Lau, Z. L. Liau, and M.-A. Nicolet, *Thin Solid Films* (in press).

42. R. Pretorius, Z. L. Liau, S. S. Lau, and M.-A. Nicolet, *Appl. Phys. Lett.*, **29**, 598 (1976).

43. R. N. Thomas and M. H. Francombe, *Solid State Electron.*, **12**, 799 (1969).

44. O. P. Pchelyakov, R. N. Lovyagin, E. A. Krivorotov, A. I. Toropov, L. N. Aleksandrov, and S. I. Stenin, *Phys. Stat. Solidi*, **A17**, 339 (1973).

45. O. P. Pchelyakov, R. N. Lovyagin, A. I. Toropov, and S. I. Stenin, *Phys. Stat. Solidi*, **A17**, 547 (1973).

46. A. G. Cullis and G. R. Booker, *J. Cryst. Growth*, **9**, 32 (1971).

47. B. A. Joyce, J. H. Neave, and B. E. Watts, *Surf. Sci.*, **15**, 1 (1969).

48. G. Dearnaley, J. H. Freeman, R. S. Nelson, and J. Stephen, *Ion Implantation*, North-Holland, Amsterdam (1973).

49. J. W. Mayer, L. Eriksson and J. A. Davies, *Ion Implantation in Semiconductors*, Academic Press, New York (1970).

50. J. W. Mayer, L. Csepregi, J. Gyulai, T. Nagy, G. Mezey, P. Revesz, and E. Kotai, *Thin Solid Films*, **32**, 303 (1976).

51. L. Csepregi, J. W. Mayer, and T. W. Sigmon, *Phys. Lett.*, **54A**, 157 (1975).

52. L. Csepregi, J. W. Mayer, and T. W. Sigmon, *Appl. Phys. Lett.*, **29**, 92 (1976).

53. L. Csepregi, E. F. Kennedy, J. W. Mayer, T. W. Sigmon, and T. R. Cass, to be published.

54. S. M. Davidson and G. R. Booker, in *Ion Implantation*, F. H. Eisen, Ed., Gordon and Breach, London (1971), p. 51.

55. L. Csepregi, W. K. Chu, H. Muller, J. W. Mayer, and T. W. Sigmon, *Rad. Eff.*, **28**, 227 (1976).

56. E. F. Kennedy, L. Csepregi, J. W. Mayer, and T. W. Sigmon, to be published.

57. L. Csepregi, E. F. Kennedy, T. J. Gallaher, J. W. Mayer, and T. W. Sigmon, in *Ion Implantation in Semiconductors and Other Materials*, edited by F. Chernow, J. A. Borders, and D. K. Price, Plenum Press, New York (1977).

58. L. Csepregi, R. P. Kullen, J. W. Mayer, and T. W. Sigmon, *Solid State Comm.*, **21**, 1019 (1977).

59. S. T. Picraux, *Rad. Eff.*, **17**, 261 (1973).

60. R. G. Hunsperger, E. D. Wolf, G. A. Shifrin, O. J. Marsh, and D. M. Jamba, in *Radiation Effects in Semiconductors*, J. W. Corbett and G. D. Watkins, Eds., Gordon and Breach, London (1971), p. 333.

61. G. Carter, W. A. Grant, J. D. Haskell, and G. A. Stephens, in *Ion Implantation*, F. H. Eisen and L. T. Chadderton, Eds., Gordon and Breach, London (1971), p. 261.

62. J. S. Harris, F. H. Eisen, B. Welch, R. D. Pashley, D. Sigurd and J. W. Mayer, *Appl. Phys. Lett.*, **21**, 601 (1972).

63. K. Gamo, T. Inada, J. W. Mayer, and F. H. Eisen, to be published.

64. T. Shimada, Y. Kato, Y. Shiraki, and K. F. Komotsubara, *J. Phys. Chem. Solids*, **37**, 305 (1976).

13

EFFECT OF INTERFACIAL REACTIONS ON THE ELECTRICAL CHARACTERISTICS OF METAL–SEMICONDUCTOR CONTACTS

E. H. Nicollian and A. K. Sinha

Bell Laboratories, Murray Hill, New Jersey

13.1   THE SCHOTTKY AND BARDEEN BARRIER

13.1.1   Introduction

The semiconductor–semiconductor, metal–semiconductor, and oxide–semiconductor interfaces are the three most important interfaces in semiconductor device technology. The semiconductor–semiconductor interface, or p–n junction, where one side of the interface is doped with acceptor-type impurities and the other side with donor-type impurities, is discussed in Chapter 2. The electrical properties of the metal–semiconductor and oxide–semiconductor interfaces are discussed in this chapter. The metal–semiconductor interface is extremely important in semiconductor device technology for two reasons. First, electrical contact must be made to the semiconductor device to operate it, and this contact almost always involves a metal–semiconductor interface. In this type, which is an ohmic contact, the metal–semiconductor interface must offer minimum resistance to current flow in either direction over a wide temperature range. Second, another type of metal–semiconductor interface can be made which offers low resistance to one direction of current flow and very high resistance to the opposite direction of current flow. Such a contact is commonly called a Schottky barrier and has wide use as a fast switch or for protecting circuit elements from high-voltage transients, for example, in integrated circuit technology.

The most important property of a metal–semiconductor interface, from the standpoint of electronic device applications, is the potential barrier, or barrier height, between the Fermi level in the metal and the majority carrier band edge of the semiconductor at the interface. Figure 13.1 shows the barrier height ϕ_B between a metal and both an n-type and a p-type semiconductor. The potential barrier between the interior of the semiconductor and the interface is called the band bending ψ_s. The barrier height plays a

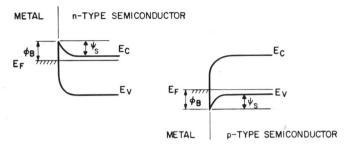

Figure 13.1. Band bending diagram of a metal–semiconductor contact showing barrier height ϕ_B and semiconductor band bending ψ_s: metal to n-type semiconductor; *left*. Metal to p-type semiconductor; *right*. E_C and E_V are the conduction and valence band edges respectively and E_F is the Fermi level.

central role in the electrical properties of the metal–semiconductor interface because it determines the I–V (current–voltage) characteristics of the structure.

The approach taken in this chapter is that barrier height is determined by the density of interface states or traps (defined later). Interface states or traps are potential wells containing energy levels in the semiconductor bandgap located at the metal–semiconductor interface. They exert their influence on barrier height because they capture and emit holes and electrons. There are four types of rectifying barrier three of which are determined by interface state or trap density. With this point of view it is confusing to call all of them Schottky barriers. For this reason we introduce a terminology to distinguish the four types of contact from each other.

There are two cases of oxide–semiconductor interfaces of interest in semiconductor device technology. The first is a thin intervening oxide layer between the metal electrode and the semiconductor thin enough to allow significant current flow. The presence of such a thin oxide layer determines the barrier height of the contact and modifies the I–V characteristics of the diode. Such an oxide layer, purposely grown, is finding use in higher-voltage metal–oxide–semiconductor solar cells. However, such an intervening oxide layer can inadvertently grow during operation of the diode, degrading its characteristics. Thus, a thin intervening oxide layer can be useful or be a reliability problem. This interface is discussed in detail in this chapter.

When the oxide layer is made thick enough so that current flow is oxide conduction limited, the diode no longer functions as a contact and the barrier height loses its importance in determining I–V characteristics. Rather, the resulting metal–oxide–semiconductor structure becomes useful in devices that utilize the field effect principle, such as the MOSFET (metal–oxide-semiconductor field effect transistor) and the CCD (charge coupled device), particularly when current flow through the thick oxide layer is negligibly small. This situation holds in the Si–SiO$_2$ system because the SiO$_2$ is a good insulator and there are large energy barriers between the metal and silicon electrodes which keep current flow through the oxide negligibly small at voltages below those required for dielectric breakdown.

The thick oxide is one of the key features of silicon integrated circuit technology. It serves the purpose of (i) a mask against impurity diffusion permitting the fabrication of p–n junctions at specified regions on a silicon slice to make an integrated circuit configuration, (ii) insulating circuit interconnections from the silicon substrate, and (iii) improving device performance and stability through passivation. A detailed description of the electrical properties of the Si–SiO$_2$ interface is beyond the scope of this chapter. However, it is pointed out that the gamut from thin to thick oxide encompasses a wide variety of semiconductor device applications.

We concentrate here on the metal–semiconductor contact and the contact with a thin intervening oxide layer between the metal and the semiconductor. Sections 13.1 and 13.2 describe the factors that influence and determine the barrier height, and Section 13.3 describes how the I–V characteristics are determined by the barrier height.

13.1.2   Interface States

Interface states play a crucial role in metal–semiconductor interfaces. In a contact between a metal and a covalent semiconductor, for example, interface states can determine the barrier height. In the literature, the term Schottky barrier has been used to represent four basic kinds of interfaces between metals and nonmetals which we shall consider. The four kinds (1) are as follows. Type 1: The nonmetal is an insulator or semiconductor, and the metal is physisorbed on the insulator surface. Type 2: The nonmetal is a highly polarizable ($\varepsilon_s > 7$) semiconductor, such as silicon, and the metal does not react with it to form a bulk compound (weak chemical bonding). Type 3: The nonmetal is a highly polarized semiconductor, and there are one or more bulk compounds which can be formed between it and the metal (strong chemical bonding). Type 4: The surface preparation of the highly polarizable semiconductor has left a very thin surface-reacted layer, possibly an oxide, between it and the metal. In this latter type the bonding is mediated by a native dielectric such as an oxide. Such an interface is produced by chemical reaction rather than cleaving under very high vacuum prior to metal deposition.

Type 1 is a true Schottky barrier, in which ϕ_B is nearly proportional to the work function difference between the metal and the semiconductor. This type of contact is the only true Schottky barrier because the barrier height depends on the work function difference between the metal and semiconductor as first described by Schottky. A typical example of a true Schottky barrier contact is the contact between a metal and an ionic semiconductor.

Type 2 is a Bardeen barrier (2) contact. In a Bardeen barrier contact the barrier height is nearly independent of the work function difference between the metal and the semiconductor. Rather, ϕ_B is determined largely by the interface states. A typical example of a Bardeen barrier contact is the contact between a metal weakly bonded to a covalent semiconductor such as the Au–Si contact.

A type 3 contact is characterized by a strong chemical reaction between the metal and the semiconductor. In this type of contact the interface is between an intermetallic compound and the semiconductor. The barrier height is still influenced by the interface states, but these have been reduced in density because of the chemical bonds between the intermetallic compound and the semiconductor. Therefore, a type 3 contact will be called a modified Bardeen barrier contact. Types 2 and 3, Bardeen barrier and modified Bardeen barrier contacts, are widely used in integrated circuits and in such microwave devices as the IMPATT diode and the GaAs FET.

A type 4 contact is characterized by a strong chemical reaction between an oxide and the semiconductor. This contact has two interfaces, one between the metal and the oxide and the other between the oxide and the semiconductor. The interface between the oxide and the semiconductor is more important from the standpoint of the electrical properties, because any charged traps at or near the oxide semiconductor interface will affect the barrier height far more than a charged trap at or near the metal–oxide interface, which will have virtually no effect on the electrical properties of the contact. The barrier height is determined largely by the properties of the oxide layer so that we shall call a type 4 contact an oxide barrier contact.

An intervening oxide layer between the metal and semiconductor can be produced deliberately, inadvertently during fabrication of the contact, or during its operating life. The oxide barrier contact has possible use in solar cells and is important in reliability studies.

Sections 13.1.3 through 13.1.7 discuss type 1 and type 2 contacts, Section 13.1.8 discusses type 3, and Section 13.2 discusses type 4.

13.1.3   The Semiconductor–Vacuum Interface

To see how interface states come about in covalent semiconductors, consider cleaving the semiconductor along a crystallographic plane under vacuum. The atoms located in the surface layer are bonded only to atoms in the layer below, the layers above having been removed during cleavage. The forces experienced by the surface atoms are different than they were before cleavage when each atom was surrounded on all sides by neighboring atoms. To minimize surface energy, the surface atoms move slightly from their former positions to share elections with neighboring surface atoms. This perturbation of the surface layer of atoms, caused by cleaving, produces potential wells at the semiconductor surface. Each potential well contains quantum states formerly in the valence or conduction band but now having an energy level in the forbidden energy gap of the semiconductor. In silicon, there would be a maximum of two states for each surface atom, taking spin degeneracy into account. The surface reconstruction only shifts the energy levels of the states. In III–V semiconductors, the states are near the band edges so that a shift in energy level caused by surface reconstruction reduces the density of states in the forbidden gap.

After cleavage, an energy barrier will exist at the semiconductor surface. An electron from the interior of the semiconductor must overcome this energy barrier to come to the surface. To see how this energy barrier is created, consider the situation immediately after the semiconductor crystal is cleaved. At this moment there will be a deficit of electrons at the surface because two surfaces are created by cleavage and on the average each surface contains half the electrons originally in the cleavage plane. As a result of the electron deficit at the cleaved semiconductor surface, electrons will diffuse from the interior of the semiconductor crystal to the surface where they become captured by the interface states. This charge in the interface states sets up an electric field that repels electrons diffusing from the bulk to the surface. A drift current of electrons is established by this electric field, which continues to increase until the drift and diffusion currents just balance. In this equilibrium condition, there will be a negative charge Q_T in the interface states, a positive space charge at the semiconductor surface which balances Q_T, and a potential barrier at the surface which repells electrons from the interior of the semiconductor.

Because the interface state levels are distributed in energy over the bandgap of the semiconductor, Q_T will be a function of bandgap energy. The slope of the Q_T-versus-ε curve is defined as the interface state density. Expressing this mathematically, the interface density is

$$N_{ss} \equiv \frac{1}{q}\left(\frac{dQ_T}{d\varepsilon}\right) = \frac{Q_{ss}}{q} \tag{1}$$

where the units of N_{ss} are $cm^{-2}\, eV^{-1}$.

13.1.4   The Semiconductor–Metal Interface (Weak Chemical Bonding)

Suppose a metal is deposited on the cleaved semiconductor surface. Suppose further that the metal makes intimate contact with the cleaved surface but either does not chemically react with it or reacts with it only weakly. The barrier height will then be determined only by the interface state density.

The barrier height will be independent of the metal and therefore will be the same for a variety of metals satisfying the conditions stated above. With an intimate contact between the metal and semiconductor, there must be an interfacial layer of atomic dimensions between the metal and semiconductor.

13.1.5   General Expression for Barrier Height

A general expression (3) can be derived for barrier height based on two realistic assumptions. The first assumption is that the interfacial layer is transparent to electrons and can withstand a potential across it. The second assumption is that the interface states are a property of the semiconductor surface and are independent of the metal. An energy band diagram of a metal *n*-type semiconductor contact in thermal equilibrium is shown in Figure 13.2. The various parameters we shall use are defined in this figure. The quantity $q\phi_0$ was the energy difference between the Fermi level and the

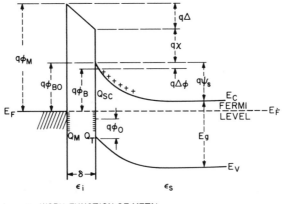

ϕ_M	=	WORK FUNCTION OF METAL
ϕ_B	=	BARRIER HEIGHT OF METAL-SEMICONDUCTOR BARRIER
ϕ_{BO}	=	ASYMPTOTIC VALUE OF ϕ_B AT ZERO ELECTRIC FIELD
ϕ_O	=	ENERGY LEVEL AT SURFACE
$\Delta\phi$	=	IMAGE FORCE BARRIER LOWERING
Δ	=	POTENTIAL ACROSS INTERFACIAL LAYER
χ	=	ELECTION AFFINITY OF SEMICONDUCTOR
ψ_s	=	BUILT-IN POTENTIAL
ϵ_s	=	PERMITTIVITY OF SEMICONDUCTOR
ϵ_i	=	PERMITTIVITY OF INTERFACIAL LAYER
δ	=	THICKNESS OF INTERFACIAL LAYER
Q_{SC}	=	SPACE-CHARGE DENSITY IN SEMICONDUCTOR
Q_{SS}	=	SURFACE-STATE DENSITY ON SEMICONDUCTOR
Q_M	=	SURFACE-CHARGE DENSITY ON METAL

Figure 13.2.   Energy band diagram of a metal–*n*-type semiconductor contact with an interfacial layer of atomic dimensions. From reference 3.

valence band edge at the surface before the metal–semiconductor contact was formed. All interface states below the energy $q\phi_0$ had to be filled with electrons to produce charge neutrality at the semiconductor surface as explained previously. Figure 13.2 shows that depositing the metal has caused a small change in interface state occupancy, as a result of charge flow between the semiconductor and metal, to line up the Fermi levels of the metal and semiconductor. The total interface state charge density is

$$Q_s = -q \int_{\phi_0}^{E_F/q} N_{ss} \, d\varepsilon \tag{2}$$

Assuming for simplicity that N_{ss} is constant over the energy range from $q\phi_0$ to the Fermi level, Eq. 2 becomes

$$Q_s = -qN_{ss}\left(\frac{E_F}{q} - \phi_0\right) \tag{3}$$

From Figure 13.2 it can be seen that

$$\frac{E_F}{q} - \phi_0 = \frac{E_g}{q} - \phi_B - \Delta\phi - \phi_0 \tag{4}$$

Thus Eq. 3 becomes

$$Q_s = -qN_{ss}[(E_g/q) - \phi_B - \Delta\phi - \phi_0] \tag{5}$$

The semiconductor surface space charge is

$$Q_{sc} = qN_D x_d = \left[2q\varepsilon_s N_D \left(\psi_s - \frac{kT}{q}\right)\right]^{1/2} \tag{6}$$

where N_D is the donor concentration in the semiconductor and x_d is the semiconductor space charge width. Using $\psi_s = \phi_B + \Delta\phi - [(E_C - E_F)/q]$ from Figure 13.2,

$$Q_{sc} = \left[2q\varepsilon_s N_D \left(\phi_B - \frac{E_c - E_F}{q} + \Delta\phi - \frac{kT}{q}\right)\right]^{1/2} \tag{7}$$

In the absence of space charge in the interfacial layer (a good assumption for intimate contact), the charge density in the metal Q_M will be equal in magnitude and opposite in polarity to the total charge density in the semiconductor which is the sum of the charge density in the interface states Q_T and the surface space charge Q_{sc}, or

$$Q_M = -(Q_T + Q_{sc}) \tag{8}$$

The potential Δ across the interfacial layer is given by applying Gauss' law to the surface charge or the metal and semiconductor:

$$\Delta = -\delta \frac{Q_M}{\varepsilon_i} \tag{9}$$

Another expression for Δ can be obtained by inspection of Figure 13.2, namely,

$$\Delta = \phi_m - (\chi + \phi_B + \Delta\phi) \tag{10}$$

Eliminating Δ from Eqs. 9 and 10 and substituting for Q_M from Eq. 8, using Eqs. 5 and 7 for Q_s and Q_{sc}, we get

$$(\phi_m - \chi) - (\phi_B + \Delta\phi) = \left\{ \frac{2q\varepsilon_s N_D \delta^2}{\varepsilon_i^2} \left(\phi_B + \Delta\phi - \frac{E_c - E_F}{q} - \frac{kT}{q} \right) \right\}^{1/2}$$
$$- q \frac{N_{ss}\delta}{\varepsilon_i} [(E_g/q) - \phi_0 - \phi_B - \Delta\phi] \tag{11}$$

Solving Eq. 11 for ϕ_B and introducing the quantities

$$c_1 = 2q\varepsilon_s N_D \delta^2 / \varepsilon_i^2 \tag{12}$$

and

$$c_2 = \varepsilon_i / (\varepsilon_i + q\delta N_{ss}) \tag{13}$$

we obtain

$$\phi_B = \left[c_2(\phi_m - \chi) + (1 - c_2)\left(\frac{E_g}{q} - \phi_0 \right) - \Delta\phi \right]$$
$$+ \left\{ \frac{c_2^2 c_1}{2} - c_2^{3/2} [c_1(\phi_m - \chi) + (1 - c_2)\left(\frac{E_g}{q} - \phi_0 \right) \frac{c_1}{c_2} \right.$$
$$\left. - \frac{c_1}{c_2}\left(\frac{E_c - E_F}{q} + \frac{kT}{q} \right) + \frac{c_2 c_1^2}{4} \right] \right\}^{1/2} \tag{14}$$

Equation 14 can be simplified by estimating c_1 and c_2. The interfacial layer will have a thickness of atomic dimensions, namely, 4 or 5 Å. The dielectric permittivity of such a thin layer can be well approximated by the free space value. For $\varepsilon_s \approx 10\varepsilon_0$, $\varepsilon_i = \varepsilon_0$, and $N_D < 10^{18}$ cm^{-3}, c_1 is small, of the order of 0.01 V, and the braced term in Eq. 14 is estimated to be less than 0.04 V. Neglecting the $\{\}$ term simplifies Eq. 14 to

$$\phi_B = c_2(\phi_m - \chi) + (1 - c_2)\left(\frac{E_g}{q} - \phi_0 \right) - \Delta\phi \tag{15}$$

There are two limiting cases of Eq. 15 which are of interest. The first case is when $N_{ss} \to 0$. From Eq. 13, $c_2 \to 1$, and Eq. 15 becomes

$$\phi_B = \phi_m - \chi - \Delta\phi \tag{16}$$

This case is type 1, the true Schottky barrier, in which the barrier height is determined by the work function of the metal and the electron affinity of the semiconductor. Because χ depends on E_F, which in turn depends on the doping concentration in the semiconductor, ϕ_B depends on ϕ_m and the doping concentration in the semiconductor. This case is most common to metal–ionic semiconductor contacts.

The second case is when $N_{ss} \to \infty$. From Eq. 13, $c_2 \to 0$, and Eq. 15 becomes

$$\phi_B = \frac{E_g}{q} - \phi_0 - \Delta\phi \tag{17}$$

This case is type 2, the Bardeen barrier, in which the Fermi level at the interface is pinned by the interface states at an energy $q\phi_0$ above the valence band edge. The barrier height is independent of the metal work function and is determined by the surface properties and doping concentration of the semiconductor. The doping concentration comes in because ϕ_0 depends on χ, which is a function of the semiconductor doping concentration.

13.1.6 Experimental Values of ϕ_B, ϕ_0, and N_{ss}

Experimental measurements of the barrier height of the metal–n-type silicon system are shown as a function of metal work function ϕ_m for a variety of different metals in Figure 13.3. The two extreme cases where $c_2 = 0$ and

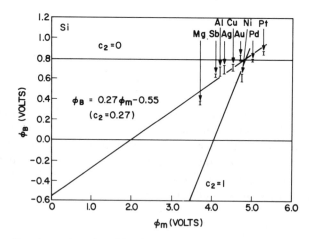

Figure 13.3. Experimental results of barrier heights for the metal–n-type Si system. From reference 3.

$c_2 = 1$ also are shown in the figure. The experimental points lie between these two extreme values, which means that interface-state density is neither zero nor infinite, in practice. That is, there is some weak chemical reaction. A least-squares straight line fit to the experimental data in Figure 13.3, yields

$$\phi_B = 0.235\phi_m - 0.352 \tag{18}$$

Comparing Eq. 18 to Eq. 15, we can obtain values of c_2, ϕ_0, and N_{ss}. To do this, we must write Eq. 15 in the same form as Eq. 18:

$$\phi_B = c_2\phi_m + c_3 \tag{19}$$

where $c_3 = (1 - c_2)[(E_g/q - \phi_0)]\Delta\phi - c_2\chi$. Solving this equation for ϕ_0 yields

$$\phi_0 = \frac{E_g}{q} - \frac{(c_2\chi + c_3 + \Delta\phi)}{1 - c_2} \tag{20}$$

and from Eq. 13

$$N_{ss} = \frac{(1 - c_2)\varepsilon_i}{qc_2\delta} \tag{21}$$

Comparing Eq. 18 to Eq. 19 and using Eqs. 20 and 21 and the previous assumptions for δ and ε_i, we find $c_2 = 0.235$, $\phi_0 = 0.33$ V, and $N_{ss} = 4 \times 10^{13}$ cm^{-2} eV^{-1}. Table 13.1 shows similar results for the widely used covalent

TABLE 13.1. SUMMARY OF BARRIER HEIGHT DATA AND CALCULATIONS FOR Si, GaP, AND GaAs[a]

Semiconductor	χ, V	N_{ss}, $\times 10^{13}$ cm^{-2} eV^{-1}	ϕ_0, V	$q\phi_0/E_g$
Si	4.05	2.7 ± 0.7	0.30 ± 0.36	0.27
GaP	4.00	2.7 ± 0.4	0.66 ± 0.20	0.29
GaAs	4.07	12.5 ± 10.0	0.53 ± 0.33	0.38

[a]After Sze (4).

semiconductors Si, GaP, and GaAs. It can be seen from the last column in the table that the values of ϕ_0 for Si, GaAs, and GaP are very close to one third the bandgap. This observation means that these semiconductors have interface-state densities which peak near one third of the gap measured from the valence band edge.

13.1.7   Influence of Interface States on Reverse Saturation Current

In addition to playing a dominant role in determining the barrier height between a metal and a covalent semiconductor, interface states also can affect the bias dependence of the reverse saturation current. Interface states influence the bias dependence of the reverse saturation current because their occupancy can be bias dependent. Interface-state occupancy is determined by the position of the Fermi level in the metal. That is, interface states are charged and discharged by electrons from the metal. The reason is that the thickness of the interfacial layer in Figure 13.2 between the metal and the interface states in the semiconductor is of atomic dimensions, so that it is in effect transparent to electrons. The electrons in the semiconductor see an energy barrier which is too thick (in a Schottky barrier) for them to tunnel through.

For a high density of interface states, the bias dependence of the reverse saturation current will be determined by image force lowering of the barrier as described in Section 13.3.4.2. The reason for this bias dependence is that the interface-state density is so large that the interface-state occupancy, determined by the Fermi level in the metal, is not changed by a detectable amount by the application of a bias to the metal–semiconductor contact. However, for intermediate interface-state densities, the occupancy of the interface states can be changed by the applied bias. In this case, only part of the bias dependence of the reverse saturation current arises from image force lowering, the remainder arising from changes in interface-state occupancy.

13.1.8   The Semiconductor–Metal Interface (Strong Chemical Bonding)

In this section we discuss type 3 contacts. Suppose we start with a cleaved covalent semiconductor surface and then chemically react a metal with it to form a contact. After the metal has chemically reacted with the semiconductor, the atoms in the surface layer of the semiconductor share their electrons with the layer of metal–semiconductor compound molecules above them so that there is no longer a maximum of two quantum energy levels associated with each surface semiconductor atom and the density of interface states is markedly reduced. It has been proposed by Andrews and Phillips (1) that the amount the interface-state density is reduced will depend on the heat of formation of the chemical reaction between them. This is a plausible proposal. To understand this model, note that the heat of formation is dependent on the bond strength between the metal and the semiconductor. Because electrons in interface states and electrons in the metal repel each other, a reduction in interface-state density results in an increase in bond strength. The lower the interface-state density, the lower the barrier height will be, so that experimental support of this picture can be provided by measuring barrier height as a function of heat of formation. The most reliable and reproducible metal–semiconductor contacts, in which a chemical reaction is involved, is the transition-metal silicide–silicon contact (see Chapter 10 for a discussion of silicide formation). The reaction causes the interface to move into the interior of the silicon lattice away from surface imperfections and contamination. Also, in this reaction the composition and therefore the heats of reaction per mole of the reaction products are accurately known.

The correlation between barrier height and heat of formation, obtained by Andrews and Phillips (1), is shown in Figure 3.5, Chapter 3. Andrews and Phillips have attempted to understand the fundamental factors that determine barrier height. Although their arguments are not conclusive, they are plausible and interesting. Their arguments to explain the correlation between barrier height and heat of formation are as follows. A significant feature of this correlation is that a simple linear relation holds. In Pauling's picture (5) of strong chemical bonding the charge transfer between the transition metal and the silicon should be proportional to $(-\Delta H_f)^{1/2}$; and in Bardeen, type 2, barriers the shift in barrier height is proportional to the transferred charge. Pauling's picture has been shown (6) to account for both ΔH_f and transfered charge in transition metal carbides, nitrides, and oxides with the (substantially) ionic NaCl structure. This work was based on measurements using electron spectroscopy for chemical analysis (ESCA). However, transition metal–metalloid compounds, with the metalloid an element from columns IV, V, or VI of the periodic table, exhibit weakly ionic structures dominated by chains or planes of strongly bonded transition metal atoms, with the transition metal–metalloid bonding being of secondary importance (7). In such cases, where the transition metal–silicon bonds are long and the interactions are weak, it seems likely that the degree of hybridized bonding between the transition metal and silicon atoms would be linear in $-\Delta H_f$. With increasing values of $-\Delta H_f$, more sp^3 orbitals are mixed into the transition-metal d orbitals; and, to first order, the degree of admixture is proportional to $-\Delta H_f$. With a sufficiently strong sp^3 admixture, the hybridized orbitals about the transition metal atom would become very similar to those about a silicon atom, and the barrier height would go to zero.

This simple argument can be used to justify the linearity of the fit in Figure 3.5 in Chapter 3 and to explain the slope and intercept as well. In the limit $\Delta H_f \to 0$, an example of which would be the Au–Si interface characterized by a small van der Waals bonding, ϕ_B equals the free silicon surface (cleaved) value of 0.83 eV.

The intercept $\phi_B = 0$ is $\Delta H_f = 4.8$ eV per transition metal atom, or 110 kcal mole^{-1}. This value of ΔH_f corresponds well with the cohesive energy of silicon, which is 108 kcal mole^{-1} (1). In elemental transition metals, the charge distribution can be accurately represented by a superposition of unhybridized, spherically symmetric charge distributions centered on each transition metal atom. Similarly, the charge distribution of free silicon atoms is spherically symmetric, and the atoms are in unhybridized states. Thus, the hybridization energy needed to make transition metal–silicon bonds indistinguishable form silicon–silicon bonds is practically identical to the hybridization energy of the silicon bonds themselves. Therefore, the slope of the line in Figure 3.5 of Chapter 3 is determined by the silicon which acts as its own fiducial point to determine the extrapolated intercept at $\phi_B=0$.

13.2   INTERVENING OXIDE LAYER

13.2.1   Introduction

In this section we discuss type 4, or oxide barrier, contacts where a relatively thin oxide exists between the metal and the semiconductor. Such an oxide layer can grow during the fabrication of the diode or during aging of the fabricated diode, or it can be grown intentionally.

For oxide thicknesses less than about 10 to 15 Å, the contact cannot be distinguished from a type 2 contact. We shall call this thickness range "thin oxides." For oxides thicker than about 15 Å, the contact will behave like a type 4 contact as long as the current is emission limited. As long as the current is emission limited, the I–V characteristics will be independent of oxide thickness. Oxide thicknesses in the emission-limited range will be called "intermediate-thickness oxides." When the oxide thickness is increased sufficiently so that current flow is limited by conduction through the oxide, the diode no longer can be considered to be an oxide barrier contact. The conduction-limited range of oxide thickness will be called "thick oxides." We have discussed type 2, contacts which include thin oxides, in Section 13.1. In this section, we discuss type 4 contacts or contacts with intervening oxide in the intermediate thickness range.

In a type 2 contact, we have seen that the barrier height is determined by interface states. A type 4 contact differs in that the barrier height is determined by the work function difference between the metal deposited on the oxide surface and the semiconductor, and by charged interface traps. Interface traps and interface states are both potential wells containing energy levels in the bandgap of the semiconductor; however, they have different origins. The origin of interface states has been discussed in the previous section.

Let us consider the origins of interface traps. By growing a native oxide on covalent semiconductors such as Si, GaP, and GaAs, the unsatisfied valence bonds on a cleaved surface, which act as interface states, are shared with oxygen atoms, that is, the heat of formation of oxides is high so that interface-state density is zero. The interface between the oxide and the semiconductor is now characterized by interface traps which arise from deviations from stoichiometry, structural defects in the bonds between the semiconductor and the oxide, and impurities at the interface. The impurities may include metal atoms that have diffused to the interface through the intermediate-thickness oxide during deposition of the metal electrode. It is known, for example, that deposition of metals by evaporation on thin and intermediate-thickness oxides, in the $Si-SiO_2$ system, produces interface traps having energy levels characteristic of that metal in silicon (8). The metal evaporation source is typically heated to a few thousand $^\circ K$, which is the average energy of the metal atoms striking the oxide surface. Some metal atoms will have considerably larger kinetic energies permitting them to penetrate the oxide all the way to the semiconductor–oxide interface. All these metal atoms can act as traps. Because in the intermediate oxide thickness range the oxide is transparent to electrons (or holes), the occupancy of the oxide traps will be determined by the postion of the Fermi level in the metal. Electrons (or holes) in the semiconductor are decoupled from the interface by the potential barrier represented by the band bending. Therefore, it is not possible to tell the position of the traps in the oxide. However, the interfacial region, being a transition region between crystalline semiconductor lattice and the amorphous oxide lattice, easily can accomodate metal atoms. Therefore, the interfacial transition region can act as a sink for impurity atoms so that a high concentration of metal atoms might be expected in the interfacial layer. For this reason and for the sake of consistency, we shall continue to refer to oxide traps as interface traps.

Interface trap density produced by the mechanism of metal atom penetration into the oxide layer decreases with increasing oxide thickness because fewer metal atoms have sufficient energy to penetrate the oxide all the way to the vicinity of the semiconductor–oxide interface (8). In the thick oxide range, no atoms penetrate all the way to the semiconductor–oxide interface and the metal deposition process no longer plays a significant role in producing interface traps.

Interface trap densities measured on type 4 contacts are usually 10 to 1000 times smaller than interface state densities measured on type 2 contacts. This is equivalent to between one interface trap per 10 to 1000 surface atoms rather than two interface states per surface atom in a type 2 contact. Another difference between type 2 and type 4 contacts, arising from the presence of the intervening oxide layer in the type 4 contact, is that interface trap densities can be changed by the passage of high current densities through the oxide layer while interface state densities are not usually susceptible to such current-induced changes. The reason for this difference is that the oxide can be considered to be an electrolyte in which electrochemical reactions can take place at room temperature (9). or in which ionic species can readily drift. Such effects are not common in the covalent semiconductors of interest here.

13.2.2   Anomalies in Oxide Barrier Contacts

Striking differences are found between Bardeen and oxide barrier-type contacts. When these differences were first observed, they were called anomalous because they were not understood. These anomalies, which result when an intermediate-thickness oxide is present between the metal and semiconductor, are the following: high minority carrier injection under forward bias; minority carrier injection increases after the flow of a high forward current; minority carrier injection decreases after the flow of avalanche current; avalanche breakdown voltage is about one half the bulk value; reverse leakage current increases after the flow of avalanche current; and reverse leakage current decreases after forward current flow. These anomalies have been observed by many workers over a long period of time on a variety of semiconductors (3, 10–23).

13.2.3   Experimental Study of the Oxide Barrier Contact

Because (*a*) interface traps play a crucial role in determining the barrier height in an oxide barrier contact and (*b*) interface trap density can be varied by passing a high current density through the oxide, the oxide barrier contact offers us the opportunity to study the effects of interface traps on barrier height and to explain the observed anomalies. A semiconductor such as nitrogen-doped GaP is particularly suitable for such a study because minority carrier injection in the oxbar contact can be easily detected by observing light emission by eye under forward bias. A stable and reproducible oxide barrier contact suitable for measurement can be made by galvanically growing (24) a 100 Å-thick oxide on GaP and also on GaAs. Figure 13.4 shows a schematic of the structure studied. Experiments on Au–GaP and Au–GaAs oxide barrier contacts (25) are described in this section. The results to be described for these oxide barrier contacts are not general and apply in detail only to oxide barrier contacts with vacuum-

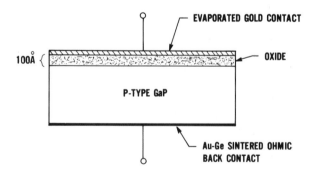

Figure 13.4.   Cross section through an experimental oxide barrier contact.

deposited gold on galvanically grown native oxides of GaP and GaAs. How the oxide is grown, how the metal is deposited, as well as the metal and semiconductor chosen have an important bearing on the experimental results obtained. However, it is our intent, using a specific oxide barrier contact, to illustrate the general principles involved in all oxide barrier contacts and to describe simple methods for measuring oxide barrier contacts which clearly reveal their properties.

It is found that the $I-V$ characteristics are the same over the oxide range from 50 to 250 Å for the galvanically grown native oxides of GaP and GaAs. Over this thickness range, there appears to be no critical voltage below which no conduction takes place. Forward and avalanche current densities of several A cm^{-2} are readily observed so that these oxides appear to be quite transparent to both electrons and holes although the exact nature of the conduction mechanism is not known. The native oxides of GaP and GaAs are better conductors over a wider range of thicknesses than thermally grown SiO_2 so that the emission-limited, intermediate oxide thickness range will be much narrower (upper limit about 30 to 40 Å) for the $Si-SiO_2$ system. However, for native oxides of GaP and GaAs, an oxide thickness of 100 Å, which was used in the experiments to be described, is well within the experimentally determined emission-limited region.

In these experiments, the initial charged interface trap density, polarity, and charged interface trap density, after current flow, will be estimated from $C-V$ measurements. Only charged interface trap density will be measured, as it is this quantity that affects barrier height. Charged interface trap density will be changed by a forward current soak which increases charged interface trap density and increases the intensity of green light emission from the nitrogen-doped GaP at a given forward current. Charged interface trap density will be changed by an avalanched current soak which decreases charged interface trap density and decreases the intensity of green light emission at a given forward current applied temporarily to observe the light emission. A lowered avalanche breakdown voltage also is observed after an avalanche current soak. Pulse $I-V$ characteristics are measured to observe changes in charged interface trap density in action. Applying voltage in short pulses with a low-duty cycle slows the rate of oxide charging, so that the lowering of the avalanche breakdown voltage can be followed as it occurs. Finally, avalanche light emission is observed in both the GaP and GaAs oxide barrier contacts, which shows that the high reverse current at low voltages (below bulk avalanche breakdown voltages) is due in fact to avalanche multiplication and is not an edge effect.

The experimental data are interpreted in terms of charge trapping in the oxide layer, and this interpretation will be used to explain the observed anomalies. However, it might be thought that pinholes, which may exist in intermediate-thickness oxides, would offer an alternative explanation of the observations. This has been considered, but it is found that all of the observations cannot be explained by a pinhole model while all of the observations can be explained by charge trapping in the oxide layer.

13.2.4   Model for High Minority Carrier Injection

In this section we present a model which shows that charged interface traps can increase the barrier sufficiently to account for the observed minority carrier injection under forward bias. Figure 13.5a shows a band bending diagram of a Au–n-type GaP Bardeen barrier contact in thermal equilibrium. Figure 13.5b shows a band bending diagram of a Au–n-type GaP oxide barrier contact in thermal equilibrium. We can draw the oxide barrier contact in thermal equilibrium because the oxide is transparent to electrons so that the interface traps can easily equilibriate.

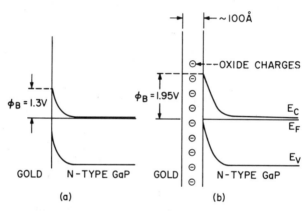

Figure 13.5.   Energy band diagram illustrating (a) a Bardeen barrier contact between gold and n-type GaP, the barrier height 1.3 V; from reference 4; (b) an oxide barrier contact between gold and n-type Gap with a galvanically grown intervening oxide layer 100 Å thick; the presence of negative charge in the oxide causes the barrier height to increase to 1.95 V; from reference 25.

By comparing Figure 13.5a to Figure 13.5b, the effect of charged interface traps on the barrier height can be seen. It is observed experimentally that the charged interface traps affect the barrier height. This means that the spatial position of the centroid of the interface trap charge distribution in the oxide must be much closer to the semiconductor than the metal so that most of the interface trap charge is imaged in the semiconductor. It will be assumed that the interface trap charge resides entirely in the interfacial region. The interface trap charge affects the barrier height because the charge in interface traps must be balanced, according to Gauss' law, by an equal and opposite charge in the semiconductor. With the polarity of interface trap charge shown in Figure 13.5, the balancing charge produced in the semiconductor results in increased band bending and therefore increased barrier height.

In Figure 13.5a, $\phi_B = 1.3$ V (4), which is too low for significant minority carrier injection. Figure 13.5b shows that negatively charged interface traps have increased ϕ_B which must be comparable to the semiconductor bandgap and the oxide must be transparent to free carriers. A negative charged interface trap density in the 10^{12} cm^{-2} range would be sufficient to make $\phi_B = 1.95$ V, which is comparable to the 2.4 V bandgap of GaP. In such a case, high minority carrier injection (1 to 10%) is to be expected. A charged interface trap density in the 10^{12} cm^{-2} range is not usually obtained immediately after oxidation; a density about a decade lower is typical, but after high forward current density flow, the oxide charge density will rise to such values.

13.2.5   Measurement of Interfacial Charge

In the next two sections, we describe experimental evidence supporting the model just described for minority carrier injection in an oxide barrier contact. Figure 13.6 shows that the net interface trap charge is initially positive when the semiconductor is p-type and initially negative when the semiconductor is negative. That is, the net interface trap charge has the same polarity as the majority carriers in the semiconductor. The solid lines in Figure 13.6 represent the measured oxide barrier contact on Au–p-type and Au–n-type GaP and the dotted lines drawn parallel to the solid lines, with intercepts at values of ϕ_B obtained from reference 4, represent Au–p-type and Au–n-type GaP Bardeen barrier contacts. Capacitance was measured in reverse bias where the leakage current was small (< 100 μA cm^{-2}) to avoid current-induced charging and errors in measured capacitance which would occur at high currents.

The acceptor concentrations of the p-type Bardeen barrier and oxide barrier contacts are the same. Similarly, the donor concentrations of the n-type Bardeen barrier and oxide barrier contacts are the same.

The slope of each line in Figure 13.6 is inversely proportional to the impurity concentration in the semiconductor as shown by Eq. 32 in Section 13.3.1. Therefore, the solid and dotted lines for p-type in Figure 13.6a are parallel to each other, while the solid and dotted lines for n-type in Figure 13.6b are parallel to each other. For the Bardeen barrier contact, the intercept on the voltage axis when $1/C^2 = 0$ gives ϕ_B. For the n-type Bardeen barrier contact, $\phi_B = 1.3$ V (4); and for the p-type Bardeen barrier contact, $\phi_B = 0.75$ V (4). The sums of these barrier heights add up to bandgap as expected. However, for the oxide barrier contact, the intercept on the voltage axis when $1/C^2 = 0$ is not equal to the semiconductor barrier height. Note that the

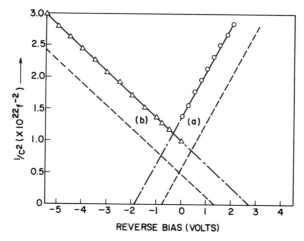

Figure 13.6. (a) Plot of $1/C^2$ vs. V for a Au-p-type GaP oxide barrier contact having a galvanically grown intervening oxide layer 100 Å thick (solid line). Metal is a 127 μm-diameter gold dot 2000 Å thick. Capacitance was measured at 1 MHz and bias swept at 0.02 V s^{-1}. Dotted line is for a Au-p-type GaP Bardeen barrier contact of the same acceptor concentration ($N_A = 1.2 \times 10^{17}$ cm^{-3}). (b) Plot of $1/C^2$ vs. V for a Au-n-type GaP oxide barrier contact with a galvanically grown intervening oxide layer 100 Å thick (solid line). Metal is a 127 μm-diameter gold dot 2000 Å thick. Capacitance was measured at 1 MHz and bias swept at 0.02 V s^{-1}. Dotted line is for a Au–n-type GaP Bardeen barrier contact of the same donor concentration ($N_D = 2.4 \times 10^{17}$ cm^{-3}). From reference 25.

intercept for the n-type oxide barrier contact alone is greater than bandgap. We can get the barrier height, the sign, and the density of the charged interface traps by considering that the voltage intercept V_i is the sum of the voltage drops across the oxide layer, Δ, and the band bending at the semiconductor surface, ψ_s:

$$V_i = \Delta + \psi_s \qquad (22)$$

The semiconductor barrier height ϕ_B is related to ψ_s by the relation

$$\phi_B = \psi_s + E_F \qquad (23)$$

In the GaP samples measured, the doping concentration obtained from the slopes of curves like Figure 13.6 are in the 10^{17} cm^{-3} range so that E_F is 0 to a good approximation. Therefore, $\phi_B = \psi_s$ and Eq. 22 becomes

$$V_i = \Delta + \phi_B \qquad (24)$$

In Figure 13.6, the intercept V_i for the p-type oxide contact is more negative than the intercept for a p-type Bardeen barrier contact, while V_i for the n-type oxide barrier contact is more positive than the intercept for the *n*-type Bardeen barrier contact. This suggests positive net interface trap charge in the *p*-type oxide barrier contact and negative net interface trap charge in the *n*-type barrier contact. This result further suggests that the interface traps are both of the donor and acceptor type. If there were only one type of trap present, the same polarity should be observed in both *n*- and *p*-type samples. Figure 13.7 illustrates a model explaining why net interface trap charge is positive with *p*-type and negative with *n*-type when both donor- and acceptor-type traps are present in nearly equal amounts. It is plausible that donor- and acceptor-type traps could be present in equal amounts if, for example, the impurity metal atoms (gold) acted as interface traps and each metal atom had associated with it two energy levels, one an acceptor level and the other a donor level.

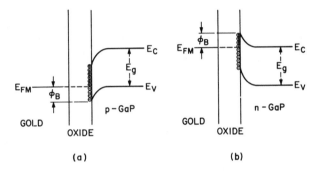

Figure 13.7. Band bending diagram of an oxide barrier contact showing interface trap model explaining the polarity of the net interface trap charge in *n*- and *p*-type samples: (*a*) Au–*p*-type GaP oxide barrier contact; (*b*) Au–*n*-type GaP oxide barrier contact.

The position of the Fermi level in the metal determines the occupancy of the interface traps because the oxide is transparent to free carriers while a potential barrier exists in the semiconductor decoupling bulk free carriers from the interface traps as described previously. The interface traps extend in energy only over the bandgap of the semiconductor, as shown in Figure 13.7. Figure 13.7*a* shows a *p*-type oxide barrier contact at zero bias so that the Fermi level in the metal is nearly in line with the valence band edge ($E_F \approx 0$). Figure 13.7*b* shows an *n*-type oxide barrier contact at zero bias so that the Fermi level in the metal is nearly lined up with the conduction band edge ($E_F \approx 0$). Acceptor-type traps below the metal Fermi level are negatively charged, while those above are neutral. Donor-type traps above the metal Fermi level are positively charged, while those below are neutral. The total acceptor trap charge in the p-type interface is $Q_A = q\phi_B N_{TA}$, and the total donor trap charge is $Q_D = (E_g - q\phi_B)N_{TD}$, where N_{TA} and N_{TD} denote total charged acceptor- and donor-type interface trap density, respectively. Because $(E_g - q\phi_B) > q\phi_B$, $Q_D > Q_A$ and the net charge is positive. Similarly, the total donor trap charge in the p-type interface is $Q_A = (E_g - q\phi_B)N_{TA}$. Again, because $(E_g - q\phi_B) > q\phi_B$, $Q_A > Q_D$ and the net charge is negative.

To estimate the magnitude of the net interface trap charge in each type, Δ must be calculated from Eq. 24. To do this, ϕ_B must be calculated from

$$\phi_B = \tfrac{1}{2}(qN_D\varepsilon_s/C_s^2) \tag{25}$$

which is just Gauss' law applied to the semiconductor surface, where N_D is the bulk donor concentration and C_s is the zero-bias semiconductor surface capacitance per unit area given by the relation

$$1/C_s = (1/C) - (1/C_{ox}) \tag{26}$$

where C is the capacitance per unit area measured at zero bias and C_{ox} is the oxide layer capacitance per unit area. Both the oxide thickness of 100 Å and oxide dielectric constant of 5.5 were measured by ellipsometry. For a 100 Å-thick oxide, $C_{ox} = 4.9 \times 10^{-7}$ F cm^{-2}; and at zero bias, $C = 7.9 \times 10^{-8}$ F cm^{-2} for a contact area of 1.27×10^{-4} cm^2.

The bulk donor concentration from the slope of Figure 13.5b is 2.4×10^{17} cm^3. Using these quantities in Eqs. 25 and 26, $\phi_B = 1.95$ V. Using $\phi_B = 1.95$ V and $V_i = 2.7$ V in Eq. 24 gives $\Delta = 0.75$ V. The voltage across the oxide is made up of two terms. The first term arises from the work function difference between the metal and semiconductor (26), and the second term arises from the net interface trap charge. Expressing this mathematically, assuming that the oxide charge is a sheet at the oxide–semiconductor interface, we obtain

$$\Delta = Q_{sc0}/2C_{ox} + qN_T/C_{ox} \tag{27}$$

where Q_{sc0} is the semiconductor surface space charge density at zero bias due to the work function difference, and qN_T is the net charged interface trap density. Equation 27 will be used to calculate N_T. The quantity Q_{sc0} is given by

$$Q_{sc0} = (2\varepsilon_s qN_D\phi_{B0})^{1/2} \tag{28}$$

where $\phi_{B0} = 1.3$ V, from reference 4, for a Bardeen barrier contact between gold and n-type GaP. From Eq. 28, $Q_{sc0}/C_{ox} = 0.3$ V; and from Eq. 27, $N_T = 1.4 \times 10^{11}$ cm^{-2}, negative. Similarly, for the p-type oxide barrier contact, the bulk acceptor concentration from the slope of Figure 13.5a is 1.2×10^{17} cm^{-3}; and at zero bias, $C = 6.7 \times 10^{-8}$ F cm^{-2}. Substituting these values into Eqs. 26 and 27, $\phi_B = 1.38$ V. Using $\phi_B = 1.38$ V and $V_i = 1.8$ V in Eq.

24 gives $\Delta = 0.42$ V. With $\phi_B = 0.75$ V, from reference 3, $Q_{sc0}/C_{ox} = 0.17$ V using Eq. 28. From Eq. 27, $N_T = 7.6 \times 10^{10}$ cm^{-2}, positive.

Changes in interface trap occupancy with bias are ignored in these calculations because the only intent is to give the reader an order-of-magnitude estimate of the net interface trap densities after oxidation.

It is emphasized that the results described for Au–GaP oxide barrier contacts may be different for different contact metals and/or different semiconductors. For example, the apparent donor- and acceptor-like interface traps observed in these oxide barriers contacts may be related to the gold used as the electrode metal because the gold may be penetrating the oxide layer all the way to the semiconductor oxide interface during vacuum deposition producing a significant number of the interface traps observed. If another metal were used, different results might be expected. For example, if the interface traps were acceptor-like in the p-type oxide barrier contact, the barrier height might be reduced rather than increased by the charged interface traps. Then, reducing the interface trap density either by a high-current soak or by making the oxide thicker so fewer metal atoms penetrate all the way to the interface would serve to increase the barrier height. Indeed, such differences are observed; therefore, the specific nature of the interface traps and their effect on the barrier height depend critically on the metal and semiconductor used as well as on the properties of the intervening oxide layer which, in turn, depends on the conditions under which it was grown.

13.2.6 Changes of Interface Trap Density Produced by Current Flow

Figure 13.8 shows how the charged interface trap density can be increased and decreased by forward and avalanche current soaks. Capacitance measured at 1 MHz is shown as a function of reverse bias, for reverse bias values below avalanche breakdown, before and after forward and avalanche current soak. Current density during these capacitance measurements, as before, was below 100 μA cm^{-2}, which means that charged interface trap density is not changed by a detectable amount during the measurement. The sample measured was a Au–p-type GaP oxide barrier contact. Curve (a) is the initial C–V curve. After this initial measurement, a hole forward current density of 16 A cm^{-2} was made to flow for about 1 min. Curve (b) was measured after this soak. Curve (b) is shifted to less positive voltages with respect to curve (a) by about 0.3 V, which means that the initial positively charged interface trap density has been increased by about 9×10^{11} charges cm^{-2}. Also, the intensity of green light emission, at a given forward current, was observed to increase. This increase of emitted light intensity suggests that increasing the positively charged interface trap density increases the barrier height. After the forward current soak, the diode was subjected to an avalanche current density of 24 A cm^{-2} for about 1 min. In this avalanche current soak, elec-

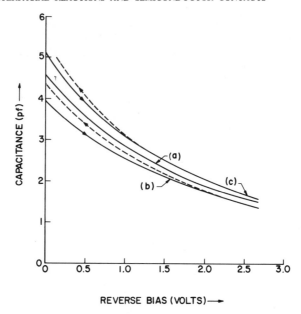

Figure 13.8.   Capacitance vs. voltage measured at 1 MHz. Bias is swept at 0.02 V s^{-1}. The diode is the same Au–p-type GaP oxide barrier contact as in Figure 13.7a. Curve (a) is the initial curve and exhibits no hysteresis. Curve (b) is measured after the passage of a hole forward current density of 15.7 A cm^{-2} through the diode for 1 min. Curve (c) is measured after an electron avalanche current density of 24 A cm^{-2} has passed through the diode for 1 min. Arrows in curves (b) and (c) indicate increasing and decreasing bias, respectively. From reference 25.

trons from the avalanche plasma in the GaP flowed through the oxide. Curve (c) was measured after the avalanche current soak. It is seen that curve (c) is shifted to more positive voltages with respect to curve (b) by about 0.46 V, which means that the positively charged interface trap density has been reduced by about 1.4×10^{12} charges cm^{-2}. Also, the intensity of green light emission, at a given forward current, was observed to decrease. This decrease of emitted light intensity suggests that decreasing the positively charged interface trap density decreases the barrier height. The induced charged interface trap density in both cases was found to be relatively stable, taking several hours to decay a detectable amount at room temperature.

The results of Figures 13.6 and 13.8 show that the barrier height is increased by the presence of charged interface traps, and that forward current soak increases the barrier height by increasing the charged interface trap density sufficiently so that significant minority carrier injection occurs. An avalanche current soak reduces the barrier height and minority carrier injection by reducing the charged interface trap density.

There is another observation from Figure 13.8 which enables us to increase our understanding of what occurs during a current soak in an oxide barrier contact. This observation is that the solid line curve (b) and the solid line curve (c) in the figure are shifted along the voltage axis parallel to each other and to curve (a). That is, at any given capacitance, in the range of capacitance shown in Figure 13.8, the voltage difference between each curve will be the same. The solid-line C–V curves were measured in the direction of increasing reverse voltage as shown by the arrows in Figure 13.8. In the direction of increasing reverse voltage, the interface traps change their occupancy by the capture of leakage current holes flowing from the metal. The capture rate is more rapid than the voltage sweep rate, so that changes in interface trap occupancy easily follow changes in reverse voltage. Changes in interface trap occupancy with bias must stretch out the C–V curve along the voltage axis. However, the figure shows that the change in interface trap occupancy with increasing reverse voltage stretches out each C–V curve the same amount along the voltage axis.

This observation is surprising because the current soaks have changed the charged interface trap density, as clearly seen in Figure 13.8 by the shifts of the C–V curves along the voltage axis after each current soak. A suggested explanation of the parallel shifts of the C–V curves, even though the current soaks have changed the charged interface trap density, is that the current soaks produce an increase in the density of one type of interface trap (donor traps for forward current soak and acceptor traps for avalanche current soak) while anihilating an equal density of the other type (acceptor traps for forward current soak and donor traps for avalanche current soak). This compensation may be reasonable if the donor and acceptor traps have a common origin, namely, gold atoms.

Consider curve (a) in Figure 13.8 before any current soak. According to this picture, the change in interface trap charge ΔQ_T resulting from interface trap occupancy change over a given voltage interval ΔV will be

$$\Delta Q_T = q(N_{TD} + N_{TA}) \Delta V \qquad (29)$$

where N_{TD} and N_{TA} are the donor and acceptor interface trap densities, respectively, and ΔQ_T is the charge which determines the degree of stretchout of the C–V curve along the voltage axis. We have assumed in Eq. 29, for simplicity, that the donor and acceptor trap densities are uniformly distributed in energy. We further assume in this model that the forward current soak produces an increased donor trap density, $N_{TD} + \Delta N_{TD}$, and a correspondingly decreased acceptor trap density, $N_{TA} - \Delta N_{TA}$. Then, if $\Delta N_{TD} = \Delta N_{TA}$, Eq. 29 will remain unchanged and curve (b) will shift parallel to curve (a) in Figure 13.8. Similarly, we assume in this model that the avalanche current soak produces a decreased donor trap density, $N_{TD} - \Delta N_{TD}$, and a correspondingly increased acceptor trap density, $N_{TA} + \Delta N_{TA}$. Again, Eq. 29 will remain unchanged and curve (c) will shift parallel to curves (a) and (b) in Figure 13.8 if $\Delta N_{TD} = \Delta N_{TA}$. It is not necessary in this model for ΔN_{TD} and ΔN_{TA} in the forward current soak to be of the same magnitude as in the avalanche current soak.

The total amount of charge that has flowed through the oxide appears to be the parameter responsible for changes in interface trap density. This can be seen from the observation that comparable changes of interface trap density occur for comparable forward and avalanche current density flowing for a given length of time while the electric field across the oxide layer is always smaller in magnitude under forward bias than it is at avalanche breakdown.

The mechanism by which interface trap density changes during a current soak may be an electrochemical reaction, the motion of an ionic species toward or away from the semiconductor–oxide interface, or the capture of free carriers by deep-lying interface traps and a subsequent slow decay. There has not been sufficient work done to distinguish these possibilities. More can be said about the mechanisms causing changes in interface trap density by current flow in the Si–SiO$_2$ system (9). In a hydrated SiO$_2$ layer, there are water-related electron-capturing centers. When an electron current is made to flow through SiO$_2$, the capture of an electron by one of these water-related centers initiates a chemical reaction. One of the products of this reaction is hydrogen and the other product is a negatively charged center which can act as an interface trap if it is near the Si–SiO$_2$ interface. By contrast, hole flow into the SiO$_2$ produces positive charge and an increase in interface trap density and occurs without a chemical reaction; thus, it appears to be simply the capture of holes by initially neutral traps in the oxide. This work (9) was done on thick oxides. Intermediate-thickness oxides may be more complicated because of the presence of metal atoms at the Si–SiO$_2$ interface from the metal deposition process; these metal atoms also may act as interface traps.

The question of the uniformity of the oxide charge induced by the two current soaks described in Figure 13.8 is important for explaining the decrease in avalanche breakdown voltage observed after an avalanche current soak. The point is that the measurements shown in Figure 13.8 do not show any nonuniformity of the charge distribution produced by the current soaks because there is no distortion in the shape of curves (b) and (c) compared to curve (a) and the shifts of the $C-V$ curves along the voltage axis are parallel. However, this permits us to place an upper limit on the extent of charge nonuniformity. To detect nonuniformity in the charge distribution in the plane of the interface from a $C-V$ curve, the extent of the nonuniformity must be comparable to the semiconductor surface space charge width. If the nonuniformity is comparable in extent to the space charge width, the shape of the $C-V$ curves compared to each other will be distorted or stretched out along the voltage axis. The fact that such distortion in the shape of the $C-V$ curves is not seen in Figure 13.8 suggests that any nonuniformities in the induced charge distribution in the plane of the interface are much smaller in extent than the semiconductor space charge width. At zero bias, the semiconductor surface space charge width, obtained from the zero bias capacitance, is about 1320 Å. At other values of reverse bias, the space charge width will be of the same order of magnitude. Therefore, nonuniformities in the oxide charge distribution must be smaller than about 1000 Å or they would have been detected. The lowered avalanche breakdown voltage cannot be ascribed to nonuniformities which cannot be detected, so that another model must be found, which is discussed later.

The dotted curves in Figure 13.8, measured in the direction of decreasing reverse voltage as indicated by the arrows, show hysteresis in curves (b) and (c) but not in curve (a). In the direction of decreasing reverse voltage, the interface traps change occupancy by the emission of electrons to the semiconductor. The electron emission rate is much slower than the voltage sweep rate so that there are more electrons remaining in interface traps than should be at a given voltage. Therefore, the dotted curves are shifted to higher reverse voltages. The observation of hysteresis in curves (b) and (c) but not in (a) means that the current soaks are producing interface traps with lower emission probabilities than the initial interface traps.

13.2.7 Pulse $I-V$ Characteristics

In this section we show, by using a pulse technique, how the decrease in charged interface trap density results in a lowering of the barrier height and a lowering of the avalanche breakdown voltage. By using short voltage pulses (50 ns) with a duty cycle of about $5 \times 10^{-4}\%$, it is possible to change the charged interface trap density under high-field conditions yet slow down the average charging rate enough to permit direct observation of the changes in the electrical characteristics of the oxide barrier contact as they gradually occur. In contrast, the change in charged interface trap density occurs too rapidly under steady-state conditions to follow its course.

The apparatus used for the pulsed $I-V$ measurements is shown schematically in Figure 13.9. This apparatus consists of a voltage pulse generator feeding the oxide barrier diode through a delay line and a trigger pulse to a sampling scope. The purpose of the delay line is to adjust the timing of the signal and trigger pulses so that the signal pulse is triggered at a convenient point on the oscilloscope screen. The pulse voltage amplitude delivered by the pulse generator was varied manually to generate the pulsed $I-V$ characteristic. A current and voltage probe feed current and voltage pulse amplitude signals to the sampling oscilloscope. The output of the sampling

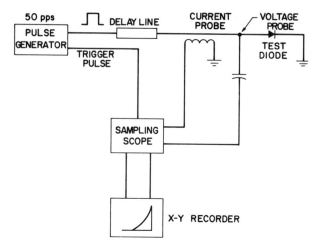

Figure 13.9. Schematic of pulsed current–voltage $(I–V)$ measuring apparatus. Pulse generator is SKL Model 503-A, delay line is HP Model 1100-A, scope is HP Model 185-B, current probe is Tektronix P60, 5052, and voltage probe is Tektronix P6035, 100×, 5 kΩ 0.5 W. Duty cycle is $5 \times 10^{-4}\%$. From reference 25.

oscilloscope was fed into an X–Y recorder to get a permanent record of the $I–V$ characteristics of the diode. The current and voltage signal pulses were sampled after the 50 ns voltage pulse was on for 30 ns.

Figure 13.10 shows a pulse $I–V$ characteristic on a Au–p-type GaP oxide barrier contact. Bulk avalanche breakdown can be seen in the figure as an abrupt increase in current at the highest voltages. Avalanche multiplication, which appears as increased leakage current, also can be seen. The basic difference between the two is that the bulk avalanche breakdown voltage is determined by the doping concentration in the semiconductor, while the low-voltage avalanche multiplication depends on the charged interface trap density. The upper curve of loop 5 in Figure 13.10 is the same curve one would obtain under steady-state conditions.

Our first purpose is to describe how this final steady-state curve develops. Initially, as pulse voltage was increased in the reverse direction, the lower curve of loop 1 was traced until bulk avalanche breakdown occurred in the semiconductor at about 11 V where the current abruptly increased. The pulse voltage was increased further until a pulse current of about 10 mA flowed through the diode, and then the pulse voltage was decreased tracing out the upper curve of loop 1. On increasing reverse pulse voltage again, the lower curve of loop 2 was followed until bulk avalanche breakdown occurred in the semiconductor. The pulse voltage was increased further until bulk avalanche breakdown occurred again and a pulse current of 12 mA flowed through the

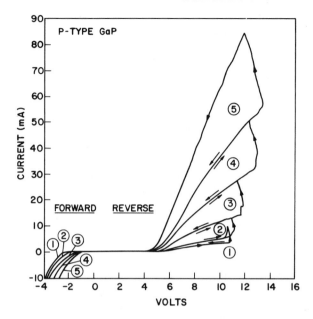

Figure 13.10.   Pulsed *I–V* characteristics measured at a sampling time of 30 ns on the same oxide barrier contact as in Figure 13.7*a*. From reference 25.

diode. Then, pulse voltage was decreased tracing out the upper curve of loop 2. Loops 3, 4, and 5 were traced similarly with bulk avalanche current going successively higher. After loop 5, the effect became saturated and no more loops could be traced even after bulk avalanche current was increased to much higher values.

The loops in Figure 13.10 can be described by noting that in bulk avalanche breakdown electrons created in the avalanche plasma, existing in the *p*-type semiconductor surface, get injected into the oxide layer. Some of the injected electrons are trapped in the oxide layer thereby decreasing the initial positive interface trap charge density. The saturation effect after loop 5 in Figure 13.10 can be interpreted to mean that all of the available traps are used up so that no further trapping can take place. This may be related to an electrochemical reaction when one of the reactants is completely used up or to the migration of a finite number of ionic species or the charging up of all the available deep-lying traps.

Another important effect of a reduced positive interface trap density is high current at very low voltages (below bulk breakdown voltage). These high currents at low voltages result from avalanche multiplication caused by impact ionization in the semiconductor. Evidence for avalanche multiplication at low voltages is given by the observation of light emission having the characteristic color for avalanche in both GaP (yellow orange) and GaAs (white). It can be seen, using a semitransparent gold electrode 50 Å thick, that the avalanche light is coming from the area of the semitransparent electrode and not from its perimeter. The problem is: How can there be avalanche multiplication in the semiconductor associated with a decrease in positive interface trap density when a decrease in positive oxide charge density results in a reduced electric field in the semiconductor? The answer must be speculative because there is not sufficient experimental evidence to prove our hypothesis.

The speculative model is as follows. As positive interface trap charge decreases by the capture of electrons flowing through the oxide, the field across the oxide increases at a given reverse voltage. Leakage current holes entering the oxide from the metal become accelerated by the field across the oxide as they cross the oxide layer. Some of these holes enter the semiconductor hot and some have been heated to a high enough energy by the oxide field to impact ionize in the semiconductor. Sufficient electron trapping can occur in the oxide to increase the voltage across the oxide to more than 1 V when the reverse voltage applied exceeds 5 V. Sufficient energy is then available to heat the holes to impact ionization energies.

A positive feedback mechanism exists in this model. Each electron trapped in the oxide decreases the positive oxide charge density, increases the oxide field, and makes it more likely that more electrons will be created by avalanche multiplication. This, in turn, means that more electrons will be trapped in the oxide, further reducing the positive oxide charge. This process can continue until all the traps are used up and the effect becomes saturated.

In summarizing this model, the decreasing positive oxide charge provides an increasing oxide field which can accelerate holes to sufficient energies to allow them to impact ionize upon entering the semiconductor. The resulting avalanche multiplication then explains the high currents observed at low voltages which are always associated with the decrease of positive interface trap charge density. A similar picture would hold when the semiconductor is n-type, the main difference being that electrons rather than holes would be accelerated by the increasing oxide field and holes rather than electrons would be trapped in the oxide.

Charge trapping is not likely to seriously interfere with the acceleration of carriers across the oxide because charge trapping is a very inefficient process. Also, the hole current flowing from the metal to the semiconductor is small compared to the multiplied electron current flowing from the semiconductor to the metal, so that the net effect will be a reduction of positive interface trap charge density. What is lacking is independent evidence such as the scattering mean free path for electrons and holes in the oxide to help decide in favor of or against the model.

Bulk avalanche at the end of the loops in Figure 13.10 and avalanche multiplication at low voltages also seen in the figure differ in the following way. Bulk avalanche occurs rapidly, in less than 30 ns, while avalanche multiplication at low voltage takes a long time to occur, of the order of ms, because it takes a relatively long time for the oxide charge density to decrease.

Figure 13.10 also shows that reducing the positive interface trap charge density reduces the barrier height. This is seen by applying a pulse voltage in the forward direction after the completion of each reverse voltage loop. The magnitude and time duration of the forward current is sufficiently small so that the resulting increase in positive interface trap charge density is insufficient to outweigh the decrease in positive interface trap density during avalanche. It is seen from Figure 13.10 that a given forward current is obtained at progressively lower forward voltages as the positive interface trap density is decreased. This is firm evidence that the semiconductor barrier height decreases with decreasing positive interface trap density.

The increased pulse voltage needed to obtain bulk avalanche breakdown in each successive loop, as seen in Figure 13.10, results from the decrease in positive interface trap charge density produced during bulk avalanche in the previous loop. This reduced positive interface trap charge results in a lower field in the semiconductor for a given pulse voltage. Therefore, a higher pulse voltage must be applied to obtain bulk avalanche breakdown as the positive interface trap charge density progressively decreases.

The bulk avalanche breakdown voltage of nearly 14 V seen in loop 5 corresponds to an acceptor concentration of 3×10^{17} to 4×10^{17} cm^{-3} (4). This result is in disagreement with the acceptor concentration of 1.2×10^{17} cm^{-3} found from the slope of the $1/C^2$-versus-V curve in Figure 13.6a. However, the slope of this curve may be in considerable error because changes in interface trap occupancy with bias have not been taken into account. The lack of hysteresis in this curve may mean that the interface traps can follow the voltage sweep, stretching out the $C-V$ curve along the voltage axis by a considerable amount.

It will be noticed further in Figure 13.10 that at high bulk avalanche currents at the end of each loop the $I-V$ characteristic has a negative slope. This negative slope results from the fact that the diode impedance during bulk avalanche breakdown becomes lower than the 50 Ω load resistance and that the characteristic walks up the 50 Ω load line as pulse voltage is increased.

Figure 13.11 shows that there can be a relatively long time delay before the onset of avalanche multiplication current at low voltages and that charge trapping is very inefficient. The observations in the figure were made by applying a step voltage of 6.3 V. This voltage produced the low-voltage avalanche condition. Current was photographed as a function of time as a

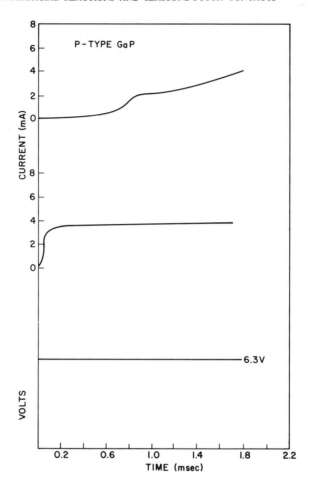

Figure 13.11. Reverse current vs. time in response to application of a step voltage of 6.3 V to the oxide barrier contact of Figure 13.7a. Upper curve is typical of the longest and lower curve is typical of the shortest delay, before the onset of current, of a random distribution of delay times. From reference 25.

single trace on an oscilloscope screen. The same step voltage was then applied several times and a distribution of delay times, before current was observed to flow, was observed. The longest and shortest delay times observed are shown in the figure.

Figure 13.11 shows that relatively long delay times, up to 1 ms, can occur before avalanche multiplication current flows. There is always a delay time before low-voltage avalanche multiplication current flows because there must be some wait for changes in interface trap charge to occur. The figure also shows that trapping efficiency is low. For a step voltage of 6.3 V, a steady-state current of 20 mA would flow. However, after 1.8 ms, only 4 mA is observed to flow. This current is only 20% of the steady-state current. The saturation value of trapped charge is in the 10^{12} cm^{-2} range when the full 20 mA would flow. However, in the 1 ms during which current flowed, about 1×10^{17} charges cm^{-2} flowed through the oxide. Thus, trapping had to be very inefficient.

The use of C–V and pulsed I–V measurements represents a powerful method of measuring the properties of oxide barrier contacts. Indeed, pulse I–V characteristics such as shown in Figure 13.10 represent a test to distinguish an oxide barrier contact from a Bardeen barrier and a type 3 contact. For an oxide barrier contact, a characteristic loop will be measured, while no such loop will be measured for a Bardeen barrier or a type 3 contact. A test such as this one would be important in reliability studies.

13.3   ELECTRICAL CHARACTERISTICS OF SCHOTTKY–BARDEEN BARRIER CONTACTS

13.3.1   Depletion Layer C–V Characteristics

In this section we derive the current–voltage characteristics of a metal–semiconductor contact and show how they depend on the barrier height and its variations. This section begins with a discussion of the C–V characteristics of a metal–semiconductor because the metal–semiconductor contact is a voltage-dependent capacitance which is useful in device applications such as a varactor, and the barrier height for Schottky barrier contact can be obtained from the C–V characteristics by extrapolating a linear $1/C^2$-versus-V curve until $1/C^2 = 0$. The intercept on the voltage axis gives the barrier height. In view of its widespread, though somewhat unfortunate, use in the literature, we retain the term Schottky barrier to describe contacts which are really a combination, in various degrees, of type 1, 2, 3, and 4 contacts defined in Section 13.1.2.

Within the depletion layer, the shape of band bending defines the electric field due to charged dopant ions stripped of their electrons/holes. The depletion layer can be characterized in terms of p–n junction theory (27) assuming a one-sided step junction with a charge density $\rho \sim qN_D$ for $x < x_d$ and $\rho \sim 0$, $dV/dx = 0$ for $x > x_d$. The properties of interest are the depletion layer width x_d, the maximum electric field E_{max} across it, and the capacitance C:

$$x_d = \frac{2\varepsilon_s V_{eff}}{qN_D} \tag{30}$$

$$E_{max} = \frac{Q_B}{\varepsilon_s} = \frac{qN_dx_d}{\varepsilon_s} \tag{31}$$

$$C = \frac{\varepsilon_s}{x_d}, \quad \frac{1}{C^2} = \frac{2}{q\varepsilon_sN_D}V_{eff} \tag{32}$$

where ε_s is the dielectric permittivity of the semiconductor, Q_B is the depletion layer charge, and V_{eff} is a function of the barrier height and the applied reverse bias V_R:

$$V_{eff} = V_B + V_R = \phi_B - \frac{E_c - E_F}{q} - \frac{kT}{q} - \Delta\phi + V_R \tag{33}$$

where $\Delta\phi$ is the image force-lowering term (see next section). For a forward bias V_f, V_R should be replaced by $-V_f$. From Eqs. 32 and 33, it may be seen that the intercept of $1/C^2$ versus V gives V_B at $V_R=0$, and hence ϕ_B. The slope is inversely proportional to the doping level.

Figure 13.12 shows typical $C-V$ characteristics of a Schottky diode formed by Ti on n-GaAs (28). From the intercept on the voltage axis, the effective diffusion voltage V_{eff} may be determined to be 0.73 V. To estimate the true barrier height ϕ_B, a net correction of about 0.11 V must be applied for the thermal energy, for effect of doping on the position of the Fermi level, and

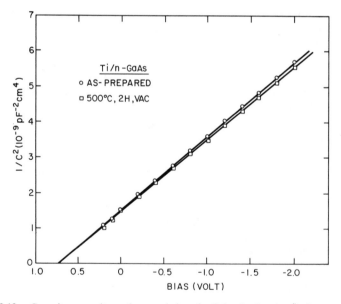

Figure 13.12. Capacitance–voltage characteristics of a Schottky barrier diode containing Ti on n-GaAs. From reference 28.

for the image force lowering of the barrier height. The resulting ϕ_B from the C–V measurements is 0.84 V, which agrees well with that (28) deduced from I–V measurements (see Section 13.3.3.4).

13.3.2 Image Force Effects

For n-type semiconductors, the depletion layer contains ionized donor atoms which are positively charged. These ions induce an image charge on the metal side. The resulting image potential energy has a sharp minimum at the interface. The superposition of this effect and that due to electric field in the depletion layer results in a peak in the conduction band edge, located a distance x_m away from the semiconductor surface (Fig. 13.13). The image force barrier lowering $\Delta\phi$ is given by

$$\Delta\phi = 2E_{\max}x_m = \sqrt{\frac{qE_{\max}}{4\pi\varepsilon_s}} \tag{34}$$

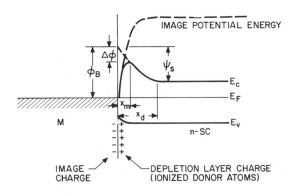

Figure 13.13. Schematic energy band diagram for a metal–n-semiconductor contact at equilibrium.

13.3.3 Forward Conduction

Under a forward bias (V_f) the quasi-Fermi level (imref) in the semiconductor moves up so that the potential barrier for electron transport from semiconductor to the metal (S→M) is decreased by V_f whereas that for electron transport from M→S remains unchanged at ϕ_B (Fig. 13.14). This gives rise to a net current flow from semiconductor to metal (S→M).

13.3.3.1 Thermionic Emission. Neglecting details of the shape or origin of the barrier, the net current flow can be predicted from simple thermodynamic arguments (29). Consider current flowing from metal to semiconductor.

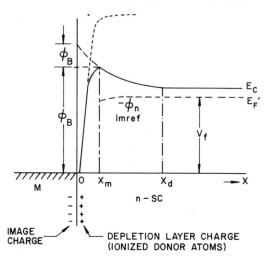

Figure 13.14.   Schematic energy band diagram for a forward biased metal–n-semiconductor contact.

The number of electrons per cm^2 per second reaching the barrier is vn_e, where n_e is the electron density and v is the electron velocity. The fraction that get across the potential barrier ϕ_B is given by the Boltzmann factor $\exp(-\phi_B/kT)$. Thus, the current density is

$$J_{M\to S} = qn_e^m v \exp\left(-\frac{\phi_B}{kT}\right) \tag{35}$$

Using a similar argument for S→M conduction

$$J_{S\to M} = qn_e^s v \exp\left[\frac{-q(V_B - V_f)}{kT}\right] \tag{36}$$

In the free electron approximation,

$$v = (kT/2\pi m^*)^{1/2} \tag{37}$$

and

$$n_e = 2\left(\frac{2\pi m^* kT}{h^2}\right)^{3/2} \exp\left[\frac{-q(E_c - E_F)}{kT}\right] \tag{38}$$

where $E_c - E_F = V_n$ for n-semiconductor and zero for metal. The net current is given by

$$J_{net} = J_{S\to M} - J_{M\to S} = A^* T^2 \exp\left(\frac{-q\phi_B}{kT}\right)\left[\exp\left(\frac{qV_f}{nkT}\right) - 1\right] \tag{39}$$

where $\phi_B = V_f + V_n$, $n = 1$ for simple thermionic emission, and the Richardson constant

$$A^* = \frac{4\pi q m^* k^2}{h^3} \tag{40}$$

A^*, in A cm^{-2} °K^{-2}, is 120 for free electrons, 264 for n-(111) Si, 252 for n-(100) Si, and 8.2 for n-GaAs under low-field conditions (4).

13.3.3.2 Recombination.   In deriving Eq. 39, we have neglected any effects taking place within the depletion layer. Under equilibrium, that is, when no applied bias is present, the depletion layer is characterized by $np = n_i^2$. However, when a forward bias is applied, electrons flow from within the semiconductor past the depletion region to the metal, and $np \gg n_i^2$. The current at very low fields can then be dominated by capture processes at traps located in the depletion layer (recombination-generation centers). The recombination rate U, in cm^{-3} s^{-1}, can be shown to be (27)

$$U \sim n_i \exp\left[\frac{-q(\phi_B - V_f)}{2kT}\right] \tag{41}$$

and the recombination current J_{rec} is

$$J_{rec} = \int_0^{x_d} qU \, dx \sim q n_i x_d \exp\left[\frac{-q(\phi_B - V_f)}{2kT}\right] \tag{42}$$

Thus, in the recombination region, log J versus V_f is linear, with a slope factor $n = 2$. With increasing V_f, x_d decreases, whereas exp $(qV_f/2kT)$ increases. The overall J_{rec} is a slowly increasing function of V_f, and eventually J_{rec} becomes dominated by the thermionic (or diffusion) current.

13.3.3.3 Charge Carrier Diffusion (30).   Another effect in the depletion layer is the diffusion of electrons (velocity v_D) in response to the imref (quasi-Fermi level) created by a forward bias in the semiconductor. As shown in Figure 13.14, the imref, ϕ_n, extends into the depletion layer until x_m, below which the electric field falls off rapidly over distances comparable to the electron mean free path, and the barrier acts as a recombination sink for electrons (velocity v_R). If $v_R \ll v_D$, the Crowell-Sze theory (30) reduces to that of thermionic emission (39). If, however, $v_D \ll v_R$, diffusion becomes the dominant conduction process:

$$J_{diff} \sim q N_c v_D \exp\left(\frac{-q\phi_B}{kT}\right)\left[\exp\left(\frac{qV_f}{kT}\right) - 1\right] \tag{43}$$

This result is very similar to the thermionic emission model except that v_D and hence saturation diffusion current density varies more rapidly with voltage and is not so sensitive to temperature.

13.3.3.4 Analysis of Experimental Results. Most of the forward I–V data on Schottky barriers are analyzed in terms of the simple thermionic emission model of Eq. 39. For $V_f \gtrsim 3kT/q$, Eq. 39 predicts a linear variation of log current versus V_f:

$$I_f = A^* T^2 S \exp\left(\frac{-q\phi_B}{kT}\right) \exp\left(\frac{qV_f}{nkT}\right) \tag{44}$$

with

$$\phi_B = \frac{kT}{q} \ln\left(\frac{A^* T^2 S}{I_0}\right) \tag{45}$$

and

$$n = \frac{q}{kT} \frac{\partial V_f}{\partial(\log I_f)} \tag{46}$$

where I_0, the saturation current, is the intercept of the linear portion of log I_f at zero bias and S is the contact area.

For most Schottky–Bardeen barrier contacts, ϕ_B as obtained from Eq. 45 is relatively insensitive to small errors in A^* or the area S. At room temperature, increasing A^*S by a factor of 2 only increases ϕ_B by ~ 0.02 V.

Figure 13.15 shows forward I–V characteristics of a Schottky barrier formed by Au on n-GaAs (31). The as-prepared diodes are characterized by two distinct transport regions; at a forward bias of less than 0.2 V, the slope n of the log I_f-versus-V_f plot is equal to ~ 2.0, signifying that the recombination current dominates. The presence of such a recombination region is believed to be the result of short minority carrier lifetime along with a relatively large barrier height ($\phi_B > E_g/2$) (32).

At larger forward bias ($V_f > 0.2$ V) the slope n changes to 1.01, which is characteristic of thermionic emission. The effect of a series resistance is apparent at the largest values of the forward voltages applied (> 0.4 V) where an additional voltage drop occurs over that due to emission current. The barrier height $q\phi_B$ associated with the thermionic emission region is found to be equal to 0.90 eV. Upon annealing Au–n-GaAs diodes at 250 or 350°C (2-h periods) there is a marked increase in the forward current as well as in the slope n. The barrier height is reduced to 0.63 and 0.60 eV, respectively, and n increases to 1.17 or 1.23. These changes are correlated with interdiffusion at the Au–n-GaAs interface (see Chapter 11).

The variation with thermal aging of the I–V characteristics of Au–W–n-GaAs diodes resembles that of W–n-GaAs diodes (32). Had any appreciable diffusion of Au atoms through the W film occurred upon thermal treatment (350 or 500°C), we would expect to observe the degradation associated with Au–n-GaAs diodes. This degradation would be manifested in the form of

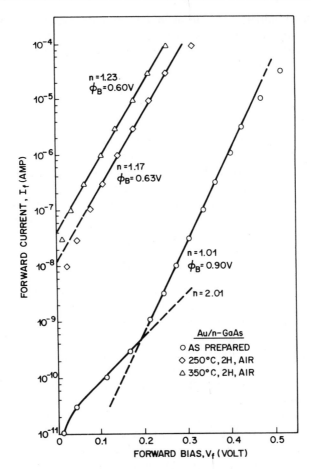

Figure 13.15. Forward I–V characteristics of a Au–n-GaAs Schottky barrier diode. From reference 31.

$n \gg 1.00$ and $q\phi_B < 0.60$ eV. In actual fact, n for the Au–W diodes remains close to unity upon thermal aging and $q\phi_B$ increases slightly. This behavior may be contrasted with that of Al–PtSi–n-Si diodes (34), whose initial barrier height was the same as that of PtSi–Si interface ($q\phi_B \sim 0.82$ eV) (35), but upon thermal treatment above 450°C, the barrier height was lowered to 0.72 eV. This change was attributed to an Al/PtSi interaction and the eventual formation of an Al–Si type of interface upon high-temperature annealing.

The parameter n describes the injection efficiency of majority carriers into the metal. It should be equal to 1.0 to 1.06 when a pure thermionic or diffusion model dominates. Deviations of n from unity have been observed by a number of investigators in the past. Padovani and Sumner (36) found that n increases with decreasing temperature. Their data on Au–n-GaAs diodes could be fitted in terms of the empirical equation

$$n = T_0 T^{-1} + 1 \qquad (47)$$

where $T_0 \approx 46°\mathrm{K}$, so that at room temperature n turns out to be 1.13, considerably higher than the value observed by us for as-prepared Au–n-GaAs diodes or by Kahng (37). Equation 47 was put on a theoretical footing by Levine (38) who has related T_0 to the distribution of interface state energy at the metal–semiconductor interface. Hackam and Harrop (39) also carried out detailed measurements of I–V characteristics of Ni–n-GaAs diodes as a function of temperature. Their data, however, show that T_0 is not a constant and that the temperature dependence of n is of the type

$$n = \alpha T^{-1/2} + \beta \qquad (47a)$$

where α and β are constants. Equations 47 and 47a represent the limiting case of Strikha's theory (40) which considers the effect of an oxide layer at the interface and predicts a temperature dependence of n of the type T^{-m}, where $\frac{1}{2} \leqslant m \leqslant 1$.

An alternative to the above theories, which mainly treat the temperature dependence of n, is provided by the thermionic-field (T-F) emission model of Padovani and Stratton (41, 42). They have considered the field-assisted tunneling through the potential barrier of electrons that have been thermionically excited to energies below the top of the barrier. The extent of T-F emission is found to be a function of the following: (a) If the donor concentration N_D becomes large enough, tunneling dominates, and the I–V characteristics become ohmic (44). (b) Temperature: the higher the temperature, the greater the proportion of thermionic diffusion current and the tendency towards $n \to 1$. According to the T-F emission theory, a plot of nT versus T has an initial slope greater than unity and it approaches unity (i.e., $n = 1$ condition) at higher temperatures (39). Note that this temperature dependence is different from Eq. 47 according to which nT versus T is linear with a slope of unity but is displaced to the right of the "ideal n" line $nT = T$.

The presently observed increase in n (> 1) for heat-treated Au–n-GaAs contacts is consistent with the following picture (see also Chapter 11). Upon thermal treatment above $250°\mathrm{C}$, Ga outdiffuses through the Au film (45), leaving vacancies in GaAs. The indiffusing Au atoms and/or vacancies may act as donors, so that alloying causes an increase in N_D and a decrease in $q\phi_B$, which favors T-F emission (31, 45). When Au–Ge contacts are used instead of Au, heat treatment above $360°\mathrm{C}$ causes the Au–Ge eutectic to form which is highly reactive and provides the Ge dopant to the GaAs near the interface, making it n^+ type; this results in ohmic behavior (46).

From the above discussion, it may be concluded that the combination of parameters ϕ_B and n represents I–V curves quite well of both ideal and non-ideal Schottky barriers. Thus, direct and meaningful comparisons can be made of I–V data on various samples if ϕ_B and n are known. However, in certain cases ϕ_B should be considered to be the true barrier height only if $n \approx 1.0$. If the detailed temperature dependence of I–V characteristics is known, the true or maximum barrier height can also be evaluated using thermionic field emission theory (see Section 13.4.1).

13.3.4   Reverse Conduction

13.3.4.1 Thermionic Model.
Under a reverse bias, the current flows mainly from the metal to the semiconductor. This direction of current flow can be seen from Eq. 39 by setting V_f negative and $|V_f| \geqslant 3kT/q$. Then,

$$J_{\text{rev}} = A^* T^2 \exp\left(\frac{-q\phi_B}{kT}\right) \tag{48}$$

One may conclude on the basis of the above equation that the reverse current should be approximately constant at the saturation values, assuming ϕ_B and A^* to be nearly constant.

13.3.4.2 Image Force Effects.
However, as the reverse bias is increased, A^* can change by up to 10% and, more importantly, the barrier height ϕ_B of Eq. 48 no longer can be assumed to be independent of bias. Image force effects begin to dominate and ϕ_B must be replaced by $\phi_B - \Delta\phi$, where, in accordance with Eq. 34

$$\Delta\phi = \sqrt{\frac{qE}{4\pi\varepsilon_0}} \tag{49}$$

and for the depletion region, in accordance with Eqs. 32 and 33,

$$E = \left[\frac{2qN_D}{\varepsilon_s}(V_B + V_R)\right]^{1/2} \tag{50}$$

Therefore, Eq. 48 should be rewritten as

$$J_{\text{rev}} = A^* T^2 \exp\left[\frac{-q}{kT}(\phi_B - \Delta\phi)\right] \tag{51}$$

and

$$\ln J_{\text{rev}} \sim (V_B + V_R)^{1/4} \tag{52}$$

That is, the reverse current slowly increases with the applied bias V_R.

13.3.4.3 Field-Induced Barrier Lowering (47). As the reverse bias is increased to relatively high preavalanche breakdown levels, an additional current component arises. This component has been ascribed to an electrostatic dipole layer at the M–S interface which leads to a field-induced barrier lowering equal to $E(\partial\phi/\partial E)$, where $\partial\phi/\partial E$ is an adjustable parameter. For many silicides on Si, this parameter is of the order of 20 Å.

13.3.4.4 Recombination-Generation Processes. In analogy with the treatment for forward conduction, we must consider nonequilibrium effects within the depletion layer. Under reverse bias, carriers are swept away from the depletion layer, so that $pn \ll n_i^2$. The dominant recombination-generation process is that of generation (emission) of electron–hole pairs at traps (27). The generation rate is

$$U \sim -\frac{n_i}{\tau_e} \qquad (53)$$

where τ_e is the effective lifetime. The generation current is

$$J_{\text{gen}} = \int_0^{x_d} q|U|dx \sim \frac{qn_ix_d}{\tau_e} \sim (V_B + V_R)^{1/2} \qquad (54)$$

Thus, compared to the thermionic component, the generation current has a stronger dependence on V_R.

13.3.4.5 Experimental Results. Unless special design features are incorporated, most Schottky barriers display abnormally large reverse currents due to leakage at the edges of the field plate. One common way to eliminate such effects is through use of a diffused guard ring of opposite polarity around the diode (Figure 13.16) (48). The p–n junction as created should be designed to have a higher breakdown voltage than the Schottky diode.

Another critical improvement consists of forming a thermodynamically stable metal silicide (or an arsenide in case of GaAs) by means of a sintering

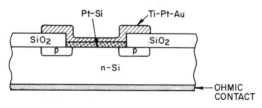

Figure 13.16. Schottky barrier contact to n-Si incorporating a diffused guard ring structure.

operation so that a clean, uniform, and intimate interface is created between the conductor and the semiconductor (49) (modified Bardeen barrier contact).

Figure 13.17 shows the reverse I–V data for a RhSi-p-Si diode. Although the agreement between experimental observations and theory improves upon going from the saturation current density model, Eq. 48, to that incorporating image force lowering, Eq. 51, the latter cannot fully account for the soft reverse characteristics. Good agreement is obtained when an effect due to field-induced barrier lowering is also included (47).

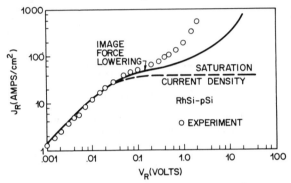

Figure 13.17. Experimentally observed soft reverse characteristics of a RhSi-p-Si diode compared with calculations of saturation current density and image force lowering effects. From reference 47.

It has been also shown that the generation current can exceed that due to thermionic emission when the M–S interface has a large barrier height and the semiconductor has an extremely small minority carrier lifetime (47). This was found to be the case with the PtSi-n-Si diode for which $\tau \sim 50$ ns and $\phi_B = 0.85$ eV.

13.4 TUNNELING AND THERMIONIC FIELD EMISSION

At higher doping levels, the depletion layer becomes very narrow (Fig. 13.18), fields become correspondingly large, and the dominant conduction mode changes from thermionic emission to quantum-mechanical tunneling (field emission) through the barrier (44). An intermediate case is that of thermionic field (T-F) emission (Fig. 13.19) (41).

13.4.1 *T-F* Emission Theory

A thermionic field emission model has been proposed (41, 42) for Schottky barriers operating at moderately high doping levels and temperatures,

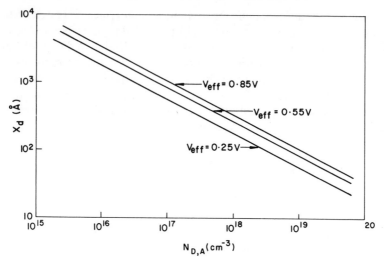

Figure 13.18. Effect of doping level N_D on the depletion layer width x_d at a metal-semiconductor junction according to Eq. 30.

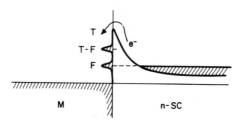

Figure 13.19. Schematic energy band diagram of a forward biased metal–n-semiconductor contact illustrating electron conduction by thermionic (T), thermionic field (T-F), and field (F) emission.

namely, when,

$$\frac{kT}{E_{00}} \approx 1, \quad \text{where } E_{00} = \frac{q\hbar}{2}\left(\frac{N_D}{m^*\varepsilon}\right)^{1/2} \tag{55}$$

T-F emission theory also predicts a linear relationship between $\log J$ and V, with

$$J = J_s \exp\left(\frac{qV}{E_0}\right) \tag{56}$$

where E_0 is a function of temperature and doping level:

$$e_0 = E_{00} \coth\left(\frac{E_{00}}{kT}\right) \tag{57}$$

$$= \left(\frac{\partial \ln J}{\partial V}\right)^{-1} \tag{58}$$

This theory has been applied to Bardeen barrier contacts whose I–V data are known for various temperatures (typically from that of liquid N_2 to $150°C$). The characteristic energy E_{00} can be obtained from the temperature dependence of the slope $[(\partial \ln J)/\partial V]^{-1}$, as shown in Figure 13.20. Next, the variation of a complex parameter incorporating the saturation current with

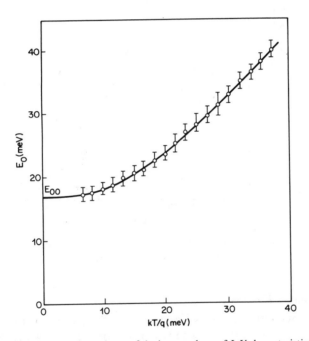

Figure 13.20. Temperature dependence of the inverse slope of I–V characteristics for a Au–n-GaAs diode. From reference 42. The intercept gives E_{00} of Eq. 57.

$1/E_0$ is plotted, and the slope gives $\phi_B + V_n$ (Fig. 13.21). The true barrier height so obtained is often significantly larger than that obtained by a straight forward application of the thermionic emission equation, especially when application of the latter yields $n > 1.1$ (45).

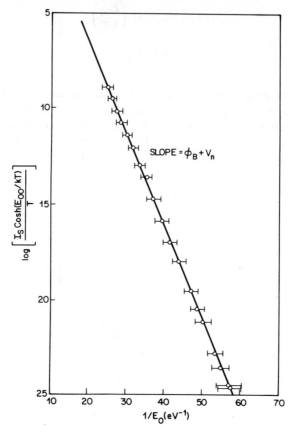

Figure 13.21. Evaluation of barrier height using T-F emission model: saturation current junction vs. $1/E_0$ for the diode of Figure 13.20.

13.4.2   Contact Resistance

The three conduction regimes are further distinguished by the functional dependence of the specific contact resistance ρ_c on the doping level, which can be described as follows (50):

$$\rho_c = \left(\frac{\partial V}{\partial J}\right)_{V \to 0} \tag{59}$$

$$\rho_c \sim \exp\left(\frac{\phi_B}{\sqrt{N_D}}\right) \text{ for field emission} \tag{60}$$

$$\rho_c \sim \exp\left\{\frac{\phi_B}{\sqrt{N_D}}\left[\coth\left(\frac{E_{00}}{kT}\right)^{-1}\right]\right\} \text{for T-F emission} \tag{61}$$

$$\rho_c \sim \exp\left[\frac{\phi_B}{kT}\right] \text{ for thermionic emission} \tag{62}$$

Figure 13.22 shows the expected $R_c(N_D)$ curves for these three cases.

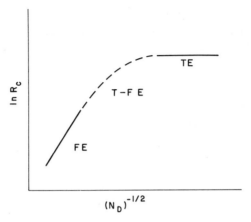

Figure 13.22. Variation of contact resistance with doping level for thermionic emission (TE), thermionic field emission (T-FE), and field emission (FE). From reference 50.

13.5 OHMIC CONTACTS TO SEMICONDUCTORS

As stated in the introduction, one of the most important applications of metal–semiconductor interfaces is that of an ohmic contact to selectively doped areas of the semiconductor device.

Ohmic contacts are electrically characterized in terms of their specific contact resistance ρ_c, which has the units of ohm cm^2, that is, the larger the contact area, the smaller the contact resistance in ohms. For the contact to be ohmic, it is desirable that the contact resistance $R_c(=\rho_c/\pi r^2)$ be smaller than or comparable to the series resistance of the semiconductor material adjacent to the contact. In many cases, the series resistance may be approximated by a spreading resistance term R_s (in ohms), given by (51)

$$R_s = \frac{\rho_s}{2\pi r} \tan^{-1}\left(\frac{2t}{r}\right) \tag{63}$$

For $t \gg r$,

$$R_s \sim \frac{\rho_s}{4r} \tag{64}$$

For $t \ll r$,

$$R_s \sim \frac{\rho_s t}{\pi r^2} \tag{65}$$

where r is the effective contact radius and t is the thickness of the semi-conductor.

Most of the experimental measurements involve determination of $R_{c,s}$ which is equal to $R_c + R_s$. In the case of sintered PtSi ohmic contacts (contact area 4×10^{-6} cm^2) to n-Si and p-Si, measured $R_{c,s}$ values were close to the estimated values for R_s, so that it was not possible to accurately estimate R_c ($= R_{c,s} - R_s$) (52). However, a comparison between $\rho_{c,s}$ ($= A R_{c,s}$) and theoretically estimated ρ_c, assuming one-dimensional tunneling (44) across the depletion layer (Fig. 13.23), showed that the experimental values for

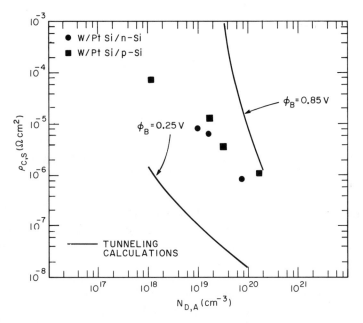

Figure 13.23. Measured specific contact resistance vs. doping level for W–PtSi contacts to n- and p-type Si. From reference 52. Solid lines show results of calculations based on a one-dimensional tunneling model. From reference 44.

n-Si samples could be orders of magnitude smaller than theoretical estimates, and correcting for speading resistance would make the discrepancies even larger. Similar disagreement was found in the case of ohmic contacts to n-GaAs where the tunneling model predicts a much steeper dependence on doping level than that experimentally found (53, 54) (see also Chapter 11).

These results are consistent with the following physical model (53). The conductor–semiconductor interface is not at all abrupt on an atomic scale.

The change from covalent bonding present in semiconductors to metallic bonding occurs over an interface that is diffuse over several interatomic distances, that is, on a scale comparable to the depletion layer width. Accordingly, it is unlikely that the idealized one-dimensional tunneling model applies to real ohmic contacts, where the current transport is most likely three-dimensional on a microscopic scale. An appropriate theory would then have to take account of this microscopic three-dimensional electron transport, nonuniform current flow, and mobility differences between n- and p-type semiconductors.

Although an exact theory has not been worked out, the above considerations have led to the following empirical guideline when searching for new ohmic contacts (53):

$$x_m > x_d \qquad (66)$$

where x_m is the width of the metallurgical junction. Ohmic contacts are favored if x_d is minimized by (a) doping the semiconductor to high levels and (b) lowering the barrier height, and x_m is maximized by metallurgically reacting the conductor–semiconductor interface.

The effect of a diffuse conductor–semiconductor interface can be illustrated with reference to the case of heat-treated Au–GaAs contacts. It has been shown that heat treatment causes outdiffusion of Ga from GaAs and indiffusion of Au. Such a diffuse interface will favor ohmic contacts. Moreover, the Ga vacancies as well as the indiffused Au atoms may be electrically active. Low-temperature heat treatment favors development of, although it does not quite achieve, Au ohmic contacts to both n^+- and p^+-GaAs. This heat treatment is effective to a much greater extent in the latter case. Pd also interdiffuses readily with GaAs, and Pd–Ge alloys have been successfully employed to form sintered ohmic contacts to n-GaAs (53). Other metals which display rapid interfusion behavior and favor ohmic I–V characteristics when in contact with $A_{III}B_V$ compound semiconductors are Sn (51) and In (55). By contrast, W contacts to both n^+- and p^+-GaAs were found to be rectifying; W does not interdiffuse with GaAs, and it is expected to have nearly the same barrier height (~ 0.7 eV) with both n- and p-type GaAs.

For the conventional "alloyed" contacts to compound semiconductors, the heat treatment is typically carried out above the eutectic temperature of metal–$A_{III}B_V$ semiconductor. Upon cooling, regrowth occurs of a thin, highly doped semiconductor layer at the interface which helps promote field emission and ohmic behavior. An example of such a contact is the Au–12 at.% Ge which is used, in conjunction with a Ni "capping" layer, to contact n-GaAs (56). At the alloying temperature ($> 500°$C, cf. eutectic point of Au–GaAs $\sim 450°$C), some GaAs dissolves into the eutectic. Ni also reacts to form NiAs; its presence improves wettability of the liquid and prevents it from "balling up" (46). On cooling, the regrown GaAs contains shallow donor atoms creating a thin, degenerately doped layer (57, 58). At the same time, Au diffuses into n-GaAs; its maximum concentration can be as high as 4×10^{21} cm^{-3}. Although there was no overwhelming tendency observed for Au to be substitutional or interstitial (59), its diffuse distribution and possible electrical activity might aid ohmic behavior as mentioned earlier. Similar principles are employed in forming ohmic "alloyed" contacts to p-GaAs, n-GaP, and p-GaP for which alloys of Au–Zn, Au–Si, and Au–Be, respectively, are utilized (60). Many other examples of ohmic contacts to semiconductors may be found in references 61 and 62.

The ohmic contact technology to silicon is much better established. In Si integrated circuits, low-resistance ohmic contacts are extensively utilized for n^+ emitter/collector and p-base regions of bipolar transistors, and for n^+ or p^+ source/drain areas and polycrystalline Si contacts in MOS (metal–oxide–semiconductor) field effect transistors. This contacting is generally accomplished using Al, PtSi, or Pd$_2$Si (see Chapter 10). With Al (63), a post-metallization anneal at 400 to 500°C is generally required. This anneal leads to interdiffusion at the interface thereby assuring an intimate contact, free from the effects of an intervening oxide layer. Ohmic contact is sometimes required to the back of the Si chip. This contact is obtained by forming a Au–Si eutectic (melting point 370°C). The eutectic, in this case, serves a dual purpose. It also helps bond the Si device chip to a Au-metallized ceramic package.

REFERENCES

1. J. M. Andrews and J. C. Phillips, *Phys. Rev. Lett.*, **35**, 56 (1975).
2. J. Bardeen, *Phys. Rev.*, **71**, 717 (1947).
3. A. M. Cowley and S. M. Sze, *J. Appl. Phys.*, **36**, 3212 (1965).
4. S. M. Sze, *Physics of Semiconductor Devices*, Wiley, New York (1969).
5. L. Pauling, *Nature of the Chemical Bond and the Structure of Molecules and Crystals: An Introduction to Modern Structural Chemistry*, Cornell University Press, Ithaca, N.Y. (1960), p. 91.
6. J. C. Phillips, *J. Phys. Chem. Solids*, **34**, 1051 (1973).
7. P. Esslinger and K. Schubert, *Z. Metallkd.*, **48**, 126, (1957).
8. S. Kar, and W. E. Dahlke, *Solid-State Electron.*, **10**, 865 (1967); and *Solid-State Electron.*, **15**, 221 (1972).
9. E. H. Nicollian, C. N. Berglund, P. F. Schmidt, and J. M. Andrews, *J. Appl. Phys.*, **42**, 5654 (1971).
10. H. P. Ladbrooke, *Solid-State Electron.*, **16**, 743, (1973).
11. H. L. Card and E. H. Rhoderick, *Solid-State Electron.*, **16**, 365 (1973).
12. C. I. Huong and S. S. Lee, *Proc. IEEE*, **61**, 477, (1973).
13. J. T. Law, in *Semiconductors*, N. B. Hannay, Ed., Reinhold, New York (1959).
14. A. M. Cowley, *J. Appl. Phys.*, **37**, 3024 (1966).
15. C. R. Crowell and G. I. Roberts, *J. Appl. Phys.*, **40**, 3726 (1969).
16. A. M. Goodman, *J. Appl. Phys.*, **34**, 329 (1963).

17. H. K. Henish, 'Rectifying Semiconductor Contacts, Oxford University Press, London (1957).
18. J. Schewchun and R. A. Clarke, Solid-State Electron., 16, 213, (1973).
19. H. C. Card and B. L. Smith, J. Appl. Phys., 42, 5863 (1971).
20. A. W. Livingstone, K. Turvey, and J. W. Allen, Solid-State Electron., 16, 357, (1972).
21. R. C. Jaklevic, D. K. Donald, J. Lambe, and W. C. Vassell, Appl. Phys. Lett., 2, 7 (1963).
22. D. J. Wheeler and D. Haneman, Salid-State Electron., 16, 875 (1973).
23. J. H. Lee and G. A. Condas, Solid-State Electron., 11, 419 (1963).
24. B. Schwartz and W. J. Sundberg, J. Electrochem. Soc., 120, 576 (1973).
25. E. H. Nicollian, B. Schwartz, D. J. Coleman, Jr., R. M. Ryder, and J. R. Brews, J. Vac. Sci. Technol., 13, 1047 (1976).
26. C. R. Crowell, H. B. Shore, and E. E. La Bate, J. Appl. Phys., 36, 3843 (1965).
27. A. S. Grove, Physics and Technology of Semiconductor Devices, Wiley, New York (1967).
28. A. K. Sinha, T. E. Smith, M. H. Read, and J. M. Poate, Solid-State Electron., 19, 489 (1976).
29. F. J. Blatt, 'Physics of Electronic Conduction in Solids, McGraw Hill, New York (1968), p. 322.
30. C. R. Crowell and S. M. Sze, Solid-State Electron., 9, 1035 (1966).
31. A. K. Sinha and J. M. Poate, Appl. Phys. Lett., 23, 666 (1973).
32. R. Williams, RCA Rev., 30, 306 (1969).
33. A. K. Sinha, Appl. Phys. Lett., 26, 171 (1975).
34. H. H. Hosack, Appl. Phys. Lett., 21, 256 (1972).
35. M. P. Lepselter and J. M. Andrews, in Ohmic Contacts to Semiconductors, B. Schwartz Ed., Electrochemical Society, New York (1969).
36. F. A. Padovani and G. G. Sumner, J. Appl. Phys., 36, 3744 (1965).
37. D. Kahng, Solid-State Electron., 6, 281 (1963).
38. J. D. Levine, J. Appl. Phys., 42, 3991 (1971).
39. R. Hackam and P. Harrop, IEEE Trans. Electron. Devices, ED19, 1231 (1972).
40. V. I. Strikha, Radio Eng. Electron. Phys. (USSR), 4, 552 (1964).
41. F. A. Padovani and R. Stratton, Solid-State Electron., 9, 695 (1966).
42. F. A. Padovani, "The Voltage Current Characteristics of Metal-Semiconductor Contacts," in Semiconductors and Semimetals, Vol. 7, Part A, R. K. Willardson and A. C. Beer, Eds., Academic Press, New York (1971), Chapter 2, pp. 75–146.
43. J. Ohura and Y. Takeishi, Jap. J. Appl. Phys., 9, 458 (1970).
44. C. Y. Chang, Y. K. Fang, and S. M. Sze, Solid-State Electron., 14, 541 (1971).
45. C. J. Madams, D. V. Morgan, and M. J. Howes, Electron. Lett., 11, 24 (1975).
46. G. Y. Robinson, Solid-State Electron., 18, 331 (1975).
47. J. M. Andrews and M. P. Lepselter, Solid-State Electron., 13, 1011 (1970).
48. J. M. Andrews, J. Vac. Sci. Technol., 11, 972 (1974).
49. M. P. Lepselter, Bell Syst. Tech. J., 40, 233 (1966).
50. A. Y. C. Yu, Solid-State Electron., 13, 239 (1970).
51. R. H. Cox and H. Strack, Solid-State Electron., 10, 1213 (1967).
52. A. K. Sinha, J. Electrochem. Soc., 120, 1767 (1973).
53. A. K. Sinha, T. E. Smith, and H. J. Levinstein, IEEE Trans. Electron. Devices, ED22, 218 (1975).
54. W. D. Edwards, W. A. Hartman, and A. B. Torrens, Solid-State Electron., 15, 387 (1972).
55. C. R. Paola, Solid-State Electron., 13, 1189 (1970).
56. N. Braslau, J. B. Gunn, and J. Staples, Solid-State Electron., 10, 381 (1967).
57. R. H. Cox and T. E. Hasty, in Ohmic Contacts to Semiconductors, B. Schwartz, Ed., Electrochemical Society, New York (1969), p. 88.
58. M. Jaros and H. L. Hartnagel, Solid-State Electron., 18, 1029 (1975).
59. J. Gyulai, J. W. Mayer, V. Rodriguez, A. Y. C. Yu, and H. C. Gopen, J. Appl. Phys., 42, 3578 (1971).

60. W. A. Brantley, B. Schwartz, V. G. Keramidas, G. W. Kammlott, and A. K. Sinha, *J. Electrochem. Soc.*, **122**, 434 (1975).
61. A. G. Milnes and D. L. Fencht, *Heterojunctions and Metal–Semiconductors Junctions*, Academic Press, New York (1972).
62. B. Schwartz, Ed., *Ohmic Contacts to Semiconductors*, Electrochemical Society, New York (1969).
63. H. C. Card, *IEEE Trans. Electron. Devices*, **ED23**, 538 (1976).

14

ION-IMPLANTED
METAL LAYERS*

S. M. Myers

Sandia Laboratories, Albuquerque, New Mexico

*This work was supported by the United States Energy Research and Development Administration, ERDA, under Contract E-(29-1)789.

14.1   INTRODUCTION

Ion implantation is emerging as a powerful tool for the study and modification of metals (1–3), with unique capabilities both for the nonthermal creation of tailored atomic mixtures and for the rapid introduction of lattice damage. Moreover, since depths up to ~ 1 μm are accessible, the technique is readily applicable to thin films. It can be used to modify existing discrete films, to produce new layers from the virgin bulk, and to conduct supportive experiments which yield pertinent metallurgical data. While implantation has not been applied extensively to metals, its considerable potential in the area has been demonstrated by a number of recent developments. These include the creation of new alloys (4–9), the modification of the corrosion and surface mechanical properties of existing materials (10, 11), and the extraction of previously unavailable data on diffusion and phase equilibrium (12–14). The purpose of this chapter is, then, to provide an introductory guide to ion-implanted metal layers: their production and characterization, their properties, and their uses.

An important feature of ion implantation is that it can produce an intimate atomic mixture of controlled composition, without many of the constraints of other alloying methods. This capability results from the injection of atoms directly into the host with kinetic energies far above the binding energy of the solid. As a consequence, the composition is not subject to thermodynamic limitations such as the solubility. Instead, implanted concentrations up to approximately the reciprocal of the sputtering yield, or typically 5 to 50 at. % (15), can be achieved. Furthermore, since the mixing can be accomplished at any temperature, it is possible to control the extent of thermal processes such as diffusion and precipitation. This versatility clearly is advantageous in the formation of metastable phases (4–9), and there are important benefits for the preparation of equilibrium alloys as well (13, 14). Thus, if the composition of an implanted layer is adjusted to correspond to some particular compound, then the formation of that compound requires minimal diffusion because the components are separated by only a few atomic jumps. This means that an equilibrium alloy can be formed at substantially lower temperatures than normal, and intermediate phases often can be bypassed.

The implantation process is accompanied by substantial lattice damage; for example, a dose of 10^{16} at. cm^{-2}, typical in metals work, might produce ~ 10 atomic displacements per host atom within the implanted region. This effect has important consequences for the final state of the resulting alloy, and under appropriate conditions it can be used to advantage. Thus, in a metal system where the amorphous state can persist, implantation provides a highly controlled way of creating this state which is independent of the properties of the melt (8, 9). In the more frequent situation where the alloy recovers rapidly to a crystalline array, transient phenomena can play a dominant role. One such effect is enhanced atomic diffusion resulting from the increased concentration of mobile point defects, which at lower temperatures can increase the migration rate by many orders of magnitude (16). This can be used to accelerate diffusion-limited processes which otherwise would require excessive times (14). Finally, it has been observed that metal atoms implanted into metals at low temperatures are often highly substitutional within the host lattice, and this has been attributed (7, 17) at least partially to the "thermal spike" associated with the ion damage cascade (18, 19). This property is very useful when the objective is to modify the composition of an ordered alloy.

The relatively new area of metals implantation is distinguished in several important respects from its extensively developed counterpart in semiconductors. One difference results simply from the wide divergence of dose levels which are of interest. Thus, semiconductors usually are implanted to introduce dopants to concentrations of <0.01 at. %, while significant modification of alloy properties typically requires several at. %. Consequently, effects related to precipitation and damage are generally more prominent in metals work. Semiconductors and metals also differ qualitatively in their recovery from a given level of damage; a semiconductor usually becomes amorphous at <1 displacement per atom (dpa), while metallic systems often remain crystalline to beyond 10^2 dpa. Because of these differences, inferences drawn for metals by analogy with semiconductors must be viewed with caution.

14.2 EXPERIMENTAL TECHNIQUES

14.2.1 Implantation Methods

The technology of ion accelerators (20) is now such that virtually any element can be implanted. Moreover, energies up to several MeV are readily available, and tens of MeV are achieved in a few facilities. The ions to be accelerated are created in a gaseous discharge from neutral vapor, which in turn is obtained either from a room-temperature gas, from heating of a solid or liquid, or from sputtering of a solid. Accelerating potentials above 0.5 MeV are produced by a Van de Graaff generator in most cases, while at lower energies it is usual to enploy an ac voltage multiplication system such as the Cockcroft-Walton supply. Beam currents generally are 1 to 1000 μA in the lower-energy regime and 0.1 to 100 μA when the Van der Graaff accelerator is used. In practice, the limit on beam current often is determined by the maximum permissible power dissipation within the target. Lateral uniformity of the implantation is achieved by sweeping of the ion beam and/or translation of the specimen; consequently, the implanted area is restricted only by available time and the dimensions of the vacuum chamber. In a typical implantation experiment, a 10 μA beam of 0.1 MeV ions might be employed to produce an average concentration of 1 at. % within the first 0.1 μm of a 1 cm^2 sample. Then, if the atomic density were 10^{23} at. cm^{-3}, the implantation would require 160 s, and 1 W of power would be dissipated within the target.

Ions implanted into a solid at constant energy will come to rest over a range of depths due to the random nature of the energy loss events. The resulting depth distribution is in general an approximately Gaussian function, centered about the average projected range R_p along the incident beam direction and characterized by the rms deviation ΔR_p (21–23). The deviation is smaller than, but comparable with, the range in most cases. There also is a depth-dependent lateral spread in the distribution (24, 25), but this usually is unimportant because it is orders of magnitude less than the dimensions of the implanted surface. The parameters R_p and ΔR_p have been calculated theoretically for a large number of ions, targets, and energies (26–28), and the values generally are correct to within 20%. This information, interpolated where necessary, aids greatly in the planning of experiments. The depth-dependent atomic displacement rate $P(x)$ also is accessible to theoretical calculation, through the energy into atomic processes per unit depth $E_A(x)$ along an individual ion track (27–29). The latter quantity has been calculated for a number of implantation conditions, and it can be used to estimate $P(x)$ via the modified Kinchin-Pease relation (30, 31):

$$P(x) = 0.8 \frac{F \cdot E_A(x)}{2E_d} \tag{1}$$

where F is the ion flux (the number of ions incident per unit area and per unit time) and E_d is the threshold energy for atomic displacement.

The factors involved in producing a desired concentration-versus-depth profile $C(x)$ can be illustrated by considering an idealized but specific example, namely, the implantation of an Al target with 0.1 MeV Al. In this case, $R_p = 0.1255$ μm and $\Delta R_p = 0.0445$ μm (32). Utilizing the Gaussian approximation

$$C(x) = C(R_p) \exp\left[(x - R_p)^2 / 2(\Delta R_p)^2 \right] \tag{2}$$

then gives the low-fluence profile shown in Figure 14.1. Also shown is the aforementioned energy deposition into atomic processes $E_A(x)$ (32). The depth distribution obtained by implantation at a single energy, such as the one given in Figure 14.1, is adequate in many instances. Where greater uniformity is desired a sequence of different energies can be used, so that several such peaks combine to give a flat profile. The distribution can be

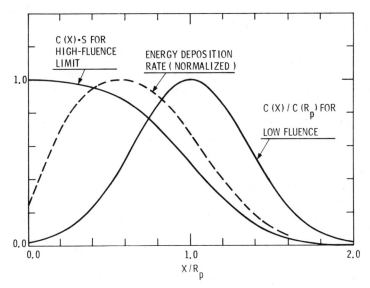

Figure 14.1. Concentration profile $C(x)$ and energy deposition rate into atomic processes for implantation of Al with 0.1 MeV Al. The depth x for the sputtering limit profile is measured from the existing surface.

adjusted further by the subsequent use of diffusion, either thermal or radiation enhanced, if other considerations do not preclude its use. Thermal diffusion generally produces a more extended, gently sloping profile, whereas radiation-enhanced diffusion tends to give an abrupt drop-off near the maximum range of the irradiating ion (32, 33).

During ion implantation the target surface recedes continuously due to sputtering. The effect on the concentration profile is negligible until the peak atomic fraction of implanted atoms approaches the reciprocal of the sputtering yield, defined as the number of atoms removed per incident ion, whereupon there are qualitative changes in shape. Thus, if $g(x)$ is the implanted profile without sputtering and normalized to unit area and the sputtering yield S and the range R_p are assumed to be independent of composition, then it is easily shown that

$$C(x, t) = \frac{1}{S} \int_{x}^{x + FSt/n_0} g(z)\, dz \qquad (3)$$

where C is in units of atomic fraction, F is the ion flux, n_0 is the atomic density, and t is the accumulated implantation time. The distance by which the surface has receded is FSt/n_0. For $C(x = R_p,\ t) \ll 1/S$, which implies $FSt/n_0 \ll R_p$, one obtains $C(x, t) = g(x)Ft/n_0$, and sputtering is not important. When $C(x = R_p, t)$ approaches $1/S$, $FSt/n_0 \sim R_p$, and the profile is skewed and shifted toward the surface; and as $t \to \infty$, the profile reaches a steady state in which $C(x, t)$ equals $1/S$ at the surface and decreases monotonically with increasing depth. This limiting high-fluence case is shown in Figure 14.1 for the conditions of that example. It should be noted, however, that the effects of sputtering may be more complicated in practice. Thus, S may vary with the continuously changing composition, invalidating one of the above assumptions. In addition, if there is diffusion, it can produce at least two effects: first, the overall depth profile is changed (34); and second, agglomeration of solute atoms can lead to nonuniform cone-like structures (35). Further discussion of sputtering can be found in Chapter 6.

An important consequence of the above considerations is that the sputtering yield determines the upper bound on the implanted concentration. This yield varies greatly with the Z of the incident ion, with its energy, and with the target composition; however, most values encountered during implantation are in the range of 2 to 20 (15), corresponding to maximum concentrations of 5 to 50 at. %. Since such concentrations are often of interest in metals work, this constitutes an important limitation. It should be noted, however, that the sputtering limit can be exceeded appreciably by suitable variation of the ion energy. Thus, the energy is reduced as a function of time so that, as the sputtered surface approaches the depths implanted earlier, the ions continue to stop in the same region of the host.

Ions incident onto a single crystal within $\sim 1°$ of a high-symmetry axis may be steered down an open channel in the lattice by the repulsive potential of the atomic cores (36). This effect, called channeling, reduces the probability of nuclear scattering by a factor of up to $\sim 10^2$. It also occurs when the beam lies within a high-symmetry plane, but the reduction in nuclear scattering is then much less. Channeling has two important consequences for ion implantation. First, it effectively removes nuclear stopping of the incident ion, leaving electronic excitation as the principal source of energy loss. This increases the ion range at a given energy by as much as an order of magnitude for mid- or high-Z ions (37). The second consequence of the channeling effect is that it substantially reduces lattice damage, which occurs only as the result of a nuclear collision in metals. Both these properties can be advantageous in certain applications.

14.2.2 Analysis

Among the most basic and pertinent data about an ion-implanted metal layer are (i) the composition-versus-depth profile, (ii) the lattice location of implanted atoms within the host matrix, (iii) the structure, composition, and morphology of second-phase precipitates, and (iv) the amount and types of lattice damage, including the degree of disorder. These factors determine the macroscopic properties of the layer such as corrosion resistance and superconducting behavior. However, the range of experimental probes which can be used is severely limited by the small amount of implanted material, typically 10^{-10} to 10^{-6} mole per cm^2 of implanted surface. Consequently, one usually is limited to measurements based on the interactions of energetic ions or electrons with the solid. Some of the more commonly used methods of characterization are discussed here. They include ion backscattering, channeling, transmission electron microscopy, and sputter profiling.

14.2.2.1 Ion Backscattering. Ion backscattering analysis, which is discussed in detail in Chapter 6, is used frequently to determine the composition-versus-depth profile of an ion-implanted layer. This is illustrated by the backscattering spectrum of Figure 14.2, which was obtained from a Be

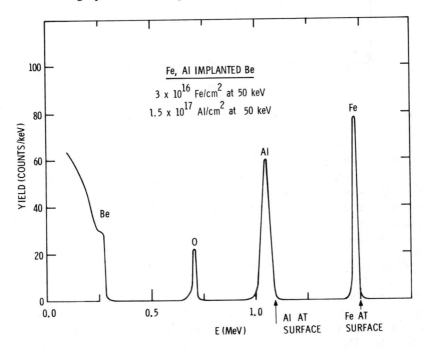

Figure 14.2.   Energy spectrum of 2 MeV He backscattered from Be which had been implanted with Al and Fe. From reference 13.

crystal implanted with both Al and Fe (13). The derived concentration profiles for the implanted species are shown in Figure 14.3. This method has a depth resolution of ~ 0.01 μm and a maximum penetration of ~ 10 μm, both of which are suited to most implantation work. Additional advantages include the essentially nondestructive nature of the measurement and the determination of absolute concentrations and depths without elaborate calibration

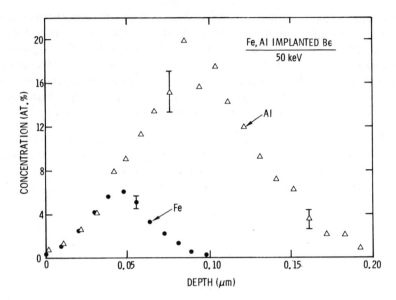

Figure 14.3. Concentration profiles of Al and Fe implanted into Be, derived from the backscattering spectrum of Figure 14.2.

procedures. Difficulties may arise, however, in detecting small concentrations of light elements in heavy substrates or in distinguishing elements of similar mass. Both these problems can be circumvented in a number of cases by the use of inelastic nuclear reactions, which are specific to one isotope, and often produce particles of energy above the yield from the host, making their detection easier.

14.2.2.2 Ion Channeling — Lattice Location. When the objective of ion backscattering is to determine composition versus depth, the channeling effect introduced at the end of Section 14.2.1 is explicitly avoided. In particular, single crystals are oriented so that neither the incident beam nor the detected portion of the backscattered flux is close to an axis or plane of high symmetry. This procedure preserves the randomness of the stopping and backscattering events which is necessary for the analysis to be quantitative. However, in determining the lattice site occupied by the implanted atom, the channeling effect can be used to advantage.

The basis for the lattice location measurement (38) is that, when the analysis beam is steered down an open channel by the repulsive cores of the lattice atoms, the probability of backscattering from a particular nucleus is sensitive to the position of that nucleus relative to the channel. Thus, for a high-symmetry axis of a single crystal, the backscattering yield from the host is generally reduced by a factor of 10 to 100. The position of an implanted element is then deduced by comparing the yield from that element with the host yield as the crystal is rotated through a channeling direction. Analysis is facilitated by the relatively weak orientational dependence of the stopping rate for the analysis beam, which determines the correspondence between backscattered energy and depth: there is typically a decrease of a factor of $\lesssim 2$ for axial channeling. This favorable circumstance results from the dominance of stopping by electronic excitation for the light ions H and He at MeV energies which are used in such measurements.

The simplest case of channeling occurs when the implanted atoms all occupy substitutional sites within the host lattice; the backscattering yields from host and implanted species then vary in direct proportion to one another as a function of orientation. This situation is illustrated in Figure 14.4, where

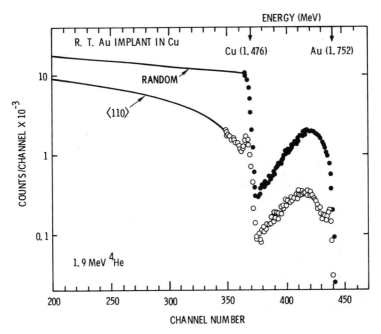

Figure 14.4. Energy spectrum of 1.9 MeV He backscattered from single-crystal Cu implanted with Au, for the incident beam along the [110] and random directions, respectively. From reference 7.

a "random," or nonchanneled, backscattering spectrum is compared to a [110] profile for single-crystal Cu implanted with Au (7). The proportional variation of the Cu and Au yields is seen more clearly in Figure 14.5, where these quantities are normalized and plotted versus the angle between the analysis beam and the [110] direction. Also shown in this figure is a similar

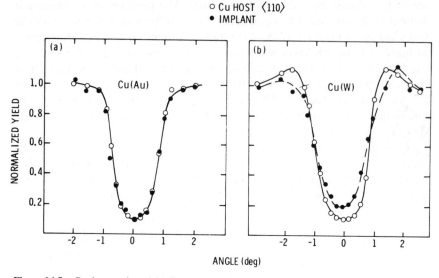

Figure 14.5. Backscattering yields for 1.9 MeV He on Cu which had been implanted with Au or W, as a function of the angle between the [110] axis and the incident beam. From reference 7.

plot for W implanted into Cu, for which the substitutional fraction is only 90% (7). When the implanted species is interstitial, the channeling behavior is more complicated but equally straightforward; the essential feature is that interstitial atoms occupy certain open channels, and consequently the backscattering yield from such atoms peaks sharply in these directions. Furthermore, the phenomenon is sufficiently well understood for different interstitial sites to be unambiguously distinguished. Finally, a nonvarying yield for the implanted species in the presence of a decrease in the host yield usually means that the implanted atoms do not occupy a well-defined site within the host lattice. In this case they generally are assumed to exist in some more complex damaged configuration.

14.2.2.3 Ion Channeling—Disorder. Since channeling requires an ordered lattice, the effect also can be used to obtain approximate information about the amount and character of implantation damage in single crystals (39–41). In general, the backscattering yield from the host crystal in a channeling direction goes up as the level of damage increases. Furthermore, the shape of the channeled backscattering spectrum depends on the type of defect. Thus, if most of the displaced atoms are well removed from lattice sites, the analysis beam tends either to be directly backscattered by these atoms or to remain within the channel. This results in a peak in the backscattering spectrum at energies corresponding to the implanted layer (40). If, on the other hand, most of the atoms are only slightly displaced, the dominant effect is to deflect the incident beam from the channel without actually backscattering it. In this case there is no distinct peak in the spectrum, but rather a monotonic increase in yield with decreasing energy. The presence of damage is then manifested by an increased slope for energies corresponding to the implanted layer. For metals, this is the more frequently encountered of the two cases, because the point defects tend to agglomerate into extended defects such as dislocations which have extensive strain fields (41). In both regimes theoretical models have been developed which allow the density of damage to be calculated from the channeling data (40, 41). However, several qualifications must be stressed. First, relatively high damage levels are necessary for detection: typical minimums are several at. % of fully displaced atoms, or $\sim 10^9$ cm of dislocation length per cm^3. It should be noted, however, that such levels are common in implanted metals. A second point is that the methods of data analysis are still the subject of some controversy, and the interpretation is semiquantitative at best. Finally, channeling measurements do not by themselves fully determine the specific defects that are present, and therefore it is preferable to combine such measurements with electron microscopy.

14.2.2.4 Transmission Electron Microscopy. Transmission electron microscopy (TEM) is a powerful and versatile tool (42) which is frequently used to characterize implanted metals (43) (see Chapter 5 for the application of this probe to deposited films). The lateral resolution is generally better than 0.002 μm, complementing the high depth resolution but poor lateral resolution of ion beam analysis. Measurements are made on samples thinned to ~ 0.1 μm, which is readily compatible with implanted layers. Ways in which transmission microscopy gives information include the following: (i) the bright-field mode provides a high-resolution image of the microstructure and the damage in particular; (ii) selected area diffraction can be used to identify the crystal structures of the phases that are present; (iii) the dark-field mode selectively images regions of the sample that are diffracting at a particular Bragg angle and is especially useful in determining the distribution and morphology of precipitates; and (iv) detection of electron-excited characteristic x-rays gives information about elemental composition in regions as small as 0.05 μm in diameter.

14.2.2.5 Sputter Profiling. The composition-versus-depth profile of an implanted metal also can be determined by monitoring surface composition in conjunction with sputtering (44). In this approach it is usual to employ Auger electron spectroscopy, secondary ion mass spectrometry, or one of the other surface probes, and to erode the sample with a beam of ions in the energy range of 0.5 to 20 keV. This method can give a depth resolution approaching a few monolayers, which is substantially better than the ~ 0.01

μm obtained with ion backscattering analysis. In addition, most elements can be detected in most metallic hosts, independent of the wider separation of atomic numbers required with backscattering. Nevertheless, the sputtering methods have significant limitations (see Chapter 6). Most importantly, both the sputtering rate and the ratio of signal strength to concentration can be sensitive to composition, leading to uncertainties in the absolute magnitudes of depth and concentration. Another disadvantage is that, unlike the ion backscattering measurement, the sputtering techniques destroy the sample, making it more difficult to follow time-varying phenomena. For these reasons, it is usually preferable to use ion backscattering analysis for profiling unless prevented by its limitations.

14.3   GENERAL BEHAVIOR OF IMPLANTED METALS

The evolution of an ion-implanted metal is logically considered in two successive stages: first, the time interval of $\sim 10^{-12}$ s during which an incident ion and its associated damage cascade reach thermal equilibrium with the surrounding lattice; and second, all modifications occurring thereafter. The first stage involves kinetic energies large compared to the binding energy of the solid, and it is controlled primarily by atomic collision kinetics. The behavior during this period is therefore nearly independent of the detailed properties of the solid, such as structure, bonding, and temperature; it makes relatively little difference whether the target is amorphous or crystalline or whether it is an insulator, a semiconductor, or a metal. (An exception to this is the importance of structure when channeling effects are present.) In contrast, processes occurring in the second stage involve energies of $\leqslant 1$ eV, and most depend at least partially on thermal activation. It is these effects which are sensitive to the detailed properties of the solid and which are most important in differentiating one implanted alloy from another. Consequently, in this section the first stage is discussed only in general terms, while the subsequent evolution is covered in more detail.

14.3.1   Atomic Displacements

The number of atomic displacements produced during ion implantation of metals is generally high. In a representative example, 0.1 MeV Al incident onto Al, an individual ion would displace on the order of one atom per lattice spacing as it traverses the lattice; expressed differently, most atoms within the implanted layer would have been displaced after a fluence of $\sim 10^{15}$ at. cm^{-2}, which is relatively low in metals work. The production rate of vacancy–interstitial pairs for a specific case can be estimated via the calculational procedure discussed in Section 14.2.1 and based on the modified Kinchin-Pease relation, Eq. 1. This approach is considered accurate to within about a factor of 2 for the majority of cases. The greatest departures have been reported in semiconductors for very massive incident species, such as the Bi_2^+ molecular ion, where the predicted damage rate can be exceeded by an order of magnitude (45). Such discrepancies have been attributed to the "thermal spike" effect (18, 19) which occurs when the individual damage cascades are sufficiently dense and long-lived for the majority of atoms within to be simultaneously in motion. In this situation the energy required per atomic displacement is reduced. In the extreme, one could envision the threshold energy being decreased from its isolated atom value of ~ 10 eV down to the melting energy, typically 0.1 eV. The principal evidence for thermal spike effects has come from experiments in which increased sputtering (46, 47) and damage rates (45) resulted from implanting molecular ions instead of atomic ions of the same elements.

14.3.2 Lattice Location

The factors controlling the lattice configuration of an implanted atom immediately after its kinetic energy has been thermalized are not well understood. This problem has been discussed most extensively in connection with lattice location measurements at temperatures where equilibrium atomic mobilities are negligible (6, 7, 17, 48, 49). The results of such experiments have varied with the implanted species, the host metal, the implanted fluence, the temperature, and other parameters. In a number of instances almost 100% of the implanted atoms are substitutional within the host lattice; in other cases a large fraction occupy one of the interstitial sites; and finally, the implanted atoms can be found in a more complex configuration, which may or may not be identified by the particular measurement technique. A notable example in the last category is the (100) dumbbell formed by an impurity and a self-insterstitial in the fcc structure (49). Of considerable significance, both theoretical and practical, in these experiments is the great frequency with which high substitutionality is obtained.

Considerations which have been invoked in the literature to interpret the lattice location data include the following:

1. The statistical probability that the implanted atom comes to rest by a replacement collision. Theoretical calculations of this probability (50) for various implanted elements in Cu do not correlate well with experimentally observed substitutional fractions (7), so that this is apparently not the dominant factor.
2. Thermal spike effects, which might lead to a transient increase in atomic mobility in the immediate vicinity of the implanted atom, permitting it to move a few atomic jumps to a more stable configuration. This quite reasonable hypothesis has not yet been confirmed directly.
3. Interaction with point defects produced by the implantation, to either

stabilize or destabilize a particular configuration. Such an effect has been shown to be important in the formation of the (100) dumbbell mentioned above (49).

4. The ease of accommodation within the host lattice, as measured by parameters such as the difference in atomic volumes between host and implanted element, the electronegativity difference, and the equilibrium solid solubility. There is indeed a qualitative correlation between the substitutional fraction and the atomic volume difference, as seen for the host Cu (6, 7) in Figure 14.6. The correlation with electronegativity is

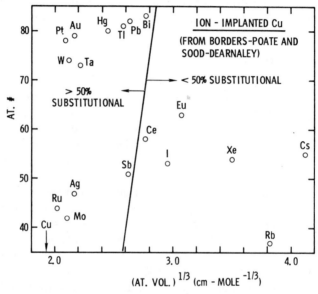

Figure 14.6. Lattice location of elements implanted into Cu at room temperature. From references 6 and 7.

considerably weaker. The equilibrium solubility is notable primarily because it is frequently exceeded by several orders of magnitude in highly substitutional implants.

Finally, caution should be exercised in equating the state observed in lattice location experiments with the condition existing immediately after thermalization of the ion cascade. For example most of the lattice location data are for implanted doses above 10^{14} at. cm^{-2}, so that there is considerable overlap among successive cascades which could change the lattice location. Modification also could result from interaction of the implanted atom with point defects, which always are present and mobile during irradiation above Stage I in the isochronal annealing profile. In summary, then, significant progress has been made in interpreting the lattice location of implanted atoms, but important questions remain.

14.3.3 Thermal Effects

As indicated above, the final states of metal-implanted metals are differentiated principally by events occurring after the $\sim 10^{-12}$ s required for thermalization of the ion cascade. Indeed, the variations in these processes are so great and depend on so many factors that only limited generalizations are possible. In the majority of cases, however, the qualitative behavior can be divided into two categories distinguished by the magnitude of the atomic mobility. Thus, a low- and a high-temperature regime are defined approximately by

$$\sqrt{D_A(T) \cdot t} < a \tag{4}$$

and

$$\sqrt{D_A(T) \cdot t} > a \tag{5}$$

respectively, where T is the temperature, t is the duration of the experiment, a is the lattice spacing, and D_A is the diffusion coefficient for the least mobile atomic species after the dissipation of the thermal spike but including possible radiation-enhanced diffusion. In the low-temperature regime, the absence of diffusion implied by Eq. 4 inhibits the evolution of the system toward thermodynamic equilibrium. As a result, metastable states produced during the implantation process can persist. Thermodynamic constraints such as solid solubilities and the Gibbs phase rule are not applicable, and precipitation of second equilibrium phases usually does not occur.

In the high-temperature regime, where there is appreciable atomic diffusion over experimental times, most metal-implanted metals move from the implanted condition toward thermodynamic equilibrium. Equilibrium phases expected from the phase diagram precipitate out, and the laws of thermodynamics are applicable. Furthermore, the approach to equilibrium tends to be quite rapid, since implantation produces an atomic mixture in which the components are separated by only a few atomic spacings. Thus, a relatively simple picture emerges and one which is consistent with most data on metal-implanted metals. This contrasts, for example, with the observed behavior of He in implanted metals (51) and various impurities in implanted semiconductors (52). There, the binding of the implanted species to lattice defects is often strong enough to retard the evolution toward equilibrium, resulting in a more complex, nonequilibrium state even when the diffusivities are high. In intermetallic systems such effects are less prominent because the impurity-defect binding energies usually are much smaller.

It should be stressed, however, that this situation is not necessarily universal, and exceptions probably exist.

Most intermetallic systems exhibit a high degree of recovery from implantation damage in both temperature regimes. Crystalline structure tends to be retained even after more than 10^2 displacements per atom (dpa). Below Stage I in the isochronal annealing profile, where all isolated point defects are immobile, the vacancy and interstitial concentrations are limited by spontaneous recombination (53). Thus, if a vacancy lies within a certain volume of spontaneous recombination about an interstitial, then the two can undergo mutual annihilation without the necessity of thermal activation. This critical volume contains $\sim 10^2$ atoms for many pure metals. At higher temperatures where the isolated interstitials (Stage I) and vacancies (Stage III) become mobile, there is additional recombination and also agglomeration into extended defects such as dislocations and, less frequently, voids. Greater detail can be found in the recent literature (54, 55). The recovery behavior of metals contrasts with that of semiconductors, which generally become amorphous at damage levels below 1 dpa (56). Factors contributing to the difference probably include the less directional bond and the absence of charge-state effects in metals.

While most alloys remain crystalline during ion implantation, recent experimental results in the low-temperature regime suggest that this is not universally true. Apparently amorphous layers were produced by implanting W into Cu (8) and Dy into Ni (9, 57) at room temperature. The characterization in both cases was based on damage peaks in the channeled backscattering spectra and diffuse bands in the transmission electron diffraction patterns. In the Dy–Ni experiment it was concluded specifically that ordered regions, if present, were smaller than 0.001 μm. Annealing at 773°K restored crystallinity in each case. The peak concentration of W in Cu was ~ 10 at. % for the disordered condition, whereas a Cu crystal implanted to only ~ 1 at. % remained crystalline. These compositions were compared with an approximate theoretical expression for the composition at which a binary alloy is most likely to become amorphous (58):

$$1.7 \approx C_1 \cdot z_1 + C_2 \cdot z_2 \tag{6}$$

where C is concentration in units of atomic fraction and z is valence. This formula gives 14 at. % for W in Cu and 35 at. % for Dy in Ni, both comparable with the respective experimental concentrations. It should be noted that amorphous metallic alloys have been extensively produced and studied using rapid cooling techniques and deposition onto low-temperature substrates (59), so that similar results for implanted metals in the low-temperature regime defined by Eq. 4 are not unreasonable. However, further

work is desirable to establish the precise nature of these implante
to determine the differences, if any, from rapidly cooled systems.

14.3.4 Enhanced Diffusion

Atomic diffusion rates can be enhanced greatly during ion implantatioι
and from the preceding discussion this has important consequences for ι
the final state of the system and the rate at which it is approached. Τ.
enhancement effect results from the concentrations of vacancies and inter
stitials exceeding their respective equilibrium values, so that the diffusion
rates by vacancy and interstitial mechanisms are proportionally greater.
The complete description of such a system requires the numerical solution
of three coupled diffusion equations for vacancies, interstitials, and atoms
(32). The atomic diffusion coefficient D_A in general will depend on both time
and position. Furthermore, if the diffusing species is other than a host atom,
additional terms in the equations can arise from impurity-defect coupling
(16, 60). However, simple, order-of-magnitude expressions for D_A will serve
the purposes of the present discussion, and the reader is referred to the
literature for greater detail (16, 32, 60).

The implanted layer is assumed here to be uniform and to have dimensions
large compared to the defect path length to annihilation. The temperature is
high enough for both vacancies and interstitials to be mobile, and the concentrations of these defects have reached the steady state. Then, if the un-
enhanced diffusion rate is negligible, it is readily shown that (61)

$$D_A \simeq \frac{\gamma P \Omega}{2\pi r_s N_s} \tag{7}$$

when the vacancies and interstitials annihilate predominantly at fixed sinks,
and

$$D_A \simeq \sqrt{\frac{\gamma P \Omega^2 D_v}{\pi r_{iv}}} \tag{8}$$

when most of them recombine. Here, P is the production rate of vacancy–
interstitial pairs as given by Eq. 1, γ is the fraction of these defects which
escape the parent cascade to migrate through the lattice, Ω is the atomic
volume, r_s is the capture radius of a fixed sink, N_s is the volume density of
these sinks, r_{iv} is the separation below which a vacancy and interstitial recombine, and D_v is the vacancy diffusion coefficient. For the fixed sink case,
the diffusion rate is proportional to the implantation flux via the defect
production rate P, but it is independent of temperature. For recombination,
D_A goes as the square root of flux and increases with temperature through
the factor $\sqrt{D_V}$.

It is instructive to consider enhanced diffusion in the example of Figure

14.1, which involved implantation of Al with 0.1 MeV Al. Here, the defect annihilation is assumed to occur primarily at fixed sinks, of which the surface is one, and the average diffusion distance to annihilation $[4\pi r_s N_s]^{-1/2}$ is set equal somewhat arbitrarily to the projected range $R_p = 0.1255$ μm. The ion flux is 10^{14} at. cm^{-2} s^{-1}, a typical value. A first approximation to D_A is then obtained using Eq. 7. The production rate P is calculated from Eq. 1, with an effective E_A equal to the total energy into atomic processes, 44 keV, divided by R_p. Choosing the threshold energy E_d to be 32 eV (62) then gives $P = 4.4 \times 10^{21}$ cm^{-3} s^{-1}. With $\gamma = \frac{1}{40}$ (32), Eq. 7 finally yields $D_A = 5.7 \times 10^{-13}$ cm^2 s^{-1}. This temperature-independent enhanced diffusion coefficient equals the thermal equilibrium value at about 560° K, and therefore the enhancement effect will be significant at temperatures below this. Some results from a full numerical calculation (32), taking account of the variations in D_A with depth and time, are shown in Figure 14.7. The steady state

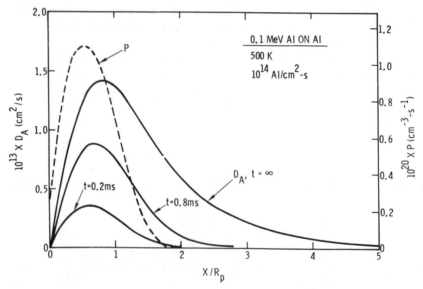

Figure 14.7. Enhanced diffusion coefficient D_A and production rate P of vacancies and interstitials for Al bombardment of Al at 0.1 MeV. From reference 32.

at $t = \infty$ is independent of temperature, whereas the profiles at shorter times are appropriate only to 500° K.

14.3.5 Precipitation

Although the precipitated phases formed in an implanted layer in the high-temperature regime defined by Eq. 5 usually are those expected from the

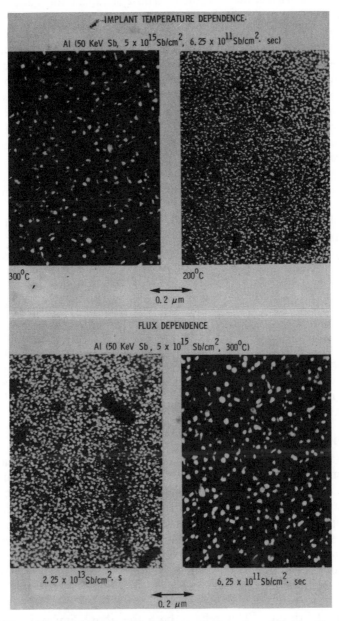

Figure 14.8. AlSb precipitation resulting from the implantation of Sb into Al, at different fluxes and temperatures. From reference 63.

phase diagram, their size and shape generally are sensitive to the experimental parameters, such as the flux, accumulated dose of implantation, and temperature. An example of this is given in Figure 14.8 which shows transmission electron dark-field micrographs of Al implanted with Sb (63). For a constant number of implanted Sb atoms, the size distribution of AlSb precipitates is seen to depend strongly on both flux and temperature. More generally, observed precipitate sizes range from the threshold of detectability by transmission electron microscopy, ~ 0.001 μm, to about the thickness of the implanted layer. The microstructure in such a system is determined in part by processes which occur in any supersaturated solid solution (64) and, in addition, by certain effects unique to the implantation environment. The latter are believed to include (63, 65, 66) the following

1. Radiation-enhanced diffusion, which was discussed above. This increases the rate of diffusion-limited precipitation.
2. Recoil dissolution, in which precipitated atoms are ejected into the host matrix by the ion displacement cascade.
3. Disordering dissolution. Thus, even though the precipitate lattice recovers from the irradiation damage, atoms near the outer surface of that precipitate may go into solution as a result of the transient disorder.
4. Precipitate fragmentation due to cutting by dislocations, which are formed by the agglomeration of the implantation damage.

These mechanisms have provided a basis for qualitative or semiquantitative interpretations of observed microstructures. However, a general theory capable of reliable prediction is not now available, so that the precipitate size and morphology always should be determined experimentally when this information is important.

14.4   APPLICATIONS

Ion implantation is used in metals for the controlled, nonthermal creation of atomic mixtures and for the efficient and controlled introduction of lattice damage, and in these respects the technique is unique. However, because of the practical limitations on surface area and depth, the relatively expensive and complicated apparatus, and the incomplete understanding of the associated physical processes, other methods generally are used unless there is a definite need for the above properties. Nevertheless, all three of these constraints are decreasing as work in the field continues. It should also be pointed out that, to date, implantation has been used in the metals area almost exclusively as a research tool, in contrast to its extensive application to semiconductor fabrication. This reflects both the above limitations and the undeveloped state of the field, and it is unlikely to continue. Indeed, as will be discussed, recent developments show promise for the fabrication of small but critical components.

In this section several current applications of implantation to metals are reviewed, ranging from the relatively well developed to the very new. An effort is made in each case to focus on the key aspects that bear on the broader applicability of the technique. The divisions among topics are somewhat arbitrary and are based largely on the work of the most active groups. Two important areas which fall outside the scope of this book will be omitted: gases in metals, notably He, and radiation damage *per se*.

14.4.1 Superconductivity

Research in superconductivity has focused increasingly on metastable alloys, especially in the search for high transition temperatures (67). A central aspect of such work is determining the properties as a function of composition, and for this purpose ion implantation would seem an ideal tool. The concentrations within a single sample could be varied at will, independent of thermodynamic constraints such as solid solubilities, and the superconducting properties could be measured repeatedly. In fact, such experiments have been performed for a number of alloys, including Nb_3Sn, the Pd–H–noble metal system, NbC, VC, and a wide array of binary alloys based on Al, Mn, Mo, and Re (4, 5, 68–72). A representative example is given in Figure 14.9,

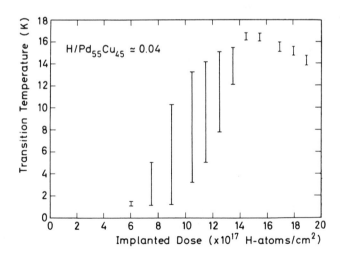

Figure 14.9. Superconducting transition temperature of a Pd–Cu alloy as a function of the fluence of H implanted at 130 keV. Vertical bars indicate the temperature interval of the transition. Prior to implantation the alloy was charged with H to a concentration of 0.04 H atom per metal atom. From reference 68.

where the transition temperature of $Pd_{0.55}Cu_{0.45}$ is plotted versus the accumulated dose of H implanted at liquid He temperature (68). Before implantation, the foil was charged with H at room temperature to about 0.04 H per metal atom, but at that concentration no superconductivity was observed. However, subsequent implantation at low temperatures where the H is immobile increased the concentration to far above its solubility and thereby produced a dramatic increase in the transition temperature. The maximum value of $17.1°K$ occurs at a fluence corresponding to a local concentration of about 0.7 H per metal atom.

Ion implantation also is very useful for studying defect-related phenomena in superconductors (70, 73–75). For example, in contrast to defected films formed by deposition onto cold substrates, the damage level in implanted layers can be varied continuously. Effects due to composition changes are avoided through self-ion bombardment or by selecting an energy such that the ions pass completely through the region of interest. Experiments of this kind have produced large changes in superconducting properties. For instance, in the A-15 system Nb_3Sn, Ar implantation at room temperature reduced the transition temperature T_c from 17.8 to $2°K$ (73); and in another A-15 compound, Nb_3Ge, He bombardment at room temperature lowered T_c from 22 to $3.5°K$ (74). Furthermore, while T_c decreases in the majority of cases, irradiation also can produce an increase. Thus, implantation of In into In at $2°K$ changed T_c from 3.7 to $4.4°K$ (75).

Aside from the mechanistic information gained about superconductivity, which is not germane here, experiments such as the above serve to illustrate both the power and the limitations of ion implantation for work with metastable alloys. Thus, when the implantation is carried out in the low-temperature regime defined by Eq. 4, the atoms remain in solution without regard to the equilibrium phase diagram. This gives unparalleled control over composition. Furthermore, recovery from the associated lattice damage usually is sufficient for retention of the superconductivity. Similarly, in studying damage effects, the high atomic displacement rate and its continuous control are ideal. However, an important difficulty arises in determining composition effects for a defect-free lattice, since in the low-temperature regime the recovery of the lattice is rarely complete. This is particularly important for superconductivity because of its relation to long-range order. The best that can be done is to compare the implantation of interest with a self-ion bombardment which produces about the same number of atomic displacements and from this to attempt a separation of composition and damage effects.

Finally, recent experimental work indicates that implantation should be applicable to fabrication of superconducting weak links in integrated circuits (76). The transition temperature of Mo is increased by implantation of N or S (5), and this effect was used to vary T_c as a function of position on sputtered

Mo films 0.08 μm thick. Thus, by using photoresist masking, regions with $T_c \simeq 4^\circ K$ were bridged by $\simeq 1$ μm sections having $T_c \simeq 2^\circ K$. These structures exhibited the ac and dc Josephson effects and were found to be stable against thermal cycling and storage at room temperature.

14.4.2 Corrosion

Ion implantation has been used widely in the study of corrosion, as is seen from the recent literature (77–80). Among the factors contributing to the usefulness of the technique in this area are:

1. The absence of thermodynamic restrictions.
2. Control over the depths at which the implanted atoms come to rest. Thus, the oxide layer and/or the underlying metal can be doped independently.
3. The tendency for implantation effects to extend much deeper than the ion range. This might be due, for example, to altered ionic mobilities in the oxide which affect its growth rate, to sweeping of the implanted element before the advancing oxide front, or to defect-assisted migration of the implanted species.
4. Modification of the near-surface corrosion characteristics without changing the bulk properties, and also without using large quantities of scarce alloying elements. These considerations are significant primarily for potential applications to fabrication. Work in the field is exemplified here by two examples.

Atomic migration during the anodic oxidation of Al was investigated using ion implantation in conjunction with ion backscattering analysis (81, 82). Initially, the mobilities of Al and O in the oxide were measured by means of an inert gas marker. Xenon was implanted into an existing Al_2O_3 layer $\simeq 0.03\,\mu$m thick, and then the oxide was grown anodically to $\simeq 0.3$ μm. At this point ion backscattering was used to determine the depth of the Xe, thereby giving its motion relative to the surface during the second oxidation. The inert gas is believed to be immobile in the Al_2O_3 matrix. Consequently, if it had remained near the surface, only O would be mobile; and conversely, if it had followed the $Al–Al_2O_3$ interface, only Al would be mobile. In fact, an intermediate result was obtained, indicating that Al and O migrate at comparable rates during anodic oxidation. Subsequently, similar methods were used to determine the direction and rate of migration for a wide range of impurities during the Al anodic oxidation.

Alloy additions of yttrium to stainless steel are known to improve its resistance to oxidation, but there is an associated degradation of mechanical properties. Consequently, implantation of Y into the near-surface region was investigated as a possible alternative to bulk alloying (83). Samples of 20 wt. % Cr–25 wt. % Ni–Nb steel were implanted with Y at a series of energies, producing an approximately uniform concentration of $\simeq 0.2$ wt. % within the first 0.2 μm. The specimens were then annealed under 1 atmosphere of CO_2 in the temperature range of 973 to 1123° K, with weight gain being measured periodically. The results were compared with measurements on unimplanted samples of the same material as well as with data for bulk alloys containing 0.13 and 0.41 wt. % Y, respectively. A representative plot of weight gain versus anneal time is shown in Figure 14.10, where the various

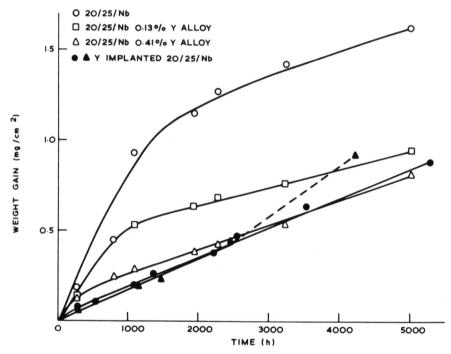

Figure 14.10.   Effect of yttrium on the oxidation of 20/25/Nb stainless steel in CO_2 at 1073°K. From reference 83.

alloys are compared at 1073° K. The addition of Y is seen to reduce consistently the amount of oxidation for anneal times up to 5000 h, with the effects of implantation and bulk alloying being similar. The latter point is particularly significant since complete oxidation of the implanted layer corresponds to only $\simeq 0.09$ mg cm^{-2} in Figure 14.10. Thus, the effect of the implanted Y extends much deeper than the implantation range. By way of explanation, it was speculated that a thin layer of the compound $YCrO_3$ forms at the metal–oxide interface and that this inhibits the ionic transport necessary for oxide growth. In any case, one may conclude that in this alloy system implantation offers considerable promise for corrosion inhibition.

14.4.3 Mechanical Properties

The near-surface mechanical properties of a metal can be modified greatly by ion implantation. Friction, wear, and hardness may change by more than an order of magnitude, as discussed in a recent review of the field (11). An example of this is given in Figure 14.11, where the reduction factor in wear rate of an oil-lubricated steel is plotted versus the fluence of implanted 30 keV N (84). The mechanisms of such mechanical effects are complicated and at

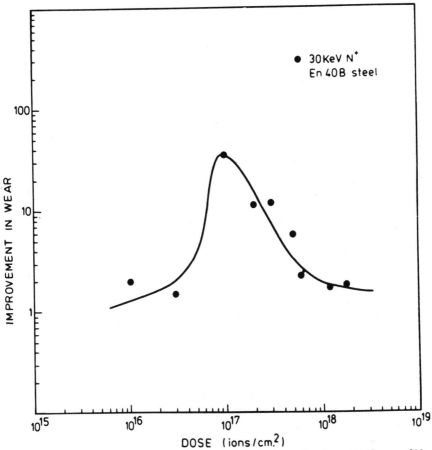

Figure 14.11.   Relative improvement in wear of En 40B steel as a function of the fluence of N implanted at 30 keV. From reference 84.

present not well understood. However, in addition to essentially chemical phenomena such as intermetallic compound formation and enhanced oxidation, the compressive stress produced by implantation also is believed to be important. For instance, in the experiment of Figure 14.11, compression effects were believed to dominate because of the relative insensitivity of the wear rate to whether the implanted species was B, N, Ar, or Mo. In addition, the peak in the wear curve coincides roughly with the dose required for maximum stress. Finally, it should be noted that, as for corrosion, the mechanical effects of implantation frequently appear to extend much deeper than the ion range. Suggested reasons for this include (i) extended strain resulting from the compressive stress in the implanted layer; (ii) migration of the implanted species before the advancing wear front; and (iii) the tendency of surface erosion not to occur simultaneously over the entire surface but rather at different points in sequence.

Experimental results such as the above suggest that implantation may be useful in improving the mechanical properties for those sections of metal components which are subject to moving contact.

14.4.4   Catalysis

Catalysts produced by ion implantation are of considerable interest because they require relatively small quantities of precious materials and because thermodynamic restrictions on composition can be bypassed. To date, published work in the area has consisted mostly of implanting the active metal Pt into various hosts and then carrying out electrochemical measurements to determine the catalytic activity (85–87). Depending on the host, its treatment, and the particular electrochemistry used, the results have varied from no change after implantation to an activity approaching that of pure Pt. Hence, in at least some cases, the use of implantation for catalyst production seems promising. Moreover, the possibilities are widened by a recent report of the successful implantation of powders (88). However, there is not now a mechanistic understanding of the observed variations in activity with experimental conditions. Among the most central questions are (i) what governs the accessibility of the implanted Pt to the reacting species and (ii) how does the activity of the Pt depend on its local environment?

14.4.5   Magnetism

Ion implantation displaces atoms essentially at random, thereby producing a microscopic rearrangement within the host lattice. This randomizing process has been used to investigate mechanisms of magnetic anisotropy in amorphous Gd–Co alloys (89–91). The growth-induced, uniaxial anisotropy observed in thin films of these materials is central to their use for magnetic memory devices (92), but the underlying mechanism has been controversial. In one mechanistic study (89), sputtered films of approximate composition $Gd_{0.2}Co_{0.8}$ and thickness ~ 1 μm were irradiated at room temperature with Ar. The ion energy, 1 to 2 MeV, was such that the Ar passed completely through the substrate, thus avoiding unwanted effects due to composition changes. In samples with existing uniaxial anisotropy, Ar doses above $\sim 5 \times 10^{13}$ at. cm^{-2} were observed to destroy the anisotropy. It was estimated that this threshold dose corresponds to a displacement level of ~ 0.2 dpa. By contrast, when samples initially without anisotropy were irradiated in a magnetic field of 1.6 kOe, uniaxial anisotropy was actually produced.

From the observations that <1 dpa is sufficient to greatly change the uniaxial anisotropy and that the final state is sensitive to the applied magnetic field during irradiation, several conclusions were drawn (89):

1. Shape anisotropy, due, for example, to preferentially oriented particles or voids, is not the dominant mechanism.
2. Stress-induced anisotropy also does not dominate. This conclusion derives specifically from the sensitivity to the applied magnetic field, which should have no bearing on stress effects.
3. The uniaxial anisotropy actually is caused by some local atomic configuration which has a preferred orientation. Exchange-coupled Co–Co pairs were suggested as one possibility. The sensitivity to the applied field would then result from the anisotropy of the Co–Co exchange coupling, which would favor energetically certain orientations of the atomic pair in the presence of the field.

14.4.6 Metallurgy of Metastable Alloys

In Section 14.3.3 a low-temperature regime is defined for implanted metals in which the atomic diffusion rates are negligible, with the result that metastable states can persist. A number of experiments, some referenced there, have been carried out to elucidate the physical processes associated with implantation in this regime. However, because of the striking success in producing both crystalline alloys and apparently amorphous structures, a new emphasis is emerging: namely, the use of ion implantation as a tool with which to study the metallurgy of metastable alloys. For example, questions being addressed include the composition range over which a particular crystalline phase can persist (8), the conditions necessary for retention of the amorphous state (8, 9), and the temperature dependences of the trapping of mobile defects at impurities (49). Such work is closely related to that discussed above for specific applications, the principal difference being its concern with the more general aspects of the microscopic behavior rather than achieving a specific macroscopic property.

The preparation of metastable alloys in the past has been accomplished primarily with rapid-cooling techniques and by deposition onto low-temperature substrates (59, 93). Ion implantation complements these methods in that both composition and the number of atomic displacements per atom can be varied continuously in a single specimen and in that the effective quench rate, if this concept is applicable, is extremely high. Possible differences in the final states of an alloy produced by the three techniques have not been explored in detail, and this is an important area for future study. However, in at least one system, namely Ag_xCu_{1-x} with $x=0$ to 0.16, all three methods have been used (94). In each case, the Ag formed a highly substitutional solution in the fcc Cu matrix, even though x greatly exceeded the solid solubility.

An example of some implantation work with metastable alloys, already discussed to some extent, is the experimental verification of the relation Eq. 6 for the composition at which a binary system is most prone to exist in the amorphous state. Thus, for W implanted into Cu, the onset of disorder occurs between 1 and 10 at. % (8), and for Dy in Ni there is disorder at 30 at. %, but Ni implanted with Pb remains crystalline up to 35 at. % (9). All of these results have been shown to be consistent with Eq. 6 (9). Another example, and one with possible implications for an existing metallurgical problem, is the study of lattice location for implanted impurities in Be (17, 95). A series of elements were introduced at room temperature to concentrations typically of $\simeq 0.5$ at. %, and their lattice sites in the Be host were determined by ion channeling. The results are summarized in Figure 14.12. When the atomic

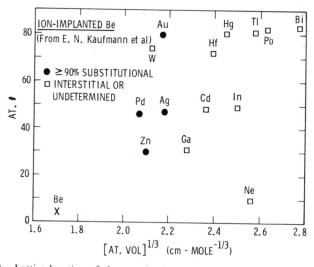

Figure 14.12.   Lattice location of elements implanted into Be at room temperature. From references 17 and 95.

volume of the implanted species is not too different from that of Be, the substitutional fraction is seen to be greater than 90%. The significance of this for Be metallurgy is that the number of elements available for solution strengthening might be increased by going to metastable alloys. Heretofore, consideration of equilibrium solubilities has limited the choice to Cu (96).

14.4.7 Metallurgy of Equilibrium Alloys

In the high-temperature regime defined by Eq. 5, atomic diffusion leads to the formation of equilibrium phases in most alloy systems. Here, ion implantation is being used to obtain information about diffusion rates and phase diagrams which was unavailable previously. The general approach is to create an alloy by implantation and then to follow its evolution during annealing by means of ion backscattering-channeling, transmission electron microscopy, or other microscopic probes discussed in Section 14.2.2. Some of the advantages of this approach are: (i) the microscopic scale of the experiment, which allows measurements at lower temperatures than with conventional techniques because of the shorter diffusion distances; (ii) the possibility of using ion irradiation to enhance the atomic diffusion rates, thereby permitting data to be obtained at still lower temperatures; (iii) production of an intimate atomic mixture of tailored composition, so that the desired phases precipitate relatively quickly and without passing through intermediate equilibrium phases, even in complex systems of higher order; and (iv) direct introduction of the solute into the host matrix, which avoids difficulties with interfacial effects such as those due to oxide layers.

Metallurgical studies in the high-temperature regime have been based principally on the use of diffusion couples, which are formed by implanting the solute or solutes into the first ~ 0.1 μm of a bulk host. The utility of this approach, as opposed to surface deposition, was demonstrated first in measurements of the diffusion rates for Fe, Mn, and In impurities in Al (97–99). There, the implanted isotopes were radioactive tracers, and profiling was accomplished by the conventional method of mechanical sectioning followed by an analysis of the radioactivity. The Fe data in particular were free of anomalies that had appeared in earlier work with solutes deposited on the Al surface (97). The improvement was attributed to the fact that, when the diffusing atoms were implanted directly into the metal, the native oxide layer on the surface could not inhibit their migration. In subsequent measurements of diffusion in Be, Al, and Fe hosts, the implanted couples were profiled by ion backscattering analysis (12, 13, 100, 101). Here, the ~ 0.01 μm depth resolution allowed the data to be extended to substantially lower temperatures. An example is given in Figure 14.13, where the diffusion coefficient D for Cu in Be is plotted in an Arrhenius format (12). The two different rates correspond to the two inequivalent directions of diffusion in the hcp Be lattice. Solid symbols are from conventional measurements, while the open characters were obtained with ion implantation and ion backscattering analysis. The ion beam data are seen to extend lower by about four orders of magnitude in D.

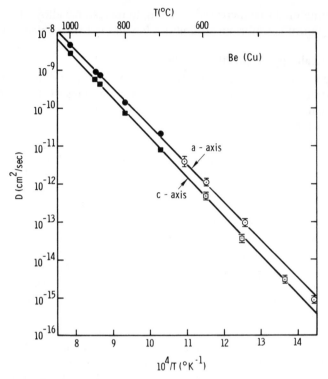

Figure 14.13.   Diffusion coefficients for Cu in Be. Open symbols are ion beam data; solid points give previous conventional results. From reference 12.

When the local concentration of the implanted species exceeds its solid solubility in the host, precipitation occurs in accordance with the phase diagram. This precipitation is quite rapid because of the small separation of the components, and its occurrence can be confirmed by transmission electron microscopy. Continued annealing then causes the implanted atoms to diffuse into solution in the underlying bulk, and eventually the precipitates within the implanted layer are completely dissolved. However, during the time interval when both phases are present, the solute concentration at the intersection of the two-phase implanted layer and the one-phase bulk is equal to the solid solubility, and this parameter is determined directly by ion backscattering analysis.

An example is given in Figure 14.14, where the solubility of Cu in Be is plotted versus reciprocal temperature (102). The open diamond symbols were obtained using ion implantation and ion backscattering analysis, while the solid points are from conventional experiments. Once again the ion beam

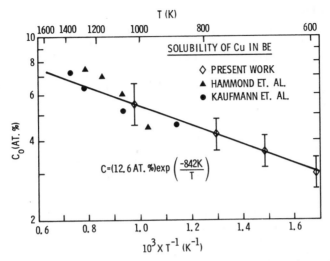

Figure 14.14.   Solubility of Cu in Be. Open diamond symbols are ion beam data; solid symbols are previous conventional results. From reference 102.

data extend to much lower temperatures. The point at 593° K is noteworthy because it was obtained in about 100 h of annealing by using ion irradiation to enhance the diffusion rate. Without the enhancement effect, the diffusion distance of ~0.1 μm necessary for the backscattering measurements would have required about six months. Alternatively, to achieve the several microns of diffusion required in the more usual electron microprobe measurement, tens of years would have been necessary.

Finally, similar methods have been used in other binary alloys of Be, Al, and Fe (13, 100, 101) and have been extended to the ternary regime in the systems Be(Al, Fe) (13), Be(Al, Cu), Be(Si, Cu) (102), and Fe(Ti, Sb) (101). In the Be(Al, Fe) work, Al was observed to reduce the solubility of Fe in Be by more than an order of magnitude.

14.5   CONCLUSIONS

Ion implantation provides a degree of control over composition and lattice damage which is unique, and its use has expanded considerably the possible manipulations of metal layers. This is illustrated by the work reviewed in the present chapter. The field is in an early stage of development and is growing rapidly, both in the understanding of physical processes and in the range and sophistication of applications. Consequently, the foregoing discussion serves to introduce the area and give some indication of its current status, but not to establish bounds. Indeed, it is probable that implantation will become a major tool in metallurgy. In particular, its use almost exclusively as a research tool is unlikely to persist.

REFERENCES

1. S. T. Picraux, E. P. EerNisse, and F. L. Vook, Eds., *Proceedings of Conference on Applications of Ion Beams to Metals*, Plenum Press, New York (1974).
2. G. Carter, J. S. Colligon, and W. A. Grant, Eds., *Proceedings of Conference on Applications of Ion Beams to Materials, 1975*, Institute of Physics, London (1976).
3. F. Chernow, J. A. Borders, and D. K. Brice, Eds., *Ion Implantation in Semiconductors and Other Materials*, Plenum Press, New York (1977).
4. B. Stritzker, in reference 2, pp. 160–167.
5. O. Meyer, in reference 2, pp. 168–175.
6. D. K. Sood and G. Dearnaley, in reference 2, pp. 196–203.
7. J. A. Borders and J. M. Poate, *Phys. Rev.*, **B13**, 969 (1976).
8. A. G. Cullis, J. M. Poate, and J. A. Borders, *Appl. Phys. Lett.*, **28**, 314 (1976).
9. R. Andrew, W. A. Grant, P. J. Grundy, J. S. Williams, and L. T. Chadderton, *Nature*, **262**, 380 (1976).
10. Reference 2, Chapter 4.
11. N. E. W. Hartley, in reference 2, pp. 210–223.
12. S. M. Myers, S. T. Picraux, and T. S. Prevender, *Phys. Rev.*, **B9**, 3953 (1974).
13. S. M. Myers and J. E. Smugeresky, *Met. Trans.*, **7A**, 795 (1976).
14. S. M. Myers, in reference 3.
15. G. Carter and J. S. Colligon, *Ion Bombardment of Solids*, Elsevier, New York (1968), Chapter 7.
16. Y. Adda, M. Beyeler, and G. Brebec, *Thin Solid Films*, **25**, 107 (1975).
17. E. N. Kaufmann, P. Raghavan, R. S. Raghaven, E. J. Ansaldo, and R. A. Naumann, *Phys. Rev. Lett.*, **34**, 1558 (1975).
18. P. Sigmund, *Appl. Phys. Lett.*, **25**, 169 (1974), and reference therein.
19. G. H. Vineyard, *Radiat. Eff.*, **29**, 245 (1976).
20. G. Dearnaley, J. H. Freeman, R. S. Nelson, and J. Stephen, *Ion Implantation*, North-Holland, Amsterdam (1973), Chapter 4.
21. D. K. Brice, *Radiat. Eff.*, **11**, 227 (1971).
22. K. B. Winterbon, *Radiat. Eff.*, **13**, 215 (1972).
23. S. Mylroie and J. F. Gibbons, in *Ion Implantation in Semiconductors and Other Materials*, B. L. Crowder, Ed., Plenum Press, New York (1973), pp. 243–253.
24. K. B. Winterbon, P. Sigmund, and J. B. Sanders, *K. Dan. Vidensk. Mat.-Fys. Medd.*, 37, No. 14 (1970).
25. H. Matsumura and S. Furukawa, *J. Appl. Phys.*, **47**, 1746 (1976).
26. J. F. Gibbons, W. S. Johnson, and S. W. Mylroie, *Projected Range Statistics*, 2nd ed., Dowden, Hutchinson, and Ross, Stroudsburg, Pa. (1975), tabulation.
27. D. K. Brice, *Ion Implantation Range and Energy Deposition Distributions*, Vol. 1, Plenum Press, New York, (1975), tabulation for high ion energies.
28. K. B. Winterbon, *Ion Implantation Range and Energy Deposition Distributions*, Vol. 2, Plenum Press, New York (1975), tabulation for low ion energies.
29. D. K. Brice, *J. Appl. Phys.*, **46**, 3385 (1975), and references therein.
30. M. T. Robinson, *Phil. Mag.*, **12**, 741 (1965); *Phil. Mag.*, **17**, 639 (1968).
31. P. Sigmund, *Appl. Phys. Lett.*, **14**, 114 (1969).
32. S. M. Myers, D. E. Amos, and D. K. Brice, *J. Appl. Phys.*, **47**, 1812 (1976).
33. R. L. Minear, D. G. Nelson, and J. F. Gibbons, *J. Appl. Phys.*, **43**, 3468 (1972).
34. R. Collins and G. Carter, *Radiat. Eff.*, **26**, 181 (1975).
35. R. S. Gvosdover, V. M. Efremenkova, L. B. Shelyakin, and V. E. Yurasova, *Radiat. Eff.*, **27**, 237 (1976), and references therein.

36. D. V. Morgan, Ed., *Channeling: Theory, Observation, and Applications*, Wiley, London (1973).
37. Reference 20, Chapter 2; J. L. Whitton, in reference 36, Chapter 8.
38. J. A. Davies, in reference 36, Chapter 13.
39. R. Behrisch and J. Roth, in *Ion Beam Surface Layer Analysis*, Vol. 2, O. Meyer, G. Linker, and F. Kappeler, Eds., Plenum Press, New York, 1976, pp. 539–565.
40. G. Foti, P. Baeri, E. Rimini, and S. U. Campisano, *J. Appl. Phys.*, **47**, 5206 (1976).
41. G. Foti, S. T. Picraux, S. U. Campisano, E. Rimini, and R. A. Kant, in reference 3.
42. P. B. Hirsch, A. Howie, R. B. Nicholson, D. W. Pashley, and M. J. Whelan, *Electron Microscopy of Thin Crystals*, Butterworth, London (1965); K. E. Easterling, *International Metals Reviews*, March 1977, pp. 1–24.
43. See, for example, references 8, 9, 63, and 65.
44. J. W. Coburn and E. Kay, *Crit. Rev. Solid State Sci.*, **4**, 561 (1974).
45. J. B. Mitchell, J. A. Davies, L. M. Howe, R. S. Walker, K. B. Winterbon, G. Foti, and J. A. Moore, in *Ion Implantation in Semiconductors*, S. Namba, Ed., Plenum Press, New York, (1975), pp. 493–500.
46. G. E. Chapman, B. W. Farmery, M. W. Thompson, and I. H. Wilson, *Radiat. Eff.*, **13**, 121 (1972).
47. H. H. Anderson and H. Bay, *Radiat. Eff.*, **19**, 139 (1973); *J. Appl. Phys.*, **45**, 953 (1974).
48. H. de Waard and L. C. Feldman, in reference 1, pp. 317–351.
49. M. L. Swanson, L. M. Howe, and A. F. Quenneville, in reference 55, pp. 316–324.
50. D. K. Brice, in reference 2, pp. 334–339.
51. E. V. Kornelsen, *Radiat. Eff.*, **13**, 227 (1972).
52. T. W. Sigmon, L. Csepregi, and J. W. Mayer, *J. Electrochem. Soc.*, **123**, 1116 (1976), and references therein.
53. H. J. Wollenberger, in reference 54, pp. 215–254.
54. A. Seeger, D. Schumacher, W. Schilling, and J. Diehl, Ed., *Proceedings of Conference on Vacancies and Interstitials in Metals*, North-Holland, Amsterdam (1970).
55. M. T. Robinson and F. W. Young, Ed., *Proceedings of Conference on Fundamental Aspects of Radiation Damage in Metals*, CONF-751006, National Technical Information Service, Springfield, Va., (1976).
56. J. W. Mayer, L. Eriksson, and J. A. Davies, *Ion Implantation in Semiconductors*, Academic Press, New York (1970), Chapter 3.
57. G. A. Stephens, E. Robinson, and J. S. Williams, in Reference 45, pp. 375–381.
58. S. R. Nagel and J. Tauc, *Phys. Rev. Lett.*, **35**, 380 (1975).
59. G. S. Cargill III, in *Solid State Physics*, Vol. 30, H. Ehrenreich, F. Seitz, and D. Turnbull, Eds., Academic Press, New York, 1975, pp. 227–320.
60. R. A. Johnson and N. Q. Lam, *Phys. Rev.*, **B13**, 4364 (1976).
61. N. Q. Lam, S. J. Rothman, K. L. Merkle, L. J. Nowicki, and D. J. Dever, *Thin Solid Films*, **25**, 157 (1975).
62. J. W. Corbett, *Solid State Phys.*, Suppl. **7**, 271 (1966).
63. R. A. Kant, S. M. Myers, and S. T. Picraux, in reference 3.
64. *Phase Transformations*, American Society for Metals, Metals Park, Ohio (1970).
65. R. S. Nelson, J. A. Hudson, and D. J. Mazey, *J. Nucl. Mater.*, **44**, 318 (1972).
66. D. I. Potter and H. A. Hoff, in reference 55, pp. 1092–1099.
67. L. R. Testardi, R. L. Meek, J. M. Poate, W. A. Royer, A. R. Storm, and J. H. Wernick, *Phys. Rev.*, **B11**, 4304 (1975).
68. B. Stritzker, *Z. Phys.*, **268**, 261 (1974).
69. O. Meyer, *J. Jap. Soc. Appl. Phys.*, Suppl. **44**, 23 (1975).

70. O. Meyer, in *New Uses of Ion Accelerators*, J. F. Ziegler, Ed., Plenum Press, New York (1975), Chapter 6.
71. G. Heim and B. Stritzker, *Appl. Phys.* (Germany), **7**, 239 (1975).
72. J. Geerk, K. C. Langguth, G. Linker, and O. Meyer, *Proceedings of Applied Superconductivity Conference*, Aug. 1976, Palo Alto, Cal., in press.
73. O. Meyer, H. Mann, and E. Phrilingos, in reference 1, pp. 15–26.
74. J. M. Poate, L. R. Testardi, A. R. Storm, and W. M. Augustyniak, in reference 2, pp. 176–182.
75. W. Bauriedl, G. Heim, and W. Buckel, *Phys. Lett.*, **57A**, 282 (1976).
76. E. P. Harris, *J. Vac. Sci. Technol.*, **12**, 1383 (1975).
77. B. L. Crowder, Ed., *Ion Implantation in Semiconductors and Other Materials*, Plenum Press, New York, (1973).
78. Reference 1, Chapter 2.
79. Reference 2, Chapter 4.
80. Reference 3, section on metals.
81. F. Brown and W. D. Mackintosh, *J. Electrochem. Soc.*, **120**, 1096 (1973).
82. W. D. Mackintosh and F. Brown, in reference 1, pp. 111–121.
83. J. E. Antill, M. J. Bennett, G. Dearnaley, F. H. Fern, P. D. Goode, and J. F. Turner, in reference 77, pp. 415–422.
84. N. E. W. Hartley, *Wear*, **34**, 427 (1975).
85. M. Grenness, M. W. Thompson, and R. W. Cahn, *J. Appl. Electrochem.*, **4**, 211 (1974).
86. M. Voinov, D. Bühler, and H. Tannenberger, in *Proceedings of Symposium on Electrocatalysis, San Francisco, May 1974*, M. W. Breiter, Ed., Electrochemical Society, New York (1974), pp. 268–274.
87. W. A. Grant, in reference 2, pp. 127–140.
88. J. H. Freeman and W. Temple, *Radiat. Eff.*, **28**, 85 (1976).
89. R. J. Gambino, J. F. Ziegler, and J. J. Cuomo, *Appl. Phys. Lett.*, **24**, 99 (1974).
90. R. Hasegawa, R. J. Gambino, J. J. Cuomo, and J. F. Ziegler, *J. Appl. Phys.*, **45**, 4036 (1974).
91. E. L. Venturini, P. M. Richards, J. A. Borders, and E. P. EerNisse, in *Magnetism and Magnetic Materials—1975*, AIP Conf. Proc. No. 29, J. J. Becker, G. H. Lander, and J. J. Rhine, Eds., American Institute of Physics, New York (1975), pp. 119–120.
92. C. H. Bajorek and R. J. Kobliska, *IBM J. Res. Dev.*, **20**, 271 (1976).
93. R. Wang and M. D. Merz, *Nature*, **260**, 35 (1976).
94. J. M. Poate, J. A. Borders, A. G. Cullis, and J. K. Hirvonen, *Appl. Phys. Lett.*, **30**, 365 (1977).
95. E. N. Kaufmann, *Bull. Amer. Phys. Soc.*, **21**, 407 (1976); E. N. Kaufmann, K. Krien, J. C. Soars, and K. Freitag, *Hyperfine Interactions*, **1**, 485 (1976); E. N. Kaufmann, private communication.
96. M. Wilhelm and F. Aldinger, *Met. Trans.*, **7A**, 695 (1976).
97. G. M. Hood, *Phil. Mag.*, **21**, 305 (1970).
98. G. M. Hood and R. J. Schultz, *Phil. Mag.*, **23**, 1479 (1971).
99. G. M. Hood and R. J. Schultz, *Phys. Rev.*, **B4**, 2339 (1971).
100. S. M. Myers and S. T. Picraux, *J. Appl. Phys.*, **46**, 4774 (1975); and unpublished results.
101. S. M. Myers and H. J. Rack, unpublished results.
102. S. M. Myers and J. E. Smugeresky, *Met. Trans.*, **8A**, 609 (1977).

MATERIALS INDEX

In this index, we list those thin film materials whose reaction and electrical properties are discussed in the text. A-B indicates a binary system, A-B-C a ternary system and A_xB_y a compound phase. We list the compound phases in alphabetical ordering except for GaAs, SiO_2 and Ta_2O_5.

SUBJECT INDEX

Accelerators, ion, 123, 535
Activation energy, of bulk diffusion, 310
  of compound formation, 334
  of electromigration, 264
  of grain boundary diffusion, 228
  of silicide formation, 376
Aluminum, barriers to diffusion, 35
  corrosion, 48
  electromigration in, 39, 264
  germanium reaction, 436
  gold reaction, 334
  grain boundary diffusion in, 230
  hafnium reactions, 36, 339
  ion implantation in, 561, 550
  metallization for devices, *see Chapters 2, 8, 9*
  metal reactions, 346
  nickel reaction, 337
  platinum silicide reaction, 25
  poly-Si reaction, 31, 442
  precipitation of Si in, 31, 443
  Schottky barriers, *see* Schottky barrier
  silicon reaction, 15, 360, 441
  spikes in Si, 15
Ambient effects, metallization, 29, 341
  silicide formation, 379
Amorphous layers, in as-deposited films, 96, 97
  epitaxial regrowth in Si and Ge, 468
  formed by implantation, 98, 435, 548, 559
  free-energy, 434
  structure, 91
Analysis techniques, Auger electron spectroscopy, 145, 207
  backscattering spectrometry, 123, 539

glancing-angle X-ray diffraction, 107
microanalysis, 114
reflection electron diffraction, 109
scanning electron microscopy (SEM), 105
secondary ion mass spectroscopy (SIMS), 149
stylus instruments, 105
transmission electron diffraction, 109, 543
X-ray topography, 108
see also Characterization of thin films
Anion of semiconductor compound, 70, 71, 73
Arrhenius equation, 310
Atomic flux divergence (in electromigration), 250
  current crowding, 253
  temperature gradients, 253
Auger electron spectroscopy (AES), 145, 207
Avalanche breakdown, 507

Backscattering spectrometry, Rutherford, 121, 123, 539
  backscattered yield, 128
  channeling effects, 460, 469, 540
  composition, concentration ratios, 129, 130
  depth resolution, 126
  energy loss factor, 126, 127
  grazing angle, 114
  heights of spectra, 129, 130
  kinematic factor, 124
  Rutherford scattering cross section, 129
  spectra (*e.g.*, Ni_2Si), 125
  straggling, Bohr theory, 127, 128